# Eighth Edition

## Maribeth Price
*South Dakota School of Mines and Technology*

# *Mastering ArcGIS*

McGraw Hill Education

MASTERING ArcGIS

Published by McGraw-Hill Education, 2 Penn Plaza, New York, NY 10121. Copyright © 2019 by McGraw-Hill Education. All rights reserved. Printed in the United States of America. No part of this publication may be reproduced or distributed in any form or by any means, or stored in a database or retrieval system, without the prior written consent of McGraw-Hill Education, including, but not limited to, in any network or other electronic storage or transmission, or broadcast for distance learning.

Some ancillaries, including electronic and print components, may not be available to customers outside the United States.

This book is printed on acid-free paper.

1 2 3 4 5 6 7 8 9 LMN 21 20 19 18

ISBN 978-1-260-09248-6
MHID 1-260-09248-8

Cover Image: ©*Anton Balazh/123RF*

All credits appearing on page or at the end of the book are considered to be an extension of the copyright page.

The Internet addresses listed in the text were accurate at the time of publication. The inclusion of a website does not indicate an endorsement by the authors or McGraw-Hill Education, and McGraw-Hill Education does not guarantee the accuracy of the information presented at these sites.

mheducation.com/highered

# Table of Contents

# Preface

Welcome to *Mastering ArcGIS*, a detailed primer on learning the ArcGIS™ software by ESRI®, Inc. This book is designed to offer everything you need to master the basic elements of GIS.

> **Notice:** ArcGIS™, ArcMap™, ArcCatalog™, ArcGIS Desktop™, ArcInfo Workstation™, and the other program names used in this text are registered trademarks of ESRI, Inc. The software names and the screen shots used in the text are reproduced by permission. For ease of reading only, the ™ symbol has been omitted from the names; however, no infringement or denial of the rights of ESRI® is thereby intended or condoned by the author.

## What's new in the eighth edition?

The primary goals for this edition included updating many of the data sets with more recent versions, particularly those utilizing population information, as well as updating the text to work with ArcGIS Desktop version 10.5.1. The interface continues to remain fairly stable, and only minor adjustments are needed. This edition should work equally well with versions 10.2 to 10.5. A few minor changes in command names or buttons are found, and a few new bugs complicate matters here and there, but most users should be able to follow along and make any necessary adjustments. It is expected that the interface will remain stable while ESRI turns its focus to developing the new desktop program, ArcGIS Pro.

I would like to thank the many people who have used and commented on this book, and I hope that it continues to serve their needs in the rapidly evolving world of GIS.

## Looking ahead

ESRI released a completely redesigned ArcGIS Desktop product in late fall 2014, called ArcGIS Pro. It has a 64-bit, multithreaded architecture, uses ribbon-style menus, integrates 2D and 3D applications, and is closely tied to ArcGIS Online. A new textbook focused on ArcGIS Pro is being developed for the *Mastering ArcGIS* series. ArcMap and ArcCatalog will exist side-by-side with this new program for many years yet, and we anticipate keeping an edition of *Mastering ArcGIS* in print as long as needed for those continuing to use them.

## Previous experience

This book assumes that the reader is comfortable using Windows™ to carry out basic tasks such as copying files, moving directories, opening documents, exploring folders, and editing text and word processing documents. Previous experience with maps and map data is also helpful. No previous GIS experience or training is necessary to use this book.

## Elements of the package

This learning system includes a textbook and web site, including

▶ Fourteen chapters on the most important capabilities of ArcGIS

▶ Comprehensive tutorials in every chapter to learn the skills, with each step demonstrated in a video clip

▶ A set of exercises, map documents, and data for practicing skills independently

▶ Reference sections on skills with video clips demonstrating each one

This book assumes that the student has access to ArcGIS Desktop Basic (formerly ArcView). A few optional topics are introduced that require an ArcGIS Desktop Standard (formerly ArcEditor) license. The Spatial Analyst extension is required for Chapter 11.

## Philosophy

This text reflects the author's personal philosophies and prejudices developed from 20 years of teaching GIS at an engineering school. The main goal is not to train geographers but to provide students in any field with GIS skills and knowledge. It is assumed that most students using this book already have a background of discipline-specific knowledge and skills upon which to draw and are seeking to apply geospatial techniques within their own knowledge domains.

- **GIS is best learned by doing it, not by studying it.** The laboratory is THE critical component of the book, and theory is introduced sparingly and integrated with experience. Hence, this book is heavy on experience and lighter on theory.

- **Independent work and projects are critical to learning GIS.** This book includes a wealth of exercises in which the student must find solutions independently without a cookbook recipe of steps. A wise instructor will also require students to develop an independent project.

## Chapter sequence

The book contains an introduction and 14 chapters. Each chapter includes roughly one week's work for a three-credit semester course. This book intentionally contains more material than the average GIS class can cover during a single semester; instructors may choose what to emphasize.

An introductory chapter describes GIS and gives some examples of how it is used. It also provides an overview of GIS project management and how to develop a project. Chapters 1–11 follow a roughly project-based sequence: data compilation, data exploration and mapping, tables and basic editing, and analysis. These chapters are the core of an introductory GIS class and, by the end of it, students should have little difficulty developing and carrying out an independent GIS project. The final chapters introduce several more advanced topics in data management.

## Chapter layout

Each chapter is organized into the following sections:

- **Concepts:** provides basic background material for understanding geographic concepts and how they are specifically implemented within ArcGIS. Most chapters have two sections, one (GIS Concepts) covering general GIS concepts and theory, and another (About ArcGIS) covering the specific implantation of those concepts within ArcGIS Desktop. A set of review questions and important terms follows the concepts section.

- **Tutorial:** contains a step-by-step tutorial demonstrating the concepts and skills. The tutorials begin with detailed instructions, which gradually become more general as mastery is built. Every step in the tutorial is demonstrated by accompanying video clips.

- **Exercises:** presents a series of problems to build skill in identifying the appropriate techniques and applying them without step-by-step help. Through these exercises, the student builds an independent mastery of GIS processes. Brief solution methods are included for all exercises, and maps of the results are shown when applicable. A full answer and methods document is available for instructors at the McGraw-Hill Instructor web site.

The web site also contains all the data needed to follow the tutorials and complete the exercises.

Instructors should use judgment in assigning exercises, as the typical class would be stretched to complete all the exercises in every chapter. Very good students can complete the entire chapter in 3–6 hours, most students would need 6–8 hours, and a few students would require 10 or more hours. Students with more computer experience generally find the material easier than others.

## Using this text

In working through this book, the following sequence of steps is suggested:

- ▶ READ through the Concepts sections to get familiar with the principles and techniques.
- ▶ ANSWER the Chapter Review Questions to test comprehension of the material.
- ▶ WORK the Tutorial section for a step-by-step tutorial and explanation of key techniques.
- ▶ REREAD the Concepts section to reinforce the ideas.
- ▶ PRACTICE by doing the Exercises.

## Using the tutorial

The tutorial provides step-by-step practice and introduces details on how to perform specific tasks. Students should be encouraged to think about the steps as they are performed and not just race to get to the end.

It is important to follow the directions carefully. Skipping a step or doing it incorrectly may result in a later step not working properly. Saving often will make it easier to go back and correct a mistake in order to continue. Occasionally, a step will not work due to differences between computer systems or software versions. Having an experienced user nearby to identify the problem can help. If one isn't available, however, just skip the step and move on without it.

## Using the videos

The web site contains two types of videos. The Tutorial Videos demonstrate each step of the tutorial. They are numbered in the text for easy reference. The Skills Reference Videos show how to perform generic tasks, such as deleting a file. The videos are intended as an alternate learning strategy. It would be tedious to watch all of them. Instead, use them in the following situations:

- ▶ When the student does not understand the written instructions or cannot find the correct menu or button
- ▶ When a step cannot be made to work properly
- ▶ When a reminder is needed to do a previously learned skill in order to complete a step
- ▶ When a student finds that watching the videos enhances learning

As you work through a chapter, keep the web site video listing on the screen, and click the appropriate link to see a video. The tutorial clips are distinguished by numbered steps and the Skills Reference Videos by their headings.

## Using video and data components

### Using the videos:

The videos are available for download from the book web site. Each chapter may be downloaded separately. The videos for Chapter 1 must be downloaded first, for they contain the instructions and folder hierarchy needed for the subsequent chapters. Please review the instructions in the Chapter 1 download before downloading the rest of the videos.

**To install the training data:**

The mgisdata.exe archive contains a folder with the documents and data needed to do the tutorials and exercises. The student must copy this folder to the computer's hard drive. If multiple people are using the same computer to complete the activities in this book, then each person should make a copy of the data in a separate folder. The data in the exe archive are a self-extracting zip file that requires approximately 350 MB of disk space. Follow these instructions to install the data:

▶ Click on the link to Install Data. If a dialog window appears asking whether to Run or Save the data, choose Run.

▶ When a dialog box appears asking whether to Open or Save the data, choose Open. Don't choose Save because it will only copy the data archive instead of extracting it.

▶ Click the Browse button to set the folder to which to extract the data (a). The data will be placed in a folder called mgisdata in whatever location you choose. In other words, if you select C:\gisclass as the target folder, then the data will be placed in C:\gisclass\mgisdata.

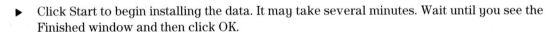

Installing the training data

Source: WinZip Self-Extractor

▶ Click Start to begin installing the data. It may take several minutes. Wait until you see the Finished window and then click OK.

## System requirements

To use the tutorials and do the exercises in this book, the student must have access to a computer with the following characteristics:

**Minimum hardware:**

PC-Intel™-platform computer with 2.2-GHz processor or better and 2 GB RAM

Suitable sound/graphics card with 24-bit color depth and 1024 × 768 minimum screen

**Software:**

Windows 10 (Home or Pro), Windows 8 (Basic, Professional, Enterprise), Windows 7™ (Ultimate, Enterprise, Professional, Home Premium), Windows Vista™ (Ultimate, Enterprise, Business, Home Premium), Windows 2000™, or Windows XP™ (Home Edition, Professional); requires Microsoft .NET framework to be installed.

A web browser, such as Edge, Chrome, Firefox, or Internet Explorer, or Microsoft® Word

A zip utility such as WinZip or 7zip

A media player that is able to display the .mp4 video format (such as Windows Media Player 12 [Windows 7 only] or QuickTime™)

ArcGIS Desktop™ 10.2 or higher (Basic Level); Standard Level and Spatial Analyst extension required for some exercises

For more information, please consult:
http://www.esri.com/software/arcgis/arcgis-for-desktop/system-requirements

Internet access is required for ArcGIS installation and for exercises requiring use of ArcGIS Online. Exercises do not require an ArcGIS Online account, although a public or organizational subscription will provide access to more capabilities and content.

For assistance in acquiring or installing these components, contact your system administrator, hardware/software provider, or local computer store.

## Acknowledgments

I would like to thank many people who made this book possible. Governor Janklow of South Dakota funded a three-month summer project in 2000 that got the book started, as part of his Teaching with Technology program. Many students in my GIS classes between 2000 and 2014 tested the text and exercises and helped immensely in making sure the tutorials were clear and worked correctly. Reviewers of previous editions, including Richard Aspinall, Joe Grengs, Tom Carlson, Susan K. Langley, Henrietta Loustsen, Xun Shi, Richard Lisichenko, John Harmon, Michael Emch, Jim Sloan, Sharolyn Anderson, Talbot Brooks, Qihao Weng, Jeanne Halls, Mark Leipnik, Michael Harrison, Ralph Hitz, Olga Medvedkov, James W. Merchant, Raymond L. Sanders, Jr., Yifei Sun, Fahui Wang, Michael Haas, Jason Kennedy, Dafna Kohn, Jessica Moy, James C. Pivirotto, Peter Price, Judy Sneller, Dave Verbyla, Birgit Mühlenhaus, Jason Duke, Darla Munroe, Wei-Ning Xiang, L. Joe Morgan, Samantha Arundel, Christopher A. Badurek, Tamara Biegas, John E. Harmon, Michael Hass, Nicholas Kohler, David Long, Jaehyung Yu, Sarah Battersby, Gregory S. Bohr, Kelly R. Dubure, Colleen Garrity, Raymond Greene, Eileen Johnson, James Leonard, and Tao Tang provided detailed and helpful comments, and the book is better than it would have been without their efforts. I also thank the reviewers who provided valuable advice for the seventh edition.

Thanks to presenters at the 2014 ESRI Educational and International Users Conference who gave me a crash review on cartography: Allen Carroll, Damien Demaj, Kenneth Field, Makram Murad-Al-Shaikh, and Larry Orman. ESRI, Inc. was prompt and generous in its granting of permission to use the screen shots, data, and other materials throughout the text. They also provided beta and prerelease versions of ArcGIS 10.1 for early development of the text. I extend heartfelt thanks to the City of Austin, Texas, for putting their fine GIS data sets in the public domain. I thank George Sielstad, Eddie Childers, Mark Rumble, Tom Junti, and Patsy Horton for their generous donations of data. I am grateful to Tom Leonard and Steve Bauer for their long-term computer lab administration, without which I could not have taught GIS courses or developed this book. I thank Linda Heindel for organizing student feedback and assisting with the initial round of edits on the first draft. I thank editors Michelle Vogler and Melissa Leick of McGraw-Hill for their unfailing encouragement and enthusiasm about the book as it took shape, as well as for their excellent feedback. I thank the McGraw-Hill team working on the eighth edition, especially Michael Ivanov and Jodi Rhomberg. I am grateful to Daryl Pope, who first started me in GIS, and to John Suppe, who encouraged me to return to graduate school and continue doing GIS on a fascinating study of Venus. I thank my partner, David Stolarz, who provided unfailing encouragement when it seemed as though the editing on this edition would never end. Last, and certainly not least, I thank Curtis Price and my daughters, Virginia and Madeleine, for their understanding and support during the many, many hours I spent working on this book.

# Introduction

## What Is GIS?

### Objectives

- ▶ Developing a basic understanding of what GIS is, its operations, and its uses
- ▶ Getting familiar with GIS project management
- ▶ Learning to plan a GIS project
- ▶ Finding resources to learn more about GIS

### Concepts

#### What is GIS?

GIS stands for Geographic Information System. In practical terms, a GIS is a set of computer tools that allows people to work with data that are tied to a particular location on the earth. Although many people think of a GIS as a computer mapping system, its functions are broader and more sophisticated than that. A GIS is a database that is designed to work with map data.

Consider the accounting department of the local telephone company. They maintain a large computer database of their customers, in which they store the name, address, phone number, type of service, and billing information for each customer. This information is only incidentally tied to where customers live; they can carry out most of the important functions (billing, for example) without needing to know where each house is. Of course, they need to have addresses for mailing bills, but it is the post office that worries about where the houses actually are. This type of information is called **aspatial data**, meaning that it is not tied, or is only incidentally tied, to a location on the earth's surface.

Employees of the service department, however, need to work with **spatial data** to provide the telephone services. When hundreds of people call in after a power outage, the service department must analyze the distribution of the calls and isolate the location where the outage occurred. When a construction company starts work on a street, workers must be informed of the precise location of buried telephone cables. If a developer builds a new neighborhood, the company must be able to determine the best place to tie into the existing network so that the services are efficiently distributed from the main trunk lines. When technicians prepare lists of house calls for the day, they need to plan the order of visits to minimize the amount of driving time. In these tasks, location is a critical aspect of the job, and the information is spatial.

In this example, two types of software are used. The accounting department uses special software called a *database management system*, or *DBMS*, which is optimized to work with large volumes of aspatial data. The service department needs access to a database that is optimized for working with spatial data, a Geographic Information System. Because these two types of software are related, they often work together, and they may access the same information. However, they do different things with the data.

A GIS is built from a collection of hardware and software components.

▶ *A computer hardware platform.* Due to the intensive nature of spatial data storage and processing, a GIS was once limited to large mainframe computers or expensive workstations. Today, it can run on a typical desktop personal computer.

▶ *GIS software.* The software varies widely in cost, ease of use, and level of functionality but should offer at least some minimal set of functions, as described in the next few paragraphs. In this book, we study one particular package that is powerful and widely used, but others are available and may be just as suitable for certain applications.

▶ *Data storage.* Some projects use only the hard drive of the GIS computer. Other projects may require more sophisticated solutions if large volumes of data are being stored or multiple users need access to the same data sets. Today, many data sets are stored in digital warehouses and accessed by many users over the Internet. Compact disc writers and/or USB portable drives are highly useful for backing up and sharing data.

▶ *Data input hardware.* Many GIS projects require sophisticated data entry tools. Digitizer tablets enable the shapes on a paper map to be entered as features in a GIS data file. Scanners create digital images of paper maps. An Internet connection provides easy access to large volumes of GIS data. High-speed connections are preferred, as GIS data sets may constitute tens or hundreds of megabytes or more.

▶ *Information output hardware.* A quality color printer capable of letter-size prints provides the minimum desirable output capability for a GIS system. Printers that can handle map-size output (36 in. × 48 in.) will be required for many projects.

▶ *GIS data.* Data come from a variety of sources and in a plethora of formats. Gathering data, assessing their accuracy, and maintaining them usually constitutes the longest and most expensive part of a GIS project.

▶ *GIS personnel.* A system of computers and hardware is useless without trained and knowledgeable people to run it. The contribution of professional training to successful implementation of a GIS is often overlooked.

GIS software varies widely in functionality, but any system claiming to be a GIS should provide the following functions at a minimum:

▶ *Data entry* from a variety of sources, including digitizing, scanning, text files, and the most common spatial data formats; ways to export information to other programs should also be provided

▶ *Data management* tools, including tools for building data sets, editing spatial features and their attributes, and managing coordinate systems and projections

▶ *Thematic mapping* (displaying data in map form), including symbolizing map features in different ways and combining map layers for display

▶ *Data analysis* functions for exploring spatial relationships in and between map layers

▶ *Map layout* functions for creating soft and hard copy maps with titles, scale bars, north arrows, and other map elements

Geographic Information Systems are put to many uses, but providing the means to collect, manage, and analyze data to produce information for better decision making is the common goal and the strength of each GIS. This book is a practical guide to understanding and using a particular Geographic Information System called ArcGIS. Using this book, you can learn what a GIS is, what it does, and how to apply its capabilities to solve real-world problems.

## A history of GIS

Geographic Information Systems have grown from a long history of cartography begun in the lost mists of time by early tribesmen who made sketches on hides or formed crude models of clay as aids to hunting for food or making war. Ptolemy, an astronomer and geographer from the second century B.C., created one of the earliest known atlases, a collection of world, regional, and local maps and advice on how to draw them, which remained essentially unknown to Europeans until the 15th century. Translated into Latin, it became the core of Western geography, influencing cartographic giants such as Gerhard Mercator, who published his famous world map in 1569. The 17th and 18th centuries saw many important developments in cartography, including the measurement of a degree of longitude by Jean Picard in 1669, the discovery that the earth flattens toward the poles, and the adoption of the Prime Meridian that passes through Greenwich, England. In 1859, French photographer and balloonist Gaspard Felix Tournachon founded the art of remote sensing by carrying large-format cameras into the sky. In an oft-cited early example of spatial analysis, Dr. John Snow mapped cholera deaths in central London in September 1854 and was able to locate the source of the outbreak—a contaminated well. However, until the 20th century, cartography remained an art and a science carried out by laborious calculation and hand drawing. In 1950, Jacqueline Tyrwhitt made the first explicit reference to map overlay techniques in an English textbook on town and country planning, and Ian McHarg was one of the early implementers of the technique for highway planning.

As with many other endeavors, the development of computers inspired cartographers to see what these new machines might do. The early systems developed by these groups, crude and slow by today's standards, nevertheless laid the groundwork for modern Geographic Information Systems. Dr. Roger Tomlinson, head of an Ottawa group of consulting cartographers, has been called the "Father of GIS" for his promotion of the idea to use computers for mapping and for his vision and effort in developing the Canada Geographic Information System (CGIS) in the mid-1960s. Another pioneering group, the Harvard Laboratory for Computer Graphics and Spatial Analysis, was founded in the mid-1960s by Howard Fisher. He and his colleagues developed a number of early programs between 1966 and 1975, including SYMAP, CALFORM, SYMVU, GRID, POLYVRT, and ODYSSEY. Other notable developers included professors Nystuen, Tobler, Bunge, and Berry from the Department of Geography at Washington University during 1958–1961. In 1970, the US Bureau of the Census produced its first geocoded census and developed the early DIME data format based on the CGIS and POLYVRT data representations. DIME files were widely distributed and were later refined into the TIGER format. These efforts had a pronounced effect on the development of data models for storing and distributing geographic information.

In 1969, Laura and Jack Dangermond founded the Environmental Systems Research Institute (ESRI), which pioneered the powerful idea of linking spatial representation of features with attributes in a table, a core idea that revolutionized the industry and launched the development of Arc/Info, a program whose descendants have captured about 90% of today's GIS market. Other vendors are still active in developing GIS systems, which include packages MAPINFO, MGA from Intergraph, IDRISI from Clark University, and the open-source program GRASS.

## What can a GIS do?

A GIS works with many different applications: land use planning, environmental management, sociological analysis, business marketing, and more. Any endeavor that uses spatial data can benefit from a GIS. For example, researchers at the US Department of Agriculture Rocky Mountain Research Station in Rapid City conducted a study of elk habitat in the Black Hills of

South Dakota and Wyoming by placing radio transmitter collars on about 70 elk bulls and cows (Fig. I.1c). Using the collars and a handheld antenna, they tracked the animals and obtained their locations. Several thousand locations were collected (Fig. I.1a), allowing the scientists to study the characteristics of the habitat where elk spend time.

The elk locations were entered into a GIS system for record keeping and analysis. Each location became a point with attached information, including the animal ID number and the date and time of the sighting. Information about vegetation, slope, aspect, elevation, water availability, and other site factors were derived by overlaying the points on other data layers, allowing the biologists to compare the characteristics of sites utilized by the elk. Figure I.1b shows a map of distances to major

Fig. I.1. Analyzing elk habitat use: (a) elk locations and study area; (b) locations on a map of distance to nearest road; (c) collaring an elk
Source: Courtesy of Mark Rumblße

roads in the central part of the study area. The elk locations clustered in the darker roadless areas, and statistical analysis demonstrated this observation empirically.

There are many applications of GIS in almost every human endeavor, including business, defense and intelligence, engineering and construction, government, health and human services, conservation, natural resources, public safety, education, transportation, utilities, and communication. In August 2016, the Industries section of the ESRI web site at www.esri.com listed 42 different application areas in these categories, each one with examples, maps, and case studies. Instead of reading through a list here, go to the site and see the latest applications.

### New trends and directions in GIS

The GIS industry has grown exponentially since its inception. Starting with the mainframe and then the desktop computer, GIS began as a relatively private endeavor focused on small clusters of specialized workers who spent years developing expertise with the software and data. Since then, the development of the Internet and the rapidly advancing field of computer hardware have been driving some significant changes in the industry.

### Proliferation of options for data sharing

Instead of storing data sets locally on individual computers or intra-organization network drives, more people are serving large volumes of data over the Internet to remote locations within an organization, across organizations, and to the general public. In the past, large collections of data were hosted by various organizations through clearinghouses sponsored by the National Spatial Data Infrastructure (NSDI) organization. The data usually existed as GIS data files in various formats and required significant expertise and the right software to download and use. Today, *GIS Servers* are designed to bring GIS data to the general public within a few clicks. Another class of data providers that give wide access to spatial data, although not specifically GIS applications, include web sites such as Google Earth, MapQuest, and Microsoft Virtual Earth.

Like many other computer industries, GIS is heading for the clouds. A cloud is a gigantic array of large computers that customers rent pieces of by the hour, instead of purchasing their own physical hardware. ArcGIS Online is a cloud-based platform for users to collaborate and share GIS data with one another, making it suddenly very easy to share data with a colleague, or with the world, even for those with no particular GIS expertise.

## Proliferation of options for working with GIS data

In the early days, people who wanted to do GIS had to purchase a large, expensive program and learn to use it. Now, a wide variety of scaled applications permit different levels of use for different levels of need. Not every user must have the full program. There are map servers for people who just need to view and print maps, free download software for viewing interactive map publications, and scaled-down versions of the full program with fewer options.

Many organizations are turning to Server GIS as a less expensive alternative to purchasing large numbers of GIS licenses. Many workers need GIS, but they only use a small subset of GIS functions on a daily basis. A Server GIS can provide both data and a customizable set of viewing and analysis functions to users without a GIS license. All they need is a web browser. Because of more simple and low-cost ways of accessing GIS data and functions, the user base has expanded dramatically.

## Expansion of GIS into wireless technology

More people are collecting and sharing data using handheld wireless devices, such as handheld computers, smartphones, and global positioning system (GPS) units. These units can now access Internet data and map servers directly so that users in the field can download background data layers, collect new data, and transmit them back to the large servers. Cell phone technology is advancing rapidly with new geolocation options, web access, and geo-applications arriving daily.

## Emphasis on open-source solutions

Instead of relying on proprietary, specialized software, more GIS functions are now implemented within open-source software and hardware. GIS data are now more often stored using engines from commercial database platforms and utilize the same development environments as the rest of the computer industry. This trend makes it easier to have the GIS communicate with other programs and computers and enhances interoperability between systems and parts of systems.

## Customization

With emphasis on open-source solutions, new opportunities have developed in creating customized applications based on a fundamental set of GIS tools, such as a hydrology tool or a wildlife management tool. Smartphone and tablet applications programming is burgeoning. These custom applications gather the commonly used functions of GIS into a smaller interface, introduce new knowledge, and formalize best practices into an easy-to-use interface. Customization requires a high level of expertise in object-oriented computer programming.

## Enterprise GIS

Enterprise GIS integrates a server with multiple ways to access the same data, including traditional GIS software programs, web browser applications, and wireless mobile devices. The goal is to meet the data needs of many different levels of users and to provide access to nontraditional users of GIS. The Enterprise GIS is the culmination of the other trends and capabilities already mentioned. The costs and challenges in developing and maintaining an Enterprise GIS are significant, but the rewards and cost savings can also be substantial.

### *What do GIS professionals do?*

It's getting so easy to create a map these days that one may wonder why one should bother to learn GIS. However, the easy solutions are based on the work of experts who provide the data and the software systems to handle them. It may not be that hard to learn to use a smartphone, but you still need the engineers and software developers to build the phones and the infrastructure to make them work. GIS is much the same. These days, GIS professionals play a variety of roles. A few broad categories can be defined.

***Primary data providers*** create the base data that form the backbone of many GIS installations. Surveyors and land-planning professionals contribute precise measurement of boundaries. Transportation agencies develop the street data used by your smartphone to give directions, or even support games like Pokemon Go. Photogrammetrists develop elevation and other data from aerial photography or the newer laser altimetry (LIDAR) systems. Remote sensing professionals extract all kinds of human-made and natural information from a variety of satellite and airborne measurement systems. Experts in global positioning systems (GPS), which provide base data, are also important.

***Applications GIS*** usually involves professionals trained in other fields, such as geography, hydrology, land use planning, business, and utilities, who utilize GIS as part of their work. Specialists in mathematics and statistics develop new ways to analyze and interpret spatial data. For these professionals, GIS is an added skill and a tool to make their work more efficient, productive, and valuable.

***Development GIS*** involves skilled software and hardware engineers who build and maintain the GIS software itself, as well as the hardware components upon which it relies—not only computers and hard drives and plotters but also GPS units, wireless devices, scanners, digitizers, and other systems that GIS could not function without. This group also includes an important class of GIS developers who create customized applications from the basic building blocks of existing GIS software.

***Distributed database GIS*** involves computer science professionals with a background in networking, Internet protocols, and/or database management systems. These specialists set up and maintain the complex server and network systems that allow data services, Server GIS, and Enterprise GIS to operate.

### *GIS project management*

A GIS project may be a small effort spanning a couple of days by a single person, or it can be an ongoing concern of a large organization with hundreds of people participating. Large or small, however, projects often follow the generalized model shown in Figure I.2, and most new users learn GIS through a project approach. A project usually begins with an assessment of needs. What specific issues must be studied? What kind of information is needed to support decision making? What functions must the GIS perform? How long will the project last? Who will be using the data? What funding is available for start-up and long-term support?

Without a realistic idea of what the system must accomplish, it is impossible to design it efficiently. Users may find that some critical data are absent or that resources have been wasted acquiring data that no one ever uses. In a short-term project the needs are generally clear-cut. A long-term organizational system will find that its needs evolve over time, requiring periodic reassessment. A well-designed system will adapt easily to future modification. The creator of a haphazard system may be constantly redoing previous work when changes arise.

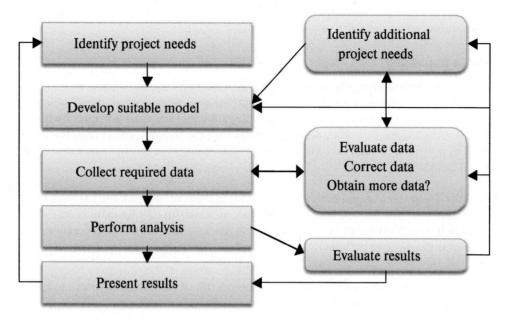

Fig. I.2. Generalized flow diagram representing steps in a GIS project

In studies seeking specific answers to scientific or managerial questions, a methodology or **model** must be chosen. Models convert the raw data of the project into useful information using a well-defined series of steps and assumptions. Creating a landslide hazards map provides a simple example. One might define a model such that, if an area has a steep slope and consists of a shale rock unit, then it should be rated as hazardous. The raw data layers of geology and slope can be used to create the hazard map. A more complex model might also take into account the dip (bedding angle) of the shale units. Models can be very simple, as in this example, or more complex with many different inputs and calculations.

Once needs are known and the appropriate models have been designed, data collection can begin. GIS data are stored as layers, with each layer representing one type of information, such as roads or soil types. The needs dictate which data layers are required and how accurate they must be. A source for each data layer must be found. In some cases data can be obtained free from other organizations. General base layers, such as elevation, roads, streams, political boundaries, and demographics, are freely available from government sources (although the accuracy and level of detail are not always what one might wish). More specific data must often be developed in-house. For example, a utility company would need to develop its own layers showing electric lines and substations—no one else would be likely to have these data.

The spatial detail and accuracy of the data must be evaluated to ensure that they are able to meet the needs of the project. For example, an engineering firm creating the site plan for a shopping mall could download elevation contours of the site for free. However, standard 10-foot contours cannot provide the detailed surface information needed by the engineers. Instead, the firm might contract with a surveying firm to measure contours at half-foot intervals.

After the data are assembled, the analysis can begin. During the analysis phase, it is not unusual to encounter problems that might require making changes to the model and/or data. Thus, the steps of model development, data collection, and analysis often become iterative, as experience gained is used to refine the process. The final result must be checked carefully against reality in order to recognize any shortcomings and to provide guidelines for improving future work.

Finally, no project is complete until the results have been communicated, whether shown informally to a supervisor, published in a scientific journal, or presented during a heated public meeting. Presentations may include maps, reports, slides, or other media.

### Learning more about GIS

As you work your way through this text, you will be amazed at all the things that a GIS can do. At times, the abundance of tasks and the flexibility of options may be overwhelming. Even so, this book gives only a bare introduction to what is possible and covers a small portion of what a GIS can do. Moreover, the industry evolves rapidly. Completing the lessons here is only the first step; you will find that you need to seek new information and training constantly as you develop your skills. How can you do this?

> **Read the Help.** Maybe you can figure out your cell phone without the manual, but if you ignore the Help files you are turning your back on a wealth of information. They don't just have instructions for doing things but also discussions, diagrams, references, and other ways to help you understand the concepts behind GIS as well as its implementation.

> **Use online courses.** ESRI has dozens of online courses and seminars to help you learn. If you have a licensed product, you have access to all of ESRI's online training for free. Talk to your GIS instructor about how to get access to these courses.

> **Build your GIS library.** GIS integrates many disciplines, including geography, surveying, cartography, statistics, computer science, spatial statistics, and so on. The more you know of these disciplines, the more you will understand, not just the *How* of doing things but also the *Why*.

> **Join a professional organization.** Many professional societies cater to GIS and remote sensing. They have newsletters and journals, conferences, and lots of professionals who can help answer your questions and share their experiences.

> **Join an online forum.** You can join an organization called GeoNet, supported by ESRI, that connects you with other GIS users. You can submit questions, look up common software problems, follow groups with similar interests, and communicate with the worldwide network of GIS professionals.

> **Join ArcGIS Online.** Not only is it easy and fun to create maps and share them, but you can also search for courses, documents, slideshows, videos, and other GIS-related materials. You can create a public account for free. If your university has a site license, you may also be given a subscription account that offers premium data content, the ability to publish data, and other benefits.

# Chapter 1. GIS Data

## Objectives

▶ Understanding how real-world features are represented by GIS data

▶ Knowing the differences between the raster and vector data models

▶ Getting familiar with the basic elements of data quality and metadata

▶ Learning the different types of GIS files used by ArcGIS

▶ Using ArcMap to view GIS data

▶ Learning about map documents, layers, and data frames

## Mastering the Concepts

### GIS Concepts

#### *Representing real-world objects on maps*

To work with maps on a computer requires developing methods to store different types of map data and the information associated with them. Map data fall into two categories: **discrete** and **continuous. Discrete** data are objects in the real world with specific locations or boundaries, such as cities, roads, or soils units. **Continuous** data represent a quantity that is measured and recorded everywhere over a surface, such as temperature or elevation.

Many different data formats have been invented to encode data for use with GIS programs; however, most follow one of two basic approaches: the **vector** model, which is designed to store discrete data, or the **raster** model, which is designed to store continuous data. In either approach, the critical task includes representing the information at a point, or over a region in space, using $x$ and $y$ coordinate values (and sometimes $z$ for height). The $x$ and $y$ coordinates are the spatial data. The information being represented, such as a soil type or a chemical analysis of a well, is called the attribute data. Raster and vector data models both store spatial and attribute data, but they do it in different ways.

Both data systems are **georeferenced,** meaning that the information is tied to a specific location on the earth's surface using $x$-$y$ coordinates defined in a standard way: a **coordinate system.** One can choose from a variety of coordinate systems, as we will see in Chapter 3. As long as the coordinate systems match, we can display any two spatial data sets together, and they will appear in the correct spatial relationship to each other.

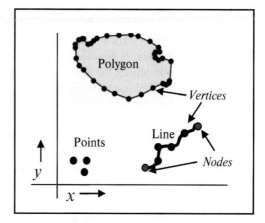

Fig. 1.1. The vector data model uses a series of $x$-$y$ locations to represent points, lines, and polygon areas.

#### *The vector model*

Vector data use a series of $x$-$y$ locations to store information (Fig. 1.1). Three basic vector objects exist: points, lines, and polygons. These objects are called **features. Point** features are used to represent objects that have no dimensions, such as a well or a sampling locality. **Line** features represent objects in one dimension, such as a road or a utility line. **Polygons** are used to represent two-dimensional areas, such as a parcel or a state. In all cases, the features are represented using one or more *x-y* coordinate locations (Fig. 1.1). A point consists of a single *x-y* coordinate pair. A line includes two or more pairs of coordinates—the end pairs of the line are termed **nodes**, and each of the intermediate pairs is called a **vertex**. A polygon is a group of **vertices** that define a closed area.

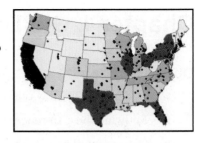

Fig. 1.2. A states feature class and a cities feature class
Source: ArcMap

The type of object used to represent features depends on the scale of the map. A river would be represented as a line on a map of the United States because at that scale it is too small for its width to encompass any significant area on the map. If one is viewing a USGS topographic map, however, the river encompasses an area and would be represented as a polygon.

In GIS, like features are grouped into data sets called **feature classes** (Fig. 1.2). Roads and rivers are different types of features and would be stored in separate feature classes. A feature class can contain only one kind of geometry—it can include point features, line features, or polygon features but never a combination. In addition, objects in a feature class have information stored about them, such as their names or populations. This information is called the **attributes** and is stored in a table (Fig. 1.3). A special field, called the Feature ID (**FID**) or ObjectID (**OID**), links the spatial data with the attributes. Each feature's attributes are stored in one row of the table, and each column has a different type of information, such as population or area. A river and a highway would not be found in the same feature class because their information would be different—flow measurements for one versus pavement type for the other—and would need to be stored in different tables with different columns.

| OBJ | Shape | STATE_NAM | SUB_REGIO | STAT |
|---|---|---|---|---|
| 1 | Polygon | Hawaii | Pacific | HI |
| 2 | Polygon | Washington | Pacific | WA |
| 3 | Polygon | Idaho | Mountain | ID |
| 4 | Polygon | Oregon | Pacific | OR |
| 5 | Polygon | Nevada | Mountain | NV |

Fig. 1.3. Each state is represented by a spatial feature (polygon), which is linked to the attributes.
Source: ArcMap

When a state is highlighted on the map, its matching attributes are highlighted in the table, and vice versa. It is this live link between the spatial and attribute information that gives the GIS system its power. It enables us, for example, to create a map in which the states are colored based on an attribute field, such as population (Fig. 1.2). This technique is called **thematic mapping** and is one example of how linked attributes can be used to analyze geographic information.

Feature classes can be stored in several different formats. Some data formats only contain one feature class. Others, called **feature datasets**, can contain multiple feature classes that are in some way related to one another. For example, a feature dataset called Transportation might contain the feature classes Roads, Traffic Lights, Railroads, Airports, and Canals.

The benefits of the vector data model are many. First, it can store individual features, such as roads and parcels, with a high degree of precision. Second, the linked attribute table provides great flexibility in the number and type of attributes that can be stored about each feature. Third, the vector model is ideally suited to mapmaking because of the high precision and detail of features that can be obtained. The vector model is also a compact way of storing data, typically requiring a tenth of the space of a raster with similar information. Finally, the vector model is ideally suited to certain types of analysis problems, such as determining perimeters and areas, detecting whether features overlap, and modeling flow through networks.

However, the vector model has some drawbacks. First, it is poorly adapted to storing continuous surfaces, such as elevation or precipitation. Contour lines (as on topographic maps) can be used to represent surfaces, but calculating derived information from contours, such as slope, flow direction, and aspect, is difficult. Finally, some types of analysis are more time-consuming to perform with vectors.

### Modeling feature behavior with topology

Two basic vector models exist: **spaghetti models** and **topological models**. A spaghetti model stores features of the file as independent objects, unrelated to each other. Simple and straightforward, this type of model is found in many types of applications that store spatial data. It is also commonly used to transfer vector features from one GIS system to another.

A topological data model stores features, but it also contains **topology**, information about how the features are spatially related to each other: for example, whether two parcels share a common boundary (**adjacency**), whether two water lines are attached to each other (**connectivity**), whether a company sprayed pesticide over the same area on two different occasions (**overlap**), or whether a highway connects to a crossroad or has an overpass (**intersection**). Although computer algorithms can determine whether these spatial relationships exist between features in a spaghetti model, storing topology can save time if the relationships must be used repeatedly.

Another application of topology involves analyzing the **logical consistency** of features. Logical consistency evaluates whether a data model or data set accurately represents the real-world relationships between features. For example, two adjacent states must share a common boundary that is exactly the same (the real-world situation), even though the states are stored in the data model as two separate features with two boundaries that coincide (Fig. 1.4). Lines representing streets should connect if the roads they represent meet. A line or a polygon boundary should not cross over itself.

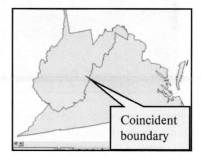

Fig. 1.4. A coincident boundary gets stored twice but is the same for both features.
Source: ArcMap

Finally, topology can be used to model the real-world behavior of features. In a network topology, for example, the connections between features are tracked so that flow through the network can be analyzed. Applications of networks include water in streams, traffic along roads, flights in and out of airline hubs, and utilities through pipes or electrical systems.

### The raster model

The raster model has the benefit of simplicity. A set of spatial data is represented as a series of small squares, called **cells** or **pixels** (Fig. 1.5). Each pixel contains a numeric code indicating a single attribute, and the raster is stored as an array of numbers.

Vector features, such as roads or land use polygons, can be converted to raster format by selecting a single attribute to be stored in the cells. In Figure 1.5a, the cells store numeric values representing a land cover type, such as 46 for conifer forest or 23 for hardwood forest. Each value is given a different color for display. The roads shown in Figure 1.5b were originally vector line features with a text attribute indicating a primary, secondary, or primitive road type. When converted to a raster, the number (1, 2, or 3) is used to represent each road that passes through a cell. The cells that don't contain roads are given a null value. Rasters that store vector features in a raster format are sometimes called **discrete** rasters.

However, rasters excel at storing continuous data, which are quantities or variables that change over the earth's surface. A **digital elevation model** (DEM), for example, stores elevation values (Fig. 1.5c). Cells are unlikely to have the same elevation as their neighbors, and the values range smoothly into one another, forming a continuous surface (or continuous field). Therefore, they are commonly called **continuous** rasters.

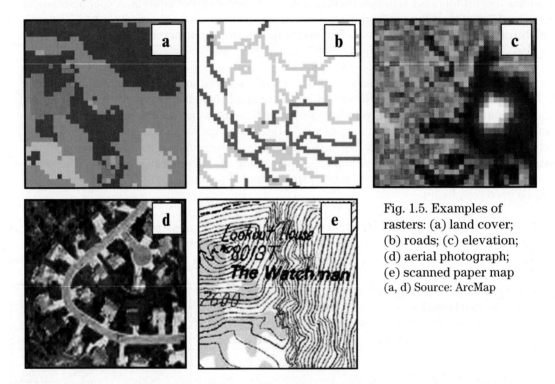

Fig. 1.5. Examples of rasters: (a) land cover; (b) roads; (c) elevation; (d) aerial photograph; (e) scanned paper map (a, d) Source: ArcMap

Image rasters (Fig. 1.5d) store brightness values and are commonly used to store aerial photographs or satellite images. Images may contain multiple arrays of values, called **bands**, to represent each pixel. Color images often contain a red band, a green band, and a blue band for each pixel, and the mixture of values in each band defines the color of the pixel. One can also do a digital scan of a paper map and store it as a raster, such as the US Geological Survey topographic map shown in Figure 1.5e, known as a **digital raster graphic** (**DRG**). Each cell stores an index code representing a different color, such as 5 for the brown contours and 1 for the white background.

A raster data set is laid out as a series of rows and columns. Each pixel has an "address" indicated by its position in the array, such as row = 3 and column = 6. Georeferencing a map in an $x$-$y$ coordinate system requires four numbers: an $x$-$y$ location for one pixel in the raster data set and the size of the pixel in the $x$ and $y$ directions. Usually the upper-left corner is chosen as the known location, and the $x$ and $y$ pixel dimensions are the same so that the pixels are square. From these four numbers, it is possible to calculate the coordinates of every other pixel based on its row and column position. In this sense, the georeferencing of the pixels in a raster data set is implicit—one need not store the $x$-$y$ location of every pixel.

The $x$ and $y$ dimensions of each pixel define the **resolution** of the raster data. The higher the resolution, the more precisely the data can be represented. Consider the 90-meter resolution roads raster in Figure 1.6. Since the raster cell dimensions are 90 meters, the roads are represented as much wider than they actually are, and they appear blocky. A 10-meter resolution raster could represent the roads more precisely; however, the file size would increase by 9 × 9, or 81 times.

Fig. 1.6. Impact of raster resolution. The 10-meter resolution raster stores the roads more precisely, but it takes 81 times as much space.
Source: ArcMap

The raster model mitigates some of the drawbacks of vectors. It is ideally suited to storing continuous information because each cell can have a value completely different from its neighbors. Many analyses are simple and rapid to perform, and an extensive set of analysis tools for rasters far outstrips those available for vectors.

The drawbacks of rasters lie chiefly in two areas. First, they suffer from trade-offs between precision and storage space to a greater extent than vectors do. The second major drawback of rasters is that they can store only one numeric attribute per raster. Vector files can store hundreds of attribute values for each spatial feature and can handle text data more efficiently.

### Coordinate systems

Both raster and vector data rely on $x$-$y$ values to locate data to a particular spot on the earth's surface. The $x$-$y$ values of the coordinate pairs can vary, however. The choice of values and units to store a data set is called its **coordinate system**. Consider a standard topographic map, which actually has three different coordinate systems marked on it. The corners are marked with degrees of latitude and longitude, which is termed a **geographic coordinate system** (**GCS**). Another set of markings indicates a scale in meters representing the UTM, or Universal Transverse Mercator, coordinate system. A third set of markings shows a scale in feet, corresponding to a State Plane coordinate system. Any location on the map can be represented by three different $x$-$y$ pairs corresponding to one of the three coordinate systems (Fig. 1.7). A global positioning system (GPS) unit also has this flexibility. It can be set to record a location in GCS degrees, UTM meters, State Plane feet, or other coordinate systems.

When creating a vector or raster data set, one must choose a coordinate system and units for storing the $x$-$y$ values. It is also important to label the data so that the user knows which coordinate system has been selected and what the units for the $x$-$y$ values are. Thus, every GIS data set must have a label that records the type of coordinate system and units used to store the $x$-$y$ data inside it.

In older GIS software, the coordinate systems of different data sets had to match in order for them to be drawn together in the same map. UTM data could only be shown with other UTM data, State Plane data with other State Plane data, and so on. If data were in different coordinate systems, they would need to be converted to the same coordinate system prior to display.

Fig. 1.7. A location can be stored using different coordinate systems and units. The x-y location of this school is shown in degrees, UTM meters, and State Plane feet.
Source: ArcMap

Although it is still true that coordinate systems must match for data to be displayed together, many GIS systems can now perform the conversion on the fly. This feature allows data to be stored in different coordinate systems yet to be drawn together. In ArcMap, the user defines a coordinate system for the map, and all of the data are converted to match (Fig. 1.8). The units defined for the map coordinate system, whether they are meters, feet, or degrees, become the **map units** and may differ from the stored units in the files. In Figure 1.8, the UTM data use meters to store the x-y coordinates, the GCS data use degrees, and the State Plane data use feet. The Oregon Statewide Lambert coordinate system uses meters, so meters become the map units.

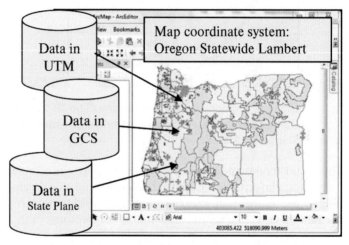

Fig. 1.8. Data in any coordinate system can be displayed together by setting the data frame coordinate system.
Source: ArcMap

### Map scale

The act of taking a set of GIS features with x-y coordinate values and drawing them on a screen or printing them on paper establishes a map scale. On a paper map, the scale is fixed at the time of printing. Within a computer system that allows interactive display, the scale changes every time the user zooms in or out of the map.

#### What is map scale?

**Map scale** is a measure of the size at which features in a map are represented. The scale is expressed as a fraction, or ratio, of the size of objects on the page to the size of the objects on the ground. Because it is expressed as a ratio, it is valid for any unit of measure. So for a common US Geological Survey topographic map, which has a scale of 1:24,000, one inch on the map represents 24,000 inches on the ground. You can use the map scale and a ruler to determine the true distance of any feature on the map, such as the width of a lake (Fig. 1.9). Measure the lake with a ruler and then set up a proportion such that the map scale equals the measured width over the actual width (x). Then solve for x. Keep in mind that the actual width and the measured width will have the same units. You can convert these units, if necessary.

Often people or publications refer to large-scale maps and small-scale maps. A large-scale map is one in which the *ratio* is large (i.e., the denominator is small). Thus, a 1:24,000 scale map is larger scale than a 1:100,000 scale map. Large-scale maps show a relatively small area, such as a quadrangle, whereas small-scale maps show bigger areas, such as states or countries.

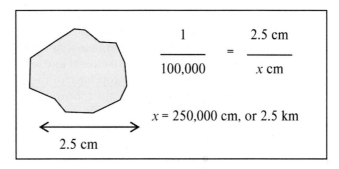

Fig. 1.9. Solving for the size of a lake

## Scales for GIS data

When data are stored in a GIS, they technically do not have a scale because only the coordinates are stored. They acquire a scale once they are drawn on the screen or on paper. However, a data set has a **source scale**, or the original scale or resolution at which it was converted to digital form. A 1:1 million scale paper map that is scanned or digitized cannot be used effectively at larger scales. The map in Figure 1.10 shows congressional districts in pink and the state outlines in thick black lines. The state boundaries are more angular and less detailed than the districts because they were digitized at a smaller scale. Thus, although it is possible to take small-scale data and zoom in to large scales, the accuracy of the data will suffer. The source scale of a data set is an important attribute and is included when documenting it.

Fig. 1.10. These two layers showing Massachusetts originated from maps with different source scales.
Source: ArcMap

One should exercise caution in using data at scales very different from the original source scale. Zooming in may give a false impression that the data are more precise than they actually are. A pipeline digitized from a 1:100,000 scale map has an uncertainty of about 170 feet in its location due to the thickness of the line on the paper. Displaying the pipeline on a city map at 1:10,000 might look all right, but it is likely to be many feet away from the actual location.

From looking at Figure 1.10, one might conclude that it is desirable to always obtain and use data at the largest possible scale. However, large-scale data require more data points per unit area, increasing data storage space, and slowing the drawing of layers. Every application has an optimal scale, and little is gained by using information at a higher scale than needed.

## *Data quality*

Representing real-world objects as points, lines, polygons, or rasters always involves some degree of **generalization**, or simplifying the data for digital storage, such as turning a house into a rectangular polygon, or even a point. No data file can capture all the spatial or attribute qualities of any object. The degree of generalization varies with the map scale. On a standard topographic map, a river has a width and can be modeled as a polygon with two separate banks. A city would be shown as a polygon area. For a national map, however, the river would simply be shown as a line, and a city would be shown as a point.

Even a detailed representation of an object is not always "true." Rivers and lakes can enlarge in size during a spring flood event or shrink during a drought. The boundary of a city changes over time as the city grows. Users of GIS data must never forget that the data they collect and use will contain flaws, and that the user has an ethical and legal responsibility to ensure that the data used for a particular purpose are appropriate for the task. When evaluating the quality of a data set, geospatial professionals consider the following aspects.

**Geometric accuracy** describes how close the $x$-$y$ values of a data set correspond to the actual locations on the earth's surface. Geometric accuracy is a function of the source scale and of how the data were captured. Surveying is one of the most accurate ways to position features. GPS units have an accuracy that ranges from centimeters to tens of meters. Maps derived from aerial photography or satellite imagery can vary widely in geometric quality based on factors such as the scale of the image, the resolution of the image, imperfections and distortions in the imaging system, and the types of corrections applied to the image. In Figure 1.11, notice that the vector road in white is offset in places from the road as it appears in the aerial photo. These differences can arise from digitizing errors, geometric distortions from the camera or satellite, or other factors.

Fig. 1.11. Aerial photo near Woodenshoe Canyon, Utah. Source: Google Earth and TeleAtlas

Moreover, not every boundary can be as precisely located as a road. Imagine that you wish to delineate the land-cover types: *forest, shrubland, grassland,* and *bare rock* in Figure 1.11. Where would you draw the line between *shrub* and *grassland*? At what point does the *shrubland* become *forest*? Six people given this photo would come up with six slightly different maps. Some boundaries would match closely; others would vary as each person made a subjective decision about where to place each boundary.

**Thematic accuracy** refers to the attributes. Some types of data are relatively straightforward to record, such as the name of a city or the number of lanes in a road. Even in this situation, the value of a feature might be incorrectly recorded. Other types of information can never be known exactly. Population data, for example, are collected through a process of surveying and self-reporting that takes many months. It is impossible to include every person. Moreover, people are born and die during the survey process, or are moving in and out of towns. Population data can never be more than an estimate. It is important to understand the limitations and potential biases associated with thematic data.

**Resolution** refers to the sampling interval at which data are acquired. Resolution may be spatial, thematic, or temporal. Spatial resolution indicates at what distance interval measurements are taken or recorded. What is the size of a single pixel of satellite data? If one is collecting GPS points by driving along a road, at what interval is each point collected? Thematic resolution can be impacted by using categories rather than measured quantities: if one is collecting information on the percent crown cover in a forest, is each measurement reported as a continuous value (32%) or as a classified range (10–20%, 20–30%)? Temporal resolution indicates how frequently measurements are taken. Census data are collected every 10 years. Temperature data taken at a climate station might be recorded every 15 minutes, but it might also be reported as a monthly or yearly average.

**Precision** refers to either the number of significant digits used to record a measurement or the statistical variation of a repeated single measurement. Many people confuse precision with accuracy, but it is important to understand the distinction. Imagine recording your body temperature with an oral digital thermometer that records to a thousandth of a degree and getting the value of 99.894 degrees Fahrenheit. This measurement would be considered precise. However, imagine that you take the reading immediately after drinking a cup of hot coffee. This action throws off the thermometer reading so that it does not record your true body temperature. Thus, the measurement is precise, but it is not accurate.

Evaluating the quality of a data set can be difficult, especially if the data were created by someone else. Professionals who create data usually also provide **metadata**, which stores information about the data set, such as where it came from, how it was developed, who assembled it, how precise it is, and whether it can be given to another person. The user can then decide whether a particular data set is suitable based on the information in the metadata.

### Citing GIS data sources

Ethical and professional considerations require that any map, publication, or report should cite the data source(s) used and give proper credit to the originators of the data. The metadata, or the site from which the data were obtained, are good sources of information for citations. The best practice is to record the citation when the data are obtained so the information is available when needed. Generally one cites only data that are publicly available (free or purchased). Data created internally within the workplace need not be cited, although often the company name or logo will appear on the map. A data set provided once in response to a personal request should be cited as a personal communication.

Keep in mind that the place you found the data may not always be its source. Your GIS administrator may have placed often-used data sets, such as ESRI Data and Maps, on a workplace fileserver for easy access, but the fact that they were obtained internally does not free you from the need to cite them, and you must cite the original source, not the local server.

#### General format

The purpose of a citation is to allow other people to find and obtain the data. It is not always possible to find all the information needed for a complete citation, but one should do one's best to make it as complete as possible. The following general format for citations may be used:

Data set name (Year published) [source type]. Producer name, producer contact information. *Resource URL: [Date accessed].*

*Data set name.* The name is assigned by the creator or provider of the data.

*Year published.* Some data sets are assembled and provided once or at long intervals, and these are considered to have a publication date. For example, the ESRI Data and Maps product is released in revised form with each version of the software and carries a publication year. Aerial photography is flown on a particular date (although mosaics such as Google Earth use multiple sources spread over several years). It may take a little hunting or a few questions to find the publication date. Some data sets are updated at shorter intervals, or are even live. These can be assigned the current year.

*Source type.* Indicate the format in which the data are available. Types might include physical media (DVD, CD-ROM), a file downloaded from the Internet, or a service that provides live data on demand. Different types of services exist, such as database service, map service, image service, map package, or layer package. New types are being added all the time.

***Producer name.*** Give the name of the person or agency that makes the data available. In some cases, this may be different from the originator of the data. For example, ESRI publishes Data and Maps using data from many different sources. ArcGIS Online serves public data.

***Producer contact information.*** For a company or small agency, the city and state should be included. For large agencies, particularly those with many offices but a unified web presence, the name itself is sufficient. Indicate a clearinghouse name, such as ArcGIS Online, if appropriate.

***Resource URL.*** This entry is optional; include when appropriate. Use only static URLs even if it means the user has to hunt for the data. (A static URL always has the same form, whereas a dynamic URL is generated automatically based on search strings or other information. Dynamic URLs generally contain gibberish and/or characters like %.)

***Date accessed.*** This entry is optional because it mainly applies to online data sets. Include the year and month when you accessed or downloaded the data.

### Examples of citations

Black Hills RIS Vegetation Database (2008) [downloaded file]. Black Hills National Forest, Custer, SD. URL: http://www.fs.usda.gov/main/blackhills/landmanagement/gis [August, 2010].

ESRI Data and Maps (2012) [DVD]. ESRI, Inc., Redlands, CA.

National Hydrology Dataset (2012) [downloaded file]. United States Geological Survey on the National Map Viewer. URL: http://viewer.nationalmap.gov/viewer/ [July 23, 2012].

USA Topo Maps (2009) [map service]. ESRI on ArcGIS Online. URL: http://server.arcgisonline.com/arcgis/services/USA_Topo_Maps/MapServer [January 1, 2012].

EIA Coalbed Methane Field Boundaries (2011) [map service]. US Department of Energy on ArcGIS Online. URL: http://arcgis.com [August, 2013].

Mineral Operations of Africa and the Middle East (2010) [layer package]. J.M. Eros and Luissette Candelario-Quintana on ArcGIS Online. URL: http://ArcGIS.com [July, 2012].

Badlands National Park GIS Database (2012) [CD-ROM]. Interior, South Dakota: National Park Service—Badlands National Park, personal communication.

## About ArcGIS

### ArcGIS overview

ArcGIS is developed and sold by Environmental Systems Research Institute, Inc. (ESRI). It has a long history and has been through many versions. Originally developed for large mainframe computers, it has metamorphosed from a system based on typed commands to a user-friendly graphical user interface (GUI). Data models, too, have changed over time, so that one is likely to encounter data sets in different formats. Knowing this background helps a student of GIS understand the nature of the ArcGIS system and its data.

The old core of the ArcGIS system was called Arc/Info and included three programs: Arc, ArcEdit, and ArcPlot (Fig. 1.12), which utilized a data model known as a **coverage** and incorporated a database program called INFO. All of the programs were command based, meaning that the user typed commands to make the program work. The difficulty of learning Arc/Info prompted ESRI to create ArcView in 1992, which was easier to use but not as powerful as Arc/Info. ArcView used a simpler data model, called the **shapefile**, although it could read coverages and convert them to shapefiles.

ArcGIS Desktop, released in 2001, combines power and ease of use. It contains two programs. ArcMap provides the means to display, analyze, and edit spatial data and data tables. ArcCatalog facilitates viewing and managing of spatial data files. ArcGIS

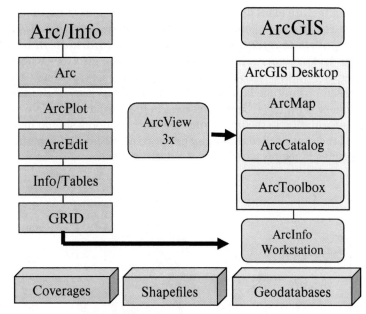

Fig. 1.12. Relationship between ESRI products and data formats

Desktop also contains ArcToolbox, a collection of tools and functions for operations in ArcCatalog and ArcMap, such as converting between data formats, managing map projections, and performing analysis (Fig. 1.13). Users may create and add their own tools or scripts for special or often-used tasks. A program called Model Builder lets users graphically arrange and run sequences of steps, and save them to be used over and over.

ArcGIS Desktop provides different levels of functionality that all use the same basic interface. Users can save money by buying only the level they need. This book primarily uses the functions available in ArcGIS Basic.

- ▶ ArcGIS Basic (formerly ArcView) provides a suite of mapping, editing, and analysis functions and is the level of functionality most users will require on a regular basis.

- ▶ ArcGIS Standard (formerly ArcEditor) adds advanced editing capabilities, such as topology and network editing, as well as additional data conversion tools.

- ▶ ArcGIS Advanced (formerly ArcInfo) provides access to the full functionality and adds more advanced tools for analysis, editing, data handling, and cartography.

Fig. 1.13. ArcToolbox and a model
Source: ArcToolbox

### *ArcMap documents*

ArcMap creates **map documents**, a collection of spatial data layers and tables and their properties. The document is designed to contain a single map page, called a **layout**, which is used to design a map for printing. The map document is also used for viewing, analyzing, editing, or managing data for a project.

When a feature class or a raster is added to ArcMap, it becomes a **layer**. The difference between a feature class and a layer is important to grasp. A feature class is a stored set of spatial data. *A layer points to a feature class and stores information about how to display it and use it*(Fig. 1.14). A layer is like a cooking recipe—it describes where to find the data file (the ingredients) and how to display it (the cooking instructions). The same feature class can be accessed by multiple layers to present different views of the same data, even within a single map document (Fig. 1.14). Layers are held in memory and stored when the map document is saved. The information in the layer, such as which symbols should be used and whether the features have labels, are called the **layer properties**. Creating a map involves manipulating these properties in menus to produce the desired result.

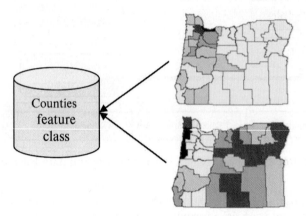

Fig. 1.14. Layers and feature classes. The counties feature class is stored on a drive and contains the spatial and attribute information. The layers point to the feature class but present the data in two different ways: as a graduated color map of population density (top) or the vacancy rate of housing units (bottom). Source: ArcMap

A layer may be saved as a **layer file**. This file stores the location of the data set upon which the layer is based and the properties of the layer. Saved layer files can be used to quickly add a layer with predefined symbols to a map document. Group layer files can be organized in multiple layers with predefined symbols into thematic maps or base maps.

Map documents organize layers using data frames, windows that contain groups of layers that are drawn together (Fig. 1.15). Each data frame has its own properties, its own data layers, its own geographic extent, its own scale, even its own coordinate system. In Figure 1.15, the top data frame shows South Dakota with its highways, counties, and rivers using a State Plane coordinate system. The second frame, showing the location of South Dakota within the United States, contains only the states, has a much smaller scale, and uses the Equidistant Conic coordinate system.

One frame at a time is designated the **active frame**. All commands and actions, such as adding layers or creating a scale bar, occur in the active frame. If you wished to add rivers to the lower data frame in Figure 1.15, first you would designate it as the active frame, and then you would add the rivers feature class to the map.

Fig. 1.15. Data frames
Source: ArcMap

### Data files in ArcGIS

ArcGIS can read a variety of GIS file formats (Table 1.1); some are native to ArcGIS and some originate from other sources.

Table 1.1. Types of files and data sources used by ArcGIS

| File type | Description |
|---|---|
| Shapefiles | Shapefiles are vector feature classes developed for the early version of ArcView and have been carried over into ArcGIS. |
| Coverages | A coverage is the vector data format developed for Arc/Info and is the oldest of the data formats. |
| Geodatabases | Geodatabases represent an entirely new model for storing spatial information with additional capabilities. |
| Database connections | Database connections permit users to log in to and utilize data from an RDBMS geodatabase. |
| Layer files | A layer file references a feature class and stores information about its properties, such as how it should be displayed. |
| Rasters | Rasters represent map data or imagery using arrays of regular cells, or pixels, containing numeric values. |
| Tables | Tables can exist as separate data objects that are unassociated with a spatial data set. |
| GIS servers | GIS servers provide GIS data over an Internet connection as maps, features, or images. |
| TINs | TINs are Triangulated Irregular Networks that store 3D surface information, such as elevation, using a set of nodes and triangles. |
| CAD drawings | Data sets created by CAD programs can be read by ArcGIS, although they cannot be edited or analyzed unless they are converted to shapefiles or geodatabases. |

Source: ArcGIS

**Shapefiles**

Shapefiles are spaghetti data models containing a feature class composed of points, lines, or polygons, but never a mixture. The attributes are stored in a **dBase** file. Shapefiles can store **multipart features**, in which a single feature includes multiple objects. For example, Hawaii requires multiple polygons to represent the different islands, but the state can be stored as a multipart feature so that it has one record in the attribute table.

Although a shapefile appears as one icon in ArcGIS, it is actually composed of multiple data files as seen in Windows Explorer (Fig. 1.16). The rivers shapefile consists of seven different files. The .shp file stores coordinate data, the .dbf file stores attribute data, and the .shx file stores a spatial index that speeds drawing and analysis. These first three files are required for every shapefile to function. Additional files may also be present: the .prj file stores projection information, the .avl file is a stored legend, and the .xml file contains metadata. To copy a shapefile, all of these files must be included. ArcCatalog takes care of this automatically, but Windows Explorer does not—one reason it is wise to manage GIS data using ArcCatalog.

Fig. 1.16. Shapefiles are groups of files but appear as single entries in ArcCatalog.
Source: Windows Explorer

In a shapefile attribute table, the first two columns of data are reserved for storing the feature identification code (FID) and the coordinate geometry (Shape) field. These fields are created and maintained by ArcGIS and cannot be modified by the user. All other fields are added by the user.

## Geodatabases

A geodatabase can contain many different objects, including feature classes, networks, tables, rasters, and topology. Figure 1.17 shows a geodatabase named oregondata. Feature classes may exist as individual objects in a geodatabase (as do the cities or counties), or they may be grouped into feature datasets, a collection of related feature classes with the same coordinate system, such as the Transportation feature dataset in Figure 1.17. For now, you can think of a feature dataset as a folder for organizing data, but they have additional capabilities that will be introduced in later chapters. Geodatabases can also store rasters like the gtopo1km raster in Figure 1.17.

Fig. 1.17. A geodatabase in ArcCatalog
Source: ArcCatalog

Three types of geodatabases are used by ArcGIS. *Personal geodatabases* are designed for use by individuals or small workgroups and are stored in a single Microsoft Access file. This file is limited to 2 GB in size and works only in the Windows operating system. *File geodatabases* are stored in a system folder, and each file can be up to 1 TB in size. File geodatabases can be accessed by multiple operating systems, including Linux or Unix. *SDE geodatabases*, also known as *enterprise geodatabases*, store GIS data within a commercial relational database management system (RDBMS), such as Oracle® or SQLServer®. SDE geodatabases are designed to meet security and management needs for large data sets accessed by many simultaneous users. In this book we use personal or file geodatabases. Which type is used does not usually matter.

## Coverages

A **coverage** is the oldest vector format, developed for Arc/Info. ArcGIS Desktop has limited functions for managing coverages, so most users will encounter them simply as an old data format that must be converted to a shapefile or exported to a geodatabase. Several things are helpful to know in this process.

Coverages contain multiple feature classes, which may store points, arcs, polygons, and polygon labels. Coverages also store topology, and the tables have several attribute fields reserved for this purpose. Figure 1.18 shows these fields for a coverage called LANDUSE (some fields use the coverage name as part of the field name). It makes sense to delete these fields during or after the conversion, for they serve no useful purpose afterwards.

Fig. 1.18. Coverages contain fields of little value once they are converted.
Source: ArcGIS Desktop

## Rasters

Rasters in ArcGIS can take a variety of formats, such as JPEG, TIF, GEOTIF, BMP, MrSID, and raw binary files (BIP, BIL, and BSQ). A list of supported raster formats can be found in the ArcMap Help under the index heading "rasters, formats." Most rasters consist of the spatial data plus a header that gives information about the file, such as its number of rows and columns and its coordinate system. This information may be stored in a separate file or as the first part of the binary raster. Rasters can also be stored inside geodatabases.

### CAD files

Data sets from CAD programs can be read by ArcGIS, although they cannot be edited or analyzed unless they are converted to shapefiles or geodatabases. A CAD file may contain multiple feature classes that correspond to layers of the drawing and can be opened separately.

### Database connections

A user can connect to a database management system (DBMS) on a network through a database connection. This connection can be open, or it can require a login ID and a password. Once inside, the user can access the tables within the database. Database connections may be available on a local network through a mounted drive, but some can be accessed online as database services.

### *Internet data services*

Many organizations make data available over the Internet. Free data services, such as Google Earth and MapQuest, provide access to huge volumes of image and map data. These services are used online in a web interface and cannot be viewed in ArcMap. Although the data quality and documentation procedures are not designed for professional-level work, the volume and popularity of these sites introduce many people to GIS techniques and data. Other sites, such as the United States Geological Survey's National Map, do allow users to download data.

### GIS servers

GIS servers provide geospatial data over Internet connections. Hosting GIS data on a server requires software that responds to requests from users for specific maps or data in the user's current window. Some GIS services are open and free, designed to provide public access to data and maps. Others may be locked down to a particular organization or group of users. Many organizations use servers to make in-house data available to personnel out in the field, or to allow employees in different locations to access the same company data sets. Servers provide data in a client-neutral format; as long as the client program knows how to use the service, the data can be used in ArcGIS Desktop, in a web application, on a tablet, or on a smartphone.

Several types of services may be offered. A **map service** renders map layers as tiles and sends them to the user as static images. This type of service is efficient and fast, but the user cannot modify how the map is symbolized. A **feature service** shares the requested data features; it tends to be slower, but the user can change how the features look and potentially edit or analyze them. A **layer package** or **map package** sends the data to the client, where they are stored locally during use. An **image service** provides access to large mosaics of satellite imagery or aerial photography. GIS servers can also provide analysis tools in addition to data. A **geoprocessing service** makes available certain computations and functions, so that users can perform the designated analysis through a web site, even if they do not have GIS software installed.

### Cloud-based services

Setting up and managing a GIS server is a complex and expensive task requiring suitable equipment, software, and expertise. Cloud-based services are lowering this barrier. A **cloud** consists of warehouses of computers and hard drives managed by a company that rents processing power and disk space to clients. Some companies offer space to individuals to back up their computers or store movies and music files in a place that is accessible to all of their devices.

Another type of cloud service is the virtual machine, or VM, created by setting aside part of a large computer with its own operating system and software, allowing it to be managed and used like an ordinary physical computer. The VM can be scaled from a standard desktop to a powerful machine capable of running intensive computations or serving thousands of web requests. The client rents the VM for an hourly fee. In web applications, cloud services can be configured to automatically add more VM units when demand is high and scale back when usage is light.

Many organizations have moved to the cloud to host their GIS data, either as a complement to or even instead of housing their own GIS servers. Advantages of cloud services include the speed and ease of deployment, the ability to scale up and down to meet changing demand, and the security benefits of hosting public information outside the organizational firewall.

### ArcGIS Online and web maps

ArcGIS Online is a cloud-based platform that provides an environment to create and share maps. It makes data available to people with little to no GIS training, yet it also addresses the needs of professional users. Anyone may use the data, and those willing to set up a free account may also share data. The user controls whether a published data set is visible to a few selected people or to the public. The platform contains easy tools for creating web pages with interactive maps.

Organizations may purchase a subscription account, which allows them to access additional data sets, manage security, designate users allowed to access or publish data, and host data in the cloud without needing to configure their own GIS servers. A member of an organizational subscription has access to many data sets, geoprocessing services, geocoding services, and groups, provided the system administrator has assigned those permissions to the account.

ArcGIS Online is designed around the **web map**, an interactive map based solely on GIS services. Web maps perform a restricted set of basic functions, such as zoom and query; however, they are device independent and can be used in ArcMap, in web browsers, on mobile devices like smartphones or tablets, and even within social media sites. They are simple to create and share.

Using servers and web maps to collect data on smartphones or tablets has become a widespread activity. It is relatively simple to set up a feature service and configure it for editing. Users with a free account can use the Explorer for ArcGIS mobile application to access web maps and collect data; some students at the South Dakota School of Mines and Technology used it to map trees on campus and rate damage caused by a blizzard (Fig. 1.19). Users with a subscription account can use the more powerful Collector application, which is capable of attaching photos and editing while out of the range of cell or wireless services, or the Survey123 application, which lets users create and fill out geotagged forms.

Fig. 1.19. Students used smartphones to map damaged trees after a blizzard.
Source: ArcGIS mobile application

The growing capabilities of ArcGIS Online are turning it into an indispensable tool. ArcGIS Desktop will remain the primary engine for creating, mapping, and analyzing data for quite some time yet, as it is far more powerful than the web services, but Desktop users certainly benefit from web services.

## Summary

▶ A GIS is a database system that uses both spatial and attribute data to answer questions about where things are and how they are related. It has many functions, including creating data, making maps, and analyzing relationships.

▶ Raster data employ arrays of values representing conditions on the ground within a square called a pixel. The array is georeferenced to a ground location using a single $x$-$y$ point.

▶ Vector data use sequences of $x$-$y$ coordinates to store discrete point, line, or polygon features. Every feature is linked to an attribute table containing information about the feature.

▶ Every GIS data set has a coordinate system defined for stored $x$-$y$ coordinate values. Many different coordinate systems are used, so each data set must be labeled with information about the coordinate system.

▶ Map scale is the ratio of the size of objects in the map to their size on the ground. The source scale affects the accuracy and precision of GIS data and for what scales it is suitable.

▶ Every GIS user has a responsibility to ensure that data are suitable for the proposed application. Data quality is measured in terms of geometric accuracy, thematic accuracy, resolution, and precision. Metadata store information about GIS data layers to help people use them properly.

▶ ArcGIS Desktop employs a menu-based interface for two programs called ArcMap and ArcCatalog. It uses many data formats, including shapefiles, coverages, geodatabases, rasters, images, TINs, CAD drawings, and Internet-based data services.

▶ ArcMap uses layers and layer properties to manipulate data in a map. Layers point to stored data sets and contain properties that indicate how they should be displayed.

▶ GIS services make data available over the Internet and are becoming an increasingly important part of GIS.

### Important Terms

| | | | |
|---|---|---|---|
| active frame | discrete | layer package | polygon |
| adjacency | feature | layer properties | precision |
| attributes | feature class | layout | query |
| band | feature dataset | Layout view | raster |
| cell/pixel | feature service | line | resolution |
| cloud | FID/OID | logical consistency | scale range |
| connectivity | generalization | map document | shapefile |
| continuous | geographic coordinate | map package | source scale |
| coordinate system | system (GCS) | map scale | spaghetti model |
| coverage | geometric accuracy | map service | Table of Contents |
| data frame | geoprocessing service | map units | thematic accuracy |
| Data view | georeferenced | metadata | thematic mapping |
| datum | image service | multipart feature | topological model |
| dBase | intersection | node | topology vector |
| digital elevation model | layer | overlap | vertex/vertices |
| digital raster graphic | layer file | point | web map |

## Chapter Review Questions

1. Explain the difference among the terms *feature*, *feature class*, and *feature dataset*.

2. Imagine you are looking at a geodatabase that contains 50 states, 500 cities, and 100 rivers. How many feature classes are there? How many features? How many attribute tables? How many total records in all the attribute tables?

3. If the following data were stored as rasters, which ones would be discrete and which would be continuous: rainfall, soil type, voting districts, temperature, slope, and vegetation type?

4. John and Mary are collecting GPS data together. John's GPS says their location is at (631058, 4885805). Mary's GPS says their location is at (1204817, 663391). Explain what is going on. What must be done to make the GPS units agree?

5. Would raster or vector be a better format for storing land ownership parcels? Give at least three reasons for your choice.

6. You measure a football field (100 yards long) on a large-scale map and find that it is 0.5 inch long. What is the scale of the map?

7. Scott is walking the boundary of a wetland area to map it. His expensive GPS records locations to the nearest 0.10 meter. Is the boundary he creates accurate? Is it precise? Explain your reasoning.

8. Imagine a feature class of agricultural fields with attributes for the crop and the organic matter content of the soil. What issues might impact the thematic accuracy of each attribute?

9. Explain some ways that GIS services are different from data that reside on your hard drive.

10. Construct an appropriate citation for the data that come with this book.

# Mastering the Skills

## *Preparing to begin*

The tutorials in this book use a data set available for download from the book web site www.mhhe.com/price7e. These data must be installed before you start using the tutorials. Each step of the tutorials and all skills learned in this book are illustrated by video clips which can also be found on the web site. See the web site for instructions on downloading and using the tutorial data and videos.

As you are working through the tutorials, you will learn better if you think carefully about each step. Be sure to follow the directions exactly, or you may find that subsequent steps look or behave differently than the book's instructions.

A **TIP** gives useful information about the program or settings that will save time or that are important to know. Pay attention to them.

A **SKILL TIP** tells about a technique that is not covered in a tutorial but that you may wish to learn about later using the Skills Reference at the back of the book. The section of the Skills Reference to consult is indicated in parentheses (ArcMap Basics).

**TIP:** In these tutorials, values you must enter are shown in this font: type this.

## Teaching Tutorial

The following examples provide step-by-step instructions for doing basic tasks and solving basic problems in ArcGIS. The steps you need to do are highlighted with an arrow →; follow them carefully. Click on the video number in the VideoIndex.htm file to view a demonstration of the steps.

**TIP:** This book comes with a folder of data called mgisdata. We suggest that you install it on the C:\ drive of your computer in a folder named C:\gisclass. If you have installed it elsewhere, move it now. If you cannot put it there for some reason, then write down where it is installed and substitute that location in the instructions: _____

### *Adding data to ArcMap*

1 → Start ArcMap.

1→ In the Getting Started window, click the New Maps text on the left side of the window, indicating that you want to start with a new, empty map document.

1→ Click on the Blank Map template icon, and click OK.

**TIP:** If using ArcGIS on a network, you may see a different configuration of toolbars. Choose Customize > Toolbars from the main menu, and you will see a list. Only these toolbars should be checked: Draw, Standard, and Tools. If you see others checked, click the names to turn them off.

2→ Click the Add Data button.

2→ Click on the *Look in* drop-down button to show available data folders (Fig. 1.20).

ArcGIS is able to work with many kinds of data, including files on your hard drive as well as online data. Data on your local computer, or on a network drive mounted on your computer, will be shown in Folder Connections, which stores shortcuts to locations containing GIS data.

Fig. 1.20. Sources of data
Source: ArcGIS

> 2→ Click the *Look in* drop-down button again, if necessary, and select Folder Connections.

If this is the first time ArcGIS has been used on the computer, this folder will be empty. Otherwise, it may contain connections, which will appear as folder entries (Fig. 1.21a). You must add a connection to your mgisdata folder to start using the data.

> 2→ To add a connection, click the Connect to Folder button in the Add Data window.
>
> 2→ Navigate to the directory containing your mgisdata folder and click once on the mgisdata <u>folder icon</u> to select it (Fig. 1.21b; yours may be in a different location). Do not select any of the subfolders in the mgisdata folder. (You are adding a connection right now; you will access the data in a moment.)
>
> 2→ Click OK to finish adding the connection.

**TIP:** Once you add a connection it will remain until you disconnect it. Connections tend to build up over time, so disconnect older ones no longer in use.

Now that the connection is added, you are ready to start adding data to the map.

> 3→ Click the *Look in* drop-down button one more time to find the connection you added. Click on it to open it.
>
> 3→ Double-click the Oregon folder to open it.
>
> 3→ Double-click the gray oregondata icon to open the geodatabase.
>
> 3→ Click the counties feature class to highlight it; then click Add to add it to the map.
>
> 4→ Click the Add Data button again. This time, double-click the Transportation feature dataset to open it.
>
> 4→ Click the airports feature class; then hold down the Ctrl key and click on the highways feature class so that both are selected. Click Add to add them.

Fig. 1.21. (a) Connection to mgisdata; (b) connecting to a folder
Source: ArcGIS

### Viewing the map

Examine the ArcMap window. The area on the left is called the **Table of Contents.** Examine the icons that appear directly below the Table of Contents title. These icons control the appearance of the Table of Contents and how it arranges the layers. The first icon, List by Drawing Order, is currently highlighted.

Source: ArcMap

5→ Hover the cursor above the highlighted icon to see the icon name.

5→ Examine the names of the other icons as well.

5→ You may click on the other options to see what they look like if you wish, but make sure afterward that the first icon is again highlighted.

The List by Drawing Order icon shows the layers in the order in which they are drawn, from bottom to top. Notice that the point and line layers that you just added were on top of the polygon layer. The draw order can be changed if needed.

6→ Click the counties layer name and drag it to the top of the list. Note that the other layers disappear from the map, as the polygons are drawn on top of them now.

6→ Move the counties layer back down to the bottom of the list.

6→ Place the airports above the highways, if they are not already there.

The three feature classes you added were placed underneath a heading called Layers. The Layers entry is a **data frame**, which is a box to contain map data to be displayed together.

We will begin by exploring the Tools toolbar. Figure 1.22 shows the tools used for zooming and panning the map.

Fig. 1.22. Part of the Tools toolbar
Source: ArcMap

 7→ Click the Zoom In tool. Place the cursor at the upper-left corner of the state, then click and hold the mouse button down to drag a box around a few counties. When finished, let go of the mouse button.

 7→ Click once in the lower-right corner of the map to zoom in again.

7→ To return to the full extent of the map (useful if you made a mistake or do not like the area you selected), click on the Full Extent button.

 7→ Click on the Zoom Out tool and click in the upper-left corner of the state again. The view zooms out with the point clicked at the center.

**TIP:** You can also draw a box using the Zoom Out tool. If a large box is drawn, the view zooms out a little bit. If a small box is drawn, the view zooms out a large amount.

 8→ Click the Pan tool; then click and drag inside the display window to move the map.

8→ Click the Go Back to Previous Extent button. It allows you to undo a zoom that you did not intend, or return to an earlier zoom level.

8→ Click the Go to Next Extent button. It reverses the effect of the Go Back button.

 8→ Click the Full Extent button to return to the view of all the counties.

**TIP:** Hover the cursor over a tool for a moment to find out the tool name and what it does.

9→ Find the scale readout on the main toolbar and click the drop-down arrow to see a list of preset scales. Choose 1:3,000,000.

Source: ArcMap

9→ Click on the Fixed Zoom In button two or three times. It zooms in to the next preset scale in the list.

9→ Click on the Fixed Zoom Out button two or three times. It zooms to the next scale.

9→ Practice these tools until you are familiar with them. When finished, click the Full Extent button to return to the full extent of the Oregon data.

The Identify tool can be used to examine the attributes of a feature.

10→ Locate the Identify tool on the Tools toolbar and click on it.

10→ Place the tool on top of one of the counties, staying away from an airport or a highway, and click on it.

10→ The county will flash on the screen, and the Identify Results box will appear. The attributes of the county, from the counties table, are displayed in the Identify window.

10→ Use the scroll bar to examine the attributes in the Identify window.

10→ Click another county and examine its attributes.

1. What is the name of the county in the northeast corner of Oregon? _____

11→ Click on a highway or airport to examine its attributes.

11→ In the Identify window, examine the entries in the *Identify from* drop-down box. You can use these to control which layer or layers will be identified.

11→ Close the Identify window by clicking the X in the upper-right corner.

**SKILL TIP:** Learn to use the Measure tool to determine areas, lengths, and perimeters in an assortment of different units (ArcMap Basics).

### Viewing attribute tables

To view the attributes for all of the features in a feature class, you can open the table.

12→ Right-click the counties layer in the Table of Contents and choose Open Attribute Table from the context menu that appears.

12→ Scroll to the right, to the end of the table and back, noting all of the fields.

2. How many records (rows) are in this table? _____

12→ Hold the cursor over the right edge of the NAME field until it turns into a double arrow bar. Click and drag the edge to increase or reduce the width of the column.

Tables have an options button with various tasks, such as finding text or exporting the table.

13→ Click the Table Options button and choose Find and Replace. Click the Find tab.

13→ Type Hood in the *Find what* box and click Find Next.

13→ The cursor jumps to the record for Hood River and subtly outlines the data cell.

13→ Examine the other Find options and then close the Find and Replace window.

A context menu gives access to several commands relating to the individual fields.

14→ Right-click the field name POP2014 to display a context menu. Choose Sort Descending. (Make sure you scroll back to the top of the table to see the most populous county, Multnomah.)

3. Which county has the smallest 2014 population? _____

14→ Right-click the POP2014 field and choose Statistics to see basic statistics and a frequency diagram of the values. Close the Statistics window when done looking.

14→ Right-click the NAME field and choose Freeze/Unfreeze Column. Scroll to the right, noting that the NAME field stays put now.

TIP: More than one field can be frozen at a time. Unfreezing allows the field to scroll again, although it remains on the left side until the field settings are reset.

TIP: You can sort, freeze, and do statistics without changing the source data.

A table **query** is used to select features meeting a certain criterion. For example, how many counties have more than 100,000 people?

 15→ Click the Select By Attributes button on the Table toolbar to open the Select By Attributes window.

15→ The top panel contains names of fields from the feature class table. Double-click the POP2014 field in the upper panel to enter it in the lower panel.

15→ Click the > button and then type the value *100000*.

15→ The lower panel should now read POP2014 > 100000. Click Apply.

16→ The selected features are highlighted in the table and on the map. Examine the bottom of the table to see how many were selected (you should see ten).

 16→ Click the Clear Selection button on the Table menu.

16→ Close the counties table by clicking the X in the upper-right corner.

### Choosing map symbols

Feature classes contain only locations and attributes and do not store symbols or labels. When a feature class is loaded into ArcMap, it becomes a **layer**. A layer is a specification for how the feature class will look and behave in the map document, and you can set properties to control it.

17→ Click the Add Data button. You last added data from the Transportation feature dataset, so click the Up arrow once to return to the main geodatabase.

17→ Click the volcanoes feature class to select it and click Add.

17→ Right-click the volcanoes layer to make the context menu appear—it contains actions or commands that can be performed on the layer.

17→ Move the cursor to the Properties entry at the bottom of the menu, and click it to open the Layer Properties window.

The tabs at the top of the window provide access to menus for setting various properties.

18→ Click on the Symbology tab.

18→ Click on the Symbol button that shows the current symbol. It opens the Symbol Selector window.

18→ Click on the Triangle 1 symbol to use it. Change the color to a reddish-brown and the size to 16. Click OK and then OK again in the Layer Properties window.

**TIP:** Clicking Apply in a window applies any changes you have made but keeps the window open. Clicking OK applies the change and closes the window.

19→ Right-click the airports layer and choose Properties.

19→ Click on the current Symbol button to open the Symbol Selector.

19→ Scroll down and find the Airplane symbol. Select it.

19→ Leave the color black, but increase the size to 25 pt, and set the angle to –45.

19→ Click OK and OK to close both windows.

**SKILL TIP:** Learn how to search for and add more symbols to a map than are listed in the Symbol Selector by default (Maps and Symbols).

You don't have to open the Properties to change the symbols for a layer, however.

20→ Right-click the highways layer symbol (not the text) to bring up a color context menu. Choose the black color.

20→ Click (left-click) on the counties symbol (not the text) to open the Symbol Selector, and choose the Beige symbol.

20→ Click the Outline Color drop-down button and hold the cursor on top of a color box until the color name appears. Set the color to the Gray 20% symbol. Click OK.

**SKILL TIP:** Use the Symbol Editor to create and manage new symbols (Maps and Symbols).

Labels are another property of a layer. You can create a set of labels and turn them on or off as you wish.

21→ Right-click the airports layer and open its properties. Click on the Labels tab.

21→ Check the box to *Label features in this layer.*

21→ The Label Field is preset to NAME, but it could be changed to something else if desired.

21→ Change the symbol to Arial 9 pt Bold font.

21→ Ignore the other options and click OK. The map should look similar to Figure 1.23.

It is a good practice to save the map periodically as you work in case the program crashes. Always be aware of where you save things, so you can find them again, and give them logical names that express what the file contains.

22→ Choose File > Save As from the main menu.

22→ Navigate to the mgisdata\MapDocuments folder, if necessary.

22→ Enter the name *Oregonmap* and save the map document.

Fig. 1.23. The map after Step 21
Source: ArcMap

### *Layer files*

Imagine that you will create a series of maps and want airports on all of them. The symbols and labels you just set will appear only in this map. Instead of setting them again each time, you can save these settings as a **layer file**. A layer file stores the location and properties used by the layer.

23→ Right-click the airports layer name and choose Save As Layer File.

23→ Navigate to the mgisdata\Oregon folder.

23→ Enter *myairports.lyr* as the name of the layer file and click Save.

Nothing obvious happens, but the layer file has been stored on disk, ready for the next time you want the airports symbolized this way on a map. What is the difference between this saved layer file and the original data set? Let's find out.

**TIP:** Removing a layer only takes it out of the map. It does not delete it from the hard drive.

24→ Right-click the airports layer and choose Remove.

24→ Click the Add Data button. Navigate inside the Transportation feature dataset and choose airports again to add it to the map.

The airplane symbol and labels are gone. You have loaded the original feature class. There are no instructions for how to display it, so the layer has a default symbol and no labels.

25→ Click the Add Data button. Navigate to the Oregon folder where you saved the layer file myairports.lyr. Select it and add it to the map.

25→ Right-click the second airports layer (that has the airplane symbol) and choose Properties. Click the Source tab.

25→ Examine the Data Source box. This layer, although saved in the Oregon folder, is accessing the feature class from the original geodatabase.

25→ Click Cancel to close the Layer Properties window.

Both of these airports layers use the <u>same</u> feature class. When you saved the layer file, it only stored the location and the display properties, not a copy of the original data.

Now let's examine some of other properties besides symbols and labels that can be set for layers. Instead of starting from scratch, we will examine properties already set and stored in a <u>group</u> layer file, which is similar to a layer file but contains properties for multiple layers.

**SKILL TIP:** Learn to create group layers and store them as group layer files (BASICS: General).

26→ Click the Save button to save the current map document.

26→ Click the New button on the main toolbar to open a new map document.

26→ Click on the Templates entry on the left, if necessary. Scroll down and select the Letter (ANSI A) Portrait template and click OK.

26→ Click on the small Data View button in the lower left map frame (see Fig. 1.25).

26→ Click the Add Data button. Navigate to the mgisdata\Usa folder.

26→ Click on the US Example layer file and choose Add.

Although only the states show in the map window, many layers are listed in the Table of Contents. The check boxes for these layers are gray and cannot be turned on or off. The creator of the layer file applied a **scale range** to these layers, defining a specific range of scales at which they appear and making sure that they appear only when appropriate.

Fig. 1.24. The scale range property sets the scales at which a layer will be displayed.
Source: ArcMap

27→ Click the Zoom In tool and zoom in to the western conterminous United States. State capitals and rivers should appear (if they do not, zoom in more).

27→ Right-click the Capitals layer and choose Properties. Click the General tab.

27→ Examine the Scale Range property (Fig. 1.24). Capitals will be shown only when the user has zoomed in below a scale of 1:40 million. Click Cancel.

1:40,000,000 ▼

Source: ArcMap

28→ Locate the Map Scale box on the main toolbar, indicating the current map scale.

28→ Click in the Map Scale box and type *40,000,000* (with or without the commas). Click Enter. The map zooms to that scale.

28→ Click the Fixed Zoom Out button. The capitals and rivers disappear.

28→ Click the Fixed Zoom In button once. They appear again.

28→ Click the Fixed Zoom In button a few more times until the labels for the capitals appear. Labels can have a different scale range than the features they represent.

29→ Click the Fixed Zoom In button until the interstates and the river labels appear.

29→ Zoom in until the counties appear, symbolized by Native American population.

29→ Place the cursor on top of a county and wait. A small label pops up to show the county name. These labels are called MapTips.

MapTips are another property of a layer. They can be set to show the contents of a specified field when the cursor is on top of the feature.

**TIP:** Double-clicking a layer's name also opens its Properties.

30→ Double-click the Counties layer to open its properties and click the Display tab.

30→ Examine the settings on this tab. Find the *Show MapTips* check box.

4. What is the name of the field that is being displayed in the MapTips? _____

30→ Click the Symbology tab for the Counties layer properties.

30→ Instead of a single symbol, this layer is set to display the percentage of Native Americans in the counties. Examine the settings but don't change anything.

30→ Click Cancel to close the Properties window.

30→ Use File > Save As to save the map document in the MapDocuments folder, naming it *USFrames.mxd*.

## Working with data frames

Now we will learn to manage data frames. Our goal is to create a map layout with two different maps, one of the United States and one of the world.

The map window has two view modes. In **Data view**, the user works with a single data frame. The **Layout view** shows how the map document will look when printed, including titles, scale bars, and so on. You are currently in Data View.

Data View
Layout View
Refresh View
Pause Drawing

Fig. 1.25. View buttons.
Source: ArcMap

31→ Click the Layout View icon in the lower left of the display area to switch to Layout view (Fig. 1.25).

31→ Click the Select Elements tool on the Tools toolbar or the Draw toolbar. This tool is used to select and work with graphic elements, such as text or data frames.

31→ Click inside the map frame. Blue square handles appear on the border of the data frame, indicating that the frame is selected by the Select Elements tool.

31→ Place the cursor on top of the middle lower blue handle, and click and drag it to size the data frame to occupy the top half of the page.

32→ In the Table of Contents, click twice slowly on the Layers name to edit it.

32→ Type *United States* for the name of the data frame and press Enter.

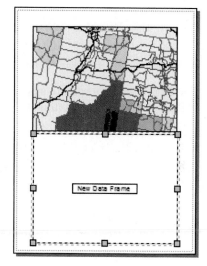

Fig. 1.26. Layout view.
Source: ArcMap

33→ Choose Insert > Data Frame from the main menu bar.

33→ When the new data frame appears, click inside it and drag it below the original data frame. Use the handles to adjust its size and shape until the layout page looks similar to Figure 1.26.

33→ Click twice slowly on the New Data Frame name in the Table of Contents, and change it to *World*.

A map may have many data frames, but one is always the active frame, indicated by boldface type in the Table of Contents. The Data View button shows the active frame. Data added to the map are placed in the active frame.

34→ Make sure that the World data frame is the active frame (its name is in boldface type).

34→ Click the Add Data button. Navigate to the mgisdata\World folder and add the country shapefile. It appears in the World data frame.

The United States data frame is currently zoomed in further than we would like. To change the scale/extent of the map within a data frame, activate the data frame and use the zoom/pan tools.

35→ Right-click on the United States data frame name in the Table of Contents and choose Activate from the context menu.

35→ Click the Full Extent button to view the entire feature class.

35→ Click the Zoom In tool and carefully draw a rectangle around the lower 48 states to fit it into the data frame.

36→ Click the Data View button in the lower-left corner of the map window to return to Data view. You should now see the United States map.

36→ Right-click the World data frame in the Table of Contents and choose Activate. The view changes to the World data frame.

36→ Click the Layout View button to return to the map layout page.

Source: ArcMap

### Data frame coordinate systems

A data frame has a coordinate system (CS), which defaults to the first layer added to the frame.

37→ Double-click the World data frame name in the Table of Contents to open its properties.

37→ Click on the Coordinate System tab and examine the coordinate system. Write its name and units (angular unit) and then close the window.
_____ Units_____

37→ Examine the coordinate system of the United States data frame. Write its name and units (linear unit) and then close the window.
_____ Units_____

The data frame coordinate system can be set to any CS desired, and will display all data sets added to the frame in that CS.

**TIP:** A plus sign by a folder expands its contents. A minus sign closes the contents. Use them to open or close folders as needed.

Source: ArcMap

38→ Open the properties for the World data frame and view the Coordinate System tab.

38→ Expand the Projected Coordinate Systems folder and then the World folder.

38→ Find the WGS 1984 Web Mercator (auxiliary sphere) coordinate system, which is used by ArcGIS Online. Click to highlight it.

38→ Examine the coordinate system. What are its units? _____

Notice that this coordinate system has both a projection (Mercator Auxiliary Sphere, in this case) and (if you scroll down) a geographic coordinate system, or datum (GCS_WGS_1984). Both items are part of the coordinate system name, WGS_1984_Web_Mercator_Auxiliary_Sphere.

38→ Click OK to accept the new coordinate system.

38→ Use the Pan tool to adjust the position of the world map in the data frame.

39→ Open Help > ArcGIS Desktop Help from the main menu bar.

39→ Click the Search tab and enter Mercator.

39→ Click the *Mercator* entry that it finds and read about the Mercator projection.

39→ Click on the Contents tab in the Help, then click *Locate topic* in the Mercator entry. Use this to find where in the Help things are located. Here we see an entire "book" about map projections.

39→ Close the Help window (or minimize it for use later).

**TIP:** You can reinforce concepts and learn new things by reading Help. An amazing amount of information is available, equivalent to many books. Make it a habit!

A data frame projects all its layers on the fly, so all layers in a data frame are not necessarily in the same coordinate system.

40→ Open the properties for the United States data frame and view the Coordinate System tab.

40→ Scroll to the bottom of the coordinate system folder tree and find the Layers folder.

40→ Click the plus sign to expand the Layers folder and again to expand its subentries.

This Layers folder is a useful way to see what coordinate systems are present in a data frame. In this frame, the Oregon Highways layer is in a coordinate system called Oregon Statewide Lambert. All other layers are in North America Equidistant Conic.

40→ Click on the Oregon Statewide Lambert coordinate system and click Apply. Move the window aside, if necessary, to see the map.

Oregon Statewide Lambert is, as the name implies, optimized for Oregon, so the United States is rotated inappropriately. If you see a rotation like this, check the data frame coordinate system and make sure it is set to a more suitable choice.

41→ Click on the North America Equidistant Conic entry and click Apply to set it to the previous option.

Let's look at a couple more coordinate systems.

41→ Scroll up on the Coordinate System tab and find the Geographic Coordinate Systems folder.

41→ Expand Geographic Coordinate Systems > North America and select the NAD 1983 entry (the simple one in a group is usually the right choice).

41→ The Current coordinate system name in the lower box changes to GCS_North _American_1983. Click Apply.

The shapes of the map features change, stretching in the east-west direction. North-south lines that were previously angled become vertical. Changing coordinate systems can dramatically affect the appearance of a map.

41→ Still on the Coordinate System tab, expand the Projected Coordinate Systems folder and navigate to the Continental > North America > USA Contiguous Albers Equal Area Conic coordinate system. Select it and click OK.

The map changes again. Every coordinate system causes distortion of areas, distances, shapes, or angles. Each one is designed to minimize or eliminate different kinds of distortion, and the name often gives a hint. An equidistant coordinate system displays accurate distances. An equal area coordinate system displays accurate areas. The best one to use depends on the purpose of the map, which is why data are stored and used in many coordinate systems instead of just one.

**TIP:** Changing the coordinate system of the data frame only changes the map display. It does not affect the coordinate systems used to store the feature classes.

### Using Internet map services

Your hard drive or office network drive is not the only source of GIS data. Many organizations host services that provide maps, imagery, and data, and individuals can share content on the ArcGIS Online platform.

42→ Open the Help window and click the Search tab.

42→ Type ArcGIS Online into the box at the top and click Ask.

42→ Click on *Adding data from ArcGIS Online* and read through the entry.

42→ Close ArcGIS Help.

**TIP:** Use the Search function in Help during a tutorial whenever you'd like to learn more about a tool or concept. Your GIS knowledge and skills will improve dramatically.

Now let's explore ArcGIS Online. You need a high-speed Internet connection for the next steps. If you are not connected, you may end the tutorial here.

43→ Save your current map.

43→ Click the New button on the main toolbar, click the New Maps entry on the left if needed, and start with a Blank Map.

43→ Click the Data View button (lower-left corner of map window).

43→ Click the Add Data drop-down triangle and choose Add Basemap.

43→ Select the Streets basemap and click Add.

Source: ArcCatalog

**TIP:** If ArcMap crashes when you try to load a basemap, start ArcMap again. Go to Customize > ArcMap Options and click the Data View tab. Uncheck the box to *Enable hardware acceleration*.

44→ Zoom to your home state and then to your home town.

44→ Zoom to your home neighborhood and find your street.

**TIP:** Be patient when using Internet services because they can take time to draw. Sometimes they may not be available, or the name may have changed. If the suggested one does not work, try another similar one, or try again later. An ESRI tag means that an ArcGIS Online subscription is required.

Notice that the map becomes more detailed as you zoom in; map services generally employ scale ranges to show more detailed layers at larger scales. The basemap is only the beginning. Many agencies and individuals provide map data for others to use.

44→ Zoom back out so that the conterminous United States is being shown.

45→ Click the Add Data drop-down again and choose Add Data from ArcGIS Online.

45→ Type *soils* in the search box and click the Search button. Examine the results.

45→ Examine the USA Soil Survey service, which should be near the top.

45→ Click on the Details link and review the information.

5. Which agency provides the data for this service? _____

45→ Click the Add link to add the service to your map.

45→ Examine the map.

You do not have as much control of a map service as your own layers, but a few display options can be set.

46→ Right-click the USA Soil Survey group layer name and choose Properties.

46→ Examine the Source tab. Scroll down to view the coordinate system.

46→ Click the Advanced tab.

46→ Move the Layer Transparency slider to about 50% and click OK.

Now you can see the underlying base map showing through the warnings.

47→ Click the Identify tool and click on one of the soil polygons.

47→ Examine the attributes of several polygons and then close the Identify window.

47→ Expand the USA Soil Survey group subheadings in the Table of Contents to see legend information.

Remember the Oregon airports layer file that you created earlier? Let's add it to this map to see the soil types each airport is located on.

48→ Click the Add Data button and navigate to the mgisdata/Oregon folder.

48→ Select the myairports.lyr layer file and click Add.

48→ A warning message about the geographic coordinate system may appear. If it does, check the box *Don't warn me again in this session* and click Close. We will learn more about this warning later.

48→ Right-click the airports layer and choose Zoom to Layer.

49→ Right-click the airports layer and choose Remove.

49→ Zoom in to your home state again.

49→ Double-click the Layers data frame name to open its properties and examine the coordinate system.

The data frame defaults to the coordinate system of the first layer added, in this case the base map, which uses WGS 1984 Web Mercator Auxiliary Sphere. Changing the coordinate system works with ArcGIS Online data also (although it may slow drawing time a bit).

50→ Scroll and collapse folders, if necessary, to find the Projected Coordinate Systems > State Plane > NAD 1983 (Meters) folder and look inside it.

50→ Choose a coordinate system for your state. In many cases there will be more than one (for example, Maine East and Maine West). Pick the one that best represents the part of the state containing your home town. Click OK.

50→ The map will redraw to a less distorted view of the state.

**TIP:** NAD 1983 stands for North American Datum of 1983. A **datum** is a correction to the coordinate system to make it more accurate. Chapter 3 has more information on datums.

51→ Click the Add Data drop-down button and choose Add Data from ArcGIS Online.

51→ Enter your home state's name in the Search box and search for data.

51→ Select an interesting map service or layer package and add it to the map.

**TIP:** Many ArcGIS Online data sets are contributed by ordinary folks, and not all of them work well. If you have trouble getting one to display, remove it and try another.

Let's view some imagery now. We will open a new map so that the data frame defaults to the coordinate system of the service and does not spend time trying to project the images on the fly.

52→ Open a new, blank map. You don't need to save the current one.

52→ Choose Add Data > Add Data from ArcGIS Online.

52→ Type the term *imagery* in the search box.

52→ Examine the details of the World Imagery map service, then add it to the map.

53→ Click the Find button on the Tools toolbar.

53→ Click the Locations tab.

53→ Choose the Esri World Geocoder locator.

A locator is an example of a **geoprocessing service**. When you click Find, the string is sent to the server, which parses it, finds the location by comparing the string to GIS data located on the server, and sends the location back to you. The work is done by the server, not by your computer.

54→ Type New York City in the *Single Line Input* box and click Find.

54→ Right-click the first New York City entry in the bottom of the window and choose Zoom To. Move the Find window out of the way but leave it open.

54→ Set the scale in the Map Scale box to 1:100,000.

54→ Zoom in more to see greater detail.

54→ Choose Bookmarks > Create Bookmark and type New York for the name. Click OK.

This map service uses a variety of imagery depending on what is available. The resolution in New York City is sub-meter, but it will not be that good everywhere. Let's try locating an address instead of a place name.

55→ Type your full home address in the Single Line Input box, including city and state (or equivalent).

55→ If it finds it, right-click it and choose Zoom To. Right-click again and choose Add Point. How well did the service do at finding your house?

56→ Create a bookmark for your house.

56→ Choose Bookmarks > New York to return to New York City.

56→ Close the Find window.

The beauty of GIS servers is that you have access to terabytes of data without having to store them, update them, or download more than the portion you need. The disadvantages are that you must be online, and you may not have access to features or be able to modify the symbols.

**TIP:** Using the *Sign in* link when searching for data in ArcGIS Online will give you access to more data, especially if your account is part of an organizational subscription. Ask your instructor about how to create or access an account.

Sign in

This is the end of the tutorial.

→ Close ArcMap. You don't need to save your changes.

## Exercises

**TIP:** If you need to capture a map or screen shot to submit to your instructor, you can do so by pressing the Alt-Prnt Screen keys on your computer, or using a program such as Jing or Snagit. Then paste the capture into a Word document. For help, ask your instructor.

Use ArcMap and the mgisdata\Usa folder to answer the following questions.

1. Is there a feature dataset in the usdata geodatabase? What is it called? List the feature classes it contains. Add them to the map.

2. What is the coordinate system of the feature classes in the Transportation feature dataset? (**Hint:** What did the data frame coordinate system default to?)

3. How many counties are there in the United States? How many of them are in California?

4. Which is the largest lake in the United States? What is its area? (The area is given in square miles. This is the type of information you find in the metadata.)

5. Which state has a county named Itawamba? (**Hint:** Use Find in the Table Options menu.)

6. In the states feature class, what is the minimum, maximum, and average 2014 population density (POP14_SQMI) of the states?

7. Use ArcGIS Help to find a discussion of feature class basics and read it to expand your knowledge. Although this text presents three basic types of feature classes (point, line, polygon), several others can be stored in a geodatabase. List the other four types.

8. Create a layout page with two data frames. The upper frame contains only Oregon counties in light yellow with light green parks. The lower frame shows the lower 48 United States in beige with blue lakes. Capture the layout.

9. Use the USA Topo Maps map service in ArcGIS Online to view your university campus. Capture the map.

10. Add the country and latlong shapefiles from the World folder to a new, blank map. Display them using the World Robinson projection. Capture the map.

Source: ArcMap

## Challenge Problem: Hey, That's Cool!

Search ArcGIS Online and find a map service that you think is interesting. Combine it with at least one data set from the mgisdata folder to create a map for a place that you know.

Capture the map in a Word document. Underneath it, construct appropriate citations for the sources of the data. Also include a statement of the coordinate system used for the map.

# Chapter 2. Managing GIS Data

## Objectives

▸ Understanding how GIS data differ from typical computer files

▸ Organizing GIS data sets and folders

▸ Creating new geodatabases by importing and exporting data

▸ Understanding how map documents work with GIS data

▸ Using ArcCatalog and ArcToolbox to manage data

▸ Documenting data sources using metadata

## Mastering the Concepts

### GIS Concepts

#### *Characteristics of GIS data*

Data files stored for use in a GIS differ in many ways from data used in other applications, such as word processing, spreadsheets, videos, or music. In order to use and manage GIS data effectively, you must understand some important characteristics.

*GIS data files may be very large.* Some data sets, for example a parcel feature class for a large city, can have hundreds of thousands of features and records. Images and rasters especially can run to hundreds of megabytes or several gigabytes.

*GIS data are often shared by multiple users.* Many data sets are collected and used across organizations rather than belonging to one individual. A parcel database must be accessed by many clerks and several city departments. Giving each a copy of the data causes problems. First, it wastes space. Second, and more important, such databases are frequently updated. Ensuring that every person has the most recent copy is a difficult task. Thus, many GIS data sets are intended to be shared; one copy is maintained on a central drive or server and everyone uses it.

*GIS data may come from many sources, yet must work together.* GIS data may come from satellites, GPS units, scanners, paper maps, tables, and other sources. They can use different coordinate systems and projections and file types. GIS requires ways to label and keep track of how files are stored and which coordinate systems are used. Some GIS systems rely on the user to keep track of everything, but more commonly today, GIS applications use headers or other label files to keep track of data set characteristics and handle the files correctly. Users need to make sure that this additional data information is present and correct.

*GIS data include local files and Internet services.* Most people work with data sets that reside on the computer being used. If a PDF is found on the Internet, the user is likely to make a copy of it on the computer for later use. Many GIS data sets work this way also, but GIS data services are becoming more prevalent. A GIS service, such as ArcGIS Online or Google Maps, allows the data to be used over the Internet, but often does not allow it to be copied to the local computer.

*GIS data formats do not always follow the latest conventions.* Computer operating systems have become more permissive in the naming of folders and files over time. Originally, computer file names were restricted to alphanumeric characters (letters or numbers only), and computers used the space character to signify the end of one file name and the beginning of another. Uppercase letters were not interchangeable with lowercase letters. Nonalphanumeric characters such as $ or % or # had special meanings. Very long names could exceed program file name limits. Today, operating systems are more flexible, but not all software applications evolve at the same rate or follow the same conventions. Even current versions of an application may contain old bits of programming code that can fail when they encounter files or folders named according to the new rules. GIS users must follow more restrictive rules about where to put data and what to name it.

### Organizing data files

Today's computers try to do our file management for us—they figure out whether we are saving writing, music, or videos and drop them into pre-organized folders. They open recently visited folders automatically. They recognize what type of file is being opened and select the right application. They even suggest what to name files. These features can be convenient for a casual user, but they cause problems in complex applications. A GIS professional must take responsibility and learn some precepts of sound file management. These precepts are common to many computing fields and are useful guidelines to follow at any time, not just for GIS.

***Be mindful when saving files.*** Don't let the computer save files with default names, because the names may not reflect what is in the file, or may be too long, or may not follow the proper naming conventions. Be careful to note where the computer is trying to save a file; in some cases this may be a default location or the last-used folder, rather than the appropriate place. Always put files in an organized place where they belong. The Desktop is a terrible place for GIS data.

***Organize files using folders.*** Finding files is much easier if you have a system of folders to organize the files that belong together. Imagine how hard it would be to find things if all the data for this class came in one giant folder, rather than organized into folders based on the geographic area (Fig. 2.1). Take the time to develop a thoughtful and logical system of folders to contain GIS data. There is no one right way to organize a data folder tree, but folder names and the structure should make sense and enable you to find things easily. However, if you are organizing data for multiple users to enjoy, then it is also beneficial to make the structure easily understandable to others.

Fig. 2.1. Folder organization is very important for GIS data.
Source: ArcGIS

In GIS, it is helpful to separate static data folders, containing finished permanent data, from working folders in which data are being downloaded, files are being added and deleted, and intermediate results clutter the folder. Separating active work folders reduces the chance of accidentally deleting a data set, or losing track of which data sets are permanent and which are works in progress. Once a project is finished, you can clean up the folder and move the data to a static location.

***Be conservative when naming files and folders.*** To make file systems compatible with any type of software, keep folder names short and use only letters, numbers, and the underscore character. Avoid spaces and never use nonalphanumeric symbols. Spaces may be used for file names if the program accepts them, but it is best to stay away from nonalphanumeric characters for all file names.

**VERY IMPORTANT TIP:** GIS folders and data set names should not contain spaces or special characters, such as #, @, &,*. Use the underscore character _ to create spaces in names if needed.

***Store GIS data in top-level drive folders.*** Windows uses personal Library and Documents folders as default locations for user data, but these locations often cause problems when used for GIS data. In a network environment, the pathnames may contain problematic characters or spaces that may prevent the GIS from working properly in all situations. It can also be difficult to find these locations, and the routes to these folders can be tedious to navigate. The best practice is to create GIS data folders at the topmost level, directly under the drive letter designation, for example, C:\gisdata or C:\mydata. From that initial point a user can create as many subfolders as needed.

***Pay attention to file extensions.*** Most files on a computer have an extension, a three- or four-letter code that indicates the type of file. A Word document has a .doc or .docx extension, and a picture might have a .jpg or .png extension. Computers use this extension to help determine which application should be used when a user double-clicks a file to open it. By default, the computer hides file extensions from the user, but it is often helpful to see them, especially when downloading files and working with data in unknown formats. You can change the default setting in the Folder Options in Windows Explorer on your computer (Fig. 2.2).

Fig. 2.2. Setting Windows to show file extensions
Source: Microsoft Windows

**TIP:** File geodatabases are simply folders with files in them, making it *possible* to store files like spreadsheets in them. However, this practice can damage the geodatabase if a user is not careful. Never use Windows or other applications to save or manage files inside a file geodatabase.

### Finding data for a project

Looking for and obtaining data for a project is an open-ended and complex activity, especially since the GIS data landscape is continually changing. New data sets emerge; old ones become unavailable or are moved to new locations. New methods for discovering and formatting information mingle with ancient, nearly forgotten file types. The volume, diversity, and complexity of available data are bewildering, but we will begin to develop the skills needed for this task. Getting proficient requires flexibility, perseverance, and willingness to experiment, and it takes time and practice. A high tolerance for frustration and an experienced mentor are valuable assets in the search for data.

Most GIS data sets are organized for a particular geographic area. As explained in Chapter 1, GIS data have an intrinsic scale for which they were designed, the source scale. When searching for data to use, the source scale must be suitable for the area and intended use of the geodatabase.

Imagine compiling data for a project focused on the state of New Jersey. A typical page-size map of the state would have a scale of about 1:2,000,000. Viewing individual counties would require a scale of about 1:300,000. Examine the three different feature classes in Figure 2.3, showing the New Jersey state boundary in the Sandy Hook area at a scale of 1:300,000. The purple, orange, and black lines represent source scales of 1:25 million, 1:5 million, and 1:50,000, respectively. The purple feature class is not appropriate for a state geodatabase at all. The orange one can be used for state- or county-wide mapping or analysis, and the black one would be fine for city-level. Sometimes one must settle for data with a smaller source scale than desired because nothing else is easily available, but the user needs to be aware of the consequent limitations on the precision of the project results.

Fig. 2.3. Three New Jersey data sets with different source scales
Source: ArcMap

Generally, the larger the scale or more specialized the project, the more difficult it can be to find preexisting data. General base data for roads, cities, and political boundaries are widely available, even at large scales. Very large-scale data, like utilities, parcels, or building footprints, will generally be found for only small areas at a time, and may not be available for download by the public because of privacy or security concerns. Specialized data sets, such as land use categories or geologic maps, are most often available either at state or national scales or as detailed maps in isolated areas.

If nothing can be found, then either the project must be revised, or the data must be created in-house. Original data can be time-consuming and expensive to create, which is why specialized large-scale data is difficult to find. For class projects, it is wise to choose topics and scales for which data may be easily located. Sometimes it is easier to find interesting data first and then develop the topic!

Searching for data can start at an agency web site or a generic clearinghouse. A list of good places to start is included at the end of this chapter, but the Internet changes so quickly that some of the links may well be obsolete by the time the book goes to press. One can sometimes have good luck with general Internet search engines by including the term "shapefile," "GIS," or "e00" along with a topic or place name—for example, "Wyoming GIS" or "geology shapefile."

ArcGIS Online can be a valuable source of data. However, two things must be remembered. First, many data sets on ArcGIS Online are designed to be used online and do not allow the user to make a local copy. Second, almost anyone can share data on ArcGIS Online. Some of it is provided by agencies and companies and is excellent, but some is worthless. As with anything on the Internet, let the searcher beware.

Data tables can also be good sources of data, such as text files, Excel spreadsheets, or even a copy/paste of tabular data found on a web page and saved in a text file or spreadsheet. If the table has fields with $x$-$y$ values in them, these rows can usually be converted to point data with attributes. If the table has attributes like county names, state names, or FIPS codes (a federal identifier of political units in the United States), then these tables can often be joined to existing feature classes and incorporated into a geodatabase as spatial data. Chapter 6 has more information about working with tables.

### Importing and exporting data

After downloading some data, the first step is to ascertain the data format, because some additional processing is usually needed before the GIS can work with, or even recognize, the data. Always start by examining the downloaded files in Windows Explorer, taking note of the file extensions in particular.

Data downloaded from the Internet is typically **zipped**, a process that combines many individual files into a single file and compresses the information to use less space. A single smaller file is easier to download than multiple larger files, but the files must be unzipped before the GIS will recognize the data. Zipped files have file extensions like .gz, .zip, .tar, .tgz, or.tar.gz. Many different zip utilities are available to "unzip" files, some free, and each one works differently. You will need one of them on your computer, so if you are not already familiar with them, ask your instructor what is typically used on your campus, or ask a coworker what is used in your organization, and learn how to install and use it.

The oldest version of ArcGIS, Arc/Info, used a proprietary zipping/conversion format to convert coverages into a single text file that could be easily downloaded, known as an **Interchange file** and bearing the file extension .e00. Files with the .e00 extension must be converted back into coverages using the *Import from E00* tool in the Conversion Tools toolset in ArcToolbox. Afterwards, the coverage can be converted to a geodatabase or a shapefile.

After unzipping, you must ascertain the data format of the new files. Sometimes data are already in a geodatabase format and can be copied directly to a geodatabase. More commonly though, the data need to be converted to the geodatabase format. The computer terms **import** and **export** usually refer to the process of copying a data file to a new location while simultaneously converting it from one data storage format to a different one. The terms also refer to the direction: an import takes external data and converts it to the current format; an export takes the current data and converts it to an external format. In practice, either may usually be used: one might right-click a shapefile and export it to a geodatabase, or right-click a geodatabase and import a shapefile—the two operations are equivalent. GIS programs typically have a set of conversion tools that allow data sets to be changed from one data format to another; which ones and how extensive the toolset is will depend on the program.

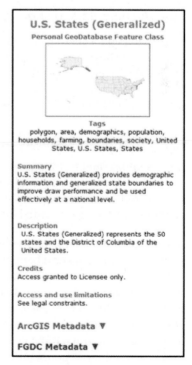

### About metadata

Metadata provide the means to discover GIS data when searching, and also store valuable information about the provenance, handling, and quality of a data set, allowing a user to determine whether the data set is fit for a given purpose (Fig. 2.4). Once created, metadata usually travels with the data set. Data downloaded from ArcGIS Online or the Internet will

Fig. 2.4. Metadata
Source: ArcGIS Online

often have metadata as part of the file (or in a separate file). When downloading, watch for links to metadata and be sure to download them along with the data set. It is a good practice to skim the metadata to learn some basic things about it, such as the originator and the source scale. The metadata may need to be updated if the user makes any changes to the original files.

Developing or updating metadata is always needed, but the level of effort can vary. Sometimes GIS data is used only by its creator. In this case, the metadata can be minimal, although it is still valuable to record some information for future reference, such as a citation documenting the source of the data and any information you found on the web site, such as the source scale or the year the data were published. In this chapter, we will learn how to quickly document a data set.

Professionals who create data that will be distributed to clients or the public incur an obligation to create more extensive metadata. They must evaluate the quality of the data and provide a report that summarizes the spatial and thematic accuracy so that users can determine whether a data set is suited to a particular purpose. Producers should also provide information on other aspects of a data set, such as what geographic area it covers, what coordinate system it uses, what the information in the attribute tables means, how a potential user can access the data, and more. If the original data were created or compiled by others, the producer must also give proper credit to the originators. This type of metadata creation is covered in Chapter 14.

Metadata also record access and use constraints. Some GIS data, including most data sets derived from federal agencies, can be freely copied and redistributed. Other data are licensed to the specific users who purchased it. Every user is responsible for understanding the access constraints placed on any data set and for abiding by them. Failure to do so can result in civil and criminal penalties against an individual or the organization.

## About ArcGIS

### *Map documents and shared data*

As previously discussed, GIS data files are often large and shared by multiple users. The map document is designed to support these characteristics. A map document (or a layer file) does not store the GIS data files within it. Instead, the map stores the name and location of the spatial data—its **source**. When it needs the data, to draw some features for example, it uses the information to find and access the data *from its original location*.

This treatment is different from what most people are accustomed to. In a word processing or slideshow document, an inserted picture is copied and saved again inside the document. If the document is e-mailed or copied to a new location, the picture goes with it. The advantage of this method is that the picture is always where it needs to be. The disadvantage is that multiple copies are created, increasing the size of the document. If you edit the original picture to improve it, the changes are not incorporated into any of the files already created.

In GIS, storing the information separately from the document and using layer properties to manipulate it has several benefits. First, it makes it easy for multiple users to share data. Each user accesses the same data but can customize how it appears. Figure 2.5 shows a counties feature class stored on a network drive; the same feature class has been used by several students in their map documents. Second, it saves space. Each student need not store a copy of the data, and the map documents will be small files no matter how much data they use. Third, it allows updates of a central database to be quickly distributed to every user. If the counties feature class is updated with new population estimates, the information will be available the next time the students open their maps.

counties feature class
F:\Dept\gisdata\Oregon\oregondata.gdb\counties

Tucker's map

Eileen's map

Ginny's map

Fig. 2.5. The same feature class can be used in multiple map documents.
Source: ArcMap

However, the practice of storing data separately has its own problems. Propagating updates to all users is wonderful, but if someone accidentally damages or deletes some data, many people are affected. Furthermore, because the map document only stores the location of the data, it can lose track of it if the data are deleted, moved, or renamed. You usually cannot e-mail a map document to someone because the data won't be sent with it. Unless the recipient has access to the same data in the original location (on a network drive, for example), the map will show up empty.

Shared data requires that data managers implement precautions to ensure that an editing mistake does not affect dozens of people. Sometimes the files are write-protected and only a small number of users are authorized to make changes. In a more complex environment where GIS data must have multiple editors, organizations may use an enterprise geodatabase.

Shared data also raises challenges for small teams working on a joint project. If each person has his or her own data, then updates cannot easily be shared. If all team members access the same data in one location, it becomes possible for two people to make simultaneous or conflicting edits to a feature class, which can potentially corrupt the database. Unless the team has an enterprise geodatabase available, it must devise a strategy for data sharing that eliminates the possibility of two people editing the same feature class at the same time.

### Map documents and pathnames

Map documents keep track of the source files by storing the location of each file as a **pathname**, which lists the successive folders traversed to a data set, each separated by backslashes. To explore the relationships between documents and pathnames, consider the file system represented in Figure 2.6. The blue box is the hard drive C:\, the green boxes are folders, the gray box is a geodatabase, and the text entries represent either map documents (.mxd) or shapefiles (.shp). The .mxd files that begin with the letters US access data in the Usa folder, and the .mxd files that begin with NJ access data from the Njdata geodatabase. The pathname C:\mgisdata\Usa\States.shp refers to the feature class States.shp, which resides in the Usa folder, which in turn resides in the mgisdata folder, which is found on drive C:\.

The pathname C:\mgisdata\Usa\States.shp is termed an **absolute pathname** because it starts at the drive letter and proceeds downward to the file. When searching for a file using an absolute pathname, the search begins at the drive level. A **relative pathname** is used to indicate that the search should begin in the folder containing the map document. A double dot indicates movement up one folder. The pathname ..\Usa\States.shp indicates a search up one folder and then down into the Usa folder and then to the States feature class (Fig. 2.6).

Map documents can store either absolute or relative pathnames as chosen by the user, with absolute pathnames as the default. Imagine that the map document ex_1.mxd in

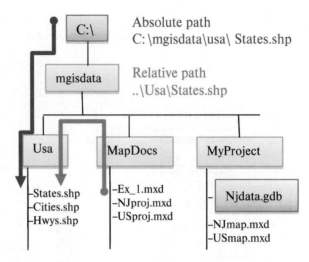

Absolute path
C:\mgisdata\usa\ States.shp

Relative path
..\Usa\States.shp

Fig. 2.6. Absolute versus relative pathnames

Figure 2.6 is being opened, and it uses the States file in the Usa folder. If the document stores absolute paths, it searches for States starting at the C:\ drive down through C:\mgisdata\Usa\States.shp in order to locate and draw the states in the map. If the document uses relative paths, it goes up one level to the mgisdata folder, then down to the Usa folder and the States data.

If a map document cannot find the data based on the stored path, red exclamation points will appear beside the lost data sets when the map document is opened (Fig. 2.7). Any time a folder or file name is changed or the location is moved, the potential exists to break a pathname link and cause the map document to lose track of the data. For example, imagine that the owner of this file system decides to rename the Usa folder USdata instead. Both types of pathnames would lose track of the data, and the map documents Ex_1, USproj, and USmap would all have red exclamation points upon being opened.

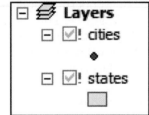

Fig. 2.7. The exclamation point indicates that the pathname link is broken.
Source: ArcMap

**TIP:** What do you suppose happens if you try to remove a space that you accidentally put in a folder name, if map documents are already referring to data inside that folder?

Absolute pathnames are best for map documents that use data residing on a permanent shared server to which everyone has access. In this situation, e-mailing map documents within the organization even works, because everyone has access to the same data in the same location. Absolute paths also work well for referring to data stored on one's own hard drive, as long as the data folders are not moved or renamed.

Relative pathnames work best for a folder containing maps and data that may be moved to different locations. If all of the map documents in Figure 2.6 used relative pathnames, they would continue to work whether the mgisdata folder resided on a drive named F:\, K:\, or D:\. The map documents in the tutorial data that come with this book are stored with relative pathnames so that the links to the data layers continue to work no matter where you install the mgisdata folder. In a university environment, students commonly use USB thumb or portable drives to save data so that it will be available on any computer. However, these units are not always assigned the same drive letter each time they are inserted in a computer. For this reason, relative paths should be used for map documents stored on thumb or portable drives.

**TIP:** A setting under Customize > ArcMap Options allows the user to set relative pathnames as the default for all new map documents.

*Pathnames should NEVER contain spaces.* ArcGIS will not permit you to put a space in a GIS data file name, but it is particularly insidious about spaces in folder names. They are permitted and will work for the most part, but occasionally one tool or function will crash if it is used on data with spaces in its pathname. The only way to proceed is to rename the folder—this causes the map documents to lose track of the data, and then they must be fixed as well. Experienced GIS users FLINCH when they see a space in a folder name, and you should learn to as well. Teach yourself never to use them anywhere on your computer, and you will be happier for it.

### Understanding file locks

When accessing a file, computers distinguish between **read access** and **write access**. Read access allows a program to take information out of a file without changing it. Write access allows changes to be made to the stored file. File systems use access restrictions to protect data. In a network file system, a folder may be designated to have read access for most users, to protect it from being accidentally (or purposely) damaged. On your own computer, you will generally have both read and write access (read-write access) to all files.

Access rights are important when multiple people use files. You may have had the experience of opening a Word document that was in use by another person (or by yourself) and received a message that the file was in use, with an offer to let you open a read-only copy. In this case, the first person who opened the file had read-write access, but the second person only had read access. If the second user made changes, they would have to be saved using a renamed copy.

Programs keep track of which documents are being used by creating a small **lock file** in the same folder as the document while it is open in an application. When the program closes the file, the lock is deleted, and another person or program is free to access the file. When you open a document, the application checks for the file lock. If it is present, you will get the "in use" message.

Because GIS data are designed to be shared, ArcGIS is more clever about locks. Most operations in ArcGIS, such as drawing features in a map, only read the data. The program recognizes this fact, so multiple users reading a data set causes no error messages. The first person to access the data will cause a lock to be placed on the file, but as long as additional users perform operations that require only read access, the presence of the lock does not matter.

A user must take specific actions to prepare a data set for write access, such as starting an editing session or trying to delete a field from a table. ArcMap checks for file locks only when write access is needed; if they are present, the user is warned and will be unable to complete the action until the lock is gone. Locks are usually deleted by removing the data from the map document or closing the program.

However, ArcGIS is prone to glitches in which it does not recognize that a lock has been removed and generates an "in use" error when a user tries a write operation, even when no other applications appear to be using the file. The best way to clear the spurious lock is to save the document and exit ArcMap. Once the program is restarted and the document reopened, the locks will be cleared and the write action can proceed. Rarely, a reboot of the computer is required to clear the locks and can be attempted if closing and reopening ArcMap does not work.

**TIP:** Like other programs, running more than one instance of ArcMap on a single computer may cause read-write access problems due to locks. For this reason, and because ArcMap is a large program that uses a great deal of memory, it is not recommended to run more than one instance of ArcMap at a time. Running ArcMap and ArcCatalog simultaneously can also cause lock issues.

### Using ArcCatalog to manage files

The ArcCatalog program shares many functions with Windows Explorer, the application used to view, copy, rename, delete, and move data files on the computer. However, ArcCatalog understands the unique characteristics of GIS data and is better at managing GIS files. For example, Chapter 1 described how a single shapefile feature class is composed of multiple files with the same prefix and different extensions (roads.dbf, roads.shp, roads.prj ...). Windows Explorer would allow a user to delete one of these files, thereby destroying the feature class and rendering it unusable. ArcCatalog, however, knows that these files form a unit. It shows them as a single icon and requires users to save, copy, or rename them as a unit. Always use ArcCatalog to manage (copy, move, rename, delete) GIS data.

**VERY IMPORTANT TIP:** Do not use Windows to copy or delete coverages, shapefiles, and geodatabases. Always use ArcCatalog to delete or copy spatial data sets.

ArcCatalog has additional functions as well. It offers the ability to preview the contents of GIS data sets, both the features and the tables, in order to ascertain what a data set contains. It includes an editor for reading and editing metadata. It performs many GIS-specific data management functions, such as creating geodatabases or feature classes, adding new columns to a table, or editing the coordinate system label of a data set.

ArcCatalog functions are frequently needed when working on a project. It may be used in two ways, as a standalone program or as the Catalog window sitting inside ArcMap. The Catalog tab is often more convenient, and it avoids issues with file locks that can occur when running ArcMap and ArcCatalog at the same time. Occasionally, using ArcCatalog is easier, especially if you are mostly searching for and managing files, or if you are focused on creating metadata. Both versions have the same functionality, so which one is used is only a matter of convenience and preference.

In this chapter we explore the functions of ArcCatalog for compiling GIS data for a project. In many cases, data already exist but need to be copied or converted to a new geodatabase. We tackle the more complicated process of creating data from scratch in Chapter 7.

An **import** function copies a data set from one location into the geodatabase, converting formats on the way if needed. An **export** function goes the other direction, copying a data set from the geodatabase to another location. In most cases they can be used interchangeably; the only difference is where to start the function. Formats that are directly supported by ArcGIS, such as shapefiles and geodatabases, can use menus to import or export them in or out of a geodatabase. External formats such as Excel tables or KML, however, usually need to be converted using tools in ArcToolbox. A perusal of the Conversion Tools toolset (Fig. 2.8) is a good way to get familiar with the different data formats that can be converted to ArcGIS files.

The import and export tools in ArcGIS can do more than simply copy files; they also allow you to modify some aspects of the file as it is copied. Imagine creating a geodatabase containing just the state of New Jersey. If we start with a feature class containing all 50 states, we can specify that only New Jersey be copied during the import process to achieve our goal.

Examine the import/export tool in Figure 2.9. The input features are a states feature class in the mgisdata\Usa folder. The output location is a NewJersey geodatabase, and the new feature class will be named statepoly. The Expression box uses a **query** to describe which features to include—the feature for which the STATE_NAME in the attribute table equals 'New Jersey.' Note that the Expression is optional; if nothing is entered, then all of the states would be copied to the new location.

### *Using ArcToolbox*

ArcToolbox contains hundreds of supporting functions and programs that expand the capabilities of ArcGIS Desktop far past what is found in the menus and toolbars. It may be used in ArcCatalog or in ArcMap. When a tool is opened (Fig. 2.10), required inputs are marked with a green dot (a). To enter data sets as inputs or outputs, use the Browse button next to the appropriate box (b). If using the tool in ArcMap, you can also select

Fig. 2.8. Conversion tools in ArcToolbox
Source: ArcToolbox

layers in the map from a drop-down list as inputs. Optional inputs are marked (c) and can usually be left to their defaults. Some tools have input parameters that extend beyond the visible pane, and the scroll bar must be used to view and enter them. To get more information about the tool or an input, click the Show/Hide Help button (d) for a brief description of the tool's function. For more detailed help, especially on the input/output data, click the Tool Help button (e).

**TIP:** When using a tool for the first time, it is strongly recommended that you read the detailed help information. It may contain critical information for making the tool work properly.

The input pane for a tool posts symbols to help you use the tool (Fig. 2.11). A green dot next to an input parameter indicates that it is required and must be entered before the tool will run. A yellow exclamation point indicates that the user should proceed with caution. Placing the cursor over the yellow sign for a moment will display a pop-up message describing the problem; in Figure 2.11, the user is attempting to define a coordinate system for a data set that already has one. A red X indicates an error that would cause the tool to fail; the tool cannot be executed until the error is cleared. Again, holding the cursor over the X will display the error message.

Fig. 2.9. Importing New Jersey to a geodatabase
Source: ArcGIS

Fig. 2.10. Preparing to run a tool
Source: ArcMap

By default, a tool runs in background mode so that you can continue working as it runs. When the tool begins running, blue scrolling text will appear at the bottom of the ArcMap or ArcCatalog window. If an error occurs, it will be reported in a dialog box. When the tool completes successfully, a blue completion message will pop up in the lower-right corner of your computer screen (Fig. 2.11). The message will go away automatically, or you can click the X to close it.

**TIP:** Background processing is slower to start and finish a task, and it may not run as reliably as foreground processing. The author ***strongly recommends*** that you turn off background processing in the Geoprocessing > Geoprocessing Options window.

Fig. 2.11. Error handling for tools
Source: ArcMap

With so many functions, finding the right one can be difficult. ArcGIS has a Search window (Fig. 2.12) that can be used to search for tools as well as other items.

### Documenting data

Compiling data for a project usually requires some work with the metadata. ArcCatalog has a Metadata Editor used to read and edit metadata for geospatial files. By default, it uses a simple format called the **Item Description** (Fig. 2.13), which shows only a few entries, even though the actual metadata may be far more extensive. The Item Description is designed for users to quickly review or enter the most important characteristics of a data set.

The fields of the Item Description, and what they need to contain, are as follows:

**Tags** record a few terms that can be used for searching for the data.

The **Summary** is a brief description, a sentence or two, describing the purpose of the data, or the intentions with which it was developed. Use it to quickly give the reader an idea of what the data set is designed for.

The **Description** can be a more extended explanation of the contents of the data set. It can include anything that is important to know, such as what the data set contains, how it was developed, what the attributes mean, and any information known about the data quality, such as the source scale. The Description may be short or it could run to several paragraphs.

The **Credits** record the source(s) of the data. It may include a formal citation if the data originally came from a book, journal article, or published map. For downloaded data, it should include the URL where the data were obtained. If the data originated from an individual or agency, the names should be listed. It is a good practice to construct a formal citation (see Chapter 1) for any data set you download and place it in the Credits box; when you need to place the citation in a report or on a map, you need only copy it.

Fig. 2.12. Searching for tools
Source: ArcGIS

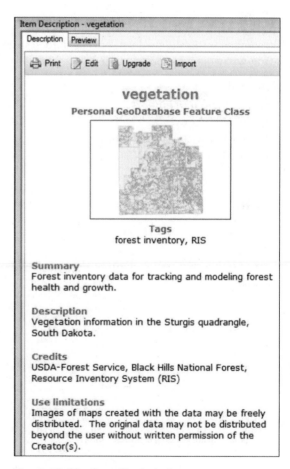

Fig. 2.13. The Item Description
Source: ArcCatalog

The *Use Limitation* provides information on any restrictions placed on the data, such as whether it can be copied or shared. This statement may be a simple sentence, but some companies use this box for an extended legal statement.

The *Appropriate Scale Range* suggests the scales at which the data are valid and should be used.

The *Bounding Box* is automatically entered by the software from the $x$-$y$ extent of the features. It is always in degrees and is used by geographic-based search engines.

**TIP:** You should make a practice of quickly documenting any data set that you intend to keep or use again. If you are developing metadata for data to be shared with others, however, it is best to learn to create the more extended forms of metadata covered in Chapter 14.

### Sources for GIS data

There is no one-stop source of GIS data, but these sites should get you started. Doing an Internet search for "sources of GIS data" will yield many more lists of sites.

**Sites with good lists of data sites**

http://en.wikipedia.org/wiki/List_of_GIS_data_sources
http://sites.tufts.edu/gis/data-sources/data-source-index/
http://www.gisinecology.com/gis_data_sources.htm

**Sites with data**

http://nationalmap.gov
http://geo.data.gov
http://www.cdc.gov/gis/data.htm
http://data.geocomm.com/

## Summary

▶ GIS data are different from more common data types such as word processing or spreadsheet documents and have special considerations that must be understood and applied.

▶ GIS data have different naming conventions than many other types of data. Folder names and file names should be short and use only numbers and letters and the underscore character.

▶ GIS users must take responsibility for setting up organized folder trees to store GIS data and making sure that files are saved in the appropriate place with descriptive names.

▶ Data compiled for a project must be taken from data with source scales suitable for the geographic area and intended purpose of the geodatabase.

▶ Finding data for a GIS project requires many general computer skills such as Internet searching, downloading, unzipping, and converting between file formats. Building these skills through practice, experimentation, and training is important.

▶ Metadata provide information about data sets and is needed for the user to evaluate its fitness for a given purpose. In-house data may use the Item Description for quick documentation; data intended for distribution needs to use a standards-based metadata format.

▶ Map documents store references to feature classes and rasters rather than storing the data. Absolute or relative pathnames may be used depending on the situation.

▶ Moving, renaming, and deleting GIS data has the potential to break links from map documents to the data. Care should be taken during these activities to avoid having to fix broken links.

▶ File locks are used by software to prevent multiple users from editing the same file. Shared GIS data require planning ahead to ensure that multiple users have the access and permissions needed without putting the data at risk.

### Important Terms

| | | | |
|---|---|---|---|
| absolute pathname | Interchange file (e00) | pathname | source |
| export | Item Description | query | write access |
| import | lock file | read access | zipped |
| | | relative pathname | |

## Chapter Review Questions

1. What characters should be avoided when naming GIS files, folders, and map documents? Explain why these restrictions are necessary.

2. What is a file extension, and what does the computer use it for?

3. Although ArcGIS Online is a good source of many types of data, what factors may interfere with using it to compile local data for a project?

4. Which kind of metadata should be used for documenting data created while studying this book? Defend your answer.

5. Which would take more storage space, a layer file showing all the US counties or a layer file showing all the US states? Explain the reasons for your answer.

6. What does it mean if you find a red exclamation point next to a map layer? How would you fix it?

7. Reexamine Figure 2.6 and its description in the text. The following table contains four actions. Below each is the list of the map documents in Figure 2.6. For each one, indicate whether the pathnames in the map document would still work (yes/no) after the action was taken, depending on whether the map document is saved with absolute (Abs) or relative (Rel) pathnames.

| Move the mgisdata folder to a USB stick | | | Copy the map documents to a USB stick to take to school and print | | | Move the Njdata geodatabase to the mgisdata folder | | | Rename the MyProject folder to NJProject | | |
|---|---|---|---|---|---|---|---|---|---|---|---|
| | Abs | Rel | | Abs | Rel | | Abs | Rel | | Abs | Rel |
| NjProj | | | NjProj | | | NjProj | | | NjProj | | |
| USProj | | | USProj | | | USProj | | | USProj | | |
| NJmap | | | NJmap | | | NJmap | | | NJmap | | |
| USmap | | | USmap | | | USmap | | | USmap | | |

8. Explain why it might cause problems to run ArcCatalog and ArcMap simultaneously.

9. What is a query? Give an example. Why might you use one when exporting data to a geodatabase?

10. What is the difference between the Summary and Description fields in the Item Description?

# Mastering the Skills

### Teaching Tutorial

The following examples provide step-by-step instructions for doing basic tasks and solving basic problems in ArcGIS. The steps you need to do are highlighted with an arrow →; follow them carefully. Click on the video number in the VideoIndex.htm file to view a demonstration of the steps.

**TIP:** Make sure your computer is set to show file extensions; in Windows this setting is generally found in the Control Panel or in the File Explorer program under Folder Options (see Fig. 2.2).

### *Using ArcCatalog*

→ Start ArcCatalog (not ArcMap!).

1→ Examine the Catalog Tree, normally on the left side of the window. This window can be used to view and explore data folders.

1→ Clicking a plus sign expands a folder to show its contents; clicking a minus sign collapses it.

1→ If any folders are expanded right now, click the minus signs to close them all.

Fig. 2.14. The Catalog Tree
Source: ArcCatalog

2→ Expand the Folder Connections entry. You will see the same connections you already added in ArcMap (Fig. 2.14). ArcMap and ArcCatalog share this information.

2→ Examine the viewing area on the right, noting that it has three tabs: Contents, Preview, and Description. Make sure the Contents tab is currently active.

Source: ArcCatalog

2→ Expand the connection to the mgisdata folder to see the subfolders.

2→ Click on the Oregon folder. The panel on the right updates to show its contents.

2→ These buttons change how the Contents tab displays the contents. Click on each one in turn and examine how the viewing panel changes. Leave it on the third setting (Details) when finished.

Source: ArcCatalog

You should see a variety of file types in the Oregon folder, including geodatabases, a raster, and some Excel spreadsheets (Fig. 2.15). If you have set Windows to show file extensions, you will see the extensions listed along with the file type.

3→ In the Catalog Tree, expand the Oregon subfolder and the oregondata geodatabase so you can see the feature classes and feature datasets.

3→ In the Catalog Tree, click on the cities feature class. The Contents tab updates to show an icon for the feature class.

3→ Click on the Preview tab. The viewing panel now shows the features.

3→ Use the Zoom and Pan tools to zoom closer to some of the cities.

3→ Use the Identify tool to get information about a city. Close the Identify window when done.

These tools help explore data sets to find out what is in them. The table can also be displayed.

| Contents | Preview | Description |

| Name | Type |
|---|---|
| GeoIProject.gdb | File Geodatabase |
| oregondata.gdb | File Geodatabase |
| L720021127av2.bil | Raster Dataset |
| L720021127av2.prj | Coordinate System |
| myairports.lyr | Layer |
| ORprecipnormals.xls | Excel File |
| ORstations.xls | Excel File |
| rocksamples.xls | Excel File |
| rocksampleskey.xls | Excel File |

Fig. 2.15. Data in the Oregon folder
Source: ArcCatalog

4→ Click the Geography drop-down button and select Table instead.

4→ Scroll to the right and examine the table attribute fields.

 4→ Click the Table Options button to view the commands available, but don't select any at this time. Just click the bar at the top of the window to close the menu.

5→ Let's preview a different data set. Click the gtoposhd raster in the Catalog Tree.

5→ The table is not that interesting, so click the Table drop-down and select Geography.

 5→ If you are still zoomed in, click the Full Extent button to see the entire raster.

5→ Click on each data set in the oregondata geodatabase in turn, previewing it so that you can quickly get an idea of the data sets available.

6→ While the last data set is still highlighted, click the Description tab. It will take a moment to open, but it shows the metadata for the data set.

6→ Click on another feature class to view its Item Description. Note the different types of information present.

6→ Click the Preview tab to return to viewing the data instead of the metadata.

Data sets have properties that can be viewed and managed in ArcCatalog. Unlike changing layer properties in ArcMap, changing properties in ArcCatalog usually makes permanent changes to the files. Be very careful when working in ArcCatalog.

7→ Right-click the parks feature class in the Catalog Tree and choose Properties. Examine each tab, but make no changes. Cancel to close the window.

7→ Right-click the gtoposhd raster in the Catalog Tree and choose Properties. Examine how raster properties differ from vector properties.

1.  What is the resolution (cell size) of this raster? How many rows and columns does it have?

**SKILL TIP:** By default only a brief description is shown. Learn how to make ArcCatalog display the complete metadata information (Metadata).

8→ Close the gtoposhd properties, if necessary, and collapse the Oregon folder.

8→ Expand the Rapidcity folder. Inside it, expand the archive folder to examine its contents (Fig. 2.16). This folder holds many types of GIS data.

2. How many coverages are there in the archive folder? _____ How many tables? _____ How many rasters? _____ How many layer files? _____ How many shapefiles?

| Name | Type |
|---|---|
| citybnd | Coverage |
| landuse | Coverage |
| buildings.shp | Shapefile |
| connects.shp | Shapefile |
| lucodes.dbf | dBASE Table |
| parcels.shp | Shapefile |
| rc_roads.shp | Shapefile |
| rceast_nw.sid | Raster Dataset |
| sdschools.shp | Shapefile |
| stategeol.lyr | Layer |
| stategeol.shp | Shapefile |
| watersheds.shp | Shapefile |

8→ Collapse the archive folder and the Rapidcity folder.

### *Creating and organizing data*

Now we will create a new geodatabase with feature classes for New Jersey, placing it outside the mgisdata folder to separate our "project work" from the book exercises. Chapter 1 suggested placing the mgisdata folder in a C:\gisclass folder. The instructions assume you are using this folder, but if not, substitute the name of whatever folder contains your mgisdata folder.

9→ Examine the Catalog Tree and find the connection to your mgisdata folder, for example, C:\gisclass\mgisdata.

Fig. 2.16. Data sets in ArcCatalog
Source: ArcCatalog

Because we created the connection directly to the mgisdata folder, we cannot access the folder above it. We need to create another folder connection.

9→ Click the Connect to Folder button on the main toolbar.

9→ Navigate through the tree to the C:\gisclass folder and select it. Click OK.

9→ Find where the C:\gisclass folder was added under Folder Connections and expand it. The mgisdata folder will be in it.

Now we see two connections to mgisdata, which might get confusing. It can also cause problems to have multiple connections to the same folder. So let's remove the first folder connection.

10→ Right-click the C:\gisclass\mgisdata connection and choose Disconnect Folder.

10→ Expand the mgisdata folder again to see the subfolders.

Now we are ready to compile data for our New Jersey project.

11→ Right-click the C:\gisclass folder in the Catalog Tree and choose New > Folder.

11→ A new folder is created, waiting for you to type its name. Enter *NJ_Project* and press Enter. (Always use an underscore instead of a space.)

11→ Right-click the NJ_Project folder and choose New > File Geodatabase. When it appears, name it *NewJersey* (no spaces!).

**TIP:** If you accidentally put a space in the name, right-click the geodatabase and choose Rename. Type in the name again without the space and press Enter.

Of course, the NewJersey geodatabase is empty, and we need to put data in it. We will start by importing some data available in our mgisdata\Usa folder and subsetting it to New Jersey.

12→ Right-click the NewJersey geodatabase and choose Import > Feature Class (single).

12→ Click the Browse button next to the empty Input Features box (Fig. 2.17).

12→ Navigate into the mgisdata\Usa\usdata geodatabase and click the states feature class. Click Add.

12→ Notice that the Output Location is already set to the NewJersey geodatabase, because you right-clicked it to start the import.

12→ In the Output Feature Class box, enter the name to be given to the new feature class, *newjerseybnd*.

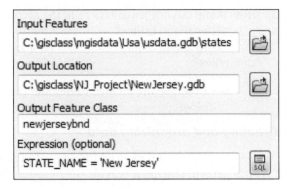

Fig. 2.17. Importing New Jersey
Source: ArcCatalog

We want only the features for New Jersey, so we must use the Expression box to select the state using values from the attribute table. The Query Builder is used to create expressions.

13→ Click the SQL button next to the Expression box to open the Query Builder.

13→ The top panel contains fields from the feature class table. Double-click the STATE_NAME field in the upper panel to enter it in the lower panel.

13→ Click the "=" button.

13→ Click the Get Unique Values button; it will scan the values in the STATE_NAME field and allow you to select from them.

13→ Find the entry for "New Jersey" and double-click it to enter it in the lower panel.

13→ The lower panel should now read STATE_NAME = 'New Jersey'.

13→ Click Verify to make sure the expression is correctly formatted. If it was, click OK to continue. If not, clear the expression and try again.

13→ When the expression is correct, click OK to close the Query Builder. The upper part of the Import window should look as shown in Fig. 2.17.

13→ Click OK to start importing.

The new feature class appears in the NewJersey geodatabase (you may need to expand the geodatabase to see it, or right-click the geodatabase and click Refresh).

14→ Click the newjerseybnd feature class in the Catalog Tree and make sure the Contents tab is selected in the viewing panel.

14→ Notice that the thumbnail image still shows all of the states; we will fix this in a moment.

14→ Click the Preview tab. The new feature class only has New Jersey.

14→ Click the Create Thumbnail button to update the icon for the data set.

14→ Click the Contents tab. The icon shows only New Jersey now.

Source: ArcCatalog

When a data set is imported or exported, any metadata it has is copied to the new feature class. However, we should update the Item Description when we change a data set.

15→ Click the Description tab in the viewing panel and click the Edit button.

15→ Modify the Tags and Summary to better reflect what is in the feature class now.

15→ In the Description box, add this statement at the end: *Subset to New Jersey from a national data set by <your name> on <current date>.*

15→ In the credits box, construct a citation to document the source of the data: *Mastering ArcGIS Tutorial Data, 8th edition, by Maribeth Price (2018) [Download]. McGraw-Hill Higher Education: Dubuque, Iowa. URL: http://www.mhhe.com/price8e.*

15→ Click Save.

**TIP:** It is good practice to update the Item Description whenever you modify a data set.

Let's import another feature class—the cities. However, every feature class table is different. The expression we used last time, STATE_NAME = 'New Jersey' may not work for the cities table. It helps to view the table first to make sure we know what expression to use.

16→ Navigate to the mgisdata\Usa\usdata geodatabase in the Catalog Tree and click on the cities feature class.

16→ Click the Preview tab and set the Preview drop-down to Table.

3.   What do you think the expression should be this time? _____

Now we will repeat the import steps, making the necessary modifications. These new instructions are abbreviated; refer back to Steps 12–13 if you need a reminder.

**TIP:** If you receive a script error when you select the Import tool in the next step, and ArcCatalog stops responding, close the program and go to page 467 in the Skills Reference.

17→ Right-click the NewJersey geodatabase and choose Import > Feature Class (single).

17→ Click the Browse button to set the Input Features to mgisdata\Usa\usdata\cities.

17→ Name the output feature class njcities.

17→ Click the SQL button and use the buttons to enter the expression ST = 'NJ'. Verify it and click OK.

We should check the output to be sure that the import worked and update the metadata.

18→ Click the njcities feature class in the NewJersey geodatabase in the Catalog Tree.

18→ Click the Preview tab to ensure that only the New Jersey cities are present.

18→ Click the Create Thumbnail button to update the icon.

18→ Click Description button and the Edit button to update the Item Description. Be sure to enter a citation in the Credits box.

Perhaps we would like to add highways to this geodatabase.

19→ Click the majroads feature class in the Transportation feature dataset of the usdata geodatabase and click the Preview tab.

19→ Set the Preview type to Table, if necessary, and examine the fields.

This table does not have a field that designates a state, because roads may not belong to a single state. However, without a state field, we cannot extract the New Jersey roads using the techniques we have learned so far. We will learn how to do it later.

→ Close ArcCatalog.

**TIP:** We have neglected one important issue when importing the data: the source data is stored in a coordinate system suitable for the United States, but a state data set needs one optimized for a smaller area. We will address this deficiency later as well.

### Using the Catalog window in ArcMap

Because of file locks, it is troublesome to run ArcMap and ArcCatalog at the same time. ArcGIS developers have created the Catalog window, a version of ArcCatalog that runs inside ArcMap.

→ Open ArcMap with a new, blank map.

20→ Click the Catalog window button on the main toolbar to open it.

20→ If it opens docked on the right side of the ArcMap window, leave it there.

20→ If it opens elsewhere, click the bar at the Catalog window's top and drag it to the blue arrow that appears on the right side of the ArcMap window. Release the mouse to dock it.

Source: ArcCatalog

**SKILL TIP:** Learn how to move and arrange the toolbars, as well as how to use the docking icons and pins to manage windows in ArcGIS (General).

In ArcMap, only the Catalog Tree is displayed. It works the same way as before in most respects.

21→ In the Catalog Tree, collapse any expanded folders, leaving only the C:\gisclass folder connection visible (as in Fig. 2.14).

21→ Expand the usdata geodatabase in the mgisdata\Usa folder.

The Preview and Description tabs from ArcCatalog are not visible, but they can be opened in a separate window.

22→ Right-click the rivers feature class and choose Item Description. It will take a few moments to open.

22→ Move/resize the Item Description so that it and the Catalog window are visible.

22→ Read through the information provided about the rivers feature class.

22→ Click the cd111 feature class in the Catalog Tree. The Item Description window updates (after a moment) to show information on this feature class.

4. What kind of data does this feature class contain? _____

23→ Click on the Preview tab in the Item Description window.

23→ Use the Zoom or Pan tool in the Item Description window to get a closer look.

23→ Change the Geography drop-down to Table to preview the attributes.

5. If we wanted to select the New Jersey districts, what would the expression be?

_____

24→ Right-click the STATE_ABBR field and choose Sort Ascending. Scroll down and find the district(s) for your state.

24→ Use the Table Options menu to open Find and Replace. Then search for *Boehner*. The cursor in the table (subtly) highlights the cell when it finds it.

24→ Close the Find/Replace and the Item Description windows.

Now that you know it's the feature class you want, it is easy to add it to the map.

25→ Click on the cd111 feature class in the <u>Catalog Tree</u> and drag it into the map window. Release the mouse button to drop it in the map.

25→ Right-click the cd111 layer symbol in the <u>Table of Contents</u> (not the Catalog Tree) and choose a pale yellow symbol.

**TIP:** Be careful to select the correct window when right-clicking an item. The <u>Table of Contents</u> window lists map layers, and making changes only affects how the map appears. The <u>Catalog Tree</u> in the Catalog tab lists folders and data sets, and it can permanently change them (usually with no Undo). Be especially careful when working in the Catalog Tree.

Since the <u>Catalog Tree</u> has the potential to change or damage our data if we are not careful, we are going to make it less easy to use it by mistake, while keeping it handy for when we need it. It will also let us see more of our map.

26→ Click the Auto Hide button on the Catalog window to hide it, creating a Catalog tab where the window was docked.

26→ Hold the cursor over the Catalog tab (no click needed) to open the Catalog window. Move off the window to hide it again.

Source: ArcCatalog

**TIP:** The Auto Hide button is a toggle switch. To keep the Catalog window open, click the Auto Hide button again. Try it now, if you like, but leave it hidden when done.

### Using ArcToolbox

By default, ArcGIS executes tools in background mode to allow the user to keep working while the tool runs. However, background processing takes longer than foreground processing, and it can cause tools to crash. The author recommends turning off background processing.

27→ Open Geoprocessing > Geoprocessing Options from the main menu bar.

27→ Under Background Processing, uncheck the Enable box and click OK.

Often tasks that we perform using menus can be done using a tool from ArcToolbox instead. Let's import another feature class as an example, this time using a tool.

28→ Click the ArcToolbox button on the main menu bar.

28→ Dock the ArcToolbox window on the right side of the ArcMap window and use the Auto Hide button to fold it away. When you want a tool, hover over the ArcToolbox tab and the toolbox will open.

Source: ArcCatalog

29→ Open ArcToolbox and expand the Conversion Tools toolset.

29→ Expand the To Geodatabase toolset.

29→ Double-click the Feature Class to Feature Class tool to open it. The tool looks exactly the same as the one we previously used (Fig. 2.17).

**TIP:** You can drag and drop data sets from ArcMap or the Catalog window into tools.

30→ Drag and drop the cd111 layer from the <u>Table of Contents</u> into the Input Features box on the tool.

30→ Drag and drop the NewJersey geodatabase from the <u>Catalog window</u> to the Output Location box on the tool (or use the Browse button to find it, if you prefer).

30→ Type njdistricts in the Output Feature Class box to name it.

30→ Click the SQL button and enter STATE_ABBR = 'NJ'.

30→ Click OK to start running the tool. Close the reporting box when it finishes.

30→ Note that the new feature class is added to the Table of Contents when a tool finishes. Zoom to New Jersey to verify that the import worked correctly.

31→ Right-click the njdistricts feature class in the <u>Catalog Tree</u> and open the Item Description window.

31→ Click the Preview tab to verify that only the New Jersey districts are present.

31→ Click the Create Thumbnail button to update the icon.

31→ Click the Description tab and Edit the Item Description to update it. Save and close the Item Description window when finished.

31→ Remove the cd111 layer from the <u>Table of Contents</u>.

6.  Notice that if the term <u>layer</u> is used in the instructions, then you are directed to right-click in the <u>Table of Contents</u>. If the term <u>feature class</u> is used, then you must right-click in the <u>Catalog Tree</u>. Explain why.

If you don't know where a tool is to be found, you can search for it. We will now use the Clip tool to extract the major roads within New Jersey.

32→ Click the Search button on the main toolbar. Dock it on the right side of the ArcMap window and click the Autohide button to fold it away.

32→ Type clip in the Search box and click the Search button.

32→ Find the Clip (Analysis) tool. Click it to open it.

32→ Click Show Help on the tool and read about what Clip does.

32→ Drag usdata\Transportation\majroads from the Catalog Tree to the Input Features in the tool.

32→ Drag njdistricts from the Catalog Tree into the Clip Features box in the tool.

32→ Click the Browse button for the Output Features and set the Save As type to *File and personal geodatabase feature classes*, if necessary.

32→ Navigate inside the NewJersey geodatabase (your other new feature classes will be visible), and name the output majroadclip. Click Save.

32→ Click OK to run the tool.

**TIP:** A bug in versions 10.4 and 10.5 prevents njdistricts from being recognized as a valid file when dragged from the Table of Contents to the tool. Work around it by dragging from the Catalog instead.

**TIP:** The majroadclip feature class should be added to the map after the tool runs, but a bug prevents that also. Add it yourself from the Catalog. To prevent the bug from recurring, save the map document and restart ArcMap.

### *Using ArcMap to export data*

A third way to move data into a geodatabase is to export it from ArcMap. This method has the advantage that you can see what is selected before exporting, and you can use the data frame to convert from one coordinate system to another. We should be saving our feature classes in the NAD 1983 State Plane New Jersey coordinate system (we will learn why in the next chapter), and using ArcMap makes it easy. We start by setting the data frame to the desired coordinate system.

33→ Right-click the Layers data frame name and choose Properties.

33→ Click the coordinate system tab and navigate from the top to Projected Coordinate Systems > State Plane > NAD 1983 (Meters). Remember to use the plain one.

33→ Find the NAD 1983 StatePlane New Jersey FIPS 2900 (Meters) coordinate system and click to highlight it. Click OK.

⊞ 📁 NAD 1983 (Intl Feet)
⊟ 📂 NAD 1983 (Meters)

Source: ArcMap

We will export earthquakes to our geodatabase next. As before, we need a query to select the New Jersey earthquakes, but this time we do it in ArcMap.

34→ Add the quakehis feature class to the map from the usdata geodatabase.

34→ Double-click the quakehis layer in the <u>Table of Contents</u> and examine the Source tab. What is the coordinate system? Close the Layer Properties window.

34→ Right-click the quakehis layer in the <u>Table of Contents</u> and choose Open Attribute Table.

35→ Click the Select By Attributes button on the Table toolbar. This opens the Select by Attributes window, very similar to the Query Builder in ArcCatalog.

35→ Enter the expression STATE = 'NJ' and click Apply.

35→ Examine the bottom of the Table window and note that six quakes are selected.

35→ Close the Table window. The selected quakes in New Jersey are highlighted.

Now that we have the quakes we need selected, we can export.

36→ Right-click the quakehis layer in the <u>Table of Contents</u> and choose Data > Export Data.

36→ Make sure it is set to Export the *Selected features*.

36→ Fill the button to *Use the same coordinate system as the data frame*.

36→ Click the Browse button. Set the Save As type *to File and Personal Geodatabase feature classes*, if necessary.

36→Navigate inside the New Jersey

| Export: | Selected features |
|---|---|

Use the same coordinate system as:

◯ this layer's source data

◉ the data frame

◯ the feature dataset you export the data into
(only applies if you export to a feature dataset in a

Output feature class:

C:\gisclass\NJ_Project\NewJersey.gdb\njquakes_sp

Fig. 2.18. Exporting in ArcMap to a new coordinate system
Source: ArcMap

geodatabase; the newjerseybnd, njcities, and njdistricts feature classes will be visible.

36→ Enter *njquakes_sp* as the output file name and click Save.

36→ Check that the Export Data window looks like Figure 2.18 and click OK.

36→ Click Yes to add the exported data to the map as a layer.

37→ Double-click njquakes_sp in the <u>Table of Contents</u> and examine the Source tab to confirm that the new feature class is in the New Jersey State Plane coordinate system.

37→ Close the properties. Right-click the original quakehis layer and choose Remove.

38→ Use the Catalog window to update the Item Description for njquakes_sp.

39→ Add the newjerseybnd feature class to the map from the NewJersey geodatabase. Examine the state boundaries closely. Why don't they match?

39→ In the <u>Catalog Tree</u>, right-click the njdistricts feature class and choose Item Description.

39→ Scroll down and examine the Scale Range section. Notice that the recommended scale for this data ranges from 1:50,000 to 1:5,000,000.

40→ Leaving the Item Description window open, click on the newjerseybnd feature class in the Catalog Tree.

40→ Examine the scale range for this data set, 1: 500,000 to 1:50,000,000. (The lower bound here is unrealistic; pay attention to the upper bound.)

40→ Close the Item Description window.

40→ Examine the scale readout on the main menu bar. If the state of New Jersey fills the screen, the scale will be about 1:2 million.

These two feature classes were designed for use at different scales, about 1:5 million for the districts and 1:50 million for the states. The New Jersey geodatabase requires larger scale data than the generalized feature class from which the state boundary was exported.

**TIP:** Pay attention to source scales when compiling data, and try not to use data sets too far outside their intended scale range.

### Using data from ArcGIS Online and the Internet

Let's see if we can find other data to add to the New Jersey geodatabase by going online. Perhaps we can find a more detailed state boundary. We will continue to use our new technique of exporting from ArcMap so that we can save data in the appropriate coordinate system.

41→ Click the small black arrow next to the Add Data button and choose *Add Data From ArcGIS Online.*

41→ Enter *detailed states* in the search window and click the Search button.

41→ Find the entry USA States, a layer package hosted by ESRI. Read about it in the Details link and write down a citation for it (see Chapter 1 for examples).

41→ Add the layer package to the map. If you see a GCS warning, check *Don't warn me again in this session* and click Close.

A layer package downloads data to ArcMap, which allows you to select features and export them to your database, just as you did previously with the quakehis feature class. Refer to the more detailed instructions in Steps 34–36 if needed.

42→ Open the attribute table for the USA States (below 1:3m) layer.

42→ Use the Select By Attributes button to select New Jersey. Close the table.

42→ Export the USA States (below 1:3m) layer, using the selected features and the data frame coordinate system, saving it with the name njstate_dtl. Add it to the map.

43→ Use the Catalog window to update the thumbnail and Item Description for the njstate_dtl feature class. Include this citation in the Credits box (did yours look like this?): *USA States (2013) [Layer package]. ESRI on ArcGIS Online, URL: http://ArcGIS.com [<today's date>].*

43→ Right-click the USA States group layer and Remove it from the map.

**TIP:** Many ArcGIS Online data sets do not allow you to export them. If you don't see the Data > Export Data function when you right-click a layer, it is not exportable. A layer package or feature service is more likely to be exportable than a map service.

Now that we have a state boundary with a suitable scale and coordinate system, we can delete our original attempt, newjerseybnd.

44→ In the <u>Catalog Tree</u>, navigate to the NewJersey geodatabase if necessary.

44→ Right-click the newjerseybnd feature class and choose Delete. Click Yes.

**TIP:** ALWAYS use ArcCatalog to delete, move, and rename GIS data sets.

### *Getting data from the Internet*

Finding and downloading data from the Internet can be challenging, but let's give it a try. A good way to start is to type a search term plus the word GIS or shapefile to find downloadable data.

45→ Open Internet Explorer and enter the search term *New Jersey shapefile.*

45→ Examine the results. Look for URLs that appear to be state-sponsored rather than commercial, for example the Bureau of GIS – State of New Jersey heading (URL: http://www.state.nj.us/dep/gis/stateshp.html). Open it.

45→ Examine the site, which appears to be an excellent source of data for New Jersey.

Before we continue, let's organize a folder for the downloads. Downloaded data can be messy, and we want to keep it separate from our main working folder for the project.

46→ Right-click the NJ_Project folder in the <u>Catalog Tree</u> and choose New > Folder. Name the folder *Downloads.*

Source: ArcCatalog

47→ Select Elevation Contours from the drop-down list on the NJDP-GIS site.

47→ Click the DOWNLOAD button and use Save As to save the .zip file in the NJ_Project/Downloads folder.

47→ Minimize (do not close!) the web browser for the moment.

**TIP:** The data file must be unzipped before ArcGIS can recognize it. The instructions assume that you have a zip utility and can use it. If you need one, you can download and install 7zip for free.

48→ Open Windows Explorer and navigate to the NJ_Project\Downloads folder.

48→ Use a zip utility to extract the files to the NJ_Project\Downloads folder.

Notice the different files with the same root name stcon and different extensions, and recall from Chapter 1 that shapefiles have this format. ArcMap can read this shapefile now.

49→ Close Windows Explorer.

49→ Add the stcon shapefile to the map from the Catalog window.

49→ Open the Item Description for stcon in the Catalog window. Click Yes if you are asked to upgrade the metadata—you will learn more about FGDC metadata in Chapter 14.

You have not changed the file, so the Item Description does not need to be updated. However, it is a good practice to create a citation in the Credits entry so you can recall where it came from.

**TIP:** You will need to search the web site and/or open the metadata link for the data set to find all the information you need for the citation. That is why we kept the site page open.

50→ Use the Preview tab to update the thumbnail.

50→ Edit the Item Description, adding this citation to the Credits box: *Elevation Contours (1987) [downloaded file]. New Jersey Department of Environmental Protection, URL: http://www.state.nj.us/dep/gis/stateshp.html [<today's date>].*

7. Where did the 1987 date for the data set come from?

51→ Double-click the stcon layer in the <u>Table of Contents</u> and examine its coordinate system.

8. Does it need to be projected to match our other New Jersey data?

51→ Right-click the stcon layer in the <u>Table of Contents</u> and export it to the NewJersey geodatabase. Use the data frame coordinate system. Save it as *njelevcon*.

Let's clean up a little and save this map document for use later.

52→ Remove all data layers from the map.

52→ Click the Add Data button and add all the data in the NewJersey geodatabase.

To ensure that moving the NJ_Project folder to a different place would not break any data links, we will set the map document to store relative pathnames.

53→ Choose File > Map Document Properties from the main menu bar.

53→ Check the box to *Store relative pathnames to data sources.*

53→ Click OK. Now as long as the map document stays in the same folder relative to its source data, the links to the data will stay valid.

54→ Click twice slowly on Layers and change the data frame name to *New Jersey*.

54→ Edit each layer name so that it is English instead of a file name, for example, *Cities, Congress Districts, Earthquakes,* and so on.

54→ Edit the symbols to make a more attractive map.

54→ Save the map document in the NJ_Project folder as *NJMap.mxd*. Be sure to save it in the right place.

Relative pathnames are usually the right choice for students working in a university setting. We will set an option to make relative pathnames the default for all new map documents.

55→ Choose Customize > ArcMap Options from the main tool bar.

55→ Click the General tab. Check the box to *Make relative paths the default for new map documents*. Click OK.

### *Using ESRI Data and Maps*

ArcGIS Desktop comes with an extensive set of GIS data called ESRI Data and Maps, either in ArcGIS Online or as a download by your GIS software administrator. Find out from your instructor if he or she has made this data available and, if so, where it is to be found. If you have the data, we will bring an additional feature class to the New Jersey geodatabase. If not, save your map now and exit ArcMap.

56→ Search the Data and Maps folders to find a US zip codes polygon feature class.

56→ Add the feature class to ArcMap.

56→ Open the table and use a query to select the zip codes for New Jersey.

56→ Export the New Jersey zip codes to the NewJersey geodatabase. Make sure you save it in the data frame coordinate system.

9.  What coordinate system do the ESRI Data and Maps feature classes use? _____

57→ Examine the citation examples in Chapter 1 and find the example for *ESRI Data and Maps*. Construct the citation.

57→ Update the Item Description for the zip codes. Be sure to put the citation in the Credits box.

This is the end of the tutorial.

→ Save the NJMap document and close ArcMap.

## Exercises

Use ArcMap, the Catalog window, and the mgisdata folder to answer the following questions. Be careful not to make any changes to properties when in the Catalog window.

1. How many feature datasets are there in the oregondata geodatabase in the mgisdata\Oregon folder? List their names. How many total feature classes does the geodatabase have? How many rasters?

2. How many rows and columns does the Landsat image L720021127 have? (**Hint:** It shows Crater Lake in Oregon.) What is the cell size (including the units)? How many bands does it have? What is its coordinate system?

3. Who created the gtopo1km raster in the oregondata geodatabase? When was it created? In what units are the elevation values?

4. In the tutorial, you saved the NJMap.mxd file in the NJ_Project folder rather than the mgisdata\MapDocuments folder. Explain why the NJ_Project folder is a better place.

5. Create a new project folder in your gisclass folder. Within it, create a new, empty file geodatabase for your home state.

6. Examine each of the feature classes in the usdata geodatabase and decide which ones can be exported to your geodatabase using a table query and which require clipping. In your answer document, list each feature class and state either "clip" or the appropriate query expression.

7. Add each of the data sets from #6 to ArcMap. Set the data frame coordinate system to an appropriate State Plane zone in the State Plane > NAD 1983 (meters) folder. If there is more than one zone for your state, just pick one, preferably the central one. State which State Plane zone you chose to use.

8. Export each of the data sets from #6 to your home state geodatabase, taking care to export only the features in your home state and using the data frame coordinate system to save it in the chosen State Plane zone.

9. Open the data frame properties and examine the Layers folder on the Coordinate System tab (as in Step 41 of the Chapter 1 tutorial). Verify that ALL of your new data sets are in the appropriate State Plane coordinate system. If not, fix them. Capture a screenshot of this window.

10. Update the Item Description for each data set that you exported.

11. **Capture** to turn in: (1) a screenshot of your project folder and geodatabase, expanded to show all of the feature classes; (2) a map of the layers, appropriately symbolized (you may omit some if the map is too cluttered); and (3) the screenshot from #9.

## Challenge Problem: My Old (insert your state name here) Home

Find three additional feature classes for your home state geodatabase, either from ArcGIS Online or a web site, choosing ones that you find interesting. Submit the same items as in Exercise 11 but only include the state boundary and the new layers in the map. Also include a list of citations, one for each data set.

# Chapter 3. Coordinate Systems

## Objectives

▶ Learning the basic properties and uses of coordinate systems

▶ Understanding the difference between geographic and projected coordinate systems

▶ Understanding different projections and the distortions they cause

▶ Choosing appropriate projections for a map or geodatabase

▶ Managing and troubleshooting coordinate systems of feature classes and images

## Mastering the Concepts

### GIS Concepts

#### *About coordinate systems*

A successful GIS system depends in large part on managing coordinate systems correctly, and a person's skill (or lack of it) can make the difference between usable and trashed data. Coordinate systems are also a vast and complex topic. Experience working with coordinate systems and regular review of these concepts will ultimately provide a sound basis for effectively working with coordinates.

In order to display and analyze maps on the screen, a GIS uses **coordinate pairs**, which specify the location and shape of a particular feature. A point is described by a single $x$-$y$ coordinate pair located in space; a line is a series of $x$-$y$ coordinates. A **coordinate space** is an agreed-upon range of coordinates in specified units used to portray the features.

Imagine that a surveyor is mapping an industrial site using typical surveying instruments that measure angles, directions, and distances. He may begin at a fence corner at the lower right of the complex and assigns that point the coordinates (0, 0). This is the **origin** of the coordinate space (Fig. 3.1). The surveyor then determines that the southwest corner of building A is 175 meters east and 200 meters north of the origin, and he assigns that corner the coordinates (175, 200). The surveyor uses meters to record the distances and determine the coordinates, thereby establishing meters as the **map units**. The combination of origin and map units becomes the coordinate space of this map. In this case the coordinate space is arbitrary and chosen for convenience.

Now imagine that the surveyor's colleague is surveying another site across town. She also chose the corner of her site as the origin, but she is measuring distances in feet. The two surveyors take their drawings back to the office and decide to create a map showing the location of both sites, so they each transfer the coordinates into a GIS. When

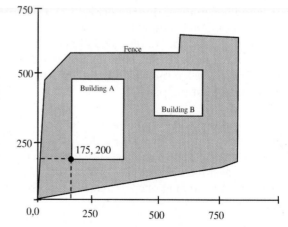

Fig. 3.1. An arbitrary coordinate space used for surveying a site

they plot the sites, however, the two different sites show up in exactly the same spot, even though in real life they are several miles apart. This problem occurs because both surveyors used an arbitrary (0, 0) origin unrelated to the other site. Moreover, the woman's buildings look about three times as large as the man's buildings because she is measuring in feet instead of meters. To plot these together the two surveyors would need to define a single common origin and convert both sites to the same map units.

To avoid problems such as this, geographers and map creators usually employ a standard **coordinate system** (CS), a coordinate space with characteristics that are defined and established according to cartographic standards. The surveyors could have used a global positioning system (GPS) instrument to establish the $x$-$y$ coordinates of their origins in a common coordinate system, such as Universal Transverse Mercator (UTM), and then entered their surveyed points relative to that base. Then their maps would have been located correctly with respect to each other.

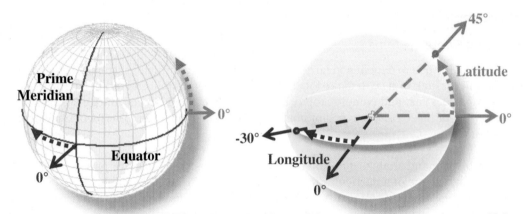

Fig. 3.2. Longitude measures E-W angles in the plane of the equator; latitude measures N-S angles above or below the plane of the equator.

### Geographic coordinate systems

One commonly used coordinate system is based on measuring angles from the center of the earth and has units of **degrees** (Fig. 3.2). **Longitude** measures horizontal angles east or west of the **Prime Meridian**, which is the line where longitude equals 0 that passes through Greenwich, England. Longitudes fall in the range –180 to +180. **Latitude** measures vertical angles above or below the equator, which has latitude equal to 0, and range from –90 at the South Pole to +90 at the North Pole. This system of measurement is called a **geographic coordinate system**, or **GCS**.

A GCS location is determined when the vector defined by a latitude-longitude pair intersects the surface of the earth. But what surface? Because the earth's shape is irregular, its surface can only be approximated, and the difference between the true shape and the approximate shape is a source of positional errors. For this reason, different approximations have been developed to meet specific mapping needs, each one representing a slightly different shape for the earth. Such an approximation is called a **datum**.

How is a datum defined? The earth's rotation causes it to bulge at the equator, so the first task is to model the earth's surface using a smooth figure called a **spheroid**, an oblong sphere with a major and a minor axis. (The bulge in Figure 3.3 is highly exaggerated.) The chosen spheroid becomes a part of the datum definition. (Some people use the term **ellipsoid** rather than spheroid.) Different spheroids have been used in the past because estimates of the earth's major and minor axes have varied over time. Clarke 1886 was a common spheroid used in North America, but it is being replaced by more recent and accurate satellite-measured spheroids.

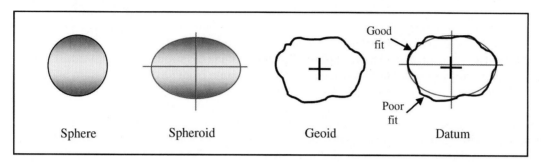

Fig. 3.3. Relationship among the spheroid, geoid, and datum

However, a spheroid does not adequately represent the shape of the earth either, so a second correction is made. The closest approximation to the earth's shape is the **geoid** (Fig. 3.3), an irregular, equipotential surface based on gravity. Sea level varies from place to place due to rotational, topographic, and compositional differences in the earth's mantle, and the geoid is often described as what the shape of the earth would be if there were only seas and no continents. The geoid can be modeled mathematically, but the equations are too computationally intensive to serve well as the basis for mapping. Instead, the spheroid is shifted relative to the geoid until a best-fit solution is obtained (Fig. 3.3). A datum, then, consists of the chosen spheroid and the location of its center relative to the geoid center.

A datum that finds the best fit overall is called an ***earth-centered datum***. The World Geodetic Survey of 1984 is an earth-centered datum based on the WGS 1984 spheroid and is commonly used for international data sets and web maps. It is the default datum for most GPS units.

Another approach is to find the best fit of the spheroid to the geoid for a particular region, such as North America, to produce a ***local datum***. The improved fit in one place, however, often means a poorer fit elsewhere, so a local datum should be used only in the regions for which it was developed. The North American Datum of 1927 (NAD 1927) is based on the Clarke 1866 spheroid. The North American Datum of 1983 (NAD 1983) is based on the more accurate GRS 1980 spheroid. Both datums use a network of surveyed benchmarks to further customize the fit and provide more accurate positions; this approach is often employed for local datums. Coordinates stored using NAD 1927 and NAD 1983 can differ by up to several hundred meters, which, although not significant for a national map, can cause serious registration issues for larger-scale maps. Figure 3.4 shows the offset between an image stored using NAD 1983 and roads stored using NAD 1927.

Adjustments of NAD 1983, based on better and better surveying of the benchmarks, take place periodically

Fig. 3.4. Offset of roads due to different datums
Source: ArcGIS

and may be used for very high-precision surveying applications when differences on the order of centimeters are important. Such adjustments have names such as CORS96, HARN, and NSRS2007. However, unless you have been told to use one of these specialized NAD adjustments, it is best to choose the standard datum, NAD 1983, which is what most data sets use.

| | |
|---|---|
| 🌐 | NAD 1983 |
| 🌐 | NAD 1983 (2011) |
| 🌐 | NAD 1983 (CORS96) |
| 🌐 | NAD 1983 (CSRS) |
| 🌐 | NAD 1983 (NSRS2007) |
| 🌐 | NAD 1983 HARN |

Source: ArcGIS

The datum forms the base definition of a geographic coordinate system (GCS) used to store locations in degrees of latitude and longitude. The same place will have slightly different values of latitude and longitude depending on which datum was used. The GCS typically takes its name from its defining datum and, to many, the terms *datum* and *GCS* are interchangeable.

At times it is necessary to convert from one datum to another; the process is called datum **transformation**. It does not always use exact mathematical formulas and may require localized estimates and fitting. Often, multiple methods are available. The transformation itself may introduce new errors in coordinate locations, on the order of several meters, so it is not desirable to transform GIS data back and forth repeatedly and multiply these errors. Usually, one picks one datum to be used for a project (or organization) and sticks with it, converting data once at the time they are brought into the geodatabase, using the best available transformation.

### *Map projections*

A GCS is a three-dimensional coordinate system, but maps need to be flat. Therefore, cartography requires converting the locations on a globe to locations on a piece of paper. This process is called **projection**. One starts with a defined GCS and applies a set of mathematical formulas that convert degrees of latitude and longitude into planar $x$-$y$ coordinates on the paper. During this conversion the three-dimensional GCS becomes a two-dimensional representation, and degrees are converted into feet or meters. The chosen earth shape (the GCS) influences exactly where locations end up on the paper, so the GCS is inherited by the projection and becomes a part of its definition.

You can imagine the projection process visually as a clear globe with the graticules of the earth printed in black. If you place a light bulb at the center of the earth and wrap a piece of paper around the equator in the shape of a cylinder (Fig. 3.5), the graticules will be cast as shadows on the paper. Tracing them gives you a map, which you can unroll and lay flat.

Fig. 3.5. Projecting a map from 3D to planar coordinates
Source: ESRI

Projections are grouped into three major classes, depending on the shape of the surface onto which the GCS locations are projected. A **cylindrical** projection uses a cylindrical surface that lies tangent to (touches) the earth at the equator along a great circle (Fig. 3.6). Rotating the cylinder sideways and making it tangent to the earth along a line of longitude produces a **transverse** projection. An **oblique** projection places the tangent at an angle. In each case, distortion is absent along the tangent and increases away from it. Cylindrical projections typically preserve shape or direction at the expense of area and distance.

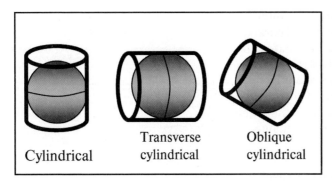

Fig. 3.6. Cylindrical projections          Fig. 3.7. Conic projections

A **conic** projection is based on setting a cone on the sphere (Fig. 3.7). If the cone is tangent to the globe along a line of latitude, then the projection is a **tangent** projection. The line of tangency is called a **standard parallel**. One can also place the cone *through* the sphere so that it touches in two places—a **secant** projection. Such projections have two standard parallels. (Cylindrical projections may also be tangent or secant.) One can also place a plane tangent or secant to the sphere, producing an **azimuthal** projection (Fig. 3.8). Other names for this method are **stereographic** and **orthographic** projection. Typically, these projections are used for displaying the earth's poles, and for that reason they are sometimes called polar projections.

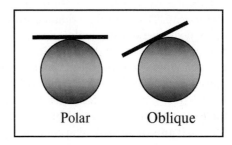

Fig. 3.8. Azimuthal projections

Projecting the earth to a plane always incurs distortion of the map. Four properties of map features may be distorted: *area, distance, shape, and direction*. Usually map projections reduce or eliminate certain distortions at the expense of the others. Maps based on cylindrical projections (Fig. 3.9a) typically preserve direction and shape at the expense of distance and area. Notice that the longitude lines point north-south, indicating the correct direction, but Alaska and Greenland are enlarged. Conic projections typically preserve area or distance at the expense of direction and shape (Fig. 3.9b). The longitude lines no longer point north-south, but the areas are better preserved. Maps based on azimuthal projections also typically preserve areas or distances (Fig. 3.9c). The inside front cover of the text summarizes some of the more common map projections along with the properties and uses of each; full descriptions are provided in the ArcGIS Help.

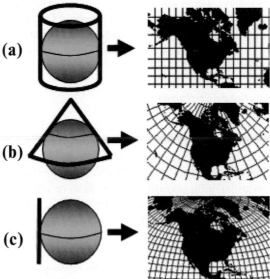

Fig. 3.9. Comparison of distortions caused by (a) cylindrical, (b) conic, and (c) azimuthal projections
Source: ArcMap

In any projection, no distortion occurs where the globe touches the cylinder, cone, or plane. The standard parallels, or the center of an azimuthal projection, are thus undistorted in all four properties, and the distortion will increase away from these locations. The standard parallels of a projection are typically placed below the upper sixth and above the lower sixth of the area of interest. For example, the region of Turkey covers approximate 36° to 42°, so the standard parallels might be set to 37° and 41°.

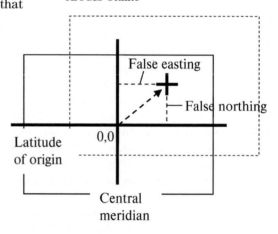

Fig. 3.10. Choosing standard parallels for Turkey
ArcGIS Online

Map projections have certain properties that are used to customize the projection for a particular geographic area and purpose; these properties are called **parameters** (Fig. 3.11). One parameter is the line of longitude that constitutes the **central meridian**, or $x = 0$ line on the map. The central meridian is chosen based on its position near the center of the area being mapped. A map of England would have a central meridian close to 0 degrees. A map of the United States would have a central meridian of about -100 degrees, because that meridian is close to the center of the continental United States. The **latitude of origin**, also sometimes called the **reference latitude**, is the $y = 0$ line. The equator is often used as the reference latitude. The **standard parallels** are also parameters. Maps may also have a **false northing** or **false easting** (Fig. 3.11). These are arbitrary numbers added to the $x$-$y$ coordinates in order to translate the map to a new location in coordinate space. Most often, the false easting and northing are used to ensure that all the coordinates of the map are positive.

Fig. 3.11. Some parameters of map projections

### Raster coordinate systems

Rasters have coordinate systems and projections just as vector feature classes do. Figure 3.12 shows a South Dakota elevation raster in a GCS and in a conic projection. The shape of the state is different, and you need to know the underlying coordinate system to be able to align this raster with other data sets. Like vectors, rasters can be projected from one coordinate system to another, although the algorithms are more complicated.

### Assigning coordinate systems to rasters

Many rasters are provided with a coordinate system at the outset. However, sometimes a user must assign the coordinate system. A raster that has been assigned a real-world coordinate system is said to be **georeferenced**. Two cases commonly occur.

**(a)**

**(b)**

Fig. 3.12. An elevation raster in (a) GCS and (b) State Plane coordinate systems
Source: ArcMap

Case I involves a raster for which the coordinate system is known, but the information has not been stored in accessible form. Some image formats store georeferencing information inside a header file or in the image itself. Other formats use a

**world file**, which is a six-line text file containing the parameters needed to georeference the raster. If the georeferencing information is missing, the user can create a world file with the appropriate parameters. A description of the contents of the world file can be found in the ArcGIS Help.

**TIP:** A world file has the first and third characters of the image file extension plus a "w" on the end. A .tif file would have a .tfw world file. A .bil file would have a .blw world file.

Case II involves a raster without a specified coordinate system, such as saved pictures from the Internet or a scanned map. The coordinate system is in pixel units from the screen or the scanner rather than real-world coordinates in feet or meters. To georeference this raster, the user must create **ground control points** by matching locations on the image to the same locations on another image or feature class that has a real-world coordinate system. Road intersections or other distinctive features are typically used. A minimum of three points are required, but additional pairs typically give a more accurate result.

Using the point pairs, a mathematical transformation is then applied (Fig. 3.13). A first-order (**affine**) transformation can translate, rotate, and skew the image to match the points. For a scanned map with a known projection, the affine transformation is usually sufficient because the shape of the map is already fine; it just needs its coordinates converted to a new set of *x-y* values.

An air photo or satellite image may have geometric distortions due to topography, focal effects, or lens distortions. A map with an unknown projection may not exactly match the shape of the coordinate system to

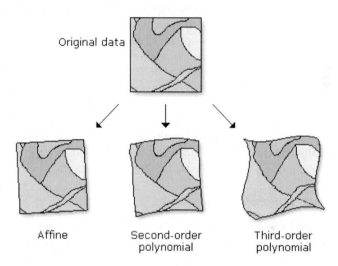

Fig. 3.13. Transformation methods for a raster
Source: ArcMap

which it is being georeferenced. In these cases, a second-order or third-order transformation allows the image to stretch or contract locally for a better fit (**rubbersheeting**). Rubbersheeting requires additional control points. Generally one should use the lowest order of transformation that produces an acceptable result. Using a higher order than needed can actually introduce new distortions.

When the transformation equations are applied to the ground control points, an offset is always found between the specified locations and the actual locations, due to slight errors in locating the points or to differences in shape between the old and new coordinate systems. This offset is evaluated by calculating a root mean square error (**RMSE**), which serves as a measure of the accuracy of the transformation. The RMS error is reported in map units of the new coordinate system and should be reported in the image metadata.

To finalize the georeferencing, the user may either save the information to a world file or **rectify** the image to create a new image. During rectification, the raster is being converted from one grid of cells to another with potentially different sizes and spacing, forcing the cells to be adjusted, or resampled, described in Chapter 11. Rasters containing categorical data should be resampled with the nearest neighbor method, and continuous rasters should be resampled with the bilinear method.

### Projecting rasters

Projecting rasters is more complicated than projecting vector data. A GCS elevation raster of South Dakota (Fig. 3.12a) has a different shape than the raster stored in a State Plane coordinate system (Fig. 3.12b); it is clear that resampling of the grid must occur if projecting from GCS to State Plane. The GCS map units are in degrees, and the cell size is 30 arc seconds, or 0.008333 degrees. During projection, the user must specify a new cell size in meters. A degree of latitude at the equator is 111.3 km. Thus, the height of the cell is

0.00833 degrees × 111.3 km/degree = 0.927 km, or about 900 meters.

The cell is not square because degrees of latitude are not equal to degrees of longitude except at the equator. In this case, the resolution in ground units is no better than 900 meters, so 900 should be the minimum cell size specified for the output, although it is fine to round the values to 900 or 1000 meters.

Resampling rasters typically degrades the accuracy of the raster and may introduce artifacts, so projecting the same raster multiple times should be avoided if possible. For best performance and quality when viewing multiple coordinate systems, the data frame should be set to match the coordinate system of the raster(s) rather than the vector data.

### *Common projection systems*

Several projections deserve special description because they are so commonly used. One, the GCS, is not actually a projection but is often treated as one. UTM is a family of coordinate systems utilized worldwide. State Plane is another such family used within the United States, and other countries have similar national or state systems used for large-scale maps.

### Geographic coordinate systems

As a three-dimensional coordinate system, a GCS cannot, strictly speaking, be displayed on a flat map or screen. However, in practice a simple equirectangular projection can be applied, which treats degrees as planar distances rather than as angles (Fig. 3.14). Unfortunately, distortions are introduced by this practice. On the globe, the latitude lines get shorter at higher latitudes, going to zero at the poles, yet in the planar display, these lines are all the same length. The further north or south an area is from the equator, the more the map gets distorted. Analysis functions also treat a GCS data set as if it were planar, and if a function uses area or distance in its calculations, then the results may be incorrect. The GCS is primarily used for distributing data; once chosen the user should project it before incorporating it into databases for mapping or analysis.

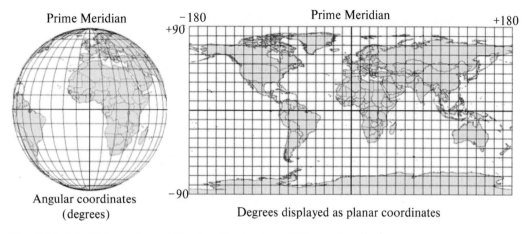

Fig. 3.14. A GCS is projected flat for display in a GIS, causing distortion.
Source: ArcGIS

**The Universal Transverse Mercator (UTM) system**

The **UTM** system is based on a secant transverse cylindrical projection. Two lines of tangency of the cylinder to the sphere fall about 180 km to each side of the central meridian (Fig. 3.15a). A zone is defined to include 3 degrees on each side of the central meridian between the 80S and 84N latitude lines. Since the projection is true along the lines of tangency, the distortion inside the zone is minimal. To map different locations, the cylinder is rotated about the globe in 6-degree intervals, producing 60 zones around the world. Areas within a single zone have negligible distortion in all four properties of shape, direction, distance, and area.

Each UTM zone has its own central meridian and splits at the equator into a north and south component. In order to eliminate negative $x$-$y$ coordinates, southern zones have a false northing of 10 million meters, and both northern and southern zones have a false easting of 500,000 meters. UTM is convenient because users need only know the zone number and hemisphere. In the United States and in ArcGIS, each zone is indicated by a number and the designation N for the northern hemisphere or S for the southern hemisphere, for example, UTM Zone 14N. (The worldwide UTM system uses letters C to X to represent different latitudinal zones.) Figure 3.15b shows the UTM zones of the conterminous United States.

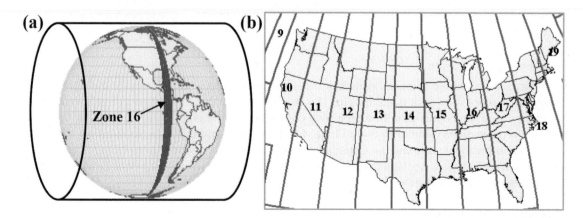

Fig. 3.15 (a) The UTM projection system has 60 north-south zones, each 6 degrees wide.
(b) This map shows the UTM zones of the conterminous United States.
Source: ArcGIS

The UTM system is used extensively for US Geological Survey 1:100,000 and 1:24,000 scale publications, for county maps, for maps of states within single zones, and for maps covering small areas. If the region of interest lies on the boundary between two zones, the user can choose the zone containing the larger area to represent the map. This choice may be acceptable if only a small area extends out of the zone. If not, State Plane zones may provide a better alternative.

### The State Plane Coordinate System

The **State Plane** Coordinate System (SPCS) includes an assortment of projections developed in the 1930s by the US Coast and Geodetic Survey for large-scale mapping in the United States. To maintain the desired level of accuracy, most states are broken into zones with different parameters used to minimize distortion in each zone (Fig. 3.16). Like the UTM system, an SPCS is identified by its name and zone, and distortion is negligible within a single zone.

The SPCS uses three different projections: Lambert Conformal Conic, Transverse Mercator, and Oblique Mercator. North-south oriented zones use a Transverse Mercator projection, as it provides the least distortion. East-west oriented zones use a Lambert Conformal Conic projection. A state may

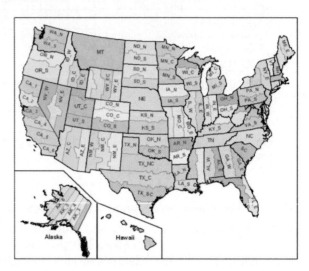

Fig. 3.16. State Plane zones of the United States
Source: ArcMap

use more than one projection depending on the size and shape of its zones. Alaska has 10 zones and uses all three projections. Zones are formally identified using a FIPS code composed of a two-digit state code and a two-digit zone code (e.g., FIPS 0407 for Zone 7 of California). Users may refer to them informally by state name and zone, for example, California State Plane Zone 7.

The original SPCS was based on NAD27 and used units of feet. Improvements in surveying necessitated the revision of the SPCS in 1983. At that time, it adopted the NAD83 datum and metric units. Certain zones were also revised and in some cases eliminated. Even though the official units changed from feet to meters, users may specify either unit for both versions of SPCS. Additional information about both the State Plane and UTM systems can be found in the ArcMap Help (index headings State Plane Coordinate System and UTM).

### State and National Grids

Some large states define a special coordinate system for statewide maps because a single State Plane zone cannot represent it accurately. The Oregon Statewide Lambert coordinate system is one example. Many countries define their own coordinate systems for national maps or their own systems of projections (similar to State Plane) for local and regional maps. Examples include the New Zealand National Grid and the Canada CSRS98 system. A variety of projections, datums, and spheroids are employed. ArcMap maintains a collection of these predefined coordinate system definitions for use with state and country data.

### *Choosing projections*

At the beginning of every GIS project, one of the first tasks is to establish the projection to be used, as best practices dictate that all data sets in a project share the same projection and datum. The choice of coordinate system impacts which map properties are preserved, and a poor choice may cause incorrect analysis results. Some guidelines for choosing projections for a project are described below. Preference should be given to one of the standard coordinate systems for the sake of interoperability, but if none of the existing systems provide the right properties, a user can define a custom projection.

***Maps of the world.*** Distortions cannot be avoided in small-scale maps, so they must be managed. Consider the applications. If distances or areas are to be analyzed as part of the project, then an equidistant or equal area map should be chosen. If navigation or wind/water currents are important, then preserving direction is critical. If portraying attractive maps is the main goal, then a compromise projection might be best.

***Maps of countries or continents.*** Many countries or regions have one or more projections defined for general mapping purposes, such as USA Contiguous Equidistant Conic or Europe Albers Equal Area Conic. Some countries have their own defined mapping systems. Like world maps, the application determines which properties of the map are important to preserve.

***Maps of states.*** Smaller states within a single UTM or State Plane zone can use them. Larger states require selecting one of the zones; if desired, one can modify the central meridian and/or parallels of the zone to work better for the entire state. Some states have a statewide projection defined, such as Oregon Statewide Lambert.

***Local and regional maps.*** Any region that fits inside a single UTM or State Plane zone can use either. Many countries, such as Canada and Australia, have defined their own mapping systems similar to the State Plane system in the United States. The preferred system would place the region to be mapped close to the center of the zone, where the distortion is least. However, other factors may take precedence, such as the existing coordinate system of one's data sources, or the decision of the agency that oversees the project.

Sometimes the region to be mapped falls close to a UTM or State Plane zone boundary or spans two zones. In this case, a customized coordinate system is desirable. Typically, one starts with a UTM or State Plane zone and adjusts the central meridian and/or the standard parallels to minimize the distortion. A good rule of thumb is that the central meridian should bisect the map extent, and the parallels for secant conic maps should keep two-thirds of the north-south extent between the parallels.

## About ArcGIS

### *Labeling coordinate systems*

ArcMap is able to display data together even when they have different projections. This capability relies on every data set having a coordinate system label, which records the characteristics of the coordinate system, including the datum and projection (Fig. 3.17). This label serves two purposes: it documents this information about the coordinate system, and it helps GIS display and manage the data. Making sure that all data sets have correct coordinate system definitions is

Fig. 3.17. Coordinate system extent and label
Source: ArcMap

a critical step in building a GIS database. Shapefiles store the labels in a separate file with a .prj extension. Geodatabases store it within the feature class. Rasters must also have coordinate system labels, storing them in various places depending on the raster format. In all cases, the coordinate system label is stored and copied as part of the data set. A coordinate system may also be stored separately as a file with a .prj extension. These files may be created, saved, copied, and used to define the spatial reference for new data sets.

Data sets may contain unprojected or projected data. If unprojected, the coordinates are stored in decimal degrees of latitude and longitude. The feature class coordinate system label would include simply the GCS (datum), such as GCS North American 1983. The stored $x$-$y$ coordinates of a data set on the disk would fall in the range −180 to +180 degrees for $x$ and −90 to +90 degrees for $y$. The map units would be degrees.

Projected data have had a map projection applied to them, converting the GCS latitude-longitude values to feet or meters in a planar coordinate system. The map units become meters or feet. Typically, the $x$-$y$ values in a projected coordinate system encompass millions or billions of meters. The feature class coordinate system definition includes the GCS/datum plus a description of the projection and its parameters, such as the central meridian, reference latitude, standard parallels, and false easting and northing, as shown in Figure 3.17.

The coordinate system is part of a larger description called the **spatial reference**, which includes the coordinate system, the X/Y domain, and the resolution. The **X/Y domain** is the range of allowable $x$-$y$ values that can be stored in a feature class. Storing a world map in degrees requires a small domain, since degree values range at most between −180 and +180. Storing a world map in feet requires a large domain, because there are more than 115 million feet in the circumference of the earth. The **resolution** represents the underlying accuracy of the values; a resolution of 0.001 meter means that coordinate values are stored to the nearest thousandth of a meter.

The **extent** of the data layer is the range of $x$-$y$ coordinates of the features actually in the feature class. You can view the extent of a data layer in ArcMap using the Source tab of the layer properties, as shown in Figure 3.17. This information can help you decide if the layer is projected or unprojected. Values between +180 and −180 indicate an unprojected GCS coordinate system with map units in degrees, whereas large values indicate a projected coordinate system with units of feet or meters.

### A note on terminology

Many people use the terms *coordinate system* and *projection* interchangeably to refer to the spatial reference (the complete label) of a data set. Using the terms this way is not strictly correct, as a coordinate system is only part of the spatial reference, and a projection simply is a mathematical conversion from a 3D to a 2D coordinate system. However, it is a common practice, and usually the intended meaning is clear in context.

### Missing coordinate system labels

Occasionally, a user will encounter a GIS feature class or raster without a coordinate system label, and the coordinate system appears as "Unknown" when viewed in ArcMap or ArcCatalog. To use the data, the user must determine the actual coordinate system of the data and create a label for it. Finding the coordinate system can be a challenge because it usually cannot be determined by looking at the data set. The user must find documentation supplied by the data provider, such as a Web page or document at the site where the data were downloaded or a separate text or document supplied with the data. The user might have to contact the data supplier to determine this information. As a last resort, the user might have to guess one of the more commonly used coordinate systems and test whether it is correct.

Once the coordinate system is known, the user may create the coordinate system label using the **Define Projection tool** in ArcToolbox or by using the XY Coordinate System tab from the feature class properties in ArcCatalog. The Define Projection tool labels a data set with a coordinate system and allows the user to specify all the coordinate system properties, including the GCS, datum, central meridian, and so on. Once defined, the data set is ready for use in ArcMap. The same tool is used to define a label for rasters.

### On-the-fly projection

A data frame may be assigned any desired coordinate system, and then ArcMap will project all data in the frame to match. To do this, ArcMap must know the coordinate system of each data set. If the data set coordinate system definition is Unknown, then ArcMap will issue a warning message and will display the data with whatever coordinates it has. Sometimes it will fortuitously match the other data in the frame, but other times it will appear totally out of place. If it matches, then you know that the Unknown layer has a coordinate system that is the same as, or at least very similar to, the data frame, and you can create a coordinate system label for it.

ArcMap can adjust differences in datums on the fly, but the user has to select an appropriate transformation method, which many users lack the advanced knowledge to do, and the transformation can cause new positional errors up to several meters. Therefore, ArcMap does not automatically perform datum transformations. The user is warned if a data set is added with a different GCS than the data already in the frame, and he or she can then select a transformation if desired. For national- or world-scale data, the differences between datums are small compared to the accuracy of the data, so no transformation is necessary. For large-scale data in situations where high accuracy is needed, the user should consult an expert on transformation methods. There is one exception: ArcMap automatically applies a transformation called NADCON if the two datums are NAD 1927 and NAD 1983.

### Projecting data

Usually, GIS projects choose a single coordinate system and convert all data to match it. Many organizations establish an official projection for their area of interest and use it consistently. Outside data from different coordinate systems are always converted to the official one at the outset, using the **Project tool** (or the export technique introduced in Chapter 2).

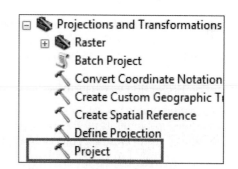

The Project tool is found under the Data Management section of the Toolbox (Fig. 3.18). It converts the $x$-$y$ coordinate values in a feature class to a different coordinate system and saves them in a new feature class with a new coordinate system label. The original

Fig. 3.18. The Projection tools
Source: ArcGIS

data remain unchanged. The original data set must have a correctly labeled coordinate system before it is projected. If the label is Unknown, the Project tool cannot work. Rasters have their own set of projection tools.

### Troubleshooting coordinate system problems

A coordinate system problem occurs when two layers should map on top of each other—but don't. They might be off by a few hundred meters or by thousands of miles. Most of these problems result from the mislabeling of a data set with the wrong coordinate system. The usual remedy is to determine the correct coordinate system and fix the label.

ArcMap can be used to troubleshoot such problems. In Figure 3.19, the user is examining the Source tab of the layer properties. The extent shows $x$-$y$ values in the range –103.6 to –103.5 for $x$ and 44.3 to 44.4 for $y$. These values are clearly degrees. However, the label underneath lists NAD 1983 UTM Zone 13N as the coordinate system, and the $x$-$y$ values are listed with meters as the map units. The label does not match the $x$-$y$ extent values. So which is correct?

***The x-y values are always correct*** because they are obtained from the values stored in the file. This data set really *is* stored in degrees, and the label is incorrect. Imagine adding this feature class to a data frame set to the

Fig. 3.19. An incorrectly labeled coordinate system
Source: ArcMap

UTM coordinate system. ArcMap reads the label and mistakenly concludes that the $x$-$y$ values are already in UTM meters and don't need to be projected on the fly. The $x$-$y$ values are plotted in degrees in the UTM coordinate space, so they don't match the other data and appear in a completely different location.

To fix the problem, you must know the correct coordinate system and use Define Projection to modify the label. You will need another data set in the same area that is known to be correct so that you can compare the results. If the data set aligns with the known one, then you set the label correctly. If it is still misaligned, then a different label is needed.

Data in a GCS are relatively easy to correct because the choice of possible datums is limited. Projected data are virtually impossible to guess unless they are in one of the common projections or you have some outside information to narrow down the possibilities.

### Define Projection versus Project

Why would a data set be mislabeled with the wrong coordinate system? The usual culprit is a user who does not understand the tools used to manage coordinate systems, and who makes the mistake of using Define Projection when he or she should use the Project tool.

Imagine Frank has a shapefile with GCS coordinates and wants to convert it to a UTM projection to match his other data. He carelessly selects the Define Projection tool from ArcToolbox and sets the coordinate system to UTM. However, the tool merely puts a UTM label on the shapefile without changing the coordinates inside. So now the shapefile is incorrectly labeled UTM and the coordinates inside are still in a GCS. (This situation is analogous to placing an albacore tuna label on a can of cat food, with disastrous results for your dinner recipe.) When the data are added to ArcMap, it will choose the wrong equations to convert the data to the data frame of the coordinate system, and the feature class will no longer appear where it should.

**WARNING:** Many people get confused about the functions of the Define Projection tool and the Project tool. This confusion can ruin data sets, so ***pay attention*** to the next two paragraphs and then be careful to select the correct tool for the task at hand.

The **Project tool** acts on the $x$-$y$ coordinates of a layer and converts them to a different coordinate system, producing a new feature class and leaving the original feature class unchanged. The new file has a new coordinate system and label. You use the Project tool if you have a layer in one coordinate system and want to convert it to a different one. The Project tool should be used only on layers that already work properly and appear in the right location.

The **Define Projection** tool changes the coordinate system label of the feature class without affecting the coordinates inside. It should only be used on a data set that has an Unknown coordinate system or on a data set that was previously mislabeled, does not appear in the right location, and needs to be fixed.

## Summary

▶ All GIS data have coordinate systems, which define the units and axes used to represent map features as $x$-$y$ coordinates. The complete definition is called the spatial reference.

▶ A geographic coordinate system, or GCS, uses angular measures of latitude and longitude with units of degrees. It is based on an approximation of the earth's shape known as a datum.

▶ A datum includes a spheroid and its location relative to the earth's geoid. It is used to reduce map errors introduced by differences between the spheroid and the earth's actual surface. Common datums in North America include NAD 1927, NAD 1983, and WGS 1984.

▶ Data for a project should be stored in a common GCS. Transforming from one datum to another should be minimized because the process may introduce new positional errors.

▶ Map projections are mathematical equations that convert degrees from a GCS into planar $x$-$y$ coordinates of meters and feet so that the map may be portrayed on a flat sheet of paper. All map projections introduce distortions of area, distance, direction, and shape.

▶ Rasters also have coordinate systems. Assigning coordinate systems to rasters may involve creating world files for them or georeferencing them based on control points.

▶ UTM and State Plane are common projection systems in the United States. Each is a family of projections designed to minimize distortion over the area covered (a zone).

▶ Every data set used in ArcGIS should have the correct coordinate system defined and stored with the map features. ArcGIS uses these definitions to display data correctly.

▶ Most coordinate system problems result from missing or incorrect labels on data sets. The Define Projection tool is used to update a data set with an undefined or incorrect label.

▶ Data may be permanently projected from one coordinate system to another using the Project tool or Project Raster tool in ArcToolbox.

**TIP:** The ArcGIS Help section also has some good material about projections.

**Important Terms**

| | | | |
|---|---|---|---|
| affine transformation | geographic coordinate system (GCS) | Project tool | reference latitude |
| central meridian | | projection, | resolution |
| coordinate pair | geoid | azimuthal | RMSE |
| coordinate space | georeferenced | conic | rubbersheeting |
| coordinate system | ground control points | cylindrical | spatial reference |
| datum | latitude | oblique | spheroid |
| Define Projection tool | latitude of origin | orthographic | standard parallel |
| degrees | longitude | secant | State Plane |
| ellipsoid | map units | stereographic | transformation |
| extent | origin | tangent | UTM |
| false easting | parameters | transverse | world file |
| false northing | Prime Meridian | rectify | X/Y domain |

## Chapter Review Questions

1. If a data set's features have $x$ coordinates between −180 and +180, what is the coordinate system likely to be? In what units are the coordinates?

2. What are the $x$-$y$ coordinates of a map's origin? _____ What is the $x$ coordinate along the central meridian? _____

3. What is the difference between a spheroid and a geoid?

4. Examine Figures 3.6 through 3.9 and explain why conic projections usually conserve area and distance but cylindrical projections typically preserve direction.

5. What extra step is performed when projecting rasters that is not needed when projecting vector data? What happens during this step?

6. What is the difference between a central meridian and the Prime Meridian?

7. You have a shapefile with an Unknown coordinate system, but a file on the web site says that the coordinate system is UTM Zone 13 NAD 1983. What is your next step?

8. True or False: A shapefile of the United States with a GCS coordinate system would have an $x$-$y$ extent that contains entirely positive values. _____ Explain your answer.

9. You have a shapefile with a UTM Zone 10 NAD 1983 coordinate system, and you want to bring it into your state database, which uses the Oregon Statewide Lambert coordinate system. What is your next step?

10. You need to create a map projection for the entire state of Idaho. What options do you have? How could you get the most accurate map projection possible?

# Mastering the Skills

## Teaching Tutorial

The following examples provide step-by-step instructions for doing basic tasks and solving basic problems in ArcGIS. The steps you need to do are highlighted with an arrow →; follow them carefully. Click on the video number in the VideoIndex.htm file to view a demonstration of the steps.

### *Displaying coordinate systems*

→ Open ArcMap and choose to start with a new, empty map.

→ Use Save As to save the document with the name of your choice. Take care to save it in the mgisdata\MapDocuments folder so it is easy to find again.

1→ Use the Add Data button to add the country and latlong shapefiles from the mgisdata\World directory.

1→ Change the data frame name from Layers to *World.*

1→ Right-click the country layer, choose Properties, and then click the Source tab.

1. What is the coordinate system (CS) for this feature class? _____
   Which units does it use? _____

Observe the values in the top of the window, which show the extent of $x$-$y$ values present in the shapefile. The extent is the range of $x$-$y$ values of features stored in the data set.

2→ Close the Layer Properties window.

2→ Observe the coordinates shown in the lower part of the window and confirm that they are in degrees. The data frame units default to the units of the first data set loaded.

2→ Zoom in to the tip of Florida in the United States.

2→ Use the cursor to hover over the most southeastern tip of Florida and observe the $x$-$y$ coordinates (Fig. 3.20).

Fig. 3.20. Hover here.
Source: ArcMap

2. What are the coordinates of the SE tip of Florida? _____

3→ Right-click the World data frame name and choose Properties.

3→ Click the General tab.

3. What are the map units of this frame? _____What are the display units? _____

The map units are dimmed because you can't set them—they are based on the data frame CS. The display units can be set to any units desired by the user.

3→ Set the display units to miles and click OK.

4. What are the coordinates for Florida's tip now? _____

The $x$ value tells you the distance Florida lies from the central meridian of the map (in this case the Prime Meridian that runs through Greenwich), and the $y$ value tells you its distance from the reference latitude (the equator).

4→ Click the Full Extent button to see the whole world again.

4→ Open the World data frame properties.

4→ Click the Coordinate System tab and notice what it says.

The data frame CS defaulted to match the first data layer added to it. However, the data frame can display maps in any CS, either geographic or projected.

> 4→ In the folder tree, scroll back to the top and collapse the Geographic Coordinate Systems folder.
>
> 4→ Expand the Projected Coordinate Systems folder and examine its contents.
>
> 4→ Expand the World folder. Click on the Mercator (world) projection. Examine the parameters and map units, and then click OK.

5.  What are the map units for this Mercator projection? _____

**TIP:** As you work with the *data frame* coordinate systems in ArcMap, keep in mind that they can be different from the coordinate systems of the *feature classes* you are working with.

> 5→ Zoom in to the tip of Florida again. The display units are still miles because you set them.
>
> 5→ Zoom to the previous extent and use the cursor and coordinate box to locate the coordinate origin, where both $x$ and $y$ equal zero (approximately—look for where the coordinates change from positive to negative).

Mercator is a cylindrical projection. The central meridian where $x = 0$ corresponds with the Prime Meridian that runs through Greenwich, England. The $y = 0$ line is the equator and is called the latitude of origin. Coordinates west of the central meridian and south of the reference latitude have negative values. Florida is west of the central meridian; its $x$ coordinates are negative.

6.  Which continent has primarily negative $x$ AND negative $y$ coordinates in this projection? _____ Which one has primarily positive $x$ and $y$ coordinates? _____

> 6→ Add the states feature class from the mgisdata\Usa\usdata geodatabase.
>
> 6→ A Geographic Coordinate Systems Warning window should appear. Read it.
>
> 6→ Click the hyperlink *About the geographic coordinate systems warning*. Read it.

7.  Why did this warning appear?

8.  Does ArcMap automatically choose a datum transformation for you. Why or why not?

> 7→ Close the Help window, if necessary, and click the Transformations button.
>
> 7→ The layers have one of the two GCS listed here (Fig. 3.21). You have the option to convert the layers in NAD83 to WGS84 or vice versa.

Fig. 3.21. Choosing a transformation
Source: ArcMap

7→ Click the *Using:* drop-down and examine the long list of transformation choices available.

Converting between datums is more complex than changing projections and requires knowledge beyond that of the typical user. This complexity is one reason it is advisable to assemble data sets using a single GCS so that the appearance of this window is rare. In this case, however, we are looking at world data. The difference between datums is never more than a few hundred meters, which would not be noticeable at this scale. Therefore, we will not apply a transformation.

7→ Click Cancel.

7→ Check the box *Don't warn me again in this session* and click Close.

8→ Double-click the states layer and view the Source tab in the layer properties.

The states feature class is stored in the North America Equidistant Conic projection. Even though the CS is different from the other layers, they all align properly because each has a correct coordinate system label, so ArcMap projects them on the fly.

9. What longitude is the central meridian? _____ What is the latitude of origin? _____ What are the standard parallels? _____ and _____.

10. Examine the standard parallels and the latitude of origin, and predict whether any areas of the United States have negative *y* coordinates in this projection. _____ Why or why not?

11. Is this Equidistant Conic projection a tangent or secant projection? _____ How can you tell? _____

8→ Close the layer properties, and open the data frame properties.

8→ Use the Coordinate System tab to set the data frame to the North America Equidistant Conic projection (in the Continental folder).

The world map looks bizarre, but remember, this CS is designed for North America.

9→ Zoom in to North America (includes Central America).

9→ Right-click the latlong layer and choose Label Features.

9→ Find the central meridian, the standard parallels, and latitude of origin on the map.

### Understanding map distortion

The circumference of the world at the equator is ~25,000 miles. The circumference at the poles is zero. Latitudes in between have intermediate lengths. Let's look again at the WGS 1984 GCS.

⊞ 🗔 Favorites
⊞ 📁 Geographic Coordinate Systems
⊞ 📁 Projected Coordinate Systems
⊟ 📂 Layers
  ⊞ 🌐 GCS_WGS_1984
  ⊞ 🌐 North_America_Equidistant_Conic

Source: ArcMap

10→ Open the World data frame properties and select the Coordinate System tab.

10→ Collapse all the folders until only the main four are shown.

10→ Expand the Layers folder.

The Layers folder shows coordinate systems of layers in the map document.

10→ Expand the + sign next to GCS_WGS_1984 and North_America_Equidistant_ Conic to see which layers have these coordinate systems.

10→ Click the GCS_WGS_1984 entry to set the data frame to that CS and click OK.

10→ Click the Full Extent button to show the whole world again.

10→ Right-click the latlong layer and choose Label Features to turn them off again.

When a GCS data set is displayed, the angular units are drawn in an equirectangular projection, as if the degrees were a planar $x$-$y$ coordinate system from $x = -180$ to $+180$ and $y = -90$ to $+90$. Distances are increasingly distorted away from the equator.

11→ Add the circles shapefile from the mgisdata\World folder.

11→ Drag the circles above the latlong layer in the Table of Contents (switch to the List by Drawing Order option, if necessary).

Source: ArcMap

Each circle has a radius of 5 degrees, and they would appear identical on a globe, but here the distance distortion of the GCS map is clearly shown. The areas and distances, if determined from these circles, would be incorrect over most of the globe. *Any GIS analysis based on distance or area measurements would be wrong when data are stored in a GCS.*

12→ Open the World data frame properties. Move the window so that you can leave it open and still see the map.

12→ Collapse/expand the folders to select the Projected Coordinate Systems > World > Mercator (world) projection. Click Apply.

Area and distance are also distorted in this projection, except at the equator, where the sphere touches the projection cylinder. However, Mercator does preserve direction and shape. The circles remain circles, and the longitude meridians point north.

13→ Change the coordinate system of the frame to Robinson (world). Click Apply.

Robinson is called a compromise projection because it minimizes distortion in all four properties but preserves none. Distortion is unavoidable when using small-scale world and national maps.

14→ Change the coordinate system to Sinusoidal (world). Click Apply.

Sinusoidal is an equal-area projection, meaning that it preserves area and, to some extent, distance. However, it does so at the expense of shape and direction. The longitude lines converge toward the poles, and the circles have equivalent areas now, but most are no longer circles. The distortion gets worse away from the equator and the central meridian.

15→ Change the coordinate system to Africa Sinusoidal in the Continental folder. Click OK to accept the change and close the window, and then zoom to Africa.

Although globally the sinusoidal project shows great distortion, when customized for the region of Africa it does a pretty good job with area and distance, although a direction distortion is apparent. For even smaller areas, UTM is a good choice.

16→ Open the data frame properties again.

16→ Collapse the Continental folder and expand the UTM folder. Navigate to UTM > WGS 1984 > Northern Hemisphere > WGS 1984 Zone 35N and select it. Click OK.

16→ Add the shapefile utmzone35 from the mgisdata\World folder. Make the symbol hollow with a thick orange outline.

The narrow strip of UTM Zone 35 runs along the tangent of a transverse cylindrical projection. Within the zone, the circles remain circles and all four properties of area, distance, shape, and direction are preserved. Areas below the equator would use Zone 35S.

### Converting coordinate systems when compiling data

When compiling data for a project, it is important to ensure that all data are converted to the same suitable coordinate system. In Chapter 2, we created a database for New Jersey, but some of the feature classes were not converted to New Jersey State Plane, a poor practice. We will take care of this conversion now.

17→ Save your current map document and open a new, blank one.

17→ In the Catalog Tree, right-click the njcities feature class in your NJ_Project/NewJersey geodatabase and choose Properties.

17→ Click the XY Coordinate System tab. Look, but DON'T change anything.

**TIP:** Be VERY careful not to modify this coordinate system. You are looking at a label that describes the *x-y* data stored in the file. *Changing the coordinate system here only makes the label incorrect, and the file will no longer display properly*.

17→ Click Cancel to close the Feature Class Properties window, just in case you changed something by accident.

The Project tool is used to convert a feature class from one CS to another by creating a copy in the new CS (Fig. 3.22).

18→ In ArcToolbox, expand the Data Management Tools > Projections and Transformations toolsets.

18→ Open the Project tool (NOT Define Projection) by double-clicking it.

18→ Drag and drop the njcities feature class from the NewJersey geodatabase to the Input Dataset or Feature Class box. The current CS will be shown in the box beneath.

Fig. 3.22. Projecting a feature class to a new CS
Source: ArcMap

**TIP:** Notice the default save location in C:\Users. Never allow ArcGIS to save data in this folder; you will have trouble finding it again. Always save data in the place you have designated for it.

18→ Click the Browse button, navigate to the NewJersey geodatabase if necessary, and enter the output name for the feature class as *njcities_sp*. Click Save.

 18→ Click the button to select the Output Coordinate System and navigate to Projected Coordinate Systems > State Plane > NAD 1983 (Meters). Use the plain one.

⊞ 🗀 NAD 1983 (Intl Feet)
⊞ 🗀 NAD 1983 (Meters)

18→ Find the NAD 1983 StatePlane New Jersey FIPS 2900 (Meters) coordinate system and click to highlight it. Click OK.

Source: ArcMap

**TIP:** Unless specifically told otherwise, always use the basic datum, NAD 1983, and not one of the specialized ones like (2011) or (CORS96) or (HARN). If you picked one of these by mistake, click the button again and select the correct one.

18→ Check your entries against Fig. 3.21 and then click OK to start projecting.

19→ Use the Project tool to convert the njdistricts and majroadclip feature classes to State Plane also. Name them appropriately and place them in the NewJersey geodatabase.

20→ In the Table of Contents, check the coordinate system of all new data sets AND the data frame to make sure they are all now set to the desired State Plane CS.

We no longer need the original files, so we should delete them. Usually all feature classes in a geodatabase should have the same coordinate system.

21→ Right-click the old njcities feature class in the Catalog Tree and choose Delete. Also delete the old njdistricts and majroadclip feature classes.

**TIP:** Recall that in Chapter 2 we used ArcMap to export and project in one step using the data frame coordinate system. Either method may be used, but exporting in ArcMap is more efficient and does not leave extra files that need deletion.

### *Troubleshooting projection problems*

The previous examples show how one is supposed to convert coordinate systems when compiling data. Unfortunately, the proper procedures are not always followed. This section will help when you encounter data sets with impaired coordinate systems caused by improper practices.

Recall that ArcMap relies on coordinate system labels to perform on-the-fly projection.

 22→ Click the New Map File button to start a new, empty map. Don't save the current map.

22→ Add the usfsrds feature class from mgisdata\BlackHills\Sturgis.mdb\Features.

22→ Open the layer properties for usfsrds, click the Source tab, and write the coordinate system and datum of the layer here: _____

22→ Also examine the Extent information, which shows the range of $x$-$y$ coordinates as stored in the file. Close the Properties window.

You are working on a project near Sturgis, SD, and have downloaded four new shapefiles to import to your geodatabase. You must ensure that each one is converted to the geodatabase coordinate system, NAD 1983 UTM Zone 13N. Because not all Internet data is trustworthy, you will do this carefully, one shapefile at a time, paying attention to the original and final coordinate systems.

23→ Add the countyline shapefile from the BlackHills\Downloads folder.

The first problem surfaces with a warning that this shapefile has an unknown spatial reference. You need to determine the coordinate system it is stored in and create a correct label for it.

23→ Click OK on the warning window. The layer appears in the Table of Contents but you can't find the line on the map. Maybe the symbol is too small?

23→ Change the symbol for the countyline layer to Highway. Still nothing.

23→ Right-click the countyline layer and choose Zoom to Layer.

Well, here is the county line, but it doesn't appear in the same coordinate space as the other roads. This error generally results from an improper or missing coordinate system label.

24→ Open the properties for countyline and click the Source tab.

24→ Examine the coordinate system and the extent carefully (Fig. 3.23).

| Extent | | |
| --- | --- | --- |
| | Top: 44.499999 ?? | |
| Left: -103.574397 ?? | | Right: -103.500014 ?? |
| | Bottom: 44.256077 ?? | |

| Data Source | |
| --- | --- |
| Data Type: | Shapefile Feature Class |
| Shapefile: | C:\gisclass\mgisdata\Black⌐ |
| Geometry Type: | Line |
| Coordinates have Z values: | No |
| Coordinates have measures: | No |
| Coordinate System: | <Undefined> |

Fig. 3.23. An unlabeled coordinate system
Source: ArcMap

This coordinate system is undefined, as you already know. The Extent shows the range of $x$-$y$ coordinates as small numbers, less than 180, which is a sure sign that you are dealing with an unprojected coordinate system in degrees, or a GCS.

So you know to assign a GCS label, but you also need to assign a datum. There is no way to know for certain, but NAD83 is the most common datum for US data, so you will try it first.

25→ Close the Layer Properties window.

25→ Open ArcToolbox > Data Management > Projections and Transformations > Define Projection.

**TIP:** You can drag and drop a layer from the Table of Contents to a tool input box, or you can use the drop-down on the tool to select a layer.

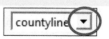

Source: ArcMap

25→ Set the input dataset to countyline.

25→ Click the coordinate system browse button to open the Spatial Reference Properties.

25→ Expand Geographic Coordinate Systems > North America and select NAD 1983.

25→ Click the Add to Favorites button, as this GCS is very common.

25→ Make sure that NAD 1983 is still highlighted and click OK.

25→ Click OK to run the Define Projection tool.

26→ Click the Full Extent button to return to the view of your study area.

Now the county line appears on the map. You think that you have solved the problem; however, to be absolutely sure, you will check the line against a reliable source.

26→ Click the Add Data drop-down button and select *Add Data From ArcGIS Online*.

26→ Search for **topo**, and add the USA Topo Maps map service (or LPK file).

26→ Zoom in to the county line and confirm that it lies on or close to the county line between Lawrence and Meade counties on the topographic map.

26→ Click the Full Extent button to return to the larger view.

Oops—the zoom goes to the full extent of the map service rather than the Sturgis area. However, we can fix this using a property of the data frame that controls how the full extent is defined.

27→ Open the Layers data frame properties and click the Data Frame tab.

27→ In the middle panel under *Extent Used By Full Extent Command*, fill the Other button and click Specify Extent.

27→ Fill the button for Outline of Features and select the usfsrds layer from the drop-down list. Click OK and OK.

27→ Click the Full Extent button again. Now it shows the area you want.

27→ Turn off the USA Topo Maps layer for now, but keep it in the map document.

So countyline is now correctly labeled, but it is still in a GCS. All the files in your geodatabase need to be in NAD 1983 UTM Zone 13N. To convert it, use the Project tool.

28→ Open the ArcToolbox > Data Management > Projections and Transformations > Feature > <u>Project</u> tool.

28→ Set the input dataset to countyline.

28→ Click the Browse button and navigate to the BlackHills\Sturgis geodatabase and inside the Features feature dataset.

28→ Name the output *lawmeadline*.

28→ Click the Browse button to set the spatial reference and expand the Layers folder.

28→ Select the NAD_1983_UTM_ Zone_13N coordinate system.

28→ Click the Add to Favorites button to be able to use it again later, then click OK to use it.

28→ Confirm that the Project tool arguments appear as in Figure 3.24, and click OK.

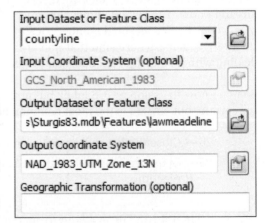

Fig. 3.24. Projecting countyline
Source: ArcMap

29→ Remove the original countyline layer and give the new lawmeadline layer the Highway symbol.

29→ Add the quadrangles shapefile from the BlackHills\Downloads folder.

29→ Open the properties for quadrangles and click the Source tab.

This shapefile is also missing the spatial reference, but in this case the quadrangles appear in the right places. Looking at the Extent, you can see that the values are far too large to be degrees, so you are dealing with a projected coordinate system. But which one is it?

Your clue is that the quadrangles are in the right places. If a data set has no label, then ArcMap does not attempt to project it on the fly but simply plots the $x$-$y$ coordinates in the file. Thus, you know that the file coordinate system matches that of the data frame, NAD 1983 UTM Zone 13N, and that is what needs to go on the label.

30→ Close the Layer Properties window.

30→ Open ArcToolbox > Data Management > Projections and Transformations > <u>Define Projection</u>.

30→ Set the input dataset to quadrangles and choose NAD_1983_UTM_Zone_13N from the Favorites folder. Click OK and OK to run the tool.

30→ Zoom in to check the quad corners against the USA Topo Maps layer, and then turn the topo layer off and zoom back to the full extent of the map.

Now the quadrangles coordinate system is labeled, and it already matches our geodatabase coordinate system, so we don't need to project it, only export it to the geodatabase.

31→ Right-click the quadrangles layer and choose Data > Export Data.

31→ Navigate to BlackHills\Sturgis\Features. Enter the output file name *fourquads*.

31→ Notice the coordinate system setting and click OK. Add the new feature class.

31→ Remove quadrangles from the Table of Contents.

It's time for the next shapefile.

> ◉ the feature dataset you export the data into
> (only applies if you export to a feature dataset

32→ Add the townships shapefile from the BlackHills\Downloads folder.

Source: ArcMap

32→ Nothing appears in the map. Open its layer properties and examine the Source tab (Fig. 3.24).

This one is interesting. According to the Data Source, the coordinate system is NAD 1983 UTM Zone 13N, but the Extent values are small and appear to be in degrees (Fig. 3.25). We have a disagreement between the extent and the label. Which one is correct?

**TIP:** The Extent *x-y* values show the values actually in the file and are the best guide to the true coordinate system.

This GCS data set is incorrectly labeled as a UTM data set. To fix it, you must re-label it with a GCS label.

Fig. 3.25. Incorrectly labeled GCS shapefile
Source: ArcMap

32→ Close the Layer Properties window.

33→ Open the <u>Define Projection</u> tool.

33→ Set the input dataset to townships. A warning icon appears on the tool because it already has a label. Ignore it—you know that the label must be corrected.

33→ Click the Browse button and select the NAD 1983 coordinate system from the Favorites folder. Click OK and OK.

33→ The townships appear in the map now. Give them a hollow, thick green symbol.

33→Check them against USA Topo Maps, These are not quite as accurate, but they are close enough. They probably had a source scale smaller than 1:24,000.

Now they are correctly located, but the townships shapefile is still in a GCS, and it needs to be projected to UTM.

34→ Right-click townships and use Data > Export data to save them in the Features feature dataset. They will be automatically projected as before. Save them with the name *townships*. Add the new feature class and remove the original one.

35→ Zoom to the full extent and turn off the townships layer.

35→ Add the benchmarks shapefile from the Downloads folder and give it the Triangle 3 symbol. You get a warning, but the benchmarks appear on the map.

35→ Open the properties for benchmarks and click the Source tab. Examine it.

It is another undefined coordinate system, but it appears to be in UTM because the benchmarks show up in the right spots. You just need to set the label as before.

36→ Close the Layer Properties window.

36→ Open the <u>Define Projection</u> tool and set the input dataset to benchmarks.

36→ Browse and select NAD 1983 UTM Zone 13N. Click OK and OK.

36→ Zoom in and compare several benchmarks to the USA Topo Maps layer (benchmarks are shown with a small triangle or *x* and an elevation).

This time you notice something odd. These benchmarks appear several hundred meters south of the benchmarks in the topo map. Something is still wrong. An offset of this magnitude is most likely to be a datum mismatch. You surmise that perhaps benchmarks was in the NAD 1927 datum, and so the label you assigned was not correct. To test your theory, you will reassign the label.

Source: ArcMap

37→ Open the <u>Define Projection</u> tool again and set the input data to benchmarks.

37→ Browse in the Projected Coordinate Systems > UTM > NAD 1927 folder and select the NAD 1927 UTM Zone 13N coordinate system. Click OK and OK.

37→ Compare the benchmark locations to the USA Topo Maps layer again.

You were right. With the NAD 1927 UTM label assigned, the benchmarks now appear where they should. However, you want all your data in the geodatabase to be in the same coordinate system, so you still need to project this file from NAD 1927 UTM to NAD 1983 UTM. This time we will use the Project tool so that we can specify the datum transformation.

38→ Open the <u>Project</u> tool and set the input to benchmarks.

38→ Save the output in the Features feature dataset as *benchmarks83*.

38→ NAD 1983 UTM Zone 13N is already selected for saving into the feature dataset.

Notice that this time a transformation method is entered to convert the NAD 1927 datum to NAD 1983. The best one is chosen for you, NADCON.

38→ Click OK to run the tool. Notice that the new locations, from the Project tool, are slightly better than the on-the-fly locations used by the data frame, because we used the appropriate transformation.

39→ Remove benchmarks and give benchmarks83 the Triangle 3 symbol.

39→ Zoom to the full extent and turn off USA Topo Maps.

You have successfully solved each coordinate system problem encountered and added four feature classes to your geodatabase in the appropriate coordinate system (Fig. 3.26). These techniques serve you well when you are downloading and compiling data for a project and encounter a data set that has coordinate system problems.

Fig. 3.26. Final layers
Source: ArcMap

### Creating a custom coordinate system

Sometimes you don't have a predefined coordinate system for the area you are working on and need to define a custom one. Imagine creating a geodatabase for Turkey to display the general features with no need for equal areas or equal distances. A search of the available coordinate systems in ArcMap yields none specific to Turkey. You begin by choosing a suitable projection.

40→ Start a new, blank map. Save changes to the previous map document.

40→ Add the country shapefile from the mgisdata\World folder.

41→ Open the attribute table for country and use the Select by Attributes button to select Turkey. Close the table.

41→ Right-click the country layer and choose Selection > Create Layer From Selected Features.

41→ The new layer appears in the Table of Contents. Rename it *Turkey* and zoom in.

42→ Open the Layers data frame properties and click the Coordinate System tab.

42→ Type *turkey* in the search box and click the Search button.

42→ Expand the Projected Coordinate Systems > Gauss Kruger folder and the Turkey subfolder. Seven different zones are suggested.

42→ Expand the National Grids folder and the Turkey subfolder. A number of coordinate systems are in here. Click several of them just to see their parameters.

42→ Close the data frame properties window.

The predefined coordinate systems in ArcMap are all based on the Gauss Kruger projection (which is another name for Transverse Mercator). We know that this projection is best for north-south-oriented areas, but Turkey is oriented east-west and is too wide for a single TM zone. Our best bet will be to design our own coordinate system based on a conic projection, with a central meridian in the center of Turkey, and standard parallels that leave a sixth of the map area above and below.

43→ Find the approximate central longitude and range of latitudes of Turkey using the cursor and reading the values from the location display.

Central longitude _____
Southern latitude _____
Northern latitude _____

Armed with this information, we can now define a custom coordinate system, which we will base on the Lambert Conformal Conic projection.

44→ Open the data frame properties and click the Coordinate System tab.

 44→ Click the Add Coordinate System drop-down button and select New > Projected Coordinate System.

44→ Enter *Turkey Lambert Conformal Conic WGS84* as the name (Fig. 3.27).

44→ Choose Lambert_Conformal_Conic from the drop-down list for the projection name.

44→ Carefully change the Central_ Meridian to 35 degrees.

44→ Set Standard_Parallel_1 to 37 degrees. Set Standard Parallel_2 to 41 degrees.

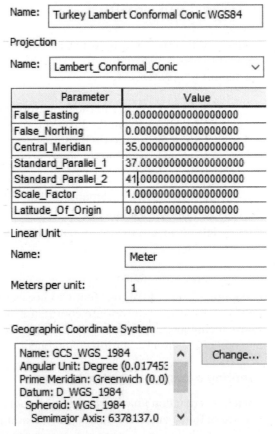

Fig. 3.27. A custom coordinate system
Source: ArcMap

44→ The Geographic Coordinate System defaulted to the data frame GCS, WGS 1984. Leave it.

44→ Click OK and OK.

The map looks fine. Now that we are sure that this coordinate system is acceptable, we can save the projection file to use again.

45→ Open the data frame properties again.

45→ Right-click the new coordinate system in the Custom folder and choose Save As.

45→ Navigate to the World folder and save the .prj file using the suggested name.

45→ Close the data frame properties.

**TIP:** To use the .prj file, click the Add Coordinate System drop-down and choose Import. You can open it in Notepad if you want to see what it looks like—it is a text file.

### Georeferencing a raster

In this section, you will download an image from the Internet and georeference it. If you don't have Internet access or if you have trouble obtaining an image, use the nwsradar_sample.gif image stored in the mgisdata\Usa folder. Your Web browser instructions might be slightly different if you are not using the Internet Explorer web browser.

46→ Open your web browser and go to http://www.nws.noaa.gov/radar_tab.php.

When georeferencing, you want to use a coordinate system that is as close as possible to the one in the image. Try to identify this one. Use the inside front cover of the book to make comparisons, if you wish.

12. On what type of projection does the map appear to be based? _____

46→ Click on Full resolution version.

46→ Right-click the enlarged image and choose Save Picture As. (If a polygon appears instead of a menu, right-click again inside the polygon to bring up the menu.)

46→ Navigate to your mgisdata\Usa folder and save the image as a GIF file. Name it *nwsradar.gif*.

47→ Create an Item Description for the raster, using the information from the web page. Remember to include a citation in the Credits section.

**TIP:** ArcGIS 10 and 10.1 had a bug in that raster Item Descriptions were not actually saved. If the Item Description goes blank again after you save it, don't worry, and don't try to do it again.

48→ Open a new map document and add the states feature class from the mgisdata\Usa\usdata geodatabase. You don't need to save the current one.

To georeference the image, you must set the data frame to the coordinate system the image uses, in this case, a GCS. You will assume it is NAD 1983, the most common US datum.

48→ Change the data frame coordinate system to NAD 1983 from the Favorites folder.

48→ Zoom to the full extent to see the states layer again, if necessary.

48→ Switch back to the browser for a moment and compare the shapes of the states in ArcMap and in the image. They appear similar.

48→ Close the browser window and return to ArcMap.

49→ Zoom in to the conterminous states so that the image and the map window are showing a similar map extent. This will facilitate finding control points.

49→ Change the symbol for the states so that it is hollow with a high-contrast outline color, such as hot pink. It needs to show up well against the image.

49→ Choose Customize > Toolbars > Georeferencing from the main menu bar.

49→ Click on Add Data and add the nwsradar.gif image from the Usa folder. Click OK if you get a warning or are asked to build pyramids.

The image does not appear yet because it is in a different coordinate system than the map. You begin by bringing the image into the same map extent as the states, which will make it easier to match control points.

50→ On the Georeferencing toolbar, choose Georeferencing > Fit to Display. The ArcMap session should look similar to Figure 3.28.

The points work best if they are spread around the map. You will first add two at diagonal corners and then two near the other corners. If the state corners or boundaries are obscured by the radar colors on your map, choose another point that is close by and clearly visible.

Fig. 3.28. Ready to begin georeferencing
Source: ArcMap

50→ Locate the southwestern corner of California and *mentally* match the corner of the state outline on the image to the corner of the pink state outline on the map.

50→ Use the Zoom In tool to zoom in to an area containing the two matching points.

50→ Click the Add Control Points tool on the Georeferencing toolbar.

50→ Carefully click the corner on the image first and then click the corresponding corner on the states map. ArcMap will help by snapping to the vertex of the state polygon. This is your first control point.

50→ Notice that the map updated after you added the control point so that the corner of California is better aligned.

**TIP:** The pairs of control points must always be added in the same order, starting with the ungeoreferenced image, and then the states.

51→ Zoom to the previous extent, and then zoom in to the New England region.

51→ Mentally locate a matching point on the Maine state boundary.

51→ Switch back to the Add Control Points tool.

51→ Click the location on the image first and then click the corresponding location on the states map. Notice that your control points are being numbered.

The match will already look much better, indicating that you chose an appropriate coordinate system. Continue switching between the Zoom and Control Point tools to add the next points, even if they are almost identical. You need at least three points for an affine transformation. Be careful not to snap to a state vertex when adding an image point.

52→ Add a control point on the Florida peninsula. Click the image first!

52→ Add a control point in the Pacific Northwest. Click the image first!

Examine the map and locate any areas where the state boundaries are still poorly aligned. These areas should receive additional control points. If you don't see any at the national scale, zoom in to a couple of locations in the middle of the country to check.

53→ Zoom to the state corner between Idaho, Wyoming, and Utah. Add a point, if necessary (image first).

53→ Zoom to the western corner of Kentucky. Add a point, if necessary (image first).

For many maps, you would continue to add more control points in misaligned areas until you no longer saw any improvement when new points were added.

 54→ Click on the View Link Table button to see the points you've added.

The points are listed in the order that you added them (Fig. 3.29), and yours will differ from the figure. The X and Y Source fields show the coordinates of each control point in the original image units (pixels). The X Map and Y Map columns show the point coordinates in map units (degrees). The 1st Order Polynomial (Affine) Transformation method is being used.

Note the residuals for each point and the total RMS error, reported in map units. In Figure 3.29, the residuals range between 0.01 and 0.03 degree, and the RMS error is about 0.02 degree. At the equator that error would be about 2 km, a reasonable value for a national map. Yours should be in the same range if you entered the control points carefully.

**TIP:** A few additional buttons on the link table are handy if you need to (from left to right) load a saved set of points, save a set of points, zoom to a point, delete a point, or add a new empty point.

Source: ArcMap

54→ On the Link Table, click the Transformation button. It shows the methods currently available with the number of control points you have.

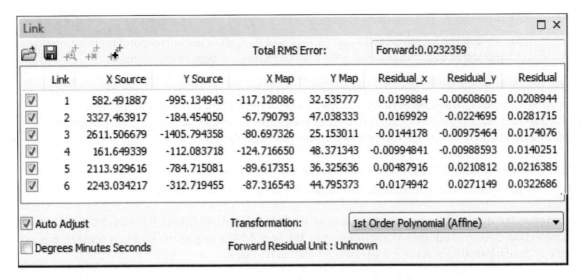

Fig. 3.29. The Link Table shows the ground control points that have been added.
Source: ArcMap

**TIP:** If the RMS error is large and does not improve when more points are added, try a second-order transformation. Second-order transformations require more control points.

54→ Once you are happy with the registration, close the Link Table. If you achieve an RMS error less than 0.02 degree, then you have done well. Write it here: _____

54→ To save the new coordinate system information with the image, click Georeferencing > Update Georeferencing.

54→ Update the Item Description with the RMSE. Include the units. You can put it in the Description box.

At this point the image is georeferenced, and the control point information has been saved with it as a world file. The next time you open the image, it will appear in the correct location.

**SKILL TIP:** Learn how to rectify an image, or permanently save it in the georeferenced coordinate system (Coordinate Systems).

### Projecting a raster

You want to place the georeferenced image in the same coordinate system as the other United States data, so we must project it. First, however, estimate the expected resolution of the projected raster. Looking at the layer properties Source tab would tell you that the cell size is 0.017979 degree. One degree of latitude is about 111 km, so the cell size in meters should be 0.017979 × 111,000 = 1995.7, or about 2000 meters.

→ Start with a new, empty map. Don't save changes to the previous map.

Remember, you can search for a tool if you don't remember where it is.

55→ Click the Search tab to open the Search window (or open and dock it from the main menu bar, if necessary).

55→ Type **project raster** in the Search box and click the Search button.

55→ Click the Project Raster (Data Management) tool to open it.

56→ Click the Browse button, navigate to the Usa folder, and set the Input raster to nwsradar.gif. Note that it has the GCS NAD 1983 coordinate system now.

56→ Name the output raster *usradar* and put it in the mgisdata\Usa\usdata geodatabase.

56→ Click the Browse button to open the Spatial Reference Properties window.

56→ Click the Add Coordinate System drop-down button and choose Import (to ensure that the raster is projected to the same coordinate system as the other US data).

56→ Select one of the data sets in the usdata geodatabase and click Add. Note that the new coordinate system will be North America Equidistant Conic.

56→ No green dot appears by the Geographic Transformation, so no change of datum is required. Leave it blank.

56→ Set the resampling technique to NEAREST, since the image contains discrete data.

56→ Set the output cell size to 2000 meters (both X and Y).

56→ Click OK to start projecting.

**TIP:** The cell size suggested in the tool is an exact conversion from the old to new cell size, but often the user rounds it to an even number.

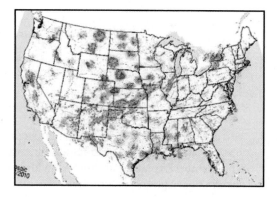

57→ Add the states feature class from usdata and give it a hollow symbol to ensure that the projection worked correctly (Fig. 3.30).

This is the end of the tutorial.

→ Close ArcMap.

Fig. 3.30. The projected radar image
Source: ArcMap

## Exercises

1. Examine the coordinate system for the streets feature class in the Austin geodatabase. What is the name of the coordinate system? Is it projected or unprojected? What are the map units?

2. What is the name of the projection (not the coordinate system) used by the feature classes in the Oregon geodatabase? What are the central meridian and the standard parallel(s)? Does it use the equator for the latitude of origin?

3. Open the map document ex_3.mxd in the mgisdata\MapDocuments folder. What is the coordinate system of the data frame in this map? _____ What are the map units of the frame? _____ What are the display units of the frame? _____ What are the feature class coordinate systems? _____

4. Choose a good projection/coordinate system from the predefined coordinate systems in ArcGIS for maps of the following areas. Explain your choice for each. (**Hint:** Use the UTM zone and State Plane zone feature classes to help locate the area relative to a zone, if needed.)

   Humboldt County, California

   Grafton County, New Hampshire

   State of Nevada

   State of New Jersey

   England

   Antarctica, true distances required

5. You are working on a statewide Illinois project and decide to define a custom coordinate system. Start with one of the State Plane zones and modify it slightly to make it better for the whole state. Explain your approach and capture the window showing the custom coordinate system description you created.

6. You are working on a statewide Colorado project and decide to define a custom coordinate system. Start with one of the State Plane zones and modify it slightly to make it better for the whole state. Explain your approach and capture the window showing the custom coordinate system description you created.

7. The Austin folder contains two shapefiles showing dog off-leash areas as points and polygons. Both have coordinate problems. Describe the problem for each; then fix them and create a map showing both the points and the polygons with a backdrop of the major transportation arteries in Austin. (**Hint:** Open ArcMap and load arteries first, in order to be able to compare the other data sets against it.) capture the map.

8. Find a map image on the Internet, save it as a GIF or JPEG, and georeference it. Display it with at least one other data set from the mgisdata folder and capture the map.

## Challenge Problem: My Country 'Tis of Thee

Create a new project folder called USProject in your gisclass folder, and create a file geodatabase in it called USContig. Find and import at least four feature classes on a common theme from the Internet to this geodatabase, including only the lower 48 states and placing them all in the USA Contiguous Albers Equal Area Conic coordinate system. Remember to create citations in the Item Description for each one. Turn in a map screenshot, a screenshot of the expanded geodatabase in the Catalog, a list of the four original coordinate systems of the data, and your four citations.

# Chapter 4. Mapping GIS Data

## Objectives

▸ Using properties of symbols to differentiate features or rasters in maps

▸ Distinguishing among nominal, categorical, ordinal, and ratio/interval data

▸ Creating maps from attributes using different map types

▸ Selecting appropriate classification methods when displaying numeric attributes

▸ Displaying thematic and image rasters

# Mastering the Concepts

## GIS Concepts

Geographic Information Systems are used for many purposes, but creating maps is one of the most common. The practice of cartography, or mapmaking, has a long history. Although a computer makes the process easier and offers the ability to explore and edit designs, the cartographer must still understand the basic principles behind portraying spatial data and the aesthetic challenges in designing an effective and attractive way to communicate ideas.

### Choosing symbols for maps

Cartographers may choose a variety of strategies for symbolizing features in a map. A data layer may simply be portrayed using one symbol, or different features can be assigned different symbols depending on the value of an attribute field. Point data are shown using marker symbols, line features with linear symbols, and polygon features with shaded area symbols (Fig. 4.1).

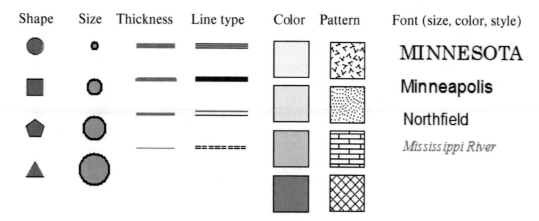

Fig. 4.1. Variations in symbols used to differentiate objects in a map

Cartographers have many ways to signify differences between features in a map: shape, size, thickness, line type, color, pattern, and font (Fig. 4.1). Traditionally these variations are used to show either changes in category (the type of thing) or in quantity. Shape, line type, pattern, and font are typically used to show changes in category, such as different types of wells (points), different classes of roads (lines), or different geologic units (polygons). Size and thickness are generally used to indicate increases in quantity, such as the population of a city (points) or the discharge of a river (lines) (Fig. 4.1). Text symbol variations usually indicate categories (towns versus rivers), although font size can indicate qualitative differences in value, such as town size, as long as not too many different sizes are employed.

Color may be used to indicate either category or quantity. Colors may be designated using one of several common methods. The **RGB** method is based on mixing red, green, and blue light on a computer screen. The **CMYK** method is often used for printing and specifies mixtures of inks used in printers or plates (cyan, magenta, yellow, black). The **HSV** method is instructive for discussing the use of color in portraying features on a map.

HSV stands for **hue**, **saturation**, **value** (Fig. 4.2). **Hue** refers to the shade of color, such as red, blue, or yellow, and is established by the wavelength of the light observed. Typically it is portrayed as a color wheel so that the color values range in degrees from 0 to 360. **Saturation** corresponds to the intensity of the color and is measured as a percentage. Imagine mixing a can of paint, starting with a white base and adding a single pigment—a small amount of pigment yields a low saturation but a large amount of pigment results in high saturation. **Value** refers to how light or dark the color is, transitioning from the full color to black. Value is also measured as a percentage. Any color can be defined using a combination of the three properties.

Fig. 4.2. Hue, saturation, and value method of defining a color
Source: ArcMap

The set of colors chosen for a map layer should follow certain guidelines. Figure 4.3 shows three different sets of five colors. A set based on variations in hue should be used to depict changes in category, such as different soil units (Fig. 4.3a). For categories, the saturation and value of the colors should be similar. Quantities are generally indicated using differences in saturation and/or value, with light or unsaturated colors indicating lower quantities and darker or more intense colors indicating higher quantities (Fig. 4.3b). Sometimes a **divergent color set** is helpful in showing variation around a significant middle value (Fig. 4.3c); for example, a climate change map showing increases or decreases of temperature from current values. Areas with no significant change are shown in the middle neutral color, colder temperatures are shown in blue, and warmer temperatures are shown in orange.

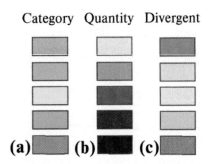

Fig. 4.3. Color set variations
Source: ArcMap

The importance of black and white symbol schemes cannot be neglected. In commercial printing or copying, color still costs more than black and white. Publishing figures in professional journals or reports may also cost significantly more in color. One should consider, too, how most viewers will see the map. Figures in a master's thesis may look wonderful in color, but some of the people reading it may receive a black and white copy version from interlibrary loan. There are many reasons why one might wish to design a map in black and white at the outset. In a black and white map, only four or five different shades of gray are typically discernible, so selecting symbols for such maps relies heavily on variations in shape, size, thickness, line type, or pattern rather than value.

Combinations of these factors may also be used when specifying symbols, such as using points that change in both shape and color, polygons that change color and pattern, or text that has differences in size and font and style. However, combinations should be used sparingly, as the more complex and varied the symbols become, the more difficult it may be to interpret them. In Figure 4.4, both volcanoes and highways are shown using combinations of color, shape, and size (thickness). The combinations serve to accentuate differences within the layer, and could be used to emphasize one type of feature, such as using a larger, saturated pink symbol to call attention to the stratovolcanoes. However, this strategy would backfire if multiple point layers were being displayed, as it would become difficult to interpret which features belong to which point layer.

Fig. 4.4. Examples of categorical data representation for points, lines, and polygons

Color choice is complicated by the fact that about 10% of men and 1% of women are color blind. Different types of color blindness are possible, although red-green color blindness is the most common. If layers are being symbolized for data exploration by one person, then the user may employ any set of colors that aids interpretation of the data. However, if the maps are intended for use by many people, it is better to avoid red-green color sets and rely on saturation, value, pattern, and shape variations instead. If divergent colors or multiple hues are needed, combinations such as brown-purple or orange-blue can be interpreted by most color-blind viewers.

Cartographers also need to be sensitive to conventions and connotations associated with different colors or symbols. A **convention** refers to the use of a particular color or symbol in a commonly understood way; for example, using blue for displaying water, blue-red for contrasting cold-heat, or a blue cross to indicate a hospital. Using conventional symbols aids interpretation of the map. **Connotation** is an emotional or psychological impact associated with a particular symbol, such as the color red indicating danger, or red-white-blue evoking feelings of patriotism. However, connotations are often culturally specific. Red may indicate danger in the United States, but is associated with joy in Thailand, or national pride in China. Making maps for an international audience demands particular care and knowledge in the choice of colors and symbols. Context can matter as well; even in the United States, a red-green map at Christmas might engender more joy than fear.

### *Types of data and types of maps*

Geographers, statisticians, and others characterize measurements or attributes as belonging to one of a set of data types: **nominal**, **categorical**, **ordinal**, **interval**, or **ratio**. The type of data affects how it should be stored in a database and what types of analyses or statistics are appropriate. In mapping, the data type also influences the kind of map used to display the feature attributes.

**Nominal data** name or identify objects, such as the name of a state. Nearly every feature will have a different name. Nominal data are often text values, but they don't have to be: a parcel number or a tax identification number also serves the purpose of uniquely identifying an object.

The US Census uses such data in its Federal Information Processing Standards Codes, or FIPS codes. Each state, county, city, block, and so on is assigned a unique identifying number that is consistent from product to product. Nominal data are usually portrayed on a map by labels. Each layer typically has one text symbol, but different text symbols may be used to accentuate different layers, such as using all uppercase for state names or italic blue text for rivers.

**Categorical data** separate features into distinct groups or classes. Figure 4.4 shows examples of categorical data for point features (volcano type), linear features (road class), and polygon features (land cover). Other examples include soil types, ethnic groups, and geological rock formations. Categorical data are often stored as text, but it is also possible to represent the categories with numeric codes, such as Commercial Services = 20 and Industrial = 40. Categorical data are represented by a **unique values map**, which gives each category a different symbol based on shape, line type, color, or pattern (Fig. 4.5).

**Ordinal data** have categories that are ranked based on some quantitative measure, although the measure may not be linear. For example, urban settlements might be classified as villages, towns, or cities. Students are assigned grades of A, B, C, D, or F. Soils are designated as A, B, C, or D depending on their infiltration. Ordinal data must be represented by unique values maps if the values are text, but value or saturation should be varied rather than hue or pattern so that the quantitative increase can be easily interpreted. Numeric ordinal data may be represented with either unique values maps or graduated color maps.

Fig. 4.5. A unique values map based on geological units
Source: ArcMap

Quantitative data represent phenomena that fall along a regularly spaced measurement scale, such as distance or rainfall. Equal changes in interval involve equal changes in the quantity being measured. For example, the energy needed to heat a thimble of water from 6 to 7 degrees is the same as heating it from 96 to 97 degrees. **Ratio data** have a meaningful zero point that indicates the absence of the thing being measured. Precipitation is an example of ratio data; zero precipitation corresponds to a total lack of rain, and two inches of rain is twice as much as one inch. Ratio data support all four arithmetic operations of addition, subtraction, multiplication, and division. **Interval data** have a regular scale but are not related to a meaningful zero point. Temperature data measured in the familiar Celsius or Fahrenheit scales are interval data because a temperature of zero does not correspond to a complete lack of temperature (heat energy). Any scale that can have negative values, such as elevation, is an interval scale. Interval scales support only addition and subtraction. Today's high temperature of 80 degrees might be hotter than yesterday's high of 40 degrees, but it cannot be said to be twice as hot, except as a figure of speech.

Quantitative numeric data take on values along a continuous scale of possibilities; every state, for example, has its own population value. In order to symbolize numeric data, the values must be partitioned into groups with specific ranges.These ranges are called classes, and the maps are called classified maps. In Figure 4.6, the cities and roads have three classes of values, and population density has four classes. Each class is symbolized using variations in symbol size, thickness, or color saturation/value.

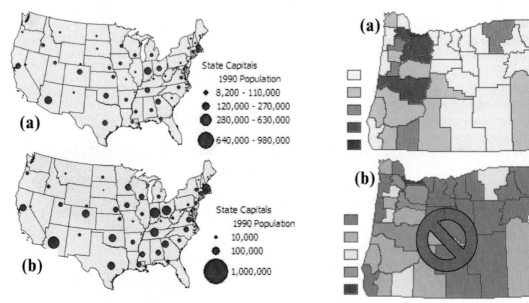

Fig. 4.6. Examples of numeric data representations for points, lines, and polygons

Point or line data are usually displayed by varying symbol size or thickness and are portrayed using a **graduated symbol map**. However, point or line data can also be portrayed using a **proportional symbol map** in which the numeric value is used to proportionally determine the size of the symbol. Instead of a few size classes, the map has a continuous range of symbol sizes. This style is often referred to as an unclassed map. Figure 4.7 compares graduated symbol and proportional symbol maps for the populations of state capitals.

Fig. 4.7. Comparison of (a) a graduated symbol map and (b) a proportional symbol map showing the populations of state capitals
Source: ArcMap

Fig. 4.8. A choropleth map showing population by varying (a) saturation and/or value (b) hue
Source: ArcMap

Numeric data classes for polygons are represented using color shaded symbols (Fig. 4.8), called a **graduated color map**, or a **choropleth map**. The maps are usually symbolized using changes in value and/or saturation (monochromatic color ramp) so that the increase in quantity is clear, as shown in the county population map in Figure 4.8a. Although rainbow color ramps are popular in software packages, using them for graduated color maps is usually a mistake, because it is difficult for the mind to interpret hues as representing larger or smaller quantities (Fig. 4.8b). Generally the eye cannot distinguish between more than 7–8 levels of the same color, so no more than that number of classes should be used.

### The Modifiable Areal Unit Problem

Data are often summarized, or aggregated, according to arbitrary area units like states or counties, and the choice of unit will influence the recorded values. In some cases, such as the number of farms or lightning strikes, units with larger areas would tend to have higher values

111

just because they are larger. In Figure 4.9a, the number of farms is greater in the bigger states like Texas and California. Certain variables, such as the number of home vacancies, are strongly linked to population, and maps made from them resemble population rather than giving insight into underlying causes. In Figure 4.9b, the most populous states have the most vacancies.

Another issue that can arise is that larger polygons receive greater prominence in the map simply because of their size. In Figure 4.9, the large western states attract more attention than the eastern states. Both of these issues are examples of a phenomenon known as the **Modifiable Areal Unit Problem**, or **MAUP**, which occurs when measurements are being aggregated over arbitrarily defined areas like states. Analyzing spatial or statistical patterns becomes more difficult, because one is rarely sure how much the patterns are affected by the aggregation scheme. However, methods to reduce the effects of MAUP on maps are available.

One approach is to **normalize** the data, dividing each value by a specified variable. If one is concerned that the size of the aggregation area is influencing the magnitude of values, one can divide by the feature area as shown in Figure 4.10a, where the number of farms is divided by state area (compare to Fig. 4.9a). If the values are being influenced by population, then one can divide by population or other suitable variable, as in Figure 4.10b, where the vacancies are divided by the number of housing units, giving a far more interesting and informative map than Figure 4.9a. Another normalization method, less commonly used, shows percentages of the total quantity falling in each feature; Figure 4.10c shows the percentage of Congress controlled by each state, accomplished by showing the number of districts in each state as a percentage of the total number of districts.

However, not all attributes need normalization. In a precipitation map, each value already represents inches of rain falling at any location. The median rent for a county does not depend directly on the area of the

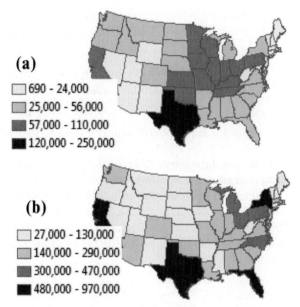

Fig. 4.9. (a) Number of farms; (b) number of vacant housing units
Source: ArcMap

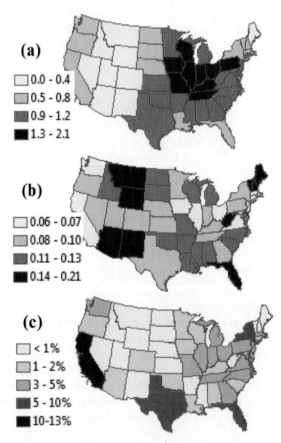

Fig. 4.10. (a) Farms normalized by state area; (b) vacancies normalized by housing units; (c) percentage of Congress controlled by each state
Source: ArcMap

county. These attributes would not be normalized. Before mapping any attribute, it is important to stop and consider how the aggregation unit might affect the map and determine whether the values should be normalized, and by what.

The visual MAUP issue, which occurs when large polygons dominate the map, can be addressed by changing the map type. A graduated symbol map applied to polygons places a symbol at the center of each polygon, minimizing the visual imbalance, as shown in this graduated symbol map of the density of farms (Fig. 4.11a; compare to Fig. 10a). A **dot density map** uses randomly placed dots to show the magnitude of a value in the attribute table (Fig. 4.11b). Each dot represents a certain number; in this map, 2000 farms. Note that the locations of the dots, which are randomly placed, do not necessarily represent the distribution of farms in the state.

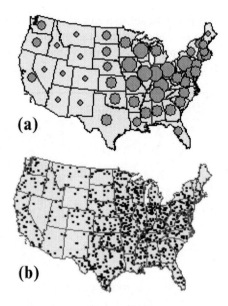

Fig. 4.11. Number of farms using an (a) area-normalized graduated symbol map or (b) dot density map
Source: ArcMap

### Chart maps

A **chart map** expands the number of attributes that can be displayed on a map by replacing a single symbol with a chart representing several attributes. The chart could be a pie chart, bar chart, or stacked bar chart. Figure 4.12 shows a pie chart map with the proportions of Caucasians, African Americans, and Hispanics in each state. The pie sizes can be all the same (Fig. 4.12a) or proportional to the sum of the three categories, thus showing the relative number of people in the state as well (Fig. 4.12b). Notice how the pie colors are chosen to facilitate recognition of the classes: cream for Caucasians, dark brown for African

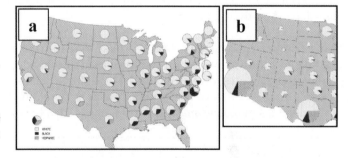

Fig. 4.12. (a) A chart map can represent several attributes, such as the proportion of ethnic groups. (b) The pie size can represent the total population.
Source: ArcMap

Americans, and reddish for Hispanics. Such details help make maps easier for the reader to interpret, and they lessen the need for the reader to look back and forth between the legend and the map.

## Displaying rasters

Recall from Chapter 1 that rasters are a cell-based data model in which an array of cells or pixels store numeric values relating to some feature or quantity on the earth's surface. Rasters may be broadly grouped into three main types, each of which is displayed differently: thematic rasters, image rasters, and indexed rasters.

### Thematic rasters

A **thematic raster** represents features or quantities, such as roads, geology, elevation, or vegetation density (Fig. 4.13). When rasters are used to store objects, such as roads or land use polygons, they are called **discrete** rasters. Typically they are created by converting features from a vector data set to a raster format, using a single attribute field as the value to be placed in the cells. Discrete rasters contain groups of adjacent cells with the same value. In contrast, **continuous** rasters store values that represent a continuously varying quantity, such as elevation.

Fig. 4.13. Thematic rasters. Discrete rasters store feature data, such as land use or road types. Continuous rasters store data that vary smoothly, such as elevation or rainfall.
Source: ArcMap

Values in a raster are always stored as numbers and may be characterized as categorical, ordinal, or interval/ratio. (Nominal data are usually not stored in rasters—it would be inefficient.) The methods used to display thematic rasters are similar to those used for displaying vector data and also depend on the data type.

A raster representing categorical or ordinal data is displayed using a unique values map. Just like the unique values map used for feature classes, each value of the raster receives its own color, as in the geology map shown in Figure 4.14a. Color schemes for unique values maps have 32 possible values and work best with data that have relatively few categories. When representing ordinal data, a monochromatic color scheme is used to communicate a sense of increase.

Fig. 4.14. Display methods for thematic rasters: (a) unique values geology; (b) classified elevation; (c) stretched elevation
Source: ArcMap

A thematic raster containing interval or ratio data must be classified into ranges in one of two ways. The **classified** display method divides the values into a small number of bins, similar to a graduated color map. A color ramp is chosen to assign the colors to each bin. The elevation map in Figure 4.14b has 12 classes represented by 12 colors from a color ramp. The **stretched** display method scales the image values to a color ramp with 256 shades (Fig. 4.14c). The raster is first subjected to a **slice**, which rescales the elevation values (ranging from 800 to 1600 meters) into 256 bins, then matches the bins to the 256 shades in the color ramp. You can think of a slice as a classification that creates 256 classes.

114

The colorful ramps, like those in the elevation rasters in Figure 4.14, ignore the guideline to use saturation/value rather than hue to represent quantities, yet the eye has no trouble interpreting the elevation increase. Continuous rasters typically represent surfaces with an underlying structure, such that low values occur near low values and high ones near high. This structure imposes a visual order and allows the eye to interpret the greater number of colors. Such ramps are not suitable for features or all rasters, but can be effective when appropriate.

### Image rasters

**Image** rasters include aerial photography and satellite data. The pixels represent degrees of brightness caused by light reflecting from materials on the surface. The brightness values are often placed on a scale of 0–255 DN (digital numbers). A dark shadow would have a low brightness and a low DN, and a white cement road would have a high brightness and a high DN.

Fig. 4.15. Raster display methods for images: (a) single-band stretched image; (b) RGB composite image (true color); (c) RGB composite with Landsat bands 7, 4, and 1
Source: ArcMap

An image raster may contain one or more bands of information. Imagine a black-and-white digital camera that measures brightness and stores a value between 0 and 255 for each pixel on its screen. This camera would produce a single-band raster that could be displayed with a grayscale color ramp from black to white, as shown in Figure 4.15a. Such images are displayed using the same **stretched** method that is applied to continuous thematic rasters. Images based on the 0–255 scale do not need to be sliced, but some images contain larger values and do require slicing.

Images typically use the RGB color model to specify colors. A color digital camera measures brightness separately for three wavelengths of light (red, green, and blue) and stores three brightness values for each pixel in three separate bands. If the value in the red band is high (close to 255) and the other two bands have low values near zero, a bright red color will result. If red and green are both high and blue is low, a mixture of red and green will create yellow. (Mixing light is a different process than mixing paint, and different colors will result.) All colors may be represented by varying proportions of red, green, and blue light, as dictated by the brightness values in the bands. Figure 4.16 shows two possible mixtures and the resulting colors.

Fig. 4.16. Computers store and identify colors as mixtures of red, green, and blue light on a scale from 0 to 255.
Source: ArcMap

To display a multiband color image, an **RGB composite** method is used. One band is assigned to each of the three color sources on the computer monitor (red, green, and blue) to produce a color image (Fig. 4.15b) from the brightness levels stored in the three bands.

Many satellites and airborne cameras measure brightness in more than three wavelengths and produce more than three bands. Landsat images, for example, have seven or more bands. Multiband images also use the RGB composite display method, but any of the bands may be applied to the RGB sources on the computer. In Figure 4.15b, the Landsat bands 3, 2, and 1 representing red, green, and blue light are used, producing a true-color image. In Figure 4.15c, bands 7, 4, and 1 are used to contrast green vegetation with dry or urban areas.

**Stretching image values**

Image DN values are often normally distributed. Figure 4.17a shows a typical **histogram** of cell values, with the horizontal axis showing the range of values with the corresponding grayscale color ramp, and the vertical axis showing the number of cells for each value (color). The image has few values near 0 or 255, with most of the pixels occurring in the middle gray range. The image is shown in Figure 4.17b, and it appears dim and featureless with little contrast.

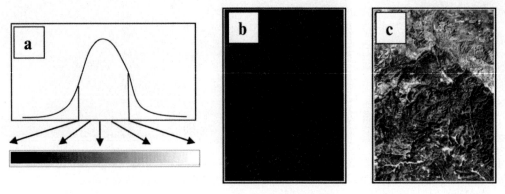

Fig. 4.17. The effects of stretching: (a) stretching an image; (b) no stretch applied; (c) a standard deviation stretch
Source: ArcMap

The **stretched** display method can improve the display of normally distributed values by ignoring the tails of the distribution. One method assigns the lowest and highest image values to 0 and 255, respectively, and stretches the remaining colors along the ramp. This is called a minimum-maximum stretch. Even greater contrast can be developed using a standard deviation stretch, which uses only the values within two standard deviations of the mean (Fig. 4.17c).

Stretches can be applied to thematic continuous rasters as well as images and can be helpful when the image values are not evenly distributed. The RGB composite display method permits the user to stretch each individual band.

**Indexed color rasters and colormaps**

The RGB color model is capable of storing millions of different colors and is best for fine rendering of images. However, it uses large amounts of memory and storage space. An alternate method for storing color identifies a restricted set of colors that actually appear in the image. It assigns an integer from 0 to 255 to each color and stores the RGB proportions needed to make it. This list of colors is called the **colormap** and is stored with the image. Each value in the raster is portrayed using its assigned color. Figure 4.18 shows part of a

Fig. 4.18. An indexed color raster stores a colormap that assigns a color to each value in the raster.
Source: ArcMap

116

digital raster graphic (DRG). A DRG is a scanned US Geological Survey topographic map. The contours are represented by pixels with the value 4 and displayed using the assigned brown color. Every pixel value and defined color used in the map is shown in the accompanying colormap: green for vegetation, black for text, and so on.

Indexed color rasters are commonly produced when scanning large color maps. Storing them in RGB could use hundreds of megabytes or more. An indexed color map typically uses less than a tenth of the space, and the eye cannot easily distinguish between the two.

### Raster pyramids

Rasters may have a much higher density of cells than can be displayed on a screen at a given scale. When this happens, the computer must skim through the raster and pick one cell from a larger cluster to display, and this takes time. A common alternative is to build **pyramids** for the raster. A pyramid is constructed by **resampling**: taking a group of adjacent cells, commonly four, merging them to form a single new pixel, and selecting one value to represent the original group (Fig. 4.19). The resampled cells are used to create a raster with one-quarter the storage space and lower resolution than the original. The procedure is repeated to create another raster of yet lower resolution, until an entire set is created, appropriate for different scales of display.

The pyramids speed drawing because the computer only needs to select the right scale of raster to display, rather than sorting through it to choose pixels. The disadvantage of rasters is that storing the additional copies increases the size of the raster by about 50%. Creating pyramids is optional, depending on whether the user is more interested in quick display or smaller file sizes.

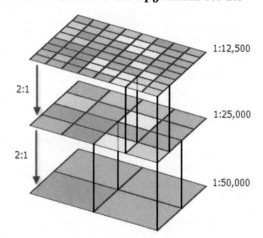

Fig. 4.19. Raster pyramids
Source: ArcGIS Desktop Help

### *Classifying numeric data*

Mapping numeric data, raster or vector, requires classifying a range of values into a small number of groups, each of which can be represented by a different color or symbol size. This process is called **classification**. There are many ways to classify data, and the choice of method affects the appearance of the map and the message that it portrays. Some common methods are compared in Figure 4.20 using the same data set of average farm size by state.

The **Jenks method** sets the class breaks at naturally occurring gaps between groups of data (Fig. 4.20a). Each class interval can have its own width, and the number of features in each class will vary. The Jenks method works well on unevenly distributed data, such as populations of the capitals shown in Figure 4.7. There are many low-population and medium-population capitals and a few very high-population capitals. The Jenks method works well with almost any data set, making it a natural choice for the default classification scheme in ArcMap.

The **equal interval** classification divides the values into a specified number of classes of equal size (Fig. 4.20b). This method is useful for ratio data, such as income or precipitation, because it gives a sense of regularity to the observed increases. However, it is hard to predict how many features will end up in each class. Notice in the farm size example that nearly all of the states fall into the first class. Compare this map to Figure 4.20a.

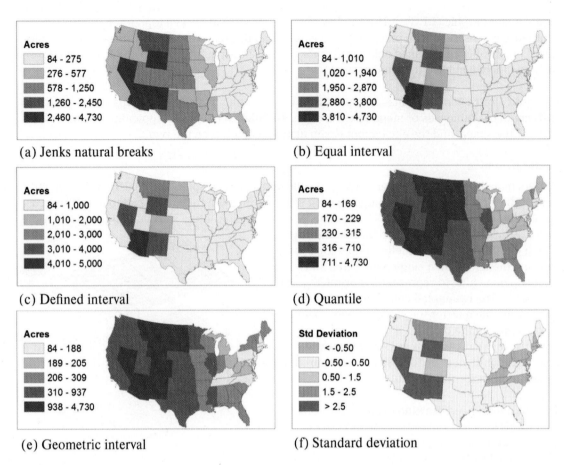

Fig. 4.20. Comparison of different classification methods using the same data set: average farm size in acres by state
Source: ArcMap

A **defined interval** classification is similar to an equal interval one, except that the user specifies the size of the class interval, and the number of classes then depends on the range of values (Fig. 4.20c). This method will create rounded values in the classes that are easy to interpret. Defined interval maps are ideal when comparing classes composed of percentages, dollars, temperatures, and other values when specific break values are desired (100, 200, etc.). It does, however, suffer the same disadvantages as the equal interval classification.

A **quantile** classification puts about the same number of features in each class (Fig. 4.20d) and enables display of groups, such as quartiles, commonly of interest in statistics. This method will create a balanced map with all classes equally well represented, but some of the features in the same class could have very different values, and features in different classes could have similar values. Quantile classifications are best applied to uniformly distributed data. Notice that a quantile classification highlights differences between the eastern states that are hidden in the Jenks, equal interval, and defined interval classifications.

A **geometric interval** classification (Fig. 4.20e) bases the class intervals on a geometric series in which each class is multiplied by a constant coefficient to produce the next higher class. A geometric interval classification is designed to work well with continuous data like precipitation and to provide about the same number of values in each class range. It works especially well with positively skewed data distributions.

A **standard deviation** classification apportions the values based on the statistics of the field. The user selects the class breaks as the number of standard deviations, and the data range determines the number of classes needed. This method excels at highlighting which values are typical and which are outliers, especially since a divergent color set is used to accentuate below versus above normal values. In Figure 4.20f, the yellow states are close to the mean farm size, several eastern states appear to have smaller than average farms, and some western states have much larger farms than average. This map is best applied to normally distributed data.

Finally, if none of the above options gives the desired map, the user can manually set class break points to any chosen values. This option works for assigning a logarithmic or exponential scale, as well.

### Choosing the classification method

The choice of classification method depends on the mapmaker's purposes and on the type of data. Jenks shows the "nearest neighbors" in a distribution, whereas a defined interval or equal interval map does a better job of portraying relative magnitude. Notice the difference between the maps in Figures 2.20a and 2.20b. The Jenks map gives the impression that many states have large farms because the class sizes increase slowly; an equal interval map shows that farms in most states actually average 1000 acres or less. Figure 4.20b makes this observation clear at first glance.

Some data possess the quality that the magnitude of the values has an intrinsic meaning. Percentage data, for example, have a physical and a psychological meaning—people have an intuitive understanding of the difference between 50% and 100%. Differences in median rent occur in dollar values to which people can attach meaning. When dealing with such data, it is wise to use a defined interval map to choose classes with logical break points, such as 20%, 5 inches of rain, or $200, rather than 12.6%, 1.47 inches of rain, or $187. The reader can more effectively interpret the classes.

The distribution of values has impact also. Jenks and geometric interval classifications are designed to work well with unevenly distributed data. Equal interval, defined interval, and quantile maps can be used with any data but show better results with evenly distributed data. The statistics behind standard deviation maps assume that the data are normally distributed.

## About ArcGIS

ArcGIS provides an extensive set of tools for symbolizing features and rasters. Within the Symbol Selector (Fig. 4.21), the user can choose from a variety of predefined symbols and modify the properties of those symbols, such as the fill color, outline width, and outline color, when the available ones are not suitable. An Edit Symbol button opens the Symbol Editor window to allow the user complete flexibility to modify individual symbol layers and create virtually any symbol desired (see the Skills Reference section for more about the Symbol Editor).

Fig. 4.21. Symbol Selector
Source: ArcMap

Symbols are organized into groups for ease of use. A **style** contains a set of symbols with a related theme and typically includes a wide range of symbol types, including point markers, line styles, polygon shades and patterns, color sets, text styles, and even north arrow or scale bar symbols. The default style the user sees is called the ESRI style. Dozens of other styles may be added by clicking the Style References button on the Symbol Selector, with themes such as Geology 24K, Environmental, Civic, Crime Analysis, or Hazmat. Users with very specific symbol needs can create their own styles and fill them with symbols copied from other styles, modified from other styles, or created *ex nihilo*. However, most users will find the standard sets quite sufficient for their needs.

Although one typically works with only a few styles in the Symbol Selector at a time, a search function in the Selector allows the user to search by keyword for matching symbols in any of the installed styles. Searching for tree, for example, yields various symbols from the 3D Trees, CADD GIS Center, Caves, Civic, Environmental, Forestry, Ordnance Survey, and Survey styles.

Once a set of symbols has been established for a layer, it is often helpful to save them as a layer file for reuse. Geologic map symbols, for example, utilize many specific colors and patterns and can take hours to set up. A graduated symbol set with manual breaks might be created to show arsenic in well water (Fig. 4.22), using specially chosen symbols to depict values below the detection limit (x), or below or above the EPA limit (blue/orange). By saving the finished layer as a layer file, the classification scheme can be quickly transferred to many map documents or used to ensure that everyone's maps use the same set of values.

**Arsenic mg/L**

✕  0.000 - 0.005

◎  0.006 - 0.009

◉  0.010 - 0.050 > EPA

⬤  0.051 - 0.100

⬤  0.101 - 0.500

Fig. 4.22. Arsenic classification
Source: ArcMap

## Summary

▶ Differentiating between features on a map requires variations in symbol shape, size, thickness, line type, color, pattern, or font.

▶ Attribute data have a data type designation: nominal, categorical, ordinal, interval, or ratio. The data type determines the kind of map and even the types of analysis that may be performed on that attribute.

▶ Nominal data name things or uniquely identify them and may be text or numbers. Each feature usually has its own value, and repeats are the exception.

▶ Categorical data group objects into smaller sets identified by a unique value. Numbers may be categorical when they are used as codes. Ordinal data consist of categories that are ranked in some way.

▶ Interval data are measured on a regular scale, and ratio data are measured on a regular scale with a meaningful zero point.

▶ Single symbol maps and labels are used to map nominal data. Unique values maps are used for categorical or ordinal data. Interval or ratio data are displayed using graduated color, graduated symbol, proportional symbol, dot density, or chart maps.

▶ Rasters may contain thematic or image data. Discrete thematic rasters can be displayed using a unique values method. Continuous thematic rasters may be classified or stretched. Image rasters use a stretched or RGB composite display.

▶ Continuous numeric data are classified before being mapped. Classification methods include Jenks natural breaks, equal interval, defined interval, quantile, geometric interval, standard deviation, and manual. The best classification method depends on the type of data and the data distribution.

▶ In ArcGIS, symbols are grouped into similar themes called styles. Tools are available to allow the user to create virtually any symbol needed for a map.

### Important Terms

| | | | |
|---|---|---|---|
| categorical data | divergent color set | image | ratio data |
| chart map | dot density map | interval data | resampling |
| choropleth map | dynamic labels | Jenks method | RGB |
| classification | equal interval | Modifiable Areal Unit | RGB composite |
| classified | geometric interval | Problem (MAUP) | slice |
| CMYK | graduated color map | nominal data | standard deviation |
| colormap | graduated symbol map | normalized data | stretched |
| connotation | histogram | ordinal data | style |
| continuous | HSV | proportional symbol map | thematic raster |
| convention | hue | quantile | unique values map |
| defined interval | saturation | pyramids | |
| discrete | value | | |

## Chapter Review Questions

1. A 1:20,000,000 scale map of the United States displays the interstates with a line symbol that is 3.4 points wide. There are 72 points to an inch. What is the uncertainty in the location of the road due to the width of the line used to represent it? Give the answer in feet and miles.

2. For each of the following types of data, state whether it is nominal, categorical, ordinal, interval, or ratio. Explain your reasoning.

   | | |
   |---|---|
   | bushels of wheat per county | pH measurement of a stream |
   | vegetation type | state rank for average wage |
   | average maximum daily temperature | number of voters in a district |
   | parcel street address (e.g., 51 Main St.) | student grade in a class (e.g., A, B+) |
   | parcel ID number (e.g., 1005690) | soil type |

3. For each of the following attributes, state whether a single symbol, graduated color, or unique values map would be most appropriate. Explain your reasoning.

   | | |
   |---|---|
   | precipitation over a county | rivers |
   | geological unit | land use |
   | acres of corn planted per county | household income by county |

4. If mapping the following attributes for counties, indicate which ones would generally be normalized and which would not.

   | | |
   |---|---|
   | average daily temperature | median rent |
   | number of Hispanics | total river miles |
   | square miles of parkland | sales tax rate |

5. State whether you would use a unique values, classified, stretched, or RGB composite display method for each of the following rasters.

   | | |
   |---|---|
   | geological units | black-and-white aerial photo |
   | 4-band satellite image | precipitation |
   | landslide hazard (high, medium, low) | slope in degrees |

6. Explain the difference between thematic rasters and image rasters.

7. For each of the following map types, find a map in ArcGIS Online or from the Internet that uses this type of symbolization: graduated color, graduated symbol (lines), graduated symbol (points), unique values. Turn in a screenshot of each map with a citation. For each one, explain how the differentiation between features is achieved, and critique the choice of symbols used.

# Mastering the Skills

## Teaching Tutorial

The following examples provide step-by-step instructions for doing basic tasks and solving basic problems in ArcGIS. The steps you need to do are highlighted with an arrow →; follow them carefully. Click on the video number in the VideoIndex.htm file to view a demonstration of the steps.

### *Exploring symbols*

The first time we opened ArcMap we started with a default blank map. This time we will use a template to specify the size and layout of the map page. We will also set a default geodatabase for the map to indicate where most of the data will come from and where it should be saved.

1→ Start ArcMap. In the Getting Started window, click on the Templates > North American entry.

1→ Scroll down to the Letter (ANSI A) Portrait entry and select it. Do not click OK yet.

1→ Click the Browse button next to the Default Geodatabase box.

1→ Navigate to the mgisdata\Oregon folder and click the oregondata geodatabase to highlight it. Click Add.

1→ Click OK to finish opening the new map.

1→ Save the map document in the MapDocuments folder, and remember to save periodically as you work.

We begin by adding some feature classes from the Oregon geodatabase.

2→ Click the Add Data button.

2→ Find the Go To Default Geodatabase button and click it to go directly to the default geodatabase, oregondata.

2→ Add the counties and volcanoes feature classes from the geodatabase.

3→ Add the airports and highways from the Transportation feature dataset.

3→ Add the rivers feature class from the Water feature dataset.

4→ In the Table of Contents, click twice slowly on the Layers data frame name and change it to *Geography*.

4→ Make sure that you are in Data view, seeing the map but not the layout page.

5→ Click on the airports symbol to open the Symbol Selector.

5→ Click the Style References button to view the list of additional styles that could be added to the Symbol Selector. Cancel without selecting any.

5→ Click the Edit Symbol button and take a moment to examine the current symbol in the Symbol Property Editor. Explore this window if you wish, but click Cancel to exit without incorporating any changes to the symbol.

**SKILL TIP:** Use the Symbol Editor to create and manage new symbols (Maps and Symbols).

6→ Still in the Symbol Selector window, type *airplane* into the search box at the top and click the magnifying glass search button.

Source: ArcMap

6→ Examine the airplane symbols from the different styles.

6→ Choose the Airplane symbol from the ESRI style; make it 25 pt. and rotate it –45 degrees. Click OK.

Airplane

Source: ArcMap

7→ Right-click the airports layer and choose Copy. Right-click the Geography data frame and choose Paste Layer(s).

7→ In the new layer, assign this Airplane symbol. View each airports layer separately, comparing the impact of the symbols.

Source: ArcMap

7→ Right-click the airports layer with the cartoon symbol and choose Remove.

**TIP:** Cartoon symbols are popular in software these days, but they are too visually complex to make good map symbols and should usually be avoided.

Recall that color may be viewed as a combination of three variations: hue, saturation, and value. Let us examine how these variations affect the color produced.

8→ Right-click on the symbol for counties and choose More Colors.

8→ If necessary, change the dropdown button from RGB or CMYK to the HSV (hue-saturation-value) color model.

Source: ArcMap

8→ Use the sliders to vary the hue, saturation, and value of the current color until you understand how these vectors impact the color produced. Click Cancel.

9→ Right-click on the symbol for counties and change the color to light yellow.

9→ Right-click on the line symbol for rivers and change the color to deep blue.

The rivers seem to be cluttering the map at this scale, so they would benefit from a scale range.

9→ Right-click the rivers layer and open its properties.

9→ Click the General tab and fill the button for *Don't show layer when zoomed out beyond* and choose the scale 1:3,000,000. Click OK.

### Labeling features

Nominal data are usually portrayed using labels. **Dynamic labels** can be created for a layer using an attribute field and can be turned on or off for the entire layer. They use an auto-placement function to avoid overlaps, and not all labels may be shown. Dynamic labels are redrawn each time the map is redrawn and may change as the scale changes or when the map is printed.

10→ Click twice slowly on the airports layer and rename it Airports.

10→ Double-click on the Airports layer to open its properties and click the Labels tab.

10→ Check the box to *Label features in this layer* and leave the method set to *Label all features the same way.*

10→ Set the Label field is set to NAME and choose Arial 9 pt. bold as the font.

10→ Click Apply to make the changes, but keep the window open. Move it aside, if necessary, to see the labels.

Basic labels are easy to set, but they often are not quite what is wanted. There are several additional options to control dynamic labels, such as adding a halo to help them stand out better.

11→ Click the Symbol button in the Label Properties window. Notice the different pre-set symbols available, but don't choose any.

11→ Click the Edit Symbol button. You can do very detailed editing here, but we just want the mask.

11→ Click the Mask tab. Choose Halo, leave the size at 2 pt., and click OK and OK to close the symbol windows. Click Apply to enact the change but keep the window open.

Some control is given over where the symbols are placed relative to the feature.

12→ Click the Placement Properties button in the Label Properties window. Make sure that the Placement tab is clicked.

12→ Examine the diagram that shows the placement priority around the point. Click Change Location to examine the other options, but don't choose one. Click Cancel.

| 2 | 2 | 1 |
|---|---|---|
| 3 |   | 2 |
| 3 | 3 | 2 |

Source: ArcMap

12→ Examine some of the other options, such as placing the label on top of a point, or placing it at an angle. Click Cancel.

12→ Click OK to close the layer properties. We are finished with the airports.

Line features have different placement options.

13→ Rename the rivers layer *Rivers*, and open its properties.

13→ Click the Labels tab, if necessary. Check the box to label the features, and make sure that the Label Field is set to the NAME field.

13→ Change the symbol to Arial 8 pt. dark blue italic, and click OK.

13→ The Rivers layer has a scale range, so set the map scale to 1:3,000,000 using the drop-down button on the main toolbar.

1:3,000,000 ▼

Source: ArcMap

Even at this scale, there are too many rivers to label, and the map is cluttered. You can set a scale range for labels separately from that for the layer.

14→ Open the Rivers properties again. On the Labels tab, click the Scale Range button.

14→ Change the settings to *Don't show labels when zoomed out beyond* 1:1,000,000. Click OK to return to the Label properties, and click OK.

14→ Set the map scale to 1:1,000,000 to examine the rivers.

It is traditional to curve labels along the rivers instead of using straight text.

15→ Open the Rivers label properties again. Click the Placement Properties button.

15→ Choose the Curved orientation button (Fig. 4.23). Check the boxes to place labels *Above* or *Below* the line.

15→ Examine the other settings, but leave them alone. Click OK to finish setting the placement options, and then OK to finish.

15→ Examine the labels, and then use the Full Extent button to zoom to all of Oregon.

**SKILL TIP:** The Label Manager makes it easier to create and edit labels for several layers at once (Labels and Annotation).

Symbols and labels may be combined when labeling.

Fig. 4.23. Setting curved labels along rivers
Source: ArcMap

16→ Rename the highways layer *Highways*, and give it a 2-pt. brown symbol.

16→ Open the Highways properties. On the Labels tab, turn on the labels and make sure the Label Field is set to HWY_SYMBOL.

16→ Click the Symbol button. In the Symbol Selector, scroll down and choose the US Route HWY symbol. Click OK to return to the Layer Properties window.

17→ Set the Scale Range to not show the symbols when zoomed out beyond 1:1,000,000. Click OK.

17→ Click the Placement Properties tab. Choose the Horizontal orientation. Click OK and OK.

17→ Zoom in to 1:1,000,000 to see the labels. Then return to the Full Extent of the map.

**SKILL TIP:** You can create classes for labels and label each class differently, such as giving the interstates one symbol and the highways another (Labels and Annotation).

18→ Rename the counties layer *Counties*.

18→ Use what you have learned so far to give the counties Arial 8 pt. bold labels with a 2-pt. halo. Don't assign a scale range.

### Creating maps from attributes for points

Currently, each feature class is displayed with a single symbol. Next, we will learn how to create maps based on attributes containing categorical, ordinal, interval, or ratio data. (If you need to review the data types, go back to the Concepts section.) Let's examine the fields in the table.

19→ Rename the volcanoes layer *Volcanoes*.

19→ Right-click the Volcanoes layer and choose Open Attribute Table.

1. What is the data type of these two attribute fields in the volcanoes feature class?

ELEVATION _____ TYPE _____

The ELEVATION field contains numeric data, so it must be portrayed using a quantities map, and the values must be classified.

20→ Close the Table window, and open the Volcanoes layer properties. Click the Symbology tab.

20→ Click the Quantities text on the left and choose Graduated colors (Fig. 4.24).

20→ Set the Value field to ELEVATION. A default set of 5 classes appears using the Jenks Natural Breaks classification.

20→ Set the color ramp to the one shown in Figure 4.24. Click OK.

The symbols now show which volcanoes have lower and higher elevations. However, the small symbol size makes the colors difficult to see—a common problem with graduated color maps for point data. We can edit the base symbol to make the map easier to interpret. We will also change the color ramp, because monochromatic color sets that vary saturation/value are better at portraying quantitive values than rainbow hues.

Fig. 4.24. The Symbology window for creating a graduated color map
Source: ArcMap

21→ Open the Volcanoes layer properties.

21→ Under the color ramp setting, click on the Symbol heading to reveal a menu. Choose Properties for All Symbols (Fig. 4.25). (If no window appears (bug in versions 10.4.1-10.5.1), hold Shift and select all the classes; then choose Properties for Selected Symbols.)

21→ Choose the Triangle 2 symbol. Click OK.

21→ Change the color ramp to a monochromatic one as in Figure 4.25. Click OK.

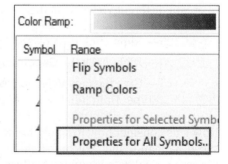

Fig. 4.25. Changing properties for graduated color symbols
Source: ArcMap

22→ Turn off the Airports layer.

22→ Right-click the Counties layer and choose Label Features to turn off the labels for now. They will still be there if you want to turn them on again later.

Chapter 4

You can now see that the eastern volcanoes are generally lower in elevation, and the highest volcanoes lie in the north. But let's try a graduated symbol map instead.

23→ Open the Volcanoes symbol properties again.

23→ Change the Quantities map type to Graduated symbols. Keep the Value field set to ELEVATION. Circles are the default symbols, but triangles would make more sense.

23→ Look right and click the Template button, which is used to set the base symbol.

23→ Choose the Triangle 2 symbol again and set its color to a reddish brown. Click OK.

23→ The smallest class symbol is barely visible. Change the minimum size value to 6 pt., the maximum size to 20 pt., and the number of classes to 4. Click OK.

It is often possible to portray the same data several different ways, and the mapmaker must choose the method that best communicates the idea being presented. In your opinion, did the graduated color or graduated symbol map do a better job of showing the differences in elevation?

24→ Rename the Volcanoes layer *Volcano Elevation*.

The elevation values in the Table of Contents have many decimal values, which is undesirable for two reasons. First, they are pointless because they are all zeros. Second, the significant figures used for labels should be consistent with the precision of the values being reported.

24→ Open the properties for the Volcano Elevation layer again.

24→ Click on the Label heading to the left of the Template button, and choose Format Labels.

24→ Under Rounding, fill the button for *Number of significant digits*, and set the number to 2. Click OK and OK.

24→ For a final touch, click twice slowly on the ELEVATION heading in the Table of Contents and type (meters).

Volcano Elevation
meters
▴ 1200 - 1500
▴ 1600 - 2000
▲ 2100 - 2500
▲ 2600 - 3400

Source: ArcMap

**TIP:** Evenly rounded class labels give a more professional look to a map.

Next, we will symbolize the volcanoes based on type. However, we don't want to lose the elevation map, so we will create a copy of the layer. One of the advantages of layers is the ability to create different views from the same feature class.

25→ Right-click the Volcano Elevation layer and choose Copy.

25→ Right-click the Geography data frame name and choose Paste Layer(s).

25→ Give the new Volcano Elevation layer (above the old one) a new name, *Volcano Type*. Drag it below the Airports layer, if necessary.

Now let's consider how to symbolize the volcano type field. It contains repeating entries like "Stratovolcano" and "Volcanic field," which represent categorical data. The proper type of map for categorical data is a unique values map.

26→ Open the Volcano Type symbol properties. Click Categories to change the map type and select the Unique values map (Fig. 4.26).

128

Fig. 4.26. Setting properties for unique values maps
Source: ArcMap

26→ Change the Value field to TYPE. Click the Add All Values button. It will not show the categories until you do.

26→ Notice the *<all other values>* entry at the top of the categories list. This offers the option of grouping small categories together. In this case, there are no features in this class, so uncheck the box to turn off display of this category.

Notice the double entry for Shield volcano(es). A data entry error has separated these two groups, which really should go together. We could edit the database, but for now we can work around it.

27→ Click one Shield volcano symbol, then hold down the Shift key and click the other symbol, so that both rows are highlighted.

27→ Right-click one of the highlighted rows and choose Group Values.

27→ Click on the Label row containing the double entry and change it to simply *Shield volcano*. Click below all of the symbols to remove the selection.

Now let's return to symbolizing the volcano type.

28→ Click the Symbol heading to access the Properties for All Symbols menu (as shown in Fig. 4.25). Set the symbol to Triangle 2 and click OK.

28→ Change the color ramp to one with dark, bold colors that will show up well.

28→ Examine the window to make sure it looks like Figure 4.26. Then click OK.

28→ Turn off the Volcano Elevation layer by unchecking it in the Table of Contents.

**TIP:** If you don't like the color assigned to a volcano, right-click its symbol in the Table of Contents to change the color. There is no map type to assign different shapes to each category, but you can do it manually using the Symbol Selector to set the symbol, size, and color for each one individually.

● Caldera
▲ Cinder cones
● Complex volcano

Source: ArcMap

129

### *Creating maps for attributes of lines*

Next we turn our attention to displaying the highways with different symbols for each type. First, we need to check the attribute table to see what fields are available.

> 29→ Right-click the Highways layer and choose Open Attribute Table.
>
> 29→ Examine in particular the HWY_TYPE field.

2. What three values are found in this field? _____ Can you decipher what they mean by examining the other fields for clues?

Again we have categorical data, and the unique values map type is the appropriate choice.

> 29→ Close the table and open the symbol properties for Highways.
>
> 29→ Choose the Categories heading and select the Unique values map type.
>
> 29→ Set the Value field to HWY_TYPE and click Add All Values.
>
> 29→ Click the Count heading above the categories to force it to count the number of features in each category. Notice the lone H value and the 58 unlabeled roads.

It is reasonable to assume that "I" stands for interstates, "S" stands for state highways, and "U" stands for US highways, but what could the "H" represent? With only one, there is a chance it is simply a data entry error. We also have no idea what to label the blank highways. However, we can edit the legend categories to create a clearer legend.

> 30→ Click on the row containing the "H" value and click the Remove button. It removes the category and places the feature in the <all other values> entry.
>
> 30→ Also click on the row with the unlabeled features and Remove it as well.
>
> 30→ Click on the "U" under the Label heading to highlight it. Type in US Highways.
>
> 30→ Click on the "S" and type in State Highways.
>
> 30→ Click on the "I" and type Interstates.
>
> 30→ Change the HWY_TYPE label to Road Class.
>
> 30→ Change the <all other values> label to Unclassified.
>
> 30→ Click on the US Highways symbol to highlight the row. Then click the Up arrow to move it below the Interstates.

Now we have symbols based on highway type, but the thin colored lines do not distinguish the types very well. This classification would benefit from using differences in symbol, thickness, and color to distinguish the different road classes.

> 31→ Double-click the symbol for the Interstates to open the Symbol Selector.
>
> 31→ Choose the Expressway symbol and click OK.
>
> 31→ Double-click the symbol for the US Highways. Select the Major Road symbol and change the color to dark green. Click OK.
>
> 31→ Double-click the symbol for the State Highways. Choose the Major Road symbol and set its color to a light brown.

31→ When finished, the legend should look as shown in Figure 4.27. Click OK.

### Creating maps for attributes for polygons

Let's create maps based on population using the counties layer.

3.  What kind of data does population represent?_____ What kind of map should be used to display it? _____

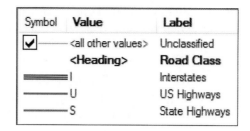

Fig. 4.27. Final legend for highways
Source: ArcMap

32→ Open the symbol properties for the Counties layer.

32→ Choose Quantities: graduated colors for map type and POP2014 for the Value field.

Remember, population data are usually normalized to the area of the aggregation unit.

32→ Choose SQMI as the Normalization field (near the bottom of the list).

32→ Change the color ramp to a monochromatic one, such as light to dark orange.

32→ Click on the Label heading above the classes and choose Format labels. Give them two significant digits and check the box to *Show thousands separators*. Click OK and OK to view the map.

We will create new layers for each new map so that we can keep the previous ones.

33→ Click twice slowly on the counties layer name and rename it *Population Density*.

33→ Right-click on the Population Density layer name and choose Copy.

33→ Right-click on the Geography data frame name and choose Paste Layer(s).

Next we will map the median age of the counties. Whereas population values may vary in part based on the size of the polygon, the median age should be independent of the area being measured. This attribute should not be normalized.

Median Age
MED_AGE

☐ 32 - 36
▨ 37 - 39
▨ 40 - 43
▨ 44 - 48
▨ 49 - 54

Source: ArcMap

34→ Rename the new layer (on top) *Median Age*.

34→ Create a graduated color map based on the MED_AGE field. Change the normalization field to none. Format the labels to have two significant digits.

### Classifying data

We have been using the default Natural Breaks classification so far, but it is not always the best method. Examine the uneven class ranges for the Median Age layer. The legend might make more sense if the age ranges are evenly separated by five years.

35→ Open the symbol properties of the Median Age layer.

35→ Click the Classify button. Take a moment to examine the histogram, and note that these data are fairly uniformly distributed.

35→ Change the Classification Method to Defined Interval. If you get an error, just close the warning and continue. It is a bug. Set the Interval Size to 5.

35→ Click OK and OK. Examine the new ranges in the Table of Contents.

Most readers would find this classification easier to understand than the uneven classes produced by the Jenks method. Next, symbolize the population of the cities.

36→ Add the majcities feature class from the mgisdata\oregondata geodatabase.

36→ Open the symbol properties for the majcities layer and choose the Quantities: graduated symbols map type.

36→ Set the Value field to POP2014 and click the Classify button.

This data set is skewed, with many small cities and one very large one. The default Jenks Natural Breaks is probably the best classification choice. Five classes are a little difficult to distinguish based on size, though, so let's reduce the number to three.

37→ Change the number of classes to three, and click OK.

37→ Click the Template button and set the color to a light one. They will show up better when the symbols overlap.

37→ Format the labels to three significant figures with thousands separators. Click OK in all of the windows to view the map.

Let's look at the activity of the volcanoes now.

38→ Make a copy of the Volcano Elevation layer and paste it into the Geography data frame. Name it *Known Eruptions*.

38→ Open the symbol properties. Keep the Graduated symbols map type but change the Value field to KNOWN_ERUP. Examine the class values.

38→ Click the Classify button.

The -999 values are flags to indicate a missing value, and they are messing up our classification statistics. We could use Exclusion to get rid of unwanted values.

39→ Click the Exclusion button. It will open a window to build a query. We will enter an expression for the values to <u>exclude</u>.

39→ Enter the expression KNOWN_ERUP < 0 in the lower panel. Verify it and click OK. You may not be able to see a difference yet.

This menu has had a bug in it for several versions of software. You have to close and reopen the layer and reset the classification to make it work.

39→ Click OK and OK to close the windows.

39→ Open the Known Eruptions symbol properties again and click the Classify button.

39→ Notice that the negative values are gone from the statistics and histogram.

39→ Set the classification method to Defined Interval and set the Interval Size to 3.

39→ Click OK and OK to finish symbolizing the layer. Turn off the other volcano layers.

The classification method can influence the appearance of a layer and thus the message it sends. An advantage of unclassed maps, such as proportional symbol maps, is that they don't require classification and present an unbiased view of the data.

40→ Copy the Known Eruptions layer and paste it in the Geography data frame.

40→ Open the new layer's symbol properties and set the map type to Proportional Symbols. Set the Value field to KNOWN_ERUP.

40→ Click the Min Value button and change it to a 10-pt. Triangle 2 symbol with a light color. Click OK.

40→ Check the Max Value symbol size to make sure that it is not unreasonably large. With a proportional symbol map, you can only set the lower size—the upper is determined by the value in the attribute field. Click OK.

**SKILL TIP:** Dot density maps are also unclassed maps and can be used for polygons.

### Displaying thematic rasters

Thematic rasters portray map data, such as soil types, roads, elevation, or precipitation. The current data frame is getting busy, so we'll create a new data frame to experiment.

41→ Click the minus sign next to the Geography data frame name to collapse it.

41→ Switch to Layout view. Arrange the data frame in the top half of the page.

41→ Choose Insert > Data Frame from the main menu bar and arrange the new one in the bottom half of the page.

41→ Change the name of the new data frame to Rasters in the Table of Contents.

We will begin with a digital elevation model (DEM), a raster that portrays an elevation surface with a pixel size of one kilometer.

42→ Activate the Rasters data frame, if necessary, by right-clicking it in the Table of Contents and choosing Activate.

42→ Add the gtopo1km raster from the oregondata geodatabase.

42→ Return to Data view. You should see the gtopo1km raster in the view.

**TIP:** The first time that a raster is displayed, the user is asked whether to build **pyramids**, which speeds the display of the raster on subsequent occasions. You can click Yes.

43→ Open the gtopo1km layer properties and click on the General tab. Note that the elevations are in meters.

43→ Click on the Source tab and examine the properties of the raster. Examine the cell size and, scrolling down, the spatial reference (coordinate system).

43→ Click on the Symbology tab and examine its settings.

4. What kind of data type is elevation? _____ What kind of raster display method(s) could be used to display it? _____

When a raster is added to a map, ArcMap chooses an initial display method based on the raster type. In this case, it chose the Stretched method with a grayscale color ramp. Notice the minimum and maximum elevation values, from 1 to 3124 meters. The Stretched method slices these values into 256 bins for use with the 256-grayscale ramp.

44→ Close the Properties window.

44→ Click the Identify tool and click on the raster to view values for a pixel.

| Field | Value |
|---|---|
| Stretched value | 241 |
| Pixel value | 2272 |
| OBJECTID | 2272 |
| COUNT | 14 |

Fig. 4.28. Identifying pixels in the DEM
Source: ArcMap

Figure 4.28 shows the results when a white pixel is clicked. The Pixel value is the elevation, 2272 meters. The Stretched value indicates the bin number, 241. This bin is displayed as nearly pure white. The COUNT field tells how many pixels have this value.

44→ Click on more pixels in the raster, choosing black, white, and gray ones. Compare the pixel values and the stretched values for each one.

44→ Close the Identify window.

45→ Open the properties for gtopo1km and examine the Symbology tab. A Standard Deviation stretch is being used with 2.5 standard deviations.

45→ Move the Properties window aside so that the map can be seen. Change the Stretch type to None and click Apply. Examine the map.

45→ Change the Stretch type back to Standard Deviations. Click Apply.

Fig. 4.29. Elevation terrain color ramp—stretched
Source: ArcMap

46→ Try experimenting with different color ramps, clicking Apply after each.

46→ End by choosing the color ramp shown in Figure 4.29. Click OK.

The Stretched method utilizes 256 equal interval classes. The classified display method gives control of the number of classes and the classification method.

47→ Rename the gtopo1km layer *Elev-Stretched*.

47→ Copy the Elev-Stretched layer and paste it into the Rasters data frame.

47→ Rename the new layer *Elev-Classified* and open its symbol properties.

48→ Change the raster display method to Classified. Find the same color ramp used for Figure 4.29 and select it.

48→ Set the number of classes to 10. Continuous data, because they often have underlying spatial patterns, can use more classes and colors. Click OK.

48→ Use the Identify tool to examine the pixel values and the class values (which go from 0 to 9 rather than 1 to 10). Close the Identify window when finished.

One advantage of using the classified display method is controlling the classification. It defaulted to Jenks Natural Breaks. However, elevations are best displayed in even classes with rounded numbers to appear like a contour map.

49→ Open the Symbology properties for the Elev-Classified layer. Click the Classify button.

49→ Change the classification method to Defined Interval. Set the interval to 200 meters.

49→ Notice the histogram. Most of the elevations fall below 2000, with very few pixels in the higher elevations. Click OK and OK.

Fig. 4.30. Elevation terrain color ramp—classified
Source: ArcMap

Notice how few white pixels show now (Fig. 4.30). The Stretched method and the Jenks classification overemphasized high elevations. The defined interval classification presents a more honest view.

50→ Add the gtoposhd raster from the oregondata geodatabase.

This type of raster is called a hillshade raster. It is derived from a DEM by modeling how the illuminated surface might appear and contains brightness values from 0 to 255. Hillshade rasters provide an intuitive and detailed view of the surface and make a useful base for transparent overlays.

50→ Click on the Elev-Classified layer and drag it above the gtoposhd layer in the Table of Contents. Remember, layers are drawn from bottom to top.

50→ Open the Elev-Classified layer properties.

50→ Click on the Display tab. Set the Transparency to 50% and click OK (Fig. 4.31).

Fig. 4.31. Transparent elevation displayed over a hillshade raster

51→ Use the minus buttons to collapse the legends for the three elevation layers.

51→ Add the slopeclass raster from the oregondata geodatabase. It shows areas of low, medium, and high slope indicated by the values 1, 2, and 3.

5.  What type of data does this raster contain? _____ What type of map should be used to display it?_____

52→ Open the properties for the slopeclass layer and click the Symbology tab.

52→ If the Value Field defaulted to the Class field, change it to the Value field.

52→ Notice that the unique values map was initially chosen. This choice is fine, but for ordinal data, a monochromatic color ramp will show the data better.

52→ Click on the Color Scheme choice and select a monochromatic ramp. Click Apply and experiment until you find one that you like. Then click OK.

53→ Add the landcover raster from the oregondata geodatabase.

53→ Right-click the landcover layer and choose Open Attribute Table.

The Value field contains the numeric values stored in the raster, and the Count field indicates how many cells contain each value. All raster attribute tables contain those two fields. The other fields were added to provide information about what the values mean.

53→ Close the Table window and open the symbol properties for the landcover layer.

53→ Scroll down and examine the colors and values.

A new Colormap method appears in the list, which assigns preselected colors to each raster value. If a raster has a colormap, then this display method will be used initially.

54→ Close the Properties window and open the Identify tool.

54→ Click on different colors of the landcover layer.

The first row shows the color index, the number identifying that color (Fig. 4.32). The second row is the color specification, indicated as mixtures of red, green, and blue. This raster has 150 categories of land cover. The raster contains another field, HABNAME, which groups the cover classes into fewer categories. Let's display the land cover using this field.

54→ Close the Identify window and open the symbol properties for landcover.

54→ Change the display method to Unique Values.

54→ Change the Value Field to HABNAME and click OK.

This raster has a colormap, so its default colors are the best choice. Transparency can be used with any raster or vector layer to enhance the display.

| Field | Value |
| --- | --- |
| Color Index | 77 |
| Color(a,r,g,b) | 255,150,210,150 |
| OBJECTID | 76 |
| Count | 581164 |
| GAPCODE | 5257 |
| ORCODE | 5257 |
| DISPLAY | 5257 Big Sagebrush Sh |
| HABNAME | Sagebrush Shrublands |

Fig. 4.32. Identifying land cover pixels
Source: ArcMap

55→ Collapse the legend for landcover layer.

55→ Click and drag the gtoposhd layer just below the landcover layer in the Table of Contents.

55→ Open the properties for the landcover layer. Click the Display tab and set the Transparency to 50%. Click OK.

### Displaying image rasters

Images represent brightness of land features as measured by a camera or spectrometer. They can have one band or multiple bands. This Landsat image over Crater Lake has eight bands.

56→ Turn off all layers in the Rasters data frame (Press the Ctrl key and click one box.).

56→ Add the LT05_Sep11_2011_P45R30clip.tif image from the mgisdata\Oregon folder. Click Close to ignore the coordinate system warning.

56→ Rename the image layer Landsat.

56→ Right-click the Landsat layer in the Table of Contents and choose Zoom to Layer.

56→ Open the Landsat layer properties and click the Symbology tab.

Notice the two ways to display the image. Stretched is appropriate for single-band images or to display one band of a multiband image at a time.

57→ Click Stretched and select Band 1 (blue light) as the band to display.

57→ Keep the default Standard Deviation Stretch and leave the other settings.

57→ Click Apply and move the Properties window aside to examine the image.

Each band collects a different wavelength of light from the surface. Different materials reflect light more or less in each band. Each band highlights different aspects of the terrain.

58→ Change the display band to Band 4 (near-infrared) and click Apply.

58→ Examine each of the other bands, noticing how different each looks.

An **RGB composite** displays three bands at a time, one in each RGB color gun of the monitor. You can experiment with assigning different bands to each color.

59→ Open the Symbol properties, if necessary, and choose RGB Composite.

59→ Choose Band 3 for the red channel, Band 2 for the green, and Band 1 for the blue. The Landsat bands will match the color guns to create a true-color image.

59→ Click Apply and examine. Notice the smoke plumes from the burning fires.

59→ Experiment with different stretches, clicking Apply after each one.

60→ Choose Band 4 for the red channel, Band 3 for the green, and Band 1 for the blue. Click OK.

Band 4 shows near-infrared light, which is strongly reflected by vegetation. This combination of bands is known as a false-color image, and it is helpful in assessing vegetation density and health. You can experiment with a few other band combinations if you have time.

ArcGIS Online has some Landsat image services which give you a subset or all of the image bands. If you have a fast Internet connection, try the following.

61→ Turn off the Landsat layer.

61→ Click the Add Data button and choose Add Data from ArcGIS Online.

61→ Search for *Landsat GLS* in the search box.

61→ Find the Landsat GLS Multispectral (or similar) image service and add it.

62→ Zoom further out to see more of the image.

62→ Continue experimenting with displaying this image if you have time. Zoom to other areas of the country to examine different ecosystems.

This is the end of the tutorial.

→ Close ArcMap. You can save the map document if you wish.

## Exercises

Use ArcMap and the data in mgisdata\Oregon\oregondata to create a map document showing various aspects of Oregon, as described below. Use what you have learned to make each map as aesthetic and legible as you can.

1. Start with a new, blank map using the Letter (ANSI A) Landscape template. Create four data frames, each of which takes up a quarter of the page. Name them Volcanic Hazards, Farms, Housing, and Physiography.

**In the Volcanic Hazards data frame:**

2. Create a map showing the population density of the counties. Also show the hospitals, marked with blue crosses.

3. Create a proportional symbol map of the volcanoes based on the KNOWN_ERUP field.

**In the Farms data frame:**

4. Create a map of counties showing the density of farms. Label the county names.

5. Create a map showing the transportation routes, symbolized by type (road type, rail, and air).

**In the Housing data frame:**

6. Create a map showing the vacancy rate of the counties. What would you normalize by in this case? Label the county names.

7. Show the major cities feature class (majcities) symbolized by population.

**In the Physiography data frame:**

8. Add the state boundary and the county boundaries, and then add the World Imagery service from ArcGIS Online. Use a hollow symbol and a contrasting outline color for the state and county boundaries so that you can see them well against the imagery.

**For all the frames:**

9. Examine the legends in the Table of Contents. Make sure you used good classification schemes, rounded label values, and appropriate significant digits or decimal places. Be sure that each layer has an informative capitalized name.

10. Capture a screen shot of each data frame in Data View, including the Table of Contents showing the corresponding legends for each frame. Also capture a screen shot of the entire layout with all four frames.

## Challenge Problem: Big and Little

Start with the map created in the Volcanic Hazards data frame and turn off the hospitals. Consult the Skills Reference section to learn how to create label classes to label features differently based on an attribute. Label the active volcanoes (KNOWN_ERUP > 0) with large red labels and the inactive volcanoes (KNOWN_ERUP = 0) with smaller green labels.

# Chapter 5. Presenting GIS Data

## Objectives

▶ Learning basic principles of map design

▶ Creating text and labels on maps

▶ Creating map layouts and printed maps

## Mastering the Concepts

### GIS Concepts

GIS analysis usually results in sharing information with others in the form of maps. Every map has a purpose established by the cartographer, and the success of a map is judged by how well the map fulfills that purpose. This chapter introduces some basic design principles for maps, but it cannot cover the full range of concepts included in the art and science of cartography. Taking a cartography class and reading map or graphic design texts are good ways to improve your skills.

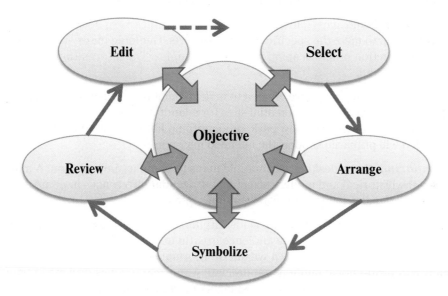

Fig. 5.1. The map design process—every step must come back to the map objective.

Map design can be envisioned as a six-step process (Fig. 5.1), but it is not linear. It begins by developing a vision of the map's objective, and this vision must inform each step along the way: selecting the data, arranging the page, symbolizing the information, reviewing the map, and editing to improve it. Each step must constantly refer back to the main objective to ensure that the final map supports the objective in every possible way.

To illustrate this process, the map used in the tutorial will be used as an example of map design. Consult these orange boxes to follow through the design considerations.

## *Determine the map objective*

Developing a clearly defined objective is the most important design step for a map. It determines the form and content of the map, and it provides the context to review a map and ask whether it works. Four issues should be considered: the purpose of the map, the audience for which it is intended, the medium in which it will be produced, and the conditions of expected use.

### Purpose

Maps serve different functions, and the function strongly influences all steps of the design. So the first question to answer is "What is the purpose?" A complete list of every map purpose is impossible, but a few of the common ones include:

- ▶ *Locating*. Some maps show the location of some place or places of interest, such as showing the location of a store or a choice of hotels on a street map, perhaps with directions to get there. A walking or biking tour map would also fall into this category. Such maps need little more than the point(s) of interest and a street matrix with labels.

- ▶ *Navigating*. In a navigation map, the location of interest is not already known, so the maps tend to be richer in data to allow the user to identify and travel to different types of locations. A highway atlas, for example, shows many cities, connecting roads, and other features of interest such as parks, rivers, or mountain ranges.

- ▶ *Compiling*. Some maps gather information, sometimes very detailed and complex information, to meet the specific needs of experts. Geologic maps, utility network maps, and construction or landscaping plans are examples. The information will be specific to the task at hand.

- ▶ *Convincing*. Many maps are created to make a point or to illustrate the conclusions of a study, such as showing the potential sites for a landfill, telling the story of historical battles, making an argument for a new park, or identifying hot spots of crime. The information included will primarily support the intended message.

- ▶ *Comparing*. Some designs use multiple maps to compare information. A map might show the historical and modern street maps, decadal differences in population or climate, or the spatial distributions of different types of cancers.

Part of determining the purpose of a map is identifying its *target audience*. The characteristics and anticipated knowledge level of the audience greatly influence the choice and presentation of the data. A traditional geologic map is a wealth of information for an expert but means little to a general citizen. A professional might not appreciate whimsical touches designed to appeal to elementary scholars. A map produced for residents of a retirement community could have larger text and symbols than usual to accommodate vision impairments common to that age group. The best maps take into account the likes, dislikes, knowledge, and abilities of the target audience and produce the best possible experience for them.

### Medium

Maps are designed with a specific medium in mind; traditionally paper or vellum, for which the size and orientation of the finished product are a main concern. Maps should be designed at the size and scale for which they are intended for production; scaling generally introduces unwanted artifacts or affects the artistic balance. The properties of the available printers—the size of paper it can handle, the type of ink used, and the printing resolution—will also influence the final product. If many copies will be produced, the creator must factor in the costs of printing in choosing the medium, the paper quality, and whether color should be used.

Today, cartographers have exciting new media to design for: electronic files like PDF or presentation slides, web sites, smartphones, and tablets. Electronic maps have different requirements than printed maps and present unique design challenges. The author of a web map has little control over the size of the product; it could be shown on a conference screen many feet across or reduced to a few inches of screen on a mobile device. The maps compete for attention in a frenzied cacophony of content, and most users will move on if they don't find what they need in less than a minute. Web maps rarely have the luxury of allowing a leisurely unfolding of complex content. They must be adaptable, obvious, and attractive.

The magic of web maps, of course, is that they can be interactive. Users can zoom, turn layers on and off, measure objects, and even perform analysis functions like getting directions or viewing an elevation profile. However, simplicity is still important. It is tempting to turn a web map into a baby GIS with dozens of buttons, but it is better to design a web map to show or do one thing really well, rather than to try to pack it with data and functions.

The **conditions of use** can be considered part of the medium as well. Posters viewed from a distance require larger fonts and symbols than maps intended to be spread on a table. Maps that would be used under dim lighting conditions should avoid complex backgrounds like imagery and use larger symbols and better contrast than a map viewed under normal room light. Maps designed for outdoor data collection on a smartphone must use large symbols and intense colors to contend with the tiny screen and bright sunlight. Being aware of special physical conditions enables the cartographer to design the map to function properly in its environment.

The purpose of the tutorial map is to portray the potential dangers of volcanoes to populations living in the Pacific Northwest. Its audience is the average adult citizen, and it is intended to be printed in color on a letter-size page.

### Select the data

Once the purpose of the map is clear, the cartographer selects the data it will contain. Think of the map as telling a story. A good story has main characters that move the action and minor characters that must not outshine the lead players. A mapmaker must know which data play the starring role, and which layers must support without intruding.

The key term in this process is **selection**. Most of us have databases full of information that could be put on maps; more often the trick is to remove what isn't needed. Simplicity is one of the guiding tenets of cartography. Every piece of data should work to support the map objective.

When selecting data for a map, one should also be aware of potential privacy issues or unintended economic consequences of presenting certain types of information. For example, laying a fault map on top of the city parcels layer might have consequences for the property values of parcels near faults. The public may not be aware that most faults are inactive and do not pose a threat. Maps of fossil sites or archeological digs must ensure that the exact locations of fossils or artifacts remain out of the public domain to protect them from poaching. One should also be cautious about putting private information on public maps, such as the names of parcel owners.

The scale of the map is an important consideration, and the data layers selected should be appropriate. Coarse-scale data should not be placed on large-scale maps. Occasionally large-scale data may be needed on a small-scale map, although often some type of simplification must be applied to individual layers. This process is called **cartographic generalization**. Eight different techniques are commonly employed (Fig. 5.2). With the exception of refinement, which can be performed using queries and selection layers, most of techniques will involve creating a new simplified feature class to be used in place of the original.

**TIP:** ArcToolbox contains many of these generalization tools in the Cartography Tools toolset.

To present information from a larger scale layer at a smaller scale, one might do the following:

▶ *Refine* a feature class by omitting certain features, especially smaller or more detailed ones, for example, removing higher-order streams and keeping the main trunk rivers.

▶ *Simplify* lines or polygons by removing vertices and/or *smoothing* to give them less detailed shapes.

▶ *Aggregate* many smaller features into larger features, such as turning many close-lying small wetland polygons into a generalized marsh area.

▶ *Collapse* features into more simple forms, such as turning detailed building footprints into simple rectangles.

▶ *Classify* a more detailed set of attributes into a simpler set, such as reducing geologic formations to a classification of "igneous, metamorphic, or sedimentary."

Some types of changes involve "falsifying" the underlying spatial data and should be used sparingly. A note on the map should indicate that such techniques have impacted the locations and/or shapes of certain layers.

▶ *Typify* features using representative rather than actual features, for example, replacing the point markers for actual homes with a smaller number of "fake" points representing scattered houses.

▶ *Displace* features slightly so they do not run over each other, for example, moving a road to avoid the appearance of an intersection with another road.

▶ *Exaggerate* the importance of features to give them greater prominence, such as enlarging the width of a water body to prevent it appearing as a line.

### Map projections

Selecting the map projection is another important task that depends on the objective of the map; for example, a map that requires distances to be measured must use a projection that preserves distance. Guidelines for choosing an appropriate projection were presented in Chapter 3; it is worth reiterating here that a geographic coordinate system is almost never appropriate for a map and should be avoided.

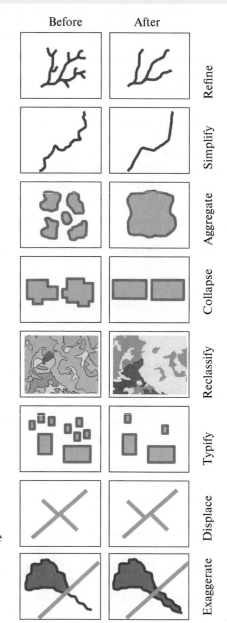

Before     After

Refine

Simplify

Aggregate

Collapse

Reclassify

Typify

Displace

Exaggerate

Fig. 5.2. Cartographic generalizations
Source: ArcMap

Sometimes the projection can influence the map story in unintended ways. The cylindrical Mercator projection has been popular for world maps and is often used in web maps (Fig. 5.3a), but critics have charged that this projection enlarges countries in Eurasia and North America, sending a subliminal message that the northern hemisphere is more important than developing countries near the equator. Attempts have been made to popularize equal-area world projections that don't exaggerate the northern hemisphere (Fig 5.3b).

The projection may also influence the elements placed in a map. A conic projection does not preserve direction, and a single arrow cannot represent north everywhere on the map. A map that distorts distance should not have a single scale bar. In these maps, a **graticule grid** is employed, which shows latitude and longitude markings (Fig. 5.4a). Two other types of grids are often used on maps. A **measurement grid** shows the map units present in the coordinate system (Fig. 5.4b). A **reference grid** shows letters and numbers delineating each grid square (Fig. 5.4c). A reference grid is commonly found in a telephone directory in conjunction with a street index, allowing streets to be located within squares of the map.

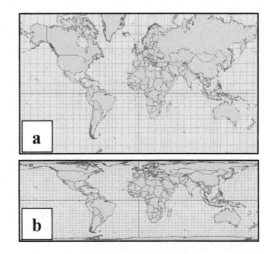

Fig. 5.3. (a) A Mercator and (b) an equal-area cylindrical projection
Source: ArcMap

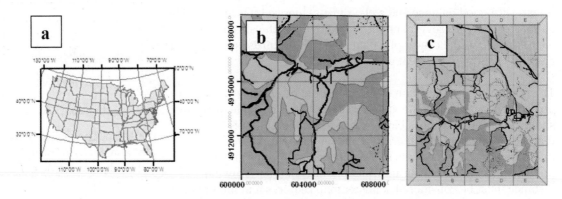

Fig. 5.4. Data frame grids: (a) graticule grid; (b) UTM measurement grid; (c) reference grid
Source: ArcMap

The Volcanic Hazards map will be focused on Washington and Oregon and have a scale of about 1:4 million in an equidistant projection. Volcanoes are the main point; counties and capitals will portray populations (census tracts would be too detailed at this scale). Highways are needed to show escape routes. Rivers will be needed to show potential lahars (volcanic mud flows).

### Arrange the map elements

The next step is to decide how the map will be arranged on the page. At this point it is a good idea to walk away from the computer and use a paper and pencil to sketch a rough mock-up of the design and decide how the parts of the map should be arranged, instead of succumbing to what is currently shown on the screen.

Return to the analogy of the map as a story. A story has main characters, supporting ones, and a defined path through the events. In a map, the direction of the story and the relative importance of the elements is achieved by establishing a **visual hierarchy**, which refers to the order in which the reader perceives the different objects in a design. A good map establishes a clear pathway through the map objects by using carefully arranged elements and appropriate symbols.

## Tools, not rules

Map layouts traditionally contain certain **map elements** in addition to the map itself. A location map is used to show where the main map extends, especially when the main map is not easily recognizable to the readers. A legend explains the symbols used. A scale bar indicates distances and a north arrow shows direction. A **graticule grid** shows latitude-longitude lines or tics. A **neatline** can be used to enclose the entire map or groups of elements within it. Titles and text explain what the map is about and cite the data used to create it. Images or graphic shapes can be added to support the story.

In grade school you might have learned that all maps must have a title, legend, scale bar, and north arrow, but that is not necessarily true. Some maps are so simple and well symbolized that a legend is not needed. Scale bars and north arrows may not be appropriate for some projections, and they may be superfluous even when allowed. A map portraying teacher salaries in the United States, for example, could omit both items; most viewers already know where north is, and distances have no bearing on the message of the map. It is better to think of this list of elements as tools, not rules, and use them only if they support the map objective.

## Balancing the elements

A map with good balance and a clear visual hierarchy is achieved by paying attention to several important guidelines: the size of the elements, the visual center of a page, the rule of thirds, alignment of elements, and the treatment of negative space.

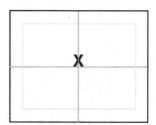

Fig. 5.5. The visual center of a map (X) is 5% higher than the geometric center. Source: ArcMap

The relative *size* of the map elements should be considered; larger objects will naturally command more attention. The scale of the map should be as large as possible to enhance the legibility of the map features. Scale bars, north arrows, and legends should be unobtrusive. Use large fonts to draw attention to titles and key points, and smaller text for informational details. The viewing distance must be considered; a poster map requires a minimum of 18-pt font to be comfortably read from several feet away, and will generally need larger symbols as well.

The **visual center** of a page is about 5% higher than the geometric center in the exact middle (Fig. 5.5). Centering the map on the geometric center makes the map appear unbalanced and heavy.

Fig. 5.6. The rule of thirds suggests putting important points at these intersections. Source: ArcMap

The **rule of thirds** is a graphic arts principle used to enliven a design; although it is primarily used as a guide to composing photographs, it can also be helpful in map design. Imagine the page cut into horizontal and vertical thirds (Fig 5.6). Important elements placed on these intersections will have greater impact than if placed in the center of the map.

***Alignment*** of the map elements is also important. This goal can be achieved in two ways. First, arrange the map elements into columns or rows, rather than letting them wander all over the page. Second, be sure that the edges of elements align with each other.

A rough sketch or mock-up is helpful in evaluating the balance and visual hierarchy of a design. In Figure 5.7, the gray boxes represent placeholders for the main map elements. Notice that two columns have been envisioned, one for the main map on the left (which is centered on the left line of thirds), and a second column for the other elements. The exterior edges of all three elements are exactly aligned at the top, left, and right.

Fig. 5.7. Use columns and align edges.
Source: ArcMap

The visual center and rule of thirds influence the view path, because the eye is typically drawn to these areas first, other things being equal. In addition, the eye tends to travel from left to right because it is used to reading that way (although this would be reversed in countries using right-to-left languages). A good map design uses these influences rather than working against them. Which way does the eye move in Figure 5.7? Probably you saw the large square on the left first, then moved to the upper right and down. Placing the map, title, and legend in these three areas, respectively, would create a logical flow through the map.

The treatment of **negative space**, or the blank areas on the page, is more important than it might first appear. Although large irregular blank areas make a map look unbalanced, too little space can make a design appear crowded. Negative space can actually be used as another design element, which can separate or group elements more subtly and naturally than neatlines. Maps often look better if you resist the temptation to put a box around every element. In Figure 5.7, the negative space serves to separate the three main map elements. Page margins are an important type of negative space.

Maps often show irregularly shaped geographic regions that require creativity in achieving balance. Imagine placing California in the layout shown in Figure 5.7. The map area is relatively small, limiting the scale and the legibility of the features. Moreover, the state's shape creates irregular negative space inside the map frame. Figure 5.8 shows three alternative treatments based on a portrait layout that maximizes the scale. The non-map elements are placed inside the data frame (a space solution that may not occur to beginning map designers, but which can be highly effective). Give each design a swift glance. Which one feels the most balanced?

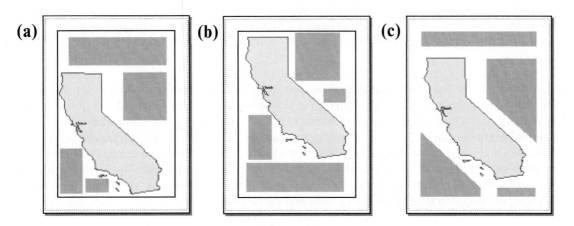

Fig. 5.8. Managing visual balance and negative space in a layout
Source: ArcMap

145

Figure 5.8a ignored the visual center and placed Californa near the bottom, creating a heavy sensation exacerbated by two more blocks added to fill the negative space. The visual center is actually blank in this layout. The upper elements draw the eye first and then force it left to the map, ignoring the left-to-right preference. Some parts of the map appear crowded, others too empty. The edges are not aligned, and the neatline crowds the map.

In Figure 5.8b, the state is now too high, although the negative space is more evenly balanced. The edges are very irregular now and the elements appear haphazardly arranged. They crowd the state in some spots and leave open areas in others. The eye is drawn to the map first and travels left-to-right across it, which is fine, but then is unsure what to view next.

Figure 5.8c is the best arrangement so far. In this case, the placeholders for the elements have been reshaped to echo the state's shape, which leaves uninterrupted negative space around California and helps draw attention to it. The north and south sections of the state fall roughly in the target areas advocated by the rule of thirds and encourage the eye to scan the state left-to-right. The exterior edges of all elements fall on the boundary of the data frame, which subtly encloses the map and allows the neatline to be removed.

In the tutorial map, Washington and Oregon form a roughly rectangular shape with little negative space, so the supporting elements can go outside the data frame. The map can be placed on the left to draw the eye first. The elements will include a location map, a legend, and some titles and supporting text. An equidistant projection and a scale bar will be used in case the reader wishes to estimate distances from the volcanoes to the vulnerable populations.

### Symbolize the data

Symbolizing the features in the layout has enormous impact on the final result. Chapter 4 introduced ways in which symbols may be manipulated by changing their shape, size, thickness, line type, font, and color. It also discussed the ways that color can be used most effectively for different data types and introduced potential issues, such as color blindness, conventions, and connotations, that should influence the choice of colors. All of these principles remain important, but in a layout we must also consider the interaction of symbols between multiple layers. Symbolization is a powerful tool for manipulating the visual hierarchy in a layout through color, contrast, foreground/background relationships, size, symbol type, patterns, and branding.

*Color* is one of the strongest influences in a map. Dark, saturated colors attract attention and help indicate the focus of a map, and so should be used sparingly. Pastel colors will tend to fade away and are perfect for the supporting layers. Color can also be used to group features or map elements—objects that are shown in similar colors will be perceived to belong with each other.

Another important aspect of color is **contrast**, or the juxtaposition of colors with very different hue, value, or saturation. Consider Figure 5.9a: How many squares do you see? Did you neglect the fifth square? The dark squares seem to stand out above the gray one, commanding the **foreground**, while the gray fades away into the **background**. This effect is achieved by the saturation and value of the colors, and even more by the contrast between them. The blue stands out more than the orange, and the outlines heighten the contrast still more, especially the orange outline that contrasts with the interior orange as well as the gray exterior. The visual hierarchy encourages the eye to start in the lower right and move clockwise.

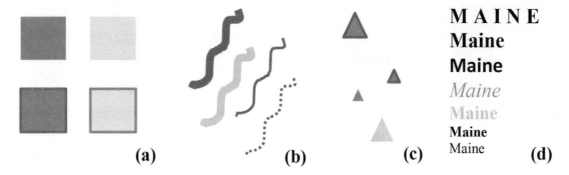

Fig. 5.9. Tools for establishing visual hierarchy: (a) color and contrast; (b) color, thickness, and line style; (c) color, size, and contrast; (d) size, spacing, font, color, contrast, and style
Source: ArcMap

The effects of color and contrast are repeated in Figure 5.9b, with the dark blue lines appearing more prominent than the gray one. *Line thickness* and *symbol style* also affect the hierarchy, with thicker and more continuous lines dominating thinner or discontinuous ones. (Which is more prominent, the thick gray line or the thin blue one?) In Figure 5.9c, the additional attribute of *size* is varied; the larger something is, the more attention it commands. Text symbols have font and style, in addition to size and color, to rank them in the hierarchy (Fig. 5.9d).

Humans do not simply perceive objects on a page; the brain works actively to organize the objects into *patterns* and interpret what it sees. Consider the four different panels in Figure 5.9—what makes them into different panels? First, the negative space between the gray squares separates them. Second, the similarity of the symbols within each panel (rectangles, lines, points, text) encourages the brain to group them together. The separation is balanced, however, by the use of repeated colors to tie the entire figure together. What is the effect of having the single purple triangle? Does it stand out more than if it were orange like the square?

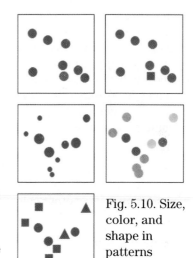

Look at each box in Figure 5.10 and pay attention to how your brain tries to group the markers. Which variation (size, color, or shape) makes the points easiest to group? Which sets are easier to separate? Which ones emphasize that the points belong to one map layer? When evaluating symbols for layers, take note of how your mind attempts to group or separate objects, and use these effects to help support the visual hierarchy.

Fig. 5.10. Size, color, and shape in patterns
Source: ArcMap

Symbols in a map should bring the important elements to the foreground and relegate less important features to the background. In Figure 5.11a, the world map has no clear hierarchy; the ocean and graticules grab as much attention as the country populations. Three slight adjustments make the hierarchy clear: (1) the oceans are displayed in light blue to displace them to the background; (2) the graticules are colored light gray to decrease the contrast; and (3) the graticules are placed behind the countries to further enforce them as background elements (Fig. 11b).

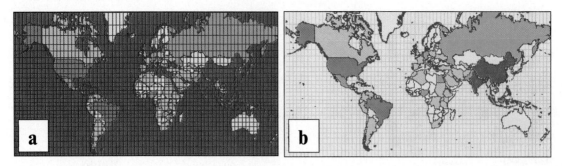

Fig. 5.11. Establishing a clear visual hierarchy by manipulating color value and contrast
Source: ArcMap

Of course, establishing the hierarchy is more difficult when many layers (and text) are involved. When in doubt how to treat a layer, one should always return to the objective of the map and ask what role the layer plays—should it be emphasized, downplayed, or even removed, to better achieve the map objective? Examine the different ArcGIS Online basemaps and study how they achieve visual hierarchy. The *Terrain With Labels* basemap is an especially good example of differentiating between multiple background layers without swamping user data placed in the foreground (Fig. 5.12).

Another factor in choosing symbols for the map is the desire to achieve **branding**. An organization may wish to impart a certain look or feel to its maps across all of its publications through the consistent use of certain colors, fonts, marker symbols, line types, or images. It is not difficult to select symbols to honor these choices. For an example of branding, examine the *National Geographic* basemap in ArcGIS Online. If you were ever a *National Geographic* reader, you will find a familiar face in this map.

Fig. 5.12. A basemap is the epitome of designing for the background.
Source: ArcGIS Online basemaps

In the tutorial map, the volcanoes and population information are most important, followed by the transportation. The county populations will cover the most map area and will be symbolized in shades of orange; other colors will be chosen to develop the appropriate contrast.

### *Review the draft*

The review step should not be skimped, any more than you would neglect to proofread a research paper before turning it in. However, we are looking for more than typographical or grammatical errors. We must return once again to the map objective and evaluate how well the map succeeds in telling its story.

#### Author's first review

Start with your own review, applying the principles in this chapter. Is it easy to determine what the map is about? Is the map balanced? Is it clear what is foreground and what is background? Does the eye go first to the more important elements, or does it dart all over, unsure what to focus on? Do the symbols clearly portray the message? Are the colors and arrangement pleasant? Hopefully you have been applying these principles all through the design, but now is the time to make sure they all work together.

Don't neglect the details. Make sure the margins are sufficient for the printer being used. Check for errors in spelling and grammar. Evaluate the legend for common mistakes, such as using file names rather than real words, using inappropriate significant figures, or allowing the neatline to crowd the text (Fig. 5.13). Many cartographers feel that the title "Legend" is superfluous and do not care to see it on maps.

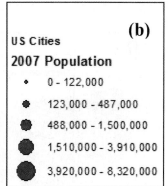

Fig. 5.13. (a) A poor legend; (b) a better legend
Source: ArcMap

Be aware of significant figures in all publications. Not only do excessive figures make the classes hard to interpret (Fig. 5.13), but also they announce that someone does not understand his or her data. Population, for example, is always an estimate, and it is ridiculous to state it down to the last place (or worse, with decimals). Classifying already reduces data to a small set of bins; variations in the hundreds and thousands make little difference.

Do not assume that every layer needs to go in the legend. If a set of features is familiar to all map viewers (such as state boundaries on a United States map) or is so apparent from the title or purpose of the map (labeled paths on a city walking tour map), consider leaving them out of the legend. Consider those basemaps again—they generally have no legends, yet most viewers have no difficulty recognizing what is being shown (Fig. 5.12).

Let's see how an initial map review is conducted by the author. The map on the left in Figure 5.14a shows a first draft based mostly on software defaults. (Hopefully a first draft would be better than this one, but unfortunately the author has seen similar student maps turned in for an assignment.) Assume that the map objective is to educate the viewer about variations in household size in the United States, with emphasis on areas with larger-than-usual households. Cover the map on the right and see if you can identify all the ways that the draft map might be improved. Then read on.

Evaluate the balance first. The map is placed in the visual center, but everything else is poorly arranged. The elements are crowded on the bottom and leave a large negative space above, from which Alaska dangles down, neither in nor out of the map. The cartographer has tried to fill the space with a box and shadow around the title, but that is a gimmicky solution. The legend is large compared to the map and contains too many significant figures. The neatlines around the legend and text boxes distract from the map. The scale and north arrow are too large, the north arrow is inappropriate in a conic projection, and the scale bar is immaterial to the map objective. The element edges are not aligned. The visual path is not clear; the eye is drawn first to the map, which is fine, but then it tends to bounce around between the other elements.

The symbolization in Figure 5.14a is also ineffective. The map uses a color scheme that resembles natural geographic features rather than human ones and, because it is based on hue rather than saturation or value, does not clearly indicate quantity. The red hues draw attention to the low values in the visual hierarchy—the opposite of the map objective. The rivers and lakes are irrelevant to the topic and distract from the message.

The revision on the right seems like a different map (Fig. 14b). A landscape orientation better fits the shape of the United States and allows a larger scale for the map to improve legibility. The

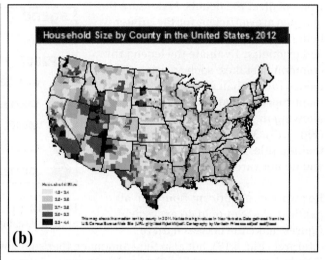

(b)

(a)

Fig. 5.14. Improving a draft layout; (a) original draft; (b) revised layout
Source: ArcMap

color scheme shows quantities and draws attention to the high values using dark, bold colors. The lakes and rivers have been removed. The legend has been reduced to the single layer that needs explaining, the significant figures have been corrected, and the redundant "Legend" heading was deleted. The distracting neatlines around the legend and text are gone, and the text has been anchored across the lower neatline. The new colors and expanded width help the title appear as part of the map, rather than floating aimlessly above it. The visual hierarchy is simple and clear.

### External review

Equally important is to have others look at the map to ensure that the intended message of the author is clear to the readers. These reviewers need not be experts; in fact, if one is designing the map for a general audience, then those are the type of people who should review it.

The first impression made by a layout is extremely important. Maps are not just rational, logical documents like spreadsheets—they can have emotional impacts, and the author must know what kinds of emotions the map elicits from the target audience. Start by giving the reviewers about 3–4 seconds to look at the map; then take it away and ask: What were the first things that you noticed? Was your overall impression positive or negative? What made you feel that way?

Let them look at the map again for a few more minutes. Then ask them what they think the map is about, and what it is telling them. Ask if any part of the map is confusing or too complicated. This feedback will help evaluate whether the map accomplishes its objective. At this point one can also get more detailed feedback on a design if the reviewers are knowledgeable users, qualified in cartography or in the field the map represents.

As you digest the review information, keep in mind that design principles sometimes conflict with each other in a particular situation, and a choice must be made based on which guideline will best achieve the map objective. On occasion, you may want to break a guideline on purpose to achieve a particular effect, as some novelists may sometimes use sentence fragments—just make sure you are doing it for a good reason and not out of ignorance or laziness. One should also remember that a map is not correct or incorrect, any more than a novel or a painting can be.

However, some maps are better than others. As you review your maps and consider alternative strategies to achieve the map objective, keep asking yourself, "Would this make the map better?"

Observing other people's maps is a great way to become a better cartographer. Spend time training your reviewing eye. When you see a map anywhere, give it the first impression test by looking at it for a few seconds. Notice what stood out and whether your first impression was positive or negative, and why you felt that way. Practice analyzing the visual pathway through a map and how it is achieved (or ignored). Then evaluate the map's balance and symbology. What works well? What doesn't? How could it be improved? When you see a great technique, incorporate it into your own maps.

We will review the Volcanic Hazards map as part of the tutorial.

### Edit and improve

Based on the review step, you will likely find some things to change. Sometimes they will be minor edits, at other times you may need to go back to one or more of the selecting, arranging, and symbolizing steps for a more drastic redesign. Regardless, there are a few principles to keep in mind when editing.

***Concentrate the message.*** Make sure that every part of the design is working toward the map objective in the clearest way possible. Include the minimum number of layers and elements needed to accomplish the map purpose. You would be surprised at how often removing a layer or reclassifying a complex symbol scheme will improve a map.

***Simplify, simplify.*** Many maps are needlessly complicated, with too many different kinds of text, symbols, or fonts. Just because the software lets you give neatlines, fill colors, and shadows to objects doesn't mean that they are always a good idea—they should serve a purpose, such as a neutral background to separate a scale bar from the map behind it. Remember that negative space is often a great way to separate or group map elements. Stay away from complicated cartoon symbols—they may be popular for web maps but they quickly start to interfere with the map interpretation. The less junk is placed on the map, the clearer the intended message will be.

Source: ArcMap

***Experiment.*** If a design just doesn't seem to be working well, experiment. Save a copy of the map under a new name and try something daring, like changing the page orientation and completely rearranging the elements. Rethink, resymbolize. That's the beauty of computer cartography—it is easy to try something different with no risk.

***Test it.*** Before pronouncing the map complete, test it under the conditions of actual use. For paper maps, print a test copy on the same printer to be used for production. Colors on a screen may differ significantly from the printed colors, and you may need to adjust to get the desired result. Margins may shift, causing the map to appear uncentered on the page. If it is a poster, print it full size and hang it on the wall, then step back and make sure that the arrangement and fonts work well from a viewing distance of two meters. If it is a web map, try it on several computers, on a tablet, and on a smartphone to ensure that it works for all media on which it might be viewed.

## About ArcGIS

### Maps in ArcGIS

ArcGIS provides an environment for cartography that spans the range from creating a simple map for a report to preparing sets of maps for commercial printing. This chapter covers the basic steps to create a printer-ready map, but the reader should be aware that many advanced cartographic tools are available, as well as tools for developing graphs and database report printouts; although they are not covered here, the Help files have more information about them.

In ArcGIS, the draft of a map intended for printing or distribution is called a **layout**. The layout mode in ArcMap facilitates map design by incorporating data frames and map elements, such as legends, north arrows, scale bars, text, and more. Figure 5.15 shows a map from a series of 1:24,000-scale maps produced to show properties of aquifers for 7.5-minute quadrangles in the Black Hills of South Dakota. The maps were intended to mimic the design of a standard USGS topographic quadrangle. They can be printed on a large-format plotter if one is available but are being distributed primarily as PDF-format files. The development of the data and the map layout took place in ArcGIS Desktop. (The layout has been drastically reduced in the figure, but originals of this and similar maps can be downloaded from http://www.sdsmt.edu/aquifers.)

Fig. 5.15. A map layout
Source: ArcGIS

Each map document can contain one map layout. Map documents may also be saved as templates, which can be used to re-create a design again and again or to give an organization's maps a common format and look. It is also possible to create map books, in which the sample map design is applied to multiple tiles covering a larger area, with the map extents, legends, titles, and other aspects generated automatically for each map using a feature called Data Driven Pages.

### Assigning map scales

GIS spatial data acquire a scale as soon as they are drawn on a computer screen, printed on paper, or saved in a PDF. The first step of creating any map is generally to define the size of the layout page, which will be influenced by the page-size capabilities of the available printers. Once the page size is set, the data frames are arranged on the layout page, and the user must decide how much of the map data will be shown within the data frame. The **map extent** is defined as the range of $x$-$y$ values displayed in the data frame. Zooming in reduces the map extent; zooming out enlarges it. The printed scale is determined by the map extent and the size of the data frame on the paper. Assigning the map scale of a data frame can follow one of three strategies.

**Automatic scaling** is used most frequently (Fig. 5.16a). The user sets the size and position of the data frame on the layout page and zooms in or out of the data until the desired map extent is achieved. The user does not try to impose a specific scale but uses his or her eye and aesthetic sense to balance the map frame with the other map elements.

**Fixed scale** is employed when the printed map must have a specific scale, such as 1:24,000. In this mode, the features always appear at the same size, and changing the data frame size changes the map extent (and may crop the map if the user is not careful) (Fig. 5.16b). Zooming in or out of the map data would change the scale and is not permitted, but panning is.

**Fixed extent** adopts the current map extent and prevents it from changing. The data frame can be resized, which changes the scale of the map but not its extent (Fig. 5.16c). The aspect ratio of the data frame is also fixed. The Zoom and Pan tools are unavailable in Fixed Extent mode.

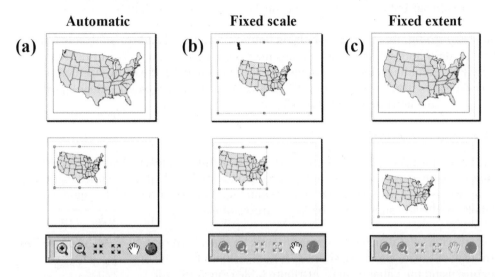

Fig. 5.16. Scaling options for data frames, shown in Layout mode. The effects of resizing the data frame differ in each case. The Zoom/Pan tools are only active in Automatic mode.
Source: ArcGIS

### Setting up scale bars

Some maps should contain either a scale bar or a statement of the scale, such as "This map is shown at a scale of 1:62,500." ArcMap can automatically construct a suitable scale and resize it when the map scale changes. The user has to set several options, which requires some knowledge of terminology.

The **divisions** are the sections into which the scale is divided. The scale bar in Figure 5.17 has four divisions. The first division may be subdivided, yielding what are called **subdivisions**. This scale bar also has four subdivisions. The **division units** are the units of measurement used to divide the scale bar (miles in this case). The **division value** is the length of each division, expressed in division units. This scale bar has a division value of 0.5 mile. Divisions should be expressed in rounded, easily divisible numbers to make them easy to use for estimating distances. The division units, the map scale, the division size, and the length of the scale bar are interrelated. Zooming changes the map scale and requires the scale bar to change also. Three options control how the scale bar responds to changes in map scale.

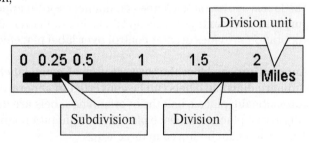

Fig. 5.17. Terminology of a scale bar
Source: ArcGIS

*Adjust width* keeps the division size and number of divisions constant. If the map scale changes, the entire scale bar gets larger or smaller. This method keeps nicely rounded numbers for the divisions but may result in an unacceptably small- or large-scale bar if the scale changes dramatically.

*Adjust division value* keeps the scale bar roughly the same length on the page and with the same number of divisions but changes the size of the divisions. If the map scale changes, the division size of the scale in Figure 5.17 might go from 0.5 mile to 0.6 or 0.4. This option preserves the width of the scale bar but may result in awkward division sizes and labels.

*Adjust number of divisions* leaves the division value constant and adds or subtracts full divisions if the scale changes. For minor changes in scale, the bar will increase or decrease slightly in size. For dramatic changes, the scale bar will remain about the same size but may have more or fewer divisions. The bar in Figure 5.17, for example, might become 1 mile long or 3 miles long, but the division size would stay 0.5.

### Labeling, text, and annotation

ArcGIS offers several ways to place text on maps. You have already learned how to create **dynamic labels**, which are derived from a feature attribute and applied to an entire layer at once. **Graphic text** is used to place a few labels manually. **Annotation** is used to gain exact control of labels. We briefly summarize the key features of each type.

**Graphic text** items are constructed by the user, one at a time. Different tools are provided to create standard, splined, callout, or multiline text or to label a feature using the value from its attribute table (Fig. 5.18). All graphic text exists on the map as simple graphic elements. Items remain where they are placed but are inefficient when many labels must be made. You must be in Layout view to create graphic text. However, the Label tool is an exception; it derives the text for the label from the attribute table and must be placed in Data view.

| | |
|---|---|
| **A** | Text |
| ᵃᵇ_c | Splined Text |
| ⊟ | Label |
| ⊔ | Callout |
| [A] | Polygon Text |
| [A] | Rectangle Text |
| Ⓐ | Circle Text |

Fig. 5.18. Text tools
Source: ArcGIS

**Dynamic labels** are created from an attribute and are redrawn every time the user changes the map view. They can be turned on or off as needed. Dynamic labels utilize a placement algorithm that omits overlapping labels. Which labels are displayed can vary, depending on the zoom scale and label size. Although largely automatic, dynamic labels provide some control options. The *Placement Properties* help manage the overlap and optimization of the labels to give the best possible result. The *Conflict Detection* options can handle priorities between different layers.

Because the labels are redrawn each time, printing maps with dynamic labels can bring surprises. The labels shown on the screen do not necessarily appear the same way on the map. For many purposes a few extra or missing labels do not matter. If they do, however, converting labels to annotation provides greater control over label placement.

**Annotation** combines the easy creation of many labels with precise control of each one. Ordinary dynamic labels can be converted to annotation. The labels that fit on the map are automatically placed, and the overlapping labels are directed to an overflow window for interactive placement. Once placed, the annotation label always appears in the same place. Annotation may be stored in three ways.

> ▶ *In the map document.* The annotation is stored as graphic text and is modified using the Draw toolbar. This annotation can be used only in a single map document.

▶ *As a feature class in a geodatabase*. The annotation can be used in multiple map documents. Editing must be used to modify it, so we cover this method in Chapter 13.

▶ *As a feature-linked annotation in a geodatabase*. The annotation is linked to the corresponding point, line, or polygon features. If the feature is deleted, so is the annotation. This type is available with ArcGIS Standard or Advanced licenses.

**Reference scale**

When labels and annotation are created, they are assigned a size in points. For a given map, the size of the text relative to the features, such as the counties shown in Figure 5.19a, is determined by the text point size and the current map scale. But what happens if you zoom in? By default, the labels (and symbols) stay the same size when the user zooms in or out (Fig. 5.19b); 12-pt. text remains 12-pt. text regardless of the scale of the map.

This behavior changes if the user sets the **reference scale**, which is a property of the data frame. The reference scale is the scale at which the labels and symbols appear at their assigned size. If a user places 12-pt. labels on a map with a scale of 1:100,000, then the labels will appear 12 pt. in size only as long as the map scale remains 1:100,000. If the user zooms in, the labels and symbols increase in size (Fig. 5.19c). If the user zooms out, they decrease in size.

Fig. 5.19. (a) Text and symbols appear at their assigned size at the reference scale. (b) If the reference scale is not set, the symbols will always appear the same size. (c) If the reference scale is set, then zooming in or out changes the size of the text and symbols.
Source: ArcGIS

**TIP:** Be sure to understand the difference among the terms *map scale*, *reference scale*, and *visible scale range*. Map scale determines the ratio of features on the map to features on the ground. The reference scale is the scale at which symbols and text appear at their assigned size. The visible scale range determines the range of scales at which a layer will appear.

## Summary

▶ Designing a map consists of six interlinked steps: determine the objective, select the data, arrange the elements, symbolize the elements, review the map, and edit/improve.

▶ Maps serve many purposes, each of which influences the content, medium, target audience, and conditions of use.

▶ Every set of data included should serve the purpose of the map and be appropriate to the map scale. Layers that are too detailed can be simplified through cartographic generalization.

▶ A suitable map projection that supports the map objective must be used; it may influence whether a north arrow or scale bar is appropriate for the map.

▶ Visual hierarchy refers to the order in which the reader perceives the different objects in a design; it is manipulated using the arrangement and symbolization of map elements.

▶ Balanced arrangement of map elements uses the size, visual center, rule of thirds, alignment, and negative space to establish a visual path through the map.

▶ Symbolization affects the visual hierarchy through size, color, contrast, foreground/background relationships, font, symbol style, and pattern.

▶ The map should be reviewed by the author and also by other to ensure that it meets its objective and is free of petty errors. First impressions are important.

▶ When editing and improving the map, work to concentrate the message and simplify the presentation. Be sure to test the map under the conditions of actual use.

▶ Three types of labeling are available. Interactive labels are used to place any kind of text. Dynamic labels are automatically generated for entire layers at a time. Annotation is created from dynamic labels but can be precisely controlled.

### Additional readings on cartography

MacEachren, Alan M. 1995. *How Maps Work: Representation, Visualization, and Design*. New York: The Guilford Press.

Monmonier, Mark. 1996. *How to Lie with Maps*. Chicago: The University of Chicago Press. Second edition. First published 1991.

Robinson, Arthur H., Joel L. Morrison, Phillip C. Muehrcke, A. Jon Kimerling, and Stephen C. Guptill. 1995. *Elements of Cartography*. New York: John Wiley & Sons. Sixth edition. First published in 1953.

### Important Terms

| | | | |
|---|---|---|---|
| annotation | division value | graticule grid | reference grid |
| automatic scaling | divisions | layout | reference scale |
| background | dynamic labels | map elements | rule of thirds |
| branding | fixed extent | map extent | Subdivisions |
| cartographic generalization | fixed scale | measurement grid | visual center |
| contrast | foreground | neatline | visual hierarchy |
| division units | graphic text | negative space | |

## Chapter Review Questions

1. Go online and find an example for each of the map purposes listed in the first section. Capture a screenshot of each one and then explain why you chose it as a good example.

Find another map to critically review, provide a large screen shot so that it can be easily seen, and answer the following questions about it.

2. Describe your first impression of this map. Was it positive or negative? What aspects of the map caused you to react that way?

3. State the map objective and explain how each layer does or does not support it. Are all of the layers necessary or could some have been omitted?

4. Describe the visual hierarchy that you see (or don't see) in this map and explain how it is created (or ignored).

5. Critique the balance and alignment of this map and describe how well it uses negative space.

6. Critique the symbols used for this map and describe how the symbols support or detract from the map objective.

7. Critique the legend of the map, if it has one.

8. Can you see any ways in which simplifying this map might improve it? List them, or explain why there are none.

# Mastering the Skills

## Teaching Tutorial

The following examples provide step-by-step instructions for doing basic tasks and solving problems in ArcGIS. The steps you need to do are highlighted with an arrow →; follow them carefully. Click on the video number in the VideoIndex.htm file to view a demonstration of the steps.

→ Start ArcMap. Choose Existing Maps > Browse for more… on the splash screen, and open the ex_5a.mxd map document from the mgisdata\MapDocuments folder.

→ Use Save As to rename the document and remember to save frequently as you work.

 1→ Switch to Layout view by clicking the Layout button in the lower-left corner of the map window.

1→ If the Layout toolbar is not visible, choose Customize > Toolbars > Layout.

 1→ Click the Zoom Whole Page button on the Layout toolbar, if necessary, to see the entire page.

### *Setting up the map page*

This map document has been set up to show volcanic hazards in Washington and Oregon, and we will use it as a starting place to learn about creating layouts. First, we are going to change the page layout and turn on the grid so we can align the objects more easily.

1→ Choose File > Page and Print Setup from the main menu bar.

1→ Examine the options to see what these are.

Look at the two sections labeled Paper and Map Page Size. The lower Map Page Size section controls the size and layout of the map itself. Always set this section first.

**TIP:** Always design and edit a map at the size at which you intend to print it.

The Paper section is used to modify the printer settings at the time of printing. It need not be set until you are ready to print. In many cases, the settings will be the same as the Map Page Size. However, after you have created an 11-inch by 17-inch map, you might decide to print it on a letter-size page for a report. Setting the Paper section makes it possible.

1→ Uncheck the box to *Use the Printer Paper settings* if it is checked.

1→ Check the box to *Scale the map elements proportionally to changes in page size* to help in resizing the frames later.

1→ Fill the button to change the Map Page Size to Landscape orientation.

1→ Also set the Paper orientation to Landscape. Click OK.

**TIP:** Checking the box to use the Printer Paper settings is not recommended unless you always use the same printer. It can cause undesired changes in the layout if you switch printers.

2→ To set the grid, choose Customize > ArcMap Options from the main menu and click the Layout View tab.

2→ Notice the options for setting rulers: the page units (currently inches), the grid, and snapping.

A grid is a set of points a specified distance apart (now set to 0.25 inch). When snapping is on, objects within the snap tolerance (currently 0.2 inch) will be automatically snapped to the closest grid location. This feature simplifies aligning map elements.

2→ Check the Show box under the *Grid* section to show the grid (Fig. 5.20).

2→ Check the Grid box under the *Snap elements to* section and make sure the Snap Tolerance is 0.2 inch.

2→ Click OK to close the Options box.

Fig. 5.20. Setting the grid options
Source: ArcMap

The Layout toolbar (Fig. 5.21) provides zoom tools that look similar to those on the regular Tools bar, but they affect the view of the page only, not the map inside the data frame. A demonstration will help show the difference.

Fig. 5.21. The Layout toolbar zoom tools
Source: ArcMap

 3→ Find the Fixed Zoom Out button on the <u>Tools</u> toolbar and click it three times. The map gets smaller within the data frame, but the page remains fixed in the map window.

 3→ Find the Fixed Zoom In button on the <u>Tools</u> toolbar and click it three times to return to the initial extent.

 3→ Now find the Fixed Zoom Out tool on the <u>Layout</u> toolbar and click it three times. Notice that the entire page now gets smaller, but the map continues to occupy a constant area of the data frame.

 3→ Use the Zoom Whole Page button on the <u>Layout</u> toolbar to fill the map window with the layout again.

**TIP:** The Zoom/Pan tools on the <u>Tools</u> toolbar control the scale and placement of the map in the active data frame. The Zoom/Pan tools on the <u>Layout</u> toolbar control the view of the layout page.

Now let's see how grid snapping works.

 4→ Click the Select Elements tool on the Draw or Tools toolbar. In the map window, click the Hazards frame to select it, and drag it to a new location. Notice that the frame corner snaps to a grid point when it gets close.

4→ Use the blue handles and snapping to give the data frame 0.75-inch margins on the top, bottom, and left sides (remember, each dot of the grid is 0.25 inch apart).

**TIP:** Always provide ample margins to avoid cropping when the map is printed.

Next, we will set up the basic layout design. Although it is simple to arrange data frames by clicking, dragging, and resizing, sometimes precise sizes and locations are needed. Set these options using the data frame properties.

5→ In the Table of Contents, right-click the Hazards data frame and choose Properties.

5→ Click the Size and Position tab.

5→ Check that the *X* and *Y* positions are set to 0.75 (inch) where you placed the corner earlier. The page has coordinates of inches, with 0.0 at the lower left.

5→ In the Size window, uncheck the *As Percentage* box if needed. Set the frame size to 6 inches wide and 7 inches high. Click OK.

**TIP:** If the map does not redraw completely, use the Refresh button, located next to the Layout and Data view buttons, to redraw it.

5→ Use the Zoom/Pan tools on the <u>Tools</u> toolbar to center the Hazards map and make it fill the frame.

**SKILL TIP:** Learn how to set the data frame to use the Fixed Scale or Fixed Extent scaling method (Layouts and Data Frames).

### Creating a location map

Many maps, especially if they show a place that is not commonly known, include a location map to inform the viewer where the map is. To add a location map, we need another data frame.

6→ Choose Insert > Data Frame from the main menu bar.

6→ Click on the Select Elements tool.

6→ Place it on the new data frame and click and drag to move the frame to the lower right of the page.

6→ Use the blue handles and the snapping function to place the new frame 0.25 inch from the Hazards frame and 0.75 inch from the edge of the page (Fig. 5.22).

6→ Find the New Data Frame name in the Table of Contents and click it twice slowly to rename it *USA*.

6→ If the name appears in boldface type, USA is already the active frame. If not, right-click the name and choose Activate.

Fig. 5.22. Arrange the frames like this.
Source: ArcMap

7→ Add the states feature class from the mgisdata\Usa\usdata geodatabase.

7→ Use the Zoom In tool from the <u>Tools</u> toolbar to zoom in to the conterminous US states.

7→ Right-click the states layer symbol and choose a light yellow color.

A location map usually includes a box showing the extent of the main map. We can add this as a property of the USA data frame.

8→ Right-click the USA data frame name and choose Properties. Click the Extent Indicators tab.

8→ Click on the Hazards name in the window to highlight it, and click the > button to move it to the right side.

8→ Check the box to *Use simple extent* (the outline of the data frame).

8→ Click the Frame button and use the drop-down to choose a 3-pt. border for the extent rectangle. Change its color to a dark orange. Click OK and OK.

We have not yet considered the issue of the map projection for this frame. The Hazards frame uses an equidistant conic projection centered on the two states, minimizing distortion.

9→ Double-click the USA frame to open its properties.

9→ Click on the Coordinate System tab and read the information. Write the coordinate system here. _____

Although the USA frame is also using an equidistant conic projection, it is centered in the middle of the country and distorts direction at the edges of the map. Notice that north does not point straight up, and the extent rectangle is skewed. This makes the location map confusing. We should choose a projection that preserves direction for the location map.

1. Which projection(s) from the front cover of the book preserves direction? _____

9→ On the data frame Coordinate System tab, expand the Projected Coordinate Systems > World folder.

9→ Choose the Mercator (world) projection and click OK.

9→ Check the box not to warn you again in this session, and then click Yes to use the coordinate system. A datum difference at this scale is not significant.

10→ Double-click the USA data frame name to open its properties. Click on the Size and Position tab.

10→ Set the data frame height to exactly 2 inches and click OK.

10→ Use the Zoom/Pan tools on the <u>Tools</u> toolbar to center the US map in the data frame.

### *Using graphic text on layouts*

You can create **graphic text** on the layout using one of the text tools on the Draw toolbar. You must be in Layout view to place graphic text.

| | |
|---|---|
| **A** | Text |
| | Splined Text |
| | Label |
| | Callout |
| | Polygon Text |
| | Rectangle Text |
| | Circle Text |

Fig. 5.23. Text tools
Source: ArcMap

 11→ Click the New Text tool on the Draw toolbar. The black triangle indicates that more text tools are accessible from this icon (Fig. 5.23).

11→ In the larger Hazards frame, click offshore of the states and type Pacific Ocean. Press Enter.

11→ Click once on the text to select it. The blue dashed line around the text indicates that it is selected.

11→ This horizontal label does not fit well, so click the Delete key to remove it.

**TIP:** To work with graphics, use the Select Elements tool from the Draw toolbar. To delete text, click the text to select it and press the Delete key. Use the Select Elements tool to draw a box around many text items, or hold down the Ctrl-key to select multiple text boxes.

12→ Let's try a different text tool. Click on the black arrow next to the New Text tool and select the Splined Text tool.

12→ Click to enter vertices of a line that curves along the coast of Oregon and Washington from south to north. Double-click to end the line.

12→ A small text box appears. Type in Pacific Ocean and press Enter. The text appears splined along the line you entered.

12→ Change the text size to 24 pt., the style to Italic, and the color to blue, using the buttons on the Draw toolbar.

12→ Right-click the Pacific Ocean text and choose Properties. Change the Character Spacing to 36 pt. to stretch the label out along the coast.

The graphic text tools require you to type the text to be displayed. The Label tool, however, can read the label from an attribute field in the feature class table. It uses the primary display field, which is one of the layer properties.

13→ Right-click the Capitals layer and choose Properties. Click the Display tab and check the Display Expression Field to verify that it is set to NAME. Click OK.

13→ Try to choose the Label tool from the drop-down button for the interactive labels. It is dimmed because the Label tool must be used in Data view.

13→ Click the Data view button. Activate the Hazards frame, if necessary.

14→ Choose the Label tool. The Label Tool Options window appears.

14→ Keep the default option to *Automatically find best placement*.

14→ Fill the button to *Choose a style*. Scroll down and click on the Capital style.

15→ Click on the capital of Oregon and watch the "Salem" label appear. If you accidentally click the wrong feature, delete the label and try again.

15→ Click on the capital of Washington (Olympia).

15→ Use the Select Elements tool to reposition the labels to be clearly seen.

15→ Return to Layout view.

**TIP:** Graphic text and other types of graphics are termed Elements. To Select All Elements or Unselect All Elements, use the functions in the Edit pull-down menu on the main menu bar.

### Working with dynamic labels and annotation

Dynamic labels are quick to set up, but annotation offers precise control. We will create two sets of annotations for this map: one inside the triangles to indicate the number of eruptions and one to label the volcano names. Annotation is created from dynamic labels.

To avoid clutter, we want labels only for the volcanoes with more than five eruptions. We will use a query to select the desired volcanoes and then create a layer from the selection.

16→ Right-click the Volcanoes layer and choose Open Attribute Table.

16→ Click the Select by Attributes button in the Table window.

16→ Enter the expression KNOWN_ERUP >= 5. Verify it and click Apply.

The selected volcanoes are highlighted in the table and on the map. Next, we will create a layer containing only the selected volcanoes.

17→ Close the Table window.

17→ Right-click the Volcanoes layer name in the Table of Contents and choose Selection > Create Layer from Selected Features.

17→ A new layer appears at the top of the Table of Contents, called Volcanoes selection. Rename it *Volcano Labels*.

17→ Right-click the Volcanoes layer and choose Selection > Clear Selected Features.

Next, we'll create dynamic labels for the Volcano Labels layer. Since we are only using the layer for the labels, we will also make the symbols invisible.

18→ Open the Volcano Labels properties and click the Labels tab.

18→ Turn on the labels, set the label field to NAME, and make the labels Arial 10 pt. bold font. Give the labels halos, if you wish.

18→ Keeping the Layer Properties window open, click on the Symbology tab.

18→ Set the symbol to Circle 1, the size to 1, and the color to No Color. Click OK twice.

**SKILL TIP:** Use the Label Manager to more easily set and control labels for multiple layers at once (Labels and Annotation).

19→ Open the Volcanoes properties and click the Labels tab.

19→ Turn on the labels, set the label field to KNOWN_ERUP, and set the symbol to Arial 8 pt.

19→ Click the Placement Properties button and fill the button to Place label on top of the point. Click OK and OK.

19→ Examine the map to make sure you set everything correctly. You should have eruption numbers in the triangles and name labels for seven volcanoes. (There are labels for the inactive volcanoes with 0 eruptions, but we will delete these later.)

**TIP:** Annotation is created for the data frame and includes all the labels currently set for the data frame. You must turn off any layer labels that you don't want converted.

20→ Right-click the Hazards data frame name in the Table of Contents and choose Convert Labels to Annotation.

20→ Fill the button to *Store annotation in the map*.

20→ Notice the reference scale value listed in the upper-right corner. This is the scale at which the labels will appear at their assigned point size.

20→ Choose to *Create annotation for features in the current extent.*

20→ Make sure that only the layers Volcano Labels and Volcanoes are listed.

20→ Keep the box checked to *Convert unplaced labels to unplaced annotation.* These labels will be put in an overflow window for individual placement later.

20→ Click Convert to create the annotation.

View the new annotation. It looks the same, but the labels are now graphic text and are individually adjustable.

**TIP:** If you made a mistake and want to remove the annotation from the map, open the properties for the data frame and click the Annotation Groups tab. Highlight the Volcanoes Anno group and click Remove.

If you had unplaced labels, they were put in an overflow window (Fig. 5.24). Your list may be different from the one in the figure. We can place these labels manually. If you had no unplaced labels, go on to Step 22.

21→ In the Overflow window, right-click one entry and choose Flash Feature. Repeat, if necessary, until you see it flash. (You may need to move the Overflow window aside.)

21→ We don't want the eruption labels for the smaller triangles. For each piece of annotation that is a number less than 5, right-click it in the Overflow Annotation window and choose Delete.

21→ Right-click each remaining entry and choose Add Annotation. The text is added to the map. The placement may not be perfect, but we will adjust it in a moment.

21→ Close the Overflow Annotation window when it is empty.

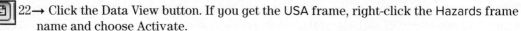

Fig. 5.24. Manually placing overflow annotation
Source: ArcMap

To adjust the annotation, we need to be in Data View. For these fine adjustments, it will be helpful to turn the snapping grid off.

22→ Open Customize > ArcMap Options on the main menu bar, click the Layout View tab, and uncheck the Grid box in the *Snap elements to* section. Click OK.

22→ Click the Data View button. If you get the USA frame, right-click the Hazards frame name and choose Activate.

2. When you returned to Data view, the labels got larger. What does this tell you about annotation? _____

22→ Click the Select Elements tool, if necessary. Click on the name label for Baker to select it. Adjust the position of the label so that it is clearly visible.

22→ Click on the 13 inside the Baker symbol to select it. Use the mouse or the arrow keys to move it to the center of the triangle.

23→ Carefully select and delete any annotation in the smaller triangles with fewer than 5 eruptions. (You can hold the Shift key to select more than one at a time.)

23→ Delete the 0 labels for the inactive volcanoes. You can draw a box around them to select them in groups—just be careful not to include any other annotation.

23→ Center the remaining labels inside their triangles.

Since we are using annotation, we should set the reference scale of the data frame to scale the symbols also when we zoom in or out.

 24→ Click the Layout button to return to Layout view.

24→ Use the Zoom/Pan tools on the <u>Tools</u> toolbar to resize and center the two states within the data frame, if necessary.

24→ Right-click the Hazards data frame name and choose Reference Scale > Set Reference Scale.

 24→ Zoom in to Mt. St. Helens and watch the symbols and labels enlarge.

24→ Return to the previous extent using the button on the Tools toolbar.

**TIP:** To set an exact reference scale, such as 1:24,000, right-click the data frame name, choose Properties, click the General tab, and type a specific reference scale in the appropriate box.

The complex interstate symbols are too prominent in the visual hierarchy, and we don't need four different road classes for this map. We can create a simpler classification.

25→ Open the Symbology tab for the Highways layer.

25→ Hold the Shift key and click on the symbols for Interstate and Freeway, so that both are highlighted.

25→ Above the highlighted rows, click on the Value heading and choose Group Values.

25→ Repeat the previous step to group the Arterial and Unclassified classes together.

25→ Click on the text under the Label heading and change them to simply *Interstates* and *Highways*.

25→ Double-click the Interstates symbol and set it to a 3.4 pt. simple brown line. Set the Highways symbol to 1.5 pt. line of the same color. Click OK to accept the changes. The final result should appear as in Figure 5.25.

Fig. 5.25. Grouping the highway symbols
Source: ArcMap

The county lines are too similar to the highways, causing confusion. We can reduce the contrast to bring the highways forward and relegate the county lines to the background.

26→ Open the Symbology properties for the Counties layer.

26→ Click on the Symbol heading above the shades and choose Properties for All Symbols. (Bug: If no menu appears, select all symbols and choose Properties for Selected Symbols instead.) Change the Outline color to Gray 10%. Click OK and OK.

### *Adding a legend to the map*

Now that the map is in good shape, it is time to continue creating the layout.

27→ Choose the Select Elements tool.

27→ Right-click the Hazards frame name and choose Activate. The legend is always based on layers in the active frame.

27→ Choose Insert > Legend from the main menu bar. A Legend Wizard will appear.

28→ Examine the list of layers and make sure you want them all. Volcano Labels has only labels, so select it on the right and click the < button to remove it from the list.

28→ Now establish the order of layers. To move a layer, click on it to select it and click the Up or Down arrow until it is in the right location. Put Volcanoes first, Counties next, then Highways, then Capitals, then Rivers, then States. Click Next.

29→ Delete the Legend title. Click Next.

29→ Choose the 1.0 Point border for the Legend. Make sure that the Gap is set to 10 pt. Click Next.

29→ To choose a different style patch, click the Rivers layer and choose the Flowing Water line. Click Next.

29→ This section gives very detailed control of the spacing between different elements of the legend. It is fine to leave the defaults on this step. Click Finish.

29→ The legend appears in the middle of the map. Click and drag it next to the map frame and even with its top (Fig. 5.26).

Fig. 5.26. Adding the legend
Source: ArcMap

**TIP:** Use Customize > ArcMap Options to turn the snap grid back on to align the boxes.

If you made a mistake creating the legend, don't worry. You can change its properties. Note that the legend is longer than the space available. We can make changes to help it fit better. The Volcanoes layer has two headings, Volcanoes and KNOWN_ERUP. We only need one.

30→ Right-click the Legend and choose Properties. Click the Items tab.

30→ Click on Volcanoes and click the Style button underneath.

30→ Choose the item style shown in Figure 5.27, called *Horizontal Single Symbol Layer Name and Label.*

30→ Click OK and OK and examine the change in the legend.

30→ Rename the Volcanoes layer *Known Eruptions* in the Table of Contents. Notice that the name changes in the legend as well.

Fig. 5.27. Choose this item style.
Source: ArcMap

31→ Open the Legend properties again and click the Items tab.

31→ Click on the Highways entry and change its item style to the same one used for the Volcanoes. Click OK and OK.

32→ Click twice slowly on the POP10_SQMI heading under the Counties layer in the Table of Contents and change it to *People/sq. mile.*

32→ Click twice on the Capitals layer name in the Table of Contents and change the name to *State Capitals.*

We have shortened the legend, but it still does not fit. Let's use two columns instead of one.

33→ Double-click on the Legend to open its Properties and click the Items tab.

33→ Move the Properties box away to be able to view the Legend.

33→ Select the Highways and check the box to *Place item(s) in new column.* Click Apply.

That looks as if it will work, but we want to align the edges of the legend with the location map.

34→ Click the Size and Position tab in the Legend Properties window.

34→ Leave the height alone but set the width to 3.275 inches. Click OK.

34→ Use ArcMap > Options to turn the snap grid back on, and use it to move the legend into its final place, 0.25 inch above the location map and right of the Hazards frame (Fig. 5.28).

Fig. 5.28. Final position of the legend
Source: ArcMap

Notice that the snap grid is helping us make the negative space equal between the elements and to ensure that the edges are perfectly aligned. Turn the snap grid off or on as needed in the remaining steps.

### *Placing a scale bar on the map*

Hazards are impacted by distance from the volcanoes, so a scale bar is appropriate on this map. The scale bar is placed in the active frame.

3. To review the scale bar terms, write the correct terms in the boxes of Figure 5.29.

35→ Make sure the Hazards frame is active.

35→ Choose Insert > Scale Bar from the main menu.

Fig. 5.29. Terminology of a scale bar
Source: ArcMap

35→ Choose the Alternating Scale Bar 1 style bar and click OK.

36→ Move the scale bar to the lower right corner of the Hazards frame, on top of Oregon.

36→ Use the Zoom In tool on the <u>Layout</u> toolbar to zoom in to the scale bar to examine it more closely.

The bar may be too long or too short, and it may have uneven divisions, such as 110 miles. If it appears unsuitable, try resizing the scale bar to improve it.

37→ Click the Select Elements tool.

37→ Click the right boundary of the scale bar (a double horizontal arrow will appear), and drag it to the left to decrease the length of the scale bar.

37→ Repeat until the scale bar is about 120 miles long.

You may have ended up with uneven labels instead of nice, round numbers. We can modify the scale bar to get exactly the properties we want.

38→ Double-click the scale bar to open its properties, and click the Scale and Units tab.

Notice that the division value is dimmed out. In order to set it explicitly, we must change the way ArcMap adjusts the bar when resizing.

38→ Under When Resizing, set the drop-down box to *Adjust number of divisions*.

38→ Set the *Division value* to *50* miles and set the *Number of subdivisions* to *2*.

38→ Click the Frame tab and set the Background color to Sand to separate the bar from the roads behind it. Click OK.

38→ Examine the new scale bar. Resize it to 100 miles long and place it neatly in the lower-right corner of the state.

### Adding other map elements

Many other elements can be added to the map using the Insert menu. Keep in mind that the elements will refer to the data frame that is currently active.

#### Adding a north arrow

 39→ Click the Zoom Whole Page button on the <u>Layout</u> toolbar.

39→ Choose Insert > North Arrow from the main menu bar.

39→ Click a slender, unobtrusive symbol like ESRI North 60.

39→ Click the Properties button and examine the settings. Click OK and OK.

39→ Click the Select Elements tool and move the north arrow to the lower-right corner of the Hazards frame. Keep it small and unobtrusive.

**TIP:** Some large-scale projections, like UTM, can have a slight direction distortion at the edges of the zone; although only a degree or two, it is still noticeable. ArcMap does not tilt the arrow automatically. Use the Properties tab to set the north arrow to the correct angle or, even better, use the data frame properties to rotate the map so that north is vertical.

The United States location map might look good with a graticules grid on it.

40→ Right-click the USA data frame name in the Table of Contents and choose Properties. Click on the Grids tab.

40→ Click New Grid. Choose the Graticule grid. Click Next.

40→ Choose Graticule and labels. Accept the default intervals. Click Next.

40→ Accept the defaults on the axes and labels. Click Next.

40→ Accept the defaults on the graticules. Click Finish and OK.

The grid appears, but the labels extend into the other map frame. A design decision is in order. You can reduce the size of the USA data frame so that the labels fit, or you can remove the grid. Since this frame merely contains a location map, having a scale or north direction or grid is not important, so we will remove it.

41→ Open the USA data frame properties and uncheck the Graticule box. Click OK.

### Adding titles and text

Titles and text provide important information about the map including data citations.

42→ Choose Insert > Title from the main menu bar. A centered text field appears with a default title.

42→ Double-click the text to open its properties.

42→ In the text box, type *Pacific Northwest*, press Enter, and type *Volcanic Hazards* on the next line. Click OK.

42→ Use the Drawing toolbar to set the font size to 18 pt. and make it bold. Center it above the legend, aligning it with the top of the Hazards data frame.

43→ Choose the Insert > Text choice in the main menu bar. Find the tiny new text box on the map, waiting for you to type the text.

43→ Type *by Your Name* and press Enter.

43→ Move the text to a centered position below the Title.

**TIP:** Did you notice the Dynamic text option under the Insert > Text menu choice? You can add times, dates, and other types of updatable text to the map.

44→ To add wrapped text, click the label drop-down on the Draw toolbar to find and select the Rectangle Text tool.

44→ Click and drag to create a box inside the blank area in the legend.

44→ Double-click inside the new text box to open its properties.

44→ Click the Text tab and type the following citation into the box without using any Enter keys: *ESRI Data and Maps [Download]. (2013) Redlands, CA: ESRI [October, 2014].*

44→ Make the text left-justified.

44→ Click the Frame tab and set the border to <None>. Click OK.

44→ Set the font size to 8 pt. Adjust the text box to fit neatly in the legend space. The legend should look similar to Figure 5.30.

### Adding neatlines

45→ Choose Insert > Neatline from the main menu bar.

45→ Fill the button to *Place around all elements*.

45→ Set the gap to 10 pt. (the gap controls the distance between the elements and the neatline).

45→ Choose the Triple Graded border from the drop-down box.

45→ Set the Background to White. A gradient border looks nice on the computer screen, but wastes ink and doesn't always print well. Click OK.

**SKILL TIP:** Learn how to add pictures or graphics to a layout (Layouts and Data Frames).

At this point, a test print can be made to make sure the colors and symbols work as well on paper as they do on screen. The map should appear similar to Figure 5.31.

46→ Save the map. It is wise to SAVE the map document before you print.

46→ Choose File > Print from the main menu bar.

46→ Make sure that the sample map and printer page settings fit each other.

46→ Click Setup to change the printer or adjust the print settings, if needed.

46→ When ready to print, click OK.

**TIP:** Examine the margins of the print. Some printers shift the map off-center, and you must adjust by moving the map elements to center them for your printer.

### Reviewing and editing the layout

Now for the final review. Go take a break, then return and look at the map at full scale, both on paper and on screen. You might notice the following issues (Fig. 5.31).

▶ The visual hierarchy established by the symbols is unclear. All of the map layers shout equally for attention. The volcanoes are the purpose, but they don't stand out. The highways and rivers are difficult to tell apart. The Pacific Ocean label is distracting.

▶ The title looks odd by itself when the other elements all have neatlines, but adding a neatline around it would probably be too much. Even the current ones create a cluttered feel.

Fig. 5.30. The final legend
Source: ArcMap

Source: ArcMap

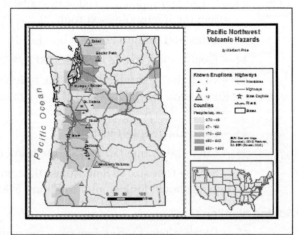

Fig. 5.31. The final layout
Source: ArcMap

As you reevaluate the design, return to the map objective: to draw attention to the fact that population centers in Washington and Oregon lie close to historically active volcanoes. Perhaps the roads and rivers are not as important as we originally decided. We can edit the map to clarify this message and make it more attractive as well, which will be your task in the first exercise.

### Exporting a layout

Maps can be saved as image files for inclusion in reports or presentations. The PDF format is a popular way to distribute maps to others, or to bring the map to a commercial printing center with better equipment.

47→ Choose File > Export Map.

47→ Choose the location and name for the image file.

47→ Set the Save As type to PDF.

47→ Click on the Options drop-down flag, if necessary, to show the Options settings.

47→ Check the resolution. The minimum resolution for printing should be 300 dpi. It could be lower for a web or screen image.

47→ Click Save to save the map.

☐ Clip Output to Graphics Extent

Source: ArcMap

**TIP:** If planning to place the map in another document, check the box to *Clip Output to Graphics Extent* and save as a JPEG or GIF. Check the resolution; the default of 96 dpi is not good enough and should be increased to 300 dpi or more.

**SKILL TIP:** Learn to use a map template to produce a quick layout of a map, or learn to create a simple graph (Layouts and Data Frames).

This is the end of the tutorial.

→ Exit ArcMap. You do not need to save your changes.

## Exercises

1.  Save the tutorial map under a new name, and then edit it to produce the version shown below, including removing the highways and rivers, and also increasing the sizes of the symbols and labels. After you finish, add the World Terrain basemap from ArcGIS Online to the main map frame to provide context (but turn off its reference layer of features and names). Make the counties about 30% transparent so that the base map shows through slightly. Save the map as a PDF to turn in.

Note: The data frame is using a Clip to Shape option to suppress the highways and rivers outside the states, and you must turn it off to see the basemap. Use the Frame tab in the data frame properties to set it back to No Clipping.

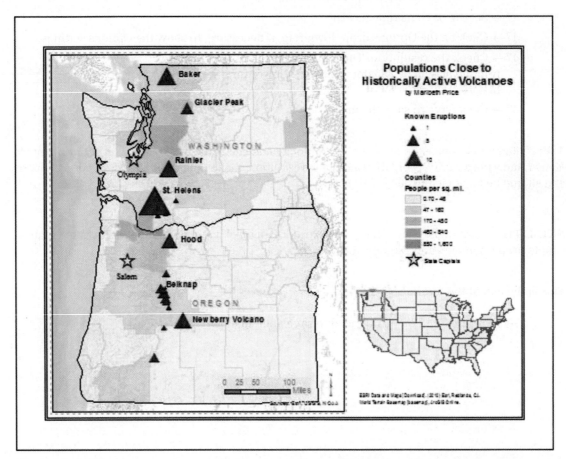

Source: ArcMap

2.  Save the tutorial map under a new name, then redesign it using a portrait orientation and without the location map. (For better performance, remove the basemap until you are finished editing.)

## Challenge Problem: Home Sweet Home

Design and create an attractive map layout showing some of the data you compiled for your home state in Chapter 2. Pick any map objective you like, but make sure the layout communicates it clearly. Export the map as a PDF file. Share it with some classmates and help each other critique your maps. Write the suggestions on the back of the PDF to turn in.

# Chapter 6. Attribute Data

## Objectives

▶ Understanding how tabular data are stored and used

▶ Using queries to select records of interest

▶ Understanding joins and cardinality concepts

▶ Exploring data in tables using statistics and summarized statistics

▶ Defining appropriate field properties

▶ Editing and calculating fields in tables

▶ Generating point layers from $x$-$y$ coordinates in tables

# Mastering the Concepts

## GIS Concepts

### Overview of tables

A **table** is a data structure for storing multiple attributes about a location or an object. It is composed of rows, called **records**, and columns, called **fields** or **attribute fields**. Figure 6.1 shows an example of a table containing attributes of counties.

Fig. 6.1. A table with information about US counties
Source: ArcGIS

In GIS, tabular data files fall into two main categories: **attribute tables** and **standalone tables**. An attribute table, such as the one shown in Figure 6.1, contains information about features in a geographic data set. In an attribute table, there is always one and only one row of information for each feature. In a shapefile, the row is linked to the spatial feature in a separate file using a unique ID number called a Feature ID, or FID. In a geodatabase, the file stores both the attributes and the $x$-$y$ coordinates in the same data file, although the coordinates are not visible in the table, and it uses an Object ID, or OID, instead of an FID (Fig. 6.1). In contrast, a standalone table simply contains information about one or more objects in tabular format instead of having information about map features. A standalone table might come from a text file, an Excel® spreadsheet, a global positioning system data file, or a database. Standalone tables exist independently of a geographic data set and may be only incidentally related to map features. They also have an OID rather than an FID.

## Database management systems

GIS tables share a history and many current properties of the database programs that are routinely used in commercial, governmental, and academic settings. Most vector-based GIS programs use an underlying database to store data, such as the ancient INFO database used for coverages, the dBase table structure used for shapefiles, the Microsoft Access engine used for personal geodatabases, or a large-scale relational database management system (RDBMS), such as SQL Server, used for enterprise geodatabases. These programs are designed to store, manipulate, analyze, and protect tabular data of all kinds. Governments use them to store information about citizens, parcels, taxes, and more. Companies use them to store information about customers. Universities use them to manage information about students, classes, and faculty. Three types of databases have traditionally been used.

A **flat file database** stores rows of information in a text or binary file. Finding information requires parsing the table and selecting the records of interest. They are simple but not efficient.

A **hierarchical database** has multiple files, each of which contains different records and fields. Parent tables can be linked to child tables through a specified field called a key. For example, a table of college classes might be linked to a table of students in each of the classes through a course ID number. Relationships between tables are fixed, which speeds the retrieval of information. However, the relationships are inflexible, designed to permit a predefined set of operations.

A **relational database** also has multiple tables stored as files. However, the relationships are not defined ahead of time. Instead, the user can temporarily associate two tables if they share a common field. This association is called a **join**, and the common field becomes the key. This database model is extremely flexible and is the preferred choice for GIS systems.

Because GIS data are intimately linked to an underlying database structure, it is not a far stretch to incorporate tables from database programs into GIS analysis. ArcGIS has capabilities for connecting to and working with RDBMS data files directly, expanding the type of data available. If a county keeps its parcel tax records in an RDBMS such as Oracle or SQL Server, for example, a GIS analyst can bring the database tables into ArcMap as standalone tables.

## Queries on tables

Often one wants information about a subset of records in a table, for example, knowing the number of parcels in the city that are designated as commercial. To determine this information, a **query** can be performed on the table. In a query, a **logical expression** is used to specify certain criteria (e.g., zoning = commercial), and then the software searches the table and finds the records (parcels) that match the criteria. Those records are returned as a **selected set**. Selected records can become the input to another action, such as printing them, exporting them to a new file (as we have already done), or executing a GIS function on them.

Most databases use a special language called Structured Query Language (SQL) to write and execute queries. A land use table might contain a field with the zoning code. If the field containing the zoning is named ZONE, and the code for commercial property is 492, the logical expression in SQL might look like this:

> SELECT * FROM landuse WHERE ZONE = 492

In this example, landuse is the name of the table, ZONE is the field name, and ZONE = 492 is the logical expression stating the criteria to be met. SQL queries can have multiple lines and include many criteria.

> SELECT * FROM landuse WHERE
> ZONE = 492 AND VALUE > 300000

This query would find all commercial parcels with a value greater than $300,000. This chapter presents methods for performing queries on tables. Chapter 8 examines queries in more detail.

### Joining and relating tables

In an RDBMS and in a GIS, tables are commonly combined using a **join**, in order to bring different sets of information together. The tables are combined using a common field called a **key**. The key field must be of the same data type in both tables. When a join is performed, the two separate tables become one and contain the information from both tables (Fig. 6.2). The join is a temporary relationship and may be removed when it is no longer needed.

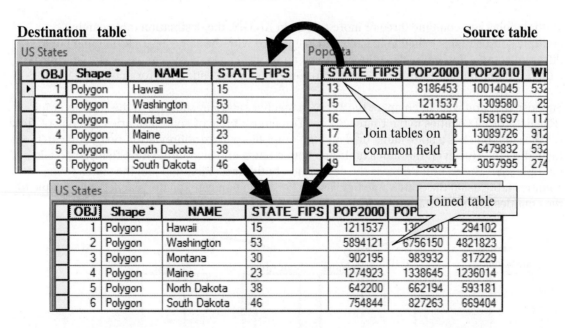

Fig. 6.2. Joining on a common field. These two tables are joined into one using the common field STATE_FIPS. The data from the source table are placed into the destination table. Source: ArcGIS

**TIP:** Joining census data is often accomplished using FIPS codes, which stands for Federal Information Processing Standard. Each state, county, city, block, and so on is assigned an identifying number, and these are combined to produce a unique identification for each census unit. For example, the state of Arizona is assigned "04" as its state FIPS code. Santa Cruz County is assigned the county FIPS code "019." The county name is not unique (California also has a Santa Cruz County), but the combination of the state and county FIPS code is. The code "04019" refers to Santa Cruz in Arizona, and "06087" refers to the one in California. A join using county names would not work because they are not unique, but a join on the FIPS code does work. Quotes are used in this discussion to show that FIPS codes are stored as text values; the leading zero on some codes would be dropped if they were stored as numbers.

Joins have a direction. The table containing the information to be appended is called the **source table**. The table that receives the appended information is called the **destination table**. In Figure 6.2, the destination table, US States, is a feature attribute table. The source table containing demographic data is a standalone table. When the two tables are joined, the demographic data are appended to the US States attribute table. Often joins are used to bring data from standalone tables into an attribute table for mapping or analysis.

Join direction matters. Once the demographics are joined to the table of the states shapefile, then the information could be used to make a map of the population data. If, however, the join had been performed in the opposite direction, with the demographics as the destination and the states as the source, then the resulting table would be a standalone table. In this case, a map showing the demographics could not be made because the demographics table is not a feature class.

Multiple joins can combine three or more tables. In ArcGIS, the destination table can have multiple source tables joined to it; each join is initiated with the same destination table but a different source. Source fields in the first join can be used as key fields in the next join. However, a table cannot simultaneously be a source for one join and a destination for another.

Suppose that you wish to compare the average teacher salary in counties to the state average (Fig. 6.3) for all counties in the United States. You have a table of county salaries and one of state salaries, but one has a state name field and the other has a state abbreviation field. You can't join them directly. However, the states table has both a name and an abbreviation field, so you do a multiple join. The counties table is the destination both times. First you join the states to the counties table using the StateName field as the key; then you can join the teacher salary table to the counties using the StateAbbr field.

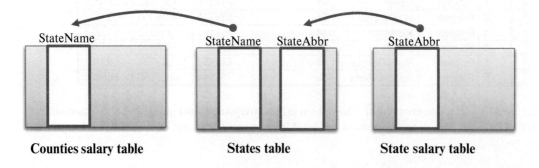

Fig. 6.3. A multiple join used to compare county teacher salaries with the state average

## Cardinality

When joining, the **cardinality** of the relationship between the tables must be considered. Cardinality is the numeric relationship between the objects in one table and their matches in the other. The simplest kind of relationship is *one-to-one*, in which each record in the destination table matches one record in the source table. In Figure 6.2, each state has one corresponding record of demographic data. In a *one-to-many* relationship, each record in the destination table could match more than one record in the source table. For example, a state typically has many school districts. In a *many-to-one* relationship, many records in the destination table would match a single record in the source table, such as many cities falling within one state. Finally, a *many-to-many* relationship indicates that multiple records can appear in both tables. For example, a student may take more than one class, and most classes have more than one student.

The direction of the join must be taken into account when ascertaining the cardinality of a relationship. The destination table is the point of reference and comes first; that is, the relationship cardinality is reported as {destination} to {source}. Imagine two tables containing *states* and *counties*. If one performs a join with *states* as the destination table and *counties* as the source, the cardinality is one-to-many because each state contains many counties (and a county may belong to only one state). If the join is reversed and *counties* is the destination table, then the cardinality becomes many-to-one because there are many counties in one state.

**TIP:** Putting the destination table first when stating cardinality is not a universal convention and may be found reversed in other data management applications. It is used here to match the convention adopted by ESRI in its publications and Help documents. It may help to always imagine the destination table on the left, as in Figure 6.4.

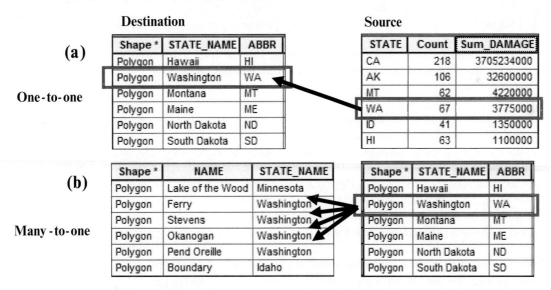

Fig. 6.4. A cardinality of one-to-one or many-to-one permits tables to be combined without violating the Rule of Joining. (a) One-to-one cardinality. (b) Many-to-one cardinality.
Source: ArcGIS

The cardinality of a relationship dictates whether the tables can be joined. ***The Rule of Joining stipulates that there must be one and only one record in the source table for each record in the destination table.*** Consider the four tables shown in Figure 6.4.

***One-to-one cardinality.*** In Figure 6.4a, a table containing the number of earthquakes and total damage in each state (source) is being joined to a states attribute table (destination), using the common field provided by the state abbreviation. Each state occurs once in each table so that there is no ambiguity about matching the source records to the destination records.

***Many-to-one cardinality.*** In Figure 6.4b, a table containing state information (source) is being joined to a table of counties (destination) using the common field provided by the state name. Although there are many counties in each state, there is only one record for each state in the source table and no ambiguity in linking the source records to the destination records. The state record does get used more than once, but it still does not violate the Rule of Joining.

***One-to-many cardinality.*** Figure 6.5 shows the reverse of the join in Figure 6.4b. Now states are the destination table and counties are the source table. Many county records in the source table match each state record in the destination table. The Rule of Joining is violated, and it becomes ambiguous which county record should be matched to the state. One cannot perform a join if a one-to-many relationship is present. Instead, we perform a different operation called a **relate**.

**Destination**

| | Shape * | STATE_NAME | STATE_FIPS |
|---|---|---|---|
| | Polygon | Hawaii | 15 |
| | Polygon | Washington | 53 |
| | Polygon | Montana | 30 |
| | Polygon | Maine | 23 |
| | Polygon | North Dakota | 38 |
| | Polygon | South Dakota | 46 |
| | Polygon | Wyoming | 56 |

**Source**

| | Shape * | NAME | STATE_NAME | FIPS |
|---|---|---|---|---|
| | Polygon | Lake of the Wood | Minnesota | 27077 |
| | Polygon | Ferry | Washington | 53019 |
| | Polygon | Stevens | Washington | 53065 |
| | Polygon | Okanogan | Washington | 53047 |
| | Polygon | Pend Oreille | Washington | 53051 |
| | Polygon | Boundary | Idaho | 16021 |
| | Polygon | Lincoln | Montana | 30053 |

Fig. 6.5. A one-to-many relationship violates the Rule of Joining because more than one record in the source table matche a record in the destination table.
Source: ArcGIS

In a relate, the two tables are still associated by a common field, but the records are not joined together. The two tables remain separate. However, if one or more records are selected in one table, then the associated records can be selected in the other table. For example, selecting the state of Washington in the states table allows the selection of all the counties in Washington in the related table, as shown by the red boxes in Figure 6.5. Many relationships in the world have cardinalities of one-to-many, such as a school to its classes or a well to its yearly water quality tests, and relates provide valuable support in dealing with those features and their attributes.

***Many-to-many cardinality.*** Many-to-many relationships exist but are difficult to deal with in a relational database. Consider a university. Each class has many students in it. However, most students are taking more than one class (Fig. 6.6). In practice, one would have to use multiple relates to represent the relationships and work with them one at a time. A one-to-many relate could be set up from the classes table to the students to allow all the students in each class to be found. Another one-to-many relate would allow the list of classes for each student to be generated.

Fig. 6.6. A many-to-many relationship
Source: ArcGIS

Sometimes a join is performed in which most records have matches, but some do not. Occasionally, a user makes a fundamental error in crafting the join, and none of the records has matches. In these cases, the join will still be performed, but the records without matches will be given zero or <Null> values in the

| Earthquakes table fields | | States table fields | | |
|---|---|---|---|---|
| STA | LOCATION | STATE_ | STATE | POP201 |
| PR | Puerto Rico Region | <Null> | <Null> | <Null> |
| PR | Puerto Rico Region | <Null> | <Null> | <Null> |
| PR | Northwest of Puerto Ri | <Null> | <Null> | <Null> |
| AL | Irondale, Alabama | Alabama | AL | 473559 |
| AL | Near Birmingham, Alab | Alabama | AL | 473559 |
| AL | Near Birmingham, Alab | Alabama | AL | 473559 |

Fig. 6.7. Result of no match for the destination records (highlighted) in the source table (white)
Source: ArcGIS

attribute fields from the source table. In Figure 6.7, an earthquakes table (highlighted) was joined to a states table (white) based on the state abbreviation and a many-to-one cardinality. Earthquakes in Puerto Rico had no match in the states table, and the source fields are given <Null> values in the joined table. It is always a good idea to view the table after a join and confirm that it went as intended. If missing values are encountered, a plan can be made to handle them. In this case, one might choose to exclude the earthquakes outside the 50 states from further consideration or to find a different states feature class that includes territories.

## Statistics on tables

### Statistics

Exploring data through statistics is a fundamental first step in data analysis, because it is important to understand the structure and properties of the data before applying more advanced techniques. A standard approach begins with calculating statistical measures such as the minimum, maximum, mean, median, and standard deviation, and viewing a histogram to evaluate whether it is potentially a normal distribution. Careful interpretation of this information can yield some important insights, especially when measured against one's expert knowledge.

Consider Figure 6.8, showing basic statistics and a frequency diagram (histogram) of earthquake magnitudes for a feature class of historical earthquakes in the United States. A geologist would recognize two odd things about this data. First, over 300 earthquakes have a magnitude of 0, which makes no sense physically and might indicate missing data. The missing values affect the current statistics, which must be recalculated without the missing values. The second oddity is more subtle. Without the zero

Fig. 6.8. Statistics and frequency diagram (histogram) for earthquakes
Source: ArcGIS

values, the histogram resembles a slightly skewed bell curve. However, a geologist knows that earthquakes follow an inverse power law, with many more small earthquakes than large ones. The creator of this feature class evidently included many small earthquakes. This choice makes sense for the purpose of the data set (to portray historic earthquakes), but it does render statistics like the minimum and average quake magnitude meaningless.

***This example illustrates the important step of exploring data and thinking about it critically before blindly calculating and accepting a result.*** Not only does the practice help prevent serious mistakes in inferring things about data sets, it can raise interesting questions. Why are there so many missing data values? Why are any magnitude 2–3 earthquakes included in this data set at all, since they typically do little damage and are rarely even felt?

A data set that passes the initial exploration can become the subject of more sophisticated statistical tests that are often provided with some GIS programs; one can also export the table for analysis in standard statistical software packages. GIS professionals can only improve their expertise by becoming more knowledgeable about statistics in all its forms.

## Summarizing tables

The **Summarize** function is a powerful form of statistics applied to tables. Summarize combines the records into groups based on a categorical attribute field and then calculates statistics separately for each group. For example, one might group a table of earthquakes according to which state the earthquake is in, and generate earthquake statistics for each state separately.

Figure 6.9 contains a table showing major historical earthquakes in the United States. Attributes for each quake include the state in which it occurred, the number of deaths it caused, the total damage caused, and the Richter scale intensity (MAG). We would like to know which states have suffered the most from historical earthquakes. In particular, we want to know the total deaths, total damage, and maximum magnitude for each state. We summarize based on the state field, dividing the earthquakes into groups by state, and specify which statistics to calculate. Separate statistics are calculated for each group (state).

| quakehis | | | | |
|---|---|---|---|---|
| **STATE** | **DEATHS** | **DAMAG** | **MAG** | **LOCATION** |
| MO | 7 | 0 | 7.88 | New Madrid, Missouri |
| SN | 51 | 0 | 7.36 | Northern Sonora, Mexico |
| AK | 0 | 0 | 8.15 | Yakutat Bay, Alaska |
| AK | 0 | 0 | 8.26 | Southeast Alaska |
| AR | 7 | 0 | 7.68 | Northeast Arkansas |
| CA | 3000 | 52400000 | 7.8 | Near San Francisco, Calif |
| CA | 12 | 6000000 | 7.48 | South of Bakersfield, Calif |

Fig. 6.9. Earthquake table
Source: ArcGIS

Because many statistics can be returned from a Summarize, this command produces a new table. The quakesum output table, shown in Figure 6.10, has one record for each state. It also has a Count field indicating the number of earthquakes in each state and one field for each of the statistics we requested. To better interpret the data, we sorted the table by the damage field in descending order to find the most grievously affected states.

| quakesum | | | |
|---|---|---|---|
| **STATE** | **Count_STATE** | **Sum_DEATH** | **Sum_DAMAG** |
| CA | 218 | 3777 | 3705234000 |
| AK | 106 | 125 | 32600000 |
| MT | 62 | 32 | 4220000 |
| WA | 67 | 15 | 3775000 |
| ID | 41 | 2 | 1350000 |
| HI | 63 | 79 | 1100000 |
| OR | 24 | 2 | 760000 |

Fig. 6.10. Summarize produces a file that contains the summary field (STATE), the number of earthquakes in each state, and the requested statistics.
Source: ArcGIS

Our next step might be to join the quakesum.dbf table to the US States layer and create a map showing the total damage or total deaths for each state (Fig. 6.11). Using Summarize, we collapsed a one-to-many relationship between states and earthquakes into a one-to-one relationship between states and earthquake statistics, allowing the data to be mapped.

Fig. 6.11. Earthquake deaths by state
Source: ArcGIS

### Field types

Creating database tables always begins with an analysis of the fields that the table will contain. Unlike a spreadsheet, in which any cell can contain any type of data, a database field (column) must contain only one type of data, perhaps text, perhaps integers, but never both. Each field must be defined, and the type of its contents must be specified, before any data are entered. Furthermore, once a **field definition** is set, it cannot be changed. The most common field types are numbers, strings, and dates. Each field may also have parameters that further specify what it may contain, such as the maximum number of characters allowed in a text field.

A field STREETNAME would be a text field because street names contain letters. The **field length** defines how many characters can be stored in the name. If the STREETNAME field is given a length of 10, then any name longer than 10 characters will be truncated to the first 10 letters. Thus, "Elm Street" fits perfectly (including the space), but "Maple Street" would be truncated to "Maple Stre" to fit in the field. It is important to ensure enough space in fields to hold the longest likely value. When designing a database, try to use the smallest field widths that will store every possible value, because extra spaces increase the size of the database.

Numeric fields have different parameters. When defining a numeric field, the user may designate a storage width (**precision**) and the number of decimal places (**scale**). For example, a field with a precision of 5 and a scale of 0 could store a number between –9999 and 99999 (the minus sign takes one space). A field with a precision of 5 and a scale of 2 could store a number between –9.99 and 99.99 (the decimal place also takes one space).

Numbers can be stored in a variety of ways. Most databases offer several formats for storing numeric values, and these formats can differ from database to database. The next section describes the common options in general terms.

### ASCII versus binary

A **byte** is the basic unit of storage space for a computer—it is composed of a string of eight digits (bits), which may be zeros or ones. These zeros and ones represent a number in base 2; such numbers are called binary numbers. A single byte can store a binary value from 0 (00000000) to 255 (11111111). Note that $255 = 2^8 - 1$. Two bytes can store values up to $2^{16} - 1$, or 65,535. The more bytes allotted to a value, the larger the number that can be stored. One million bytes constitute a megabyte (MB), and one billion bytes equal a gigabyte (GB), terms that probably sound familiar as measurements of data storage capacity.

All text is stored as sequences of characters using a special code called **ASCII** in which every number, every letter, and every symbol (such as $) is assigned a single-byte code between 0 and 255. To store the word *cat*, the computer stores the code for *c*, the code for *a*, and the code for *t*. Thus, it takes three bytes to store *cat*. The word *horse* requires five bytes. Numbers may also be stored using ASCII by storing the ASCII one-byte code for each numeral. Thus, it requires three bytes of data to store *147* and five bytes to store *147.6*. This scheme is a simple and standard way to store information. Text files and HTML files, among others, are stored in ASCII.

181

Unlike text, numbers have another storage option called **binary**. In this case, a number is stored in base 2 directly rather than being assigned one byte per character. The number *16* would be stored in binary as *00010000* in a single byte of information. Binary is a more efficient way to store numbers than ASCII. The number *14456* would require five bytes in ASCII but only two bytes in binary. It is also faster to compute with binary values because the computer by design does all its calculations in base 2. If the number is already stored in base 2, the computer does not need to convert it before calculating. Thus, it is often advantageous to use binary storage. Many types of files use a binary encoding scheme, including spreadsheets, word processing documents, and shapefiles. Raster and image data are also stored as binary data.

### Precision

Very large numbers, such as 1,000,000,000,000, require many bytes to store, as do very small numbers, such as 0.0000000000001. People often use scientific notation when dealing with very large or very small numbers; these two would be written as $1.0 \times 10^{12}$ and $1.0 \times 10^{-13}$. Computers can also use scientific notation, usually called exponential or floating-point data. This notation stores values composed of a **mantissa** (the decimal part of the number) and an exponent. For example, the number 123456789 stored as an exponential number would consist of a mantissa of 1.23456789 and an exponent of 8, yielding $1.23456789 \times 10^{8}$ (often written as 1.23456780e08). The computer truncates the mantissa at a certain level of precision—when storing values in the trillions, differences of tens or hundreds are often of little interest. The number might become 1.2345e08. In ArcGIS, a **single-precision** floating-point field stores eight significant digits of information in the mantissa. A **double-precision** field stores 16 significant digits. Floating-point data types are much more flexible than numeric or binary types because they can store either very large or very small values using the same field.

### Different types for different data formats

Every attribute must be defined before use; that is, the field type must be specified (text or numeric) and the field properties set. Once a field is defined, the definition cannot be changed. If a mistake is made defining the field, one must delete the field and redefine it. Different GIS systems may have different allowable data types, although most use the basic categories of text, integers, and floating-point values. Table 6.1 shows the specific data types used in ArcGIS. Geodatabases also have a Raster and GUID type for advanced users.

Table 6.1. Field data types available for feature classes (*geodatabases only)

| Field type | Explanation | Examples |
|---|---|---|
| Short | Integers stored as 2-byte binary numbers<br>*Range of values – 32,000 to +32,000* | 255<br>12001 |
| Long | Integers stored as 10-byte binary numbers<br>*Range of values – 2.14 billion to +2.14 billion* | 156000<br>457890 |
| Float | Floating-point values with eight significant digits in the mantissa | 1.289385e12<br>1.5647894e–02 |
| Double | Double-precision floating-point values with 16 significant digits in the mantissa | 1.12114118119141e13 |
| Text | Alphanumeric strings | 'Maple St'<br>'John H. Smith' |
| Date | Date/time format for calendar dates and times | 07/12/2008<br>10/17/1963 13:24:06 |
| BLOB* | Binary large object; any complex binary data, including images, documents, etc. | |

## About ArcGIS

### Tables in ArcGIS

Tables in ArcGIS may come from any of the underlying database programs supported by ArcGIS. The tables may be in different formats for storage, but the table itself always looks the same and has the same functions so that users don't need to learn different commands for working with different file types.

Tables are viewed and manipulated in the Table window in ArcMap (Fig. 6.12). This window contains a Table Options menu and several tools at the top. More commands are accessible from a context menu that appears if a field is right-clicked. If multiple tables are opened, they all appear in the Table window. Tabs at the bottom of the window are used to switch between tables. Tables may also be viewed side by side or vertically using a menu under the Table Options.

Fig. 6.12. Parts of a table
Source: ArcGIS

In a shapefile, the table is stored in a dBase format file and includes a unique feature identification number (FID) that links the spatial and the attribute data. In a geodatabase and in any standalone table, each record contains an Object ID (OID) analogous to the row number in a spreadsheet. Both shapefiles and geodatabase feature classes contain a Shape field that represents the $x$-$y$ coordinate values of the feature. Geodatabase tables for line and polygon feature classes also have Shape_Area and/or Shape_Length fields, which keep track of the areas and lengths of features.

Figure 6.12 shows some terminology associated with tables. A table consists of rows and columns. A row is called a record, and it contains information about a single object or feature. A column is called a field, and it stores one type of information. Each field has a name shown in the top row. Field names must contain 13 or fewer characters and should contain only letters, numbers, and underscores. An alternative name, called an **alias**, can be set in the layer properties to give a field a more descriptive name that does not have to follow the naming rules. For example, the somewhat perplexing field MEDREN could be given a more understandable alias, such as MedianRent. Aliases form part of a layer definition. If defined in ArcMap, they persist only inside that map document. If defined as part of a layer file, they can be used in different map documents by loading the layer file instead of the feature class. Settings in the table properties control whether the actual field name or alias is shown in tables and windows as you work.

In an attribute table, some fields are required, and they are created and updated by the program that creates the geographic data set. For example, a shapefile has a Feature ID (FID) field and a Shape field. The geodatabase feature class has an Object ID and a Shape field. *These fields should never be altered by the user.* Likewise, users of shapefiles must make sure that they never delete records in an attribute table unless the accompanying features in the spatial data set are also deleted. Editing features only within an editing session in ArcMap will meet this criterion, as does deleting rows in a geodatabase feature class.

As we learned in Chapter 1, different GIS data formats use different underlying databases. The coverage model uses a database called INFO. Shapefiles use a dBase format file. Personal geodatabases use the database underlying Microsoft Access, called Jet. File geodatabases use their own format. SDE geodatabases use a large-scale commercial RDBMS, such as Oracle or SQL Server. ArcGIS can read comma-delimited text files (fields separated by commas) and tab-delimited text files (fields separated by tabs), as well as Excel™ spreadsheets if they are correctly formatted.

### Editing and calculating fields

ArcGIS offers two ways to change the values in a table: by typing the information directly into the fields or by calculating the value of a field. Typing information into fields must be done during an edit session in ArcMap. Calculating fields can be done inside or outside an edit session and uses the Field Calculator to generate arithmetic expressions using fields as variables. For example, the percentage of Hispanics in each state could be calculated from the two fields containing the total state population and the number of Hispanics (Fig. 6.13).

Calculating within an edit session is safest, because a mistake can be undone. However, it can be inconvenient to start and stop an edit session each time you want to run a calculation. If you choose to calculate outside an edit session, you must be VERY careful to ensure that the field you are calculating either is empty or contains data that are easy to replace (by a different calculation, for example).

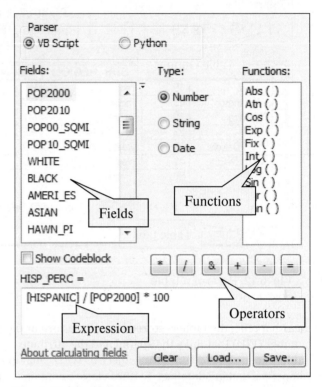

Fig. 6.13. The Field Calculator
Source: ArcGIS

The first window in the Field Calculator (Fig. 6.13) contains the fields in the table that can be used to create the expression. The functions box contains different functions that can be used in expressions. The functions displayed will depend on whether Number, String, or Date is selected. The expression used to calculate is entered in the large box at the bottom. The operators (\*, /, &, +, −, =) appear to the lower right. The Advanced button allows more complex VB Script or Python expressions to be entered. A complicated expression can be saved and loaded for use another time.

The Calculate Geometry tool (Fig. 6.14) can add information on feature areas, perimeters, lengths, or $x$-$y$ centroids to a table. The field to contain the information must already exist. The user may choose which coordinate system and output units to use in the calculation.

Fig. 6.14. The Calculate Geometry tool
Source: ArcGIS

### *Importing tables*

Many GIS projects encounter situations when external tabular data must be imported for use in ArcGIS. Several different formats may be used, but one must always take care that the layout of the records and fields follows the requirements of ArcGIS tables. Field names, for example, must be present in the first row and must follow the ArcGIS naming conventions. Columns that are supposed to hold numbers must hold only numbers and avoid NoData or xx or Null or N/A to indicate missing values. Blank rows, merged cells, and formulas are not allowed. Most import formats are read-only within ArcGIS. They can be viewed, but not edited or changed, and some tools and functions may not work on them. Usually, the import file is exported to a dbf file or geodatabase table if changes to it are needed.

Dbf files, or dBASE files, are commonly produced by database programs. These files can be directly imported to ArcGIS and even modified and returned to the database program. This is the only import format that can be edited within ArcGIS.

Text files, such as those shown in Figure 6.15, come from a variety of sources. They might come from text copied from a web site, data copied from a word processing document, or values produced from a statistical program. Nearly any program that handles tabular data has the option to save them as text, and it is one of the most basic and standard formats for transferring data between programs. All text files are stored using ASCII characters. Three primary formats are found. Figure 6.15 shows the same tabular information stored in all three formats.

A tab-delimited file has the field values separated by the ASCII tab character (Fig. 6.15a). A comma-delimited file has the field values separated by commas (Fig. 6.15b). The fixed-column format (also known as a formatted or space-delimited file) is most often encountered as the output of a computer program, such as FORTRAN. It appears similar to a tab-delimited file, but no tabs are present. Instead, each row is composed of a string of characters, and each field starts and ends at a specified character location (column). In Figure 6.15c, the first field includes columns 1–5; the station field goes from columns 6 to 30, and so on. (The red lines are not actually present in the file but have been added to the figure to help you visualize where each field starts and stops.)

**TIP:** ArcGIS can read either tab-delimited or comma-delimited text files. It cannot read fixed-column text files. (However, Excel™ can read fixed-column files and save them as a comma-delimited CSV file.)

Fig. 6.15. Text file formats: (a) tab-delimited; (b) comma-delimited; (c) fixed-column
Source: ArcGIS

Text files have no way to store what type of data is stored in a field. One problem with importing tabular data and saving them as a dbf file or geodatabase table is that ArcGIS must interpret the data in the field and assign a data type as best it can, and the results may be less than satisfactory. For example, a column containing zip codes may be interpreted as numbers, and any leading zeros will be eliminated (05575 becomes 5575). Putting quotes around them in the text table will enforce that they are to be interpreted as text. Text fields with only a few characters may be given a 255-character field in the new table. Often, too, a text file will contain formatting problems, such as field names with spaces, or an xx in a numeric column to indicate missing data. Such issues will cause the import to fail and may need to be fixed before the import can be successful.

Simple editing of text files can be done using system tools, such as the Notepad program. However, more advanced and powerful editing is available using the Microsoft Excel™ program. It can open and save a variety of text formats and can be used to translate one type to another, such as converting a fixed-column text file to a tab-delimited file. (The files shown in Figure 6.15 were all saved from an Excel™ worksheet.) Since ArcGIS can open Excel™ files directly, it can also be helpful to assign specific text or numeric formatting to the worksheet columns to aid ArcGIS in translating the data types.

**TIP:** To read the newer .xlsx spreadsheet format you may need to install the 2007 Office System Driver on your computer. To work around it, save the spreadsheet in the older .xls format.

## *Mapping x-y coordinates from a table*

Another useful GIS function, Add XY, takes $x$-$y$ coordinates from a table and converts them to locations on a map (Fig. 6.16). The $x$-$y$ values must be given in a real-world coordinate system, such as latitude-longitude or UTM meters. The coordinate system details must also be known, including the geographic coordinate system/datum and the projection, and must be explicitly set in the Add XY window. Moreover, it should match the $x$-$y$ values in the file, rather than the coordinate system of the map. For example, if the map is in UTM and the table $x$-$y$ coordinates are given as longitude-latitude values, the coordinate system must be set to GCS.

| ORstations$ | | | |
|---|---|---|---|
| COOPID | LAT | LON | STATION I |
| 350036 | 42.183333 | -119.900000 | ADEL |
| 350118 | 42.966667 | -120.000000 | ALKALI LAKE |
| 350145 | 44.400000 | -123.750000 | ALSEA F H FALL |
| 350197 | 44.916667 | -120.733333 | ANTELOPE 1 NW |
| 350217 | 42.250000 | -123.166667 | APPLEGATE |

Fig. 6.16. Climate stations from a spreadsheet displayed as points on a map
Source: ArcGIS

Global positioning systems (GPS) frequently provide tables of $x$-$y$ data. A GPS receiver can calculate its location by traingulating distances and directions from a constellation of orbiting GPS satellites and may produce tables of $x$-$y$ locations as output. Other examples of $x$-$y$ data sources include benchmarks, surveyed points, and locations measured on a map. Regardless of the source, the procedure for mapping $x$-$y$ locations from a table is identical.

The layer added from $x$-$y$ points is called an **event layer**. This terminology comes from considering each entry in the table to be an event, such as an earthquake or a traffic accident. Not all tables contain events, of course, but the terminology remains. Event layers appear similar to point feature classes, but they do not constitute an actual feature class, for they remain a table of objects represented by spatial points. Some geoprocessing functions will not accept event layers as input. In many cases, users will export the event layer to a shapefile or a geodatabase feature class to make it permanent. When exported, all the attribute fields in the original table, including joined fields, become attributes of the feature class.

Latitudes and longitude outside GIS systems are commonly reported in degrees-minutes-seconds and must be converted to decimal degrees before they can be used to create an event layer. This conversion is most easily performed in Excel™ using the formula:

$$X \text{ in decimal degrees} = \text{degrees} + \text{minutes}/60 + \text{seconds}/3600$$

For example, 103° 30' 15" would convert to $103 + 30/60 + 15/3600 = 103.504167$. Keep in mind that degrees are large units, and six or seven decimal places should be kept to ensure reasonable accuracy. At the equator, a meter is equal to approximately $10^{-5}$ degrees. Also, longitudes west of the Prime Meridian in England are negative, but the minus sign is often omitted for convenience. If this longitude is reported by a GPS unit in Nebraska, for example, the true value needed by the GIS is $-103.504167$.

## Summary

▶ Tables consist of rows and columns of information. A row is associated with one feature and includes columns of information called fields.

▶ Tables associated with spatial data sets are called attribute tables and contain records, one for each feature in the data set. Standalone tables are not associated with map features.

▶ Relational database management systems construct temporary links between data tables and are the preferred model for GIS software.

▶ Queries allow the user to select certain records based on one or more criteria. Once selected, these records may be viewed, exported, or analyzed.

▶ Tables may be joined or related on a common field in order to access information in one table from another. Joins may be performed on tables with one-to-one or many-to-one cardinality. Relates must be used on tables with one-to-many or many-to-many cardinality.

▶ The Statistics function calculates basic statistical values for all selected records in a table. The Summarize command generates statistics about groups of features defined by a categorical field.

▶ Fields must be defined to contain a specific type of data, such as text, integers, dates, or floating point values. Once defined, the field type cannot change.

▶ ArcGIS tables can display and manipulate tabular data from a variety of sources, including dBase files, INFO files, geodatabases, SQL queries, or comma-delimited text files.

▶ New fields may be added to tables by defining names and field types. New tables may have data entered by typing in values or by using the Calculate function. The Calculate Geometry function can put areas and lengths in attribute tables for the corresponding features.

**TIP:** Field names must have 13 or fewer characters; may include letters, numbers, and the underscore character (_); and should not contain spaces or special characters, such as @, #, !, $, or %. Field names must also start with a letter, not a number.

### Important Terms

| | | | |
|---|---|---|---|
| alias | double-precision | key | Rule of Joining |
| ASCII | event layer | logical expression | scale |
| attribute field | field | mantissa | selected set |
| attribute table | field definition | precision | single-precision |
| binary | field length | query | source table |
| byte | flat file database | record | standalone table |
| cardinality | hierarchical database | relate | Summarize |
| destination table | join | relational database | table |

## Chapter Review Questions

1. Describe the difference between an attribute table and a standalone table.

2. Which type of databases are GIS systems based on? How does this type of system differ from other DBMS types?

3. List the types of data sources from which tables may display data.

4. Describe how storing the number 255 in ASCII differs from storing it as a binary representation.

5. Choose the best field type for each of the following types of data in a geodatabase:

    populations of countries in the world

    precipitation in inches

    number of counties in a state

    highway name

    distances between US cities, in meters

    birthdays

6. What is the cardinality of each of the following relationships? Assume that the destination is on the left.

    students to college classes

    states to governors

    students to grades

    counties to states

7. Describe the differences between a join and a relate.

8. You have a table of states and a table of airports, both with a state abbreviation field. Can you join them if states is the destination table? If airports is the destination table? Explain your answer.

9. Describe the difference between using Statistics and using Summarize functions on a field.

10. For each of the following problems, using data sets for the United States, state whether using a query, the Statistics function, or the Summarize function would be the best approach to solving it.

    _____ Find all towns with more than 20,000 people.
    _____ Find the total number of volcanoes in each state.
    _____ Determine the total damage caused by earthquakes in the United States.
    _____ Find the states in which Hispanics exceed the number of African Americans.
    _____ Find out which subregion of the country has the most Hispanics.

# Mastering the Skills

## Teaching Tutorial

The following examples provide step-by-step instructions for doing basic tasks and solving basic problems in ArcGIS. The steps you need to do are highlighted with an arrow →; follow them carefully. Click on the video number in the VideoIndex.htm file to view a demonstration of the steps.

→ Start ArcMap, if necessary. Navigate to the MapDocuments folder in the mgisdata directory and open the map document ex_6.mxd.

→ Use Save As to give the document a new name and save frequently as you work.

### Viewing tables

Let's begin by exploring some tables and learning some basic skills for working with them. Many table settings affect only how the tables appear and do not change the underlying data.

1→ Right-click on the US States layer and choose Open Attribute Table.

1→ Narrow the STATE_NAME field to a more suitable width by clicking and dragging the right border of the field.

The OBJECTID field simply counts the rows, so hide it from sight.

1→ Right-click the OBJECTID field name and choose Turn Field Off.

Notice that the STATE_ABBR field is also too wide for its data but that narrowing it will cut off the field name. Each field has display properties that can be set.

2→ Right-click on STATE_ABBR and choose Properties.

2→ Change the Alias to *ABBR*, and click OK.

2→ Right-click the STATE_NAME field, open Properties, and give it the alias *NAME*.

Scroll to the right and look at the fields of population data from the US census. It is difficult to match the values to the right state, however, because the state names quickly scroll out of sight.

3→ Right-click on the NAME field and choose Freeze/Unfreeze Column.

3→ Now scroll again and see how much easier it is to read the data.

3→ Right-click on the NAME field and choose Freeze/Unfreeze Column again to unfreeze the field.

**TIP:** An unfrozen column will not automatically go back to its original place.

Now let's use the Sort function to obtain information about the largest and smallest states.

3→ Right-click on the POP2014 heading and choose Sort Ascending from the menu.

**TIP:** Right-click a field name and choose Advanced Sorting to sort on more than one field.

Notice that the POP14_SQMI field has decimal values in it, which makes it hard to compare the population densities of different states. We can format the field so that no decimals are displayed.

4→ Right-click on the POP14_SQMI field and choose Properties.

4→ Click on the button next to Numeric to open the Number Format window.

4→ Fill the button that says *Number of decimal places* and enter 0. Click OK and OK.

We can even modify which fields are shown and the order in which they appear.

5→ Right-click the US States layer and choose Properties. Click the Fields tab.

5→ Click the Options menu and make sure that Show Field Aliases is checked.

5→ Click on the NAME field in the list on the left. Notice the details about its appearance and definition on the right (Text data type with a length of 25 characters).

5→ Click the Turn All Fields Off button to turn all the fields off. Click Apply and move the Properties window aside so that the table is visible.

5→ Turn the check boxes on for the NAME and POP2014 fields, and click Apply. These two fields are the only ones shown.

5→ Click the Turn All Fields On button to turn all the fields back on. Click Apply.

6→ Click on the NAME field in the Properties window to highlight it.

6→ Click on the Move Down button to move the NAME field back to its original position below the Shape field. Click Apply.

6→ Click on the Options button and examine the menu items. Then choose Sort Ascending, and click Apply.

6→ Click Options again and choose Reset Field Order and click Apply.

1. Examine the properties for the POP2014 and POP14_SQMI fields. Why does one use an integer field type and the other a floating point field type?

**TIP:** Use Reset Field Order to move an unfrozen field back to its original place.

7→ Close the US States Layer Properties window.

7→ Turn your attention again to the US States table in the Table window.

7→ Examine the text at the bottom that says (0 out of 51 Selected). It tells you that there are 51 records in this table (50 states plus the District of Columbia).

7→ Right-click the US Counties layer in the Table of Contents and choose Open Attribute Table. It opens as a new tab in the same Table window.

8→ Click on the US States tab in the lower-left corner of the Table window to switch back to the US States table.

8→ To see both tables at once, choose Table Options > Arrange Tables > New Vertical Tab Group. The tables now appear side by side.

8→ Choose Table Options > Arrange Tables > Move to Previous Tab Group to return to the original layout.

8→ Right-click the US Counties tab at the bottom of the Table window and choose Close, leaving only the US States table.

**TIP:** The changes to the table demonstrated in this section only affect the properties of the layer in this map document. They do not change the source file.

## Using queries and statistics on tables

Often it is necessary to isolate a subset of records from a table and work with them alone. GIS handles this task using a query. Imagine that you are exploring the politics of the larger states. Let's select the states that had more than 5 million people in the year 2014.

9→ Click the Select By Attributes button in the Table window.

9→ Enter the expression POP2014 > 5000000 in the query box.

9→ Click the Verify button to make sure that you entered the expression correctly. If it is correct, click OK. (If not, click OK, then Clear, and try again.)

9→ Click Apply to execute the query, then close the Select By Attributes window.

**TIP:** Field names in queries may appear with brackets or quotes or nothing around them (like [POP2014] or "POP2014") depending on the version of ArcMap and the format used to store the table. The window will always select the right style if you double-click the field names rather than entering them by typing.

The selected states are highlighted in the map and in the Table window. Also, the text at the bottom of the Table window changes to read *22 out of 51 Selected*, indicating that 22 states have more than 5 million people.

10→ Click on the Show Selected Records button at the bottom of the Table window to view only the selected records.

10→ Click the Show All Records button to show them all again.

Suppose you now want to look at the states with fewer than 5 million people. The Switch Selection button is faster than another query.

10→ Click the Switch Selection button in the Table window toolbar (or choose Table Options > Switch Selection).

10→ Clear the selected set by clicking the Clear Selection button on the Table window.

Now it is time to view some population statistics for the states.

11→ Right-click on the POP2014 field name and choose Statistics from the menu.

11→ Review the statistics values and examine the frequency distribution.

2. Is this a normal distribution? _____ What is the population of the largest state? _____ (Be sure to report it using an appropriate number of significant figures.)

11→ Use the drop-down list in the Statistics box to select the POP10_SQMI field.

11→ Close the Statistics window.

The Statistics command uses all records, unless a subset of records has been selected, in which case it uses only the selected subset to calculate the statistics.

12→ Click the Select By Attributes button in the Table window again.

12→ Click Clear to delete the previous expression, if necessary.

12→ Enter the expression SUB_REGION = 'New England'. Click Verify to check it, and then click Apply and close the Select By Attributes window.

**TIP:** Beginners often find it easier to enter expressions using double-clicks and buttons, rather than trying to type the expressions manually. The program helps you get the syntax right.

13→ Right-click the POP2014 field in the Table window and choose Statistics. Examine the Count statistic and observe that only six states are included—the New England states just selected.

13→ Close the Statistics window.

13→ Click the Clear Selection button in the Table window to clear the selected records.

13→ Close the Table window.

14→ Turn off US States in the Table of Contents and turn on 114th Congress.

14→ Open the table for the 114th Congress layer.

114→ Examine the fields, noting the field PARTY.

15→ Open the layer properties for 114th Congress layer and click the Symbology tab.

15→ Create a Categories: unique values map based on the PARTY field. (Remember to click the Add All Values button to show the categories). Click OK.

15→ Right-click the symbols in the Table of Contents to change the symbol for the Democratic counties to blue, the Republican counties to red, and the Vacant districts to light gray (Fig. 6.17).

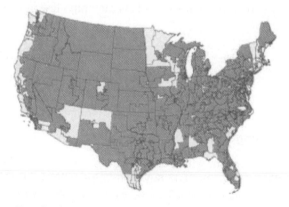

Fig. 6.17. 114th Congress districts
Source: ArcMap

How many Democratic districts are there? We can find out using a query.

16→ Click the Select By Attributes button in the Table window.

16→ Enter the expression PARTY = 'Democrat'. Use the Get Unique Values button so that you can select the 'Democrat' value instead of typing it.

16→ Click Apply and close the Select By Attributes window.

3.  How many Democratic districts are there? _____

Finding the number of Democratic districts using a single query is fine. But suppose you wanted to know the number of districts for all three categories: Republican, Democrat, and Vacant? The Summarize command can generate statistics about multiple groups defined by a field.

16→ Use the Clear Selection button in the table window to clear the selected records.

17→ Right-click the Party field and choose Summarize. Note that the Party field is already entered in box #1 because you right-clicked it.

17→ You are only interested in the number of districts, which is always counted by default, so you need not enter any statistics. Do not change anything in box #2.

17→ Click the Browse button. Make sure that the *Save as type* is set to dBASE table and that you are currently in the Usa folder. Give the new file the name cd114numdist.dbf.

17→ Click Save and OK, and answer Yes to add the resulting table to the map.

**TIP:** Because this table was created to answer a transitory question, we saved it in the folder instead of making it a permanent addition to the geodatabase. Be sure to give your saved tables descriptive names that indicate what they contain.

Examine the Table of Contents window. The icons at the top change how layers are listed. Originally, it was set to the first icon on the left, List By Drawing Order. Because a standalone table was added, it switched to the second view style, List By Source. Each layer is now listed under its folder or geodatabase name. This is the only view in which standalone tables are visible.

Source: ArcMap

18→ Right-click the new cd114numdist.dbf table and choose Open.

18→ See that all three categories are listed, with the Count_PARTY field showing the number of districts in each category.

18→ Right-click the cd114numdist tab in the Table window and choose Close.

### Joining tables

Let's find out which districts may have changed their party affiliation since the 113th Congress. We need to compare the party in two different tables using a query, so we must combine the two tables using a join. To join, we need to find a key field.

19→ Add the cd113 feature class from the mgisdata\Usa\usdata geodatabase.

19→ Right-click the cd113 layer and open the table. Examine the fields and try to identify the appropriate key field that might work to combine the tables. The key must be unique to each record and have the same value in both tables.

4. What is the best potential key field in this table? _____

19→ Click the 114th Congress tab in the Table window and confirm that this field is present.

We want 114th Congress to be the destination table. It is currently the active table in the Table window, so it will become the destination table as we intend.

20→ Open the Table Options menu and choose Joins and Relates > Join.

20→ Make sure that the top drop-down reads *Join attributes from a table* (Fig. 6.18).

20→ Choose DISTRICTID as the field that the join will be based on.

20→ Choose cd113 as the table to join to the layer.

20→ Choose DISTRICTID as the field in the table to base the join on.

20→ You don't need to validate the join. Make sure that your window reads as shown in Figure 6.18, and then click OK. Choose Yes if asked to create an index.

20→ Scroll to the right, noticing the additional fields now in the table.

**What do you want to join to this layer?**

Join attributes from a table

1. Choose the field in this layer that the join will be based on:

   DISTRICTID

2. Choose the table to join to this layer, or load the table from

   ◇ cd113

   ☑ Show the attribute tables of layers in this list

3. Choose the field in the table to base the join on:

   DISTRICTID

Join Options
   ⦿ Keep all records

Fig. 6.18. Joining the tables
Source: ArcMap

The fields from cd113 have been appended to the right of the fields for 114th Congress. However, it can get confusing to remember which fields came from which table.

21→ Choose Table Options and select Show Field Aliases to uncheck it.

Now each field name is prefixed with the name of its table (the original name, not the title in the Table of Contents). Hence, the 114th Congress table fields are prefixed with the name cd114, and the cd113 fields are prefixed with cd113.

Now we can use a query to find the districts that have changed affiliation so that the cd114.PARTY field is different from the cd113.PARTY field.

22→ Click the Select By Attributes button in the Table window.

22→ Clear the previous expression, if necessary.

22→ Double-click the cd114.PARTY field to enter it in the expression box; then click the "< >" (not equal to) button, and double-click the cd113.PARTY field.

22→The final expression should look like cd114.PARTY <> cd113.PARTY. Verify it and then click Apply. Close the Select By Attributes window.

22→ Click the Table button to Show selected records.

5.   Which party received most of the changed seats in the 114th Congress? _____

Let's save the selection as a new layer for future reference and to symbolize it differently.

23→ Right-click the 114th Congress layer in the Table of Contents, and choose Selection > Create Layer From Selected Features.

23→ The new layer appeared near the top as 111th Congress selection. Rename it *Changed Party 114*.

23→ We can turn off the join now. Click the Table Options menu, and choose Joins and Relates > Remove Join(s) > cd113.

23→ Clear the selected records, click the button to Show All Records, and close the Table window.

24→ Turn off the cd113 layer.

24→ Click on the Changed Party 114 layer symbol to open the Symbol Selector and change it to 10% Crosshatch, which shows the changed districts without obscuring the party color (Fig. 6.19).

24→ You may need to click the Refresh button (next to the Layout View button) to make the feature selection go away.

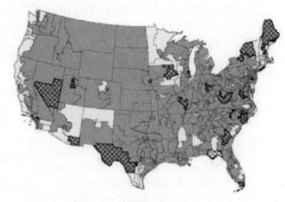

Fig. 6.19. Map of changed districts
Source: ArcMap

### Summarizing tables

Some states have only one district; others have more. Suppose that we want to know how many districts each state has. The Summarize function will yield this information. It will group the districts by state and calculate statistics for each group.

25→ Open the attribute table of 114th Congress. Make sure no records are selected.

25→ Right-click on the STATE_ABBR field and choose Summarize.

The Summarize command always counts the number of features in each group and will report it in the output table, so you don't need to request any other statistics.

25→ Enter the name of the new table to be created, *dists_per_state*, making sure it will be saved as a dBase file in the Usa folder. Click OK.

25→ Click Yes to add the table to the map document.

6. Is this new table an attribute table or a standalone table? _____

26→ Find the new dists_per_state.dbf table in the Table of Contents and open it.

26→ Right-click the Count_STATE_ABBR field and choose Sort Descending.

7. Use a query to determine how many states have only one district: _____. Clear the selection before going on.

Sometimes the goal is to generate statistics about each group in addition to counting the records. Perhaps you would like to know the total area represented by each party.

27→ Click the 114th Congress tab in the Table window.

27→ Right-click the PARTY field and choose Summarize.

27→ Verify that the field to summarize is set to PARTY (Fig. 6.20).

27→ Locate the SQMI field in the list, expand it, and choose Sum as the statistic to calculate.

27→ Enter a location and name for the output table, calling it *partyarea.dbf* and placing it in the Usa folder.

27→ Click OK and Yes.

28→ Find the new partyarea table in the Table of Contents and open it.

28→ Right-click the Sum_SQMI field and choose Sort Descending.

8. Which party represents the greater area, and by how much? _____

28→ Close the Table window before going on.

A map showing the number of districts in each state would be nice, but the dist_per_state table is a standalone table and can't be mapped. However, it can be joined to the US States feature class. In this join, the US States table must be the destination.

9. What is the cardinality of this join?

_____

29→ Right-click on the US States layer and choose Joins and Relates > Join.

29→ Choose ABBR as the field in US States to base the join on.

29→ Choose dists_per_state as the table to join.

29→ Choose STATE_ABBR as the field in the source table to join on.

29→ You don't need to validate the join. Click OK to finish.

29→ Open the US States table, and scroll all the way to the right to find the new fields added as a result of the join.

The number of districts in each state is contained in the Count_STATE_ABBR field.

30→ Close the US States table.

30→ Turn off all of the layers in the map, and turn on the US States layer.

30→ Open the US States layer properties, click the Symbology tab, and create a Quantities: graduated color map based on the Count_STATE_ABBR field (Fig. 6.21).

| 1. Select a field to summarize: |
| PARTY |

2. Choose one or more summary statistics to be incl output table:

⊞ STFIPS
⊞ STATE_ABBR
⊞ POP2008
⊟ SQMI
  ☐ Minimum
  ☐ Maximum
  ☐ Average
  ☑ Sum
  ☐ Standard Deviation
  ☐ Variance

3. Specify output table:

C:\gisclass\mgisdata\Usa\partyarea.dbf

Fig. 6.20. Summarizing statistics
Source: ArcMap

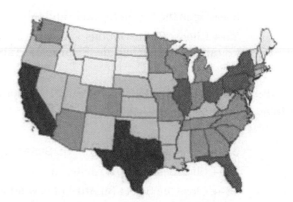

Fig. 6.21. The number of districts
Source: ArcMap

**TIP:** Hold the Ctrl key and check/uncheck one box to turn all layers on/off at once.

31→ Right-click the US States layer and choose Copy.

31→ Right-click the Layers data frame and choose Paste Layer(s).

31→ Rename the new layer *Number of Districts*.

### Relating tables

It would be interesting to group the representatives according to subregion, as they would have similar issues to address with their constituencies. However, the SUB_REGION field is in the US States table, not the districts table. Setting up a relate between the US States table and the 114th Congress table will allow us to select a subregion and obtain a list of representatives from it. The common key between the tables is the state abbreviation.

10. What is the cardinality between subregions and districts? _____

11. Which is the source table, and which is the destination table? _____

32→ First remove the previous join by right-clicking on the US States layer and choosing Joins and Relates > Remove Join(s) > Remove All Joins.

32→ Right-click on the US States layer and choose Joins and Relates > Relate.

32→ Enter ABBR as the field to base the relate on (Fig. 6.22).

32→ Enter 114th Congress as the source table to be related.

32→ Enter STATE_ABBR as the key field from the related table.

32→ Enter *Congress* for the name of the relate. Click OK.

1. Choose the field in this layer that the

ABBR

2. Choose the table or layer to relate to

◈ 114th Congress

3. Choose the field in the related table c

STATE_ABBR

4. Choose a name for the relate:

Congress|

Now let's use the relate to find out which representatives are from New England states.

Fig. 6.22. Setting up a relate
Source: ArcMap

33→ Open the US States table and the 114th Congress table.

33→ Choose Table Options > Arrange Tables > New Vertical Tab Group so that both tables are side by side. Enlarge the Table window if needed to see both tables well.

**TIP:** When more than one table is visible, the one with the darker bar across the top is the active table, and choices made from the Table Options menu will be applied to the active table.

33→ Click the title bar of the US States table to make sure it is the active table.

33→ Click Select By Attributes and select the states in the New England subregion.

33→ Close the Select By Attributes window.

Now that the New England states are selected and highlighted, the relate can access the representatives from these states.

34→ Click on the Table Options button and choose Related Tables > Congress: cd114.

34→ The other table view highlights the related records, now also selected.

The Changed Party 114 layer is based on a selection from the 114th Congress layer, and so both of them show up in the relate. We're not interested in the changed representatives right now, and there are few matches, so let's close that table.

34→ Right-click the Changed Party 114 tab in the Table window and choose Close.

12. How many representatives come from New England states? _____

Exporting the selected records from the representatives table will provide a new file that can be saved permanently and given to interested constituents.

35→ Click the bar of the 114th Congress table to make sure it is active.

35→ Choose Table Options > Export. Make sure it says to export the *Selected records*.

35→ Click the Browse button.

35→ Change the *Save as type* to dBASE Table. Navigate inside the Usa folder.

35→ Type in *NE_reps.dbf* as the name of the table and choose Save and OK.

35→ Say Yes to add the table to the map.

35→ Open the new NE_reps table and verify that it contains the correct representatives.

36→ Choose Selection > Clear Selected Features from the main menu to clear the selection in all tables.

36→ Click the US States table to make it active and choose Table Options > Joins and Relates > Remove Relate(s) > Remove All Relates.

36→ Close the Table window.

### Editing values in tables

Sometimes fields in a table must be created or updated manually. Imagine that you wish to add more information to the partyarea table you created earlier, indicating status as the majority or minority party.

37→ Open the partyarea table.

37→ Click Table Options and choose Add Field. Name it *STATUS*.

37→ Set the Type to Text and set the Length to 10. Click OK.

**TIP:** Always think about how many characters or decimal places you need before creating a field.

To type values into a table, you must be in an edit session. Editing can be performed in only one folder or geodatabase at a time, so you must specify which source you want to edit.

38→ Click the Editor Toolbar button on the Standard toolbar, to open the Editor toolbar.

38→ Choose Editor > Start Editing from the Editor toolbar.

38→ Click the partyarea table to highlight it and click OK.

Notice that the field headings turn white, indicating that this table is now editable.

39→ Click in the STATUS field in the Democrat row and type Minority. Click Enter.

39→ Type Majority in the Republican row and click Enter.

39→ Type Vacant in the Vacant row and click Enter.

39→ Choose Editor > Save Edits.

39→ Choose Editor > Stop Editing to close the edit session.

39→ Close the Table window.

**TIP:** You cannot add a new field to a table during an edit session.

### Calculating fields in tables

Imagine that we are writing a report and need a table showing the percentage of Hispanics in each state. The US States table has only the number of Hispanics. Creating a new field and calculating the percentage of Hispanics solves the problem.

40→ Open the US States attribute table.

40→ Click on the Table Options button and choose Add Field.

40→ Name the field HISP_PERC and select Float as its type. Click OK.

Now calculate the percentage of Hispanics. The minority population fields in this table represent 2010 census data, not 2014, so we will use the POP2010 field for calculating percentages.

**TIP:** Be VERY careful when calculating, because you can overwrite existing data and won't be able to undo it unless you are in an edit session. Be sure a field is empty before you calculate it.

41→ Right-click on the heading of the new, _empty_ HISP_PERC field (located to the far right of the table), and choose Field Calculator from the context menu. Click Yes.

41→ Double-click the field names and the operator buttons to the expression HISPANIC / POP2010 * 100 and click OK.

**TIP:** Do all of the records equal zero, or null? You might have accidentally selected a single record by clicking on the table. Check the bottom status bar to see if any records are selected. If so, choose Table Options > Clear Selection from the table and try the calculation again.

Finally, for a report you need to determine the density of Hispanics per square kilometer in the counties. You'll need to create and calculate two new fields, the area in kilometers and the Hispanic density. Because you're calculating areas, make sure the data frame coordinate system is set to an equal-area projection.

42→ Close the Table window.

42→ Open the Layers data frame properties and click the Coordinate System tab.

42→ Set the coordinate system to the Projected Coordinate Systems > Continental > North America > USA Contiguous Albers Equal Area Conic projection.

43→ Open the US Counties attribute table.

43→ Click the Table Options button and choose Add Field.

43→ Name the field AREA_KM and set its Type to Float. Click OK.

43→ Add another new field named HISP_SQKM, also making it a Float type.

**TIP:** It is safer, but less convenient, to open an edit session before doing field calculations. Then you can undo a mistake if you make one.

Now we will calculate the new fields. Areas are calculated with a special function.

44→ Right-click the <u>empty</u> AREA_KM field heading and choose Calculate Geometry. Click Yes to calculate outside an edit session.

44→ Set the Property to Area, fill the button to *Use the coordinate system of the data frame,* and set the units to square kilometers. Click OK.

45→ Right-click the <u>empty</u> HISP_SQKM field and choose Field Calculator. Enter the expression HISPANIC / AREA_KM and click OK.

45→ Close the Table window. Turn off all layers and then turn on the US Counties.

46→ Open the symbology properties for US Counties and create a Quantities: graduated colors map using the HISP_SQKM field.

The population density of Hispanics has an extremely skewed distribution due to the presence of densely populated areas like Los Angeles. The Jenks classification method does not show the distribution of Hispanics well.

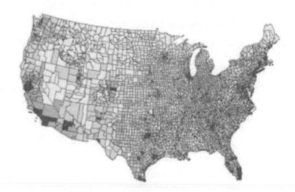

46→ Open the US Counties symbology properties again.

46→ Use the Classify button and change the classification scheme to Geometrical Interval with five classes. Click OK and OK and view the new map (Fig. 6.23).

46→ Save the map document.

Fig. 6.23. Density of Hispanics
Source: ArcMap

**SKILL TIP:** Learn how to create new, empty tables and how to edit values in tables (Tables).

### *Working with Excel and XY data*

We will do a brief exercise on using Excel data in ArcGIS and displaying locations in a table as points on a map.

47→ Click on the New Map File button.

47→ Change the default geodatabase to mgisdata\Oregon\oregondata. Click OK to open the new blank map.

48→ Add the gtoposhd raster from the mgisdata\Oregon\oregon geodatabase.

48→ Click the Add Data button, navigate up to the mgisdata\Oregon folder, and double-click the ORstations.xlsx workbook. Select the ORstations$ worksheet and click Add.

**TIP:** An Excel .xls or .xlsx file is a workbook that may contain multiple worksheets. Only single worksheets may be opened in ArcMap.

**TIP:** If you receive a "Cannot connect" error when you open a spreadsheet, then your system requires the 2007 Office System Driver to be installed. Or use an .xls version of the spreadsheet.

48→ Open the ORstations$ table and examine the fields. Note the LAT and LON fields containing the station location in decimal degrees.

Any table with $x$-$y$ coordinate locations can be displayed as a point layer. You need to know the coordinate system of the $x$-$y$ locations in the table. In this case the units are degrees of latitude and longitude, so we know that it is a geographic coordinate system (GCS). A look at the web site where these data were downloaded from the National Climatic Data Center (NCDC) would tell us that the GCS is based on the NAD 1983 datum.

49→ Close the Table window.

49→ Right-click the ORstations$ table name and choose Display XY Data.

49→ Set the X Field to LON and the Y Field to LAT.

49→ Click the Edit button to set the coordinate system.

49→ Choose Geographic Coordinate Systems > North America > NAD 1983 (be sure to choose the plain option). Click OK and OK. Click OK in the notice.

Observe the station points, called an event layer. The precipitation values are stored in a different worksheet. We will add it and join it to the event layer so that we can map the precipitation.

50→ Click Add Data. Navigate up to the Oregon folder and double-click the ORprecipnormals.xlsx file. Add the Sheet1$ worksheet.

50→ Open the Sheet1$ table and examine the monthly and annual precipitation values for each station.

50→ Open the ORstations$ table and compare it with the Sheet1$ table in the Table window.

**TIP:** Always open an Excel table after adding it to make sure that it was interpreted correctly.

13. What is the common field in these two tables? _____

51→ Close the Table window.

51→ Right-click the new ORstations$ Events layer and choose Joins and Relates > Join. Click Yes to continue.

51→ Choose STATION NAME as the field to base the join on.

51→ Choose the Sheet1$ as the table to be joined, and STATION as the second field. Click OK.

51→ Open the attribute table for the ORstation$ Events layer and examine it to make sure it contains the precipitation values. Close it when finished.

An event layer is a temporary layer in a map document. You can export it to save it in permanent form in the oregondata geodatabase.

> 52→ Right-click the ORstations$ Events layer and choose Data > Export Data.
>
> 52→ Choose to use the same coordinate system as the data frame so that the stations will be stored in a projection (Oregon Statewide Lambert) rather than the original GCS.
>
> 52→ Click the Browse button. Change the *Save as type* to File and Personal Geodatabase feature classes.
>
> 52→ Navigate inside the mgisdata\Oregon\oregondata geodatabase and enter *precip* for the name of the output feature class. Click Save, OK, and Yes to add the feature class.

**TIP:** Exporting a joined table or layer will place the attribute fields of the source table in the output along with those from the destination table.

> 53→ Right-click the ORstations$ Events layer, which you no longer need, and choose Remove.
>
> 53→ Open the symbology properties of the new precip layer and create a Quantities: graduated symbol map based on the ANN field (Fig. 6.24).

This is the end of the tutorial.

> → Close ArcMap. Save the map document if you wish.

Fig. 6.24. Oregon precipitation
Source: ArcMap

**SKILL TIP:** Not all Excel files can be successfully opened in ArcMap. Learn more about which files are suitable and how to set up a file that can be read (Tables).

## Exercises

Open the ex_6.mxd map document in the MapDocuments folder of the mgisdata directory and answer the following questions.

1. (a) Recalculate the earthquake statistics without the missing data values and compare the results to the values reported in Figure 6.8.
   (b) Perhaps many quakes have missing magnitude values because they occurred before 1960, when seismic networks for measuring magnitudes were not yet common. Test this hypothesis and report what you find.
   (c) Examine the quake magnitudes on a map and propose an explanation for why the smaller quakes (magnitude less than 4) were included in the data set. (**Hint:** Use the query expression MAG > 0 AND MAG < 4.)

2. How many counties in the United States have the name Washington? What is their total 2014 population? Which one has the largest area?

3. Calculate the percentage of the population in each state that is African American. (Use the BLACK field for the African Americans and the POP2010 field for the total population in each state.) Which state has the largest *number* of African Americans? Which state has the largest *percentage* of African Americans?

4. Which subregion of the country has the greatest number of African Americans? How many does it have?

5. Add the table dating from the mgisdata\Usa\usdata geodatabase to the map document. This table contains information about marital status by county. Examine the table. Which fields could you use to join this table to the US Counties table? Why would the county name field NOT work?

   In the US Counties table, examine the three fields STATE_FIPS, CNTY_FIPS, and FIPS. How are these fields related?

6. Join the dating table to the US Counties table. Then select the counties that *lost* population in the 20-year period from 1990 to 2010. How many are there?

   In what area of the country are these counties predominantly located? Make a map showing the "losing" counties. Capture the map. (**Hint:** Create a new layer from the selected features.)

7. Use Summarize to determine the number of counties in each state that lost population. Which three states had the most losing counties, and how many losing counties did each have?

8. Imagine that you are the PR director of a group seeking aid for Native American artists, and you wish to contact the *Democratic* representatives in the 114th Congress who are from states in which the number of Native Americans and Eskimos (AMERI_ES field in US States) exceeds 150,000. Which states are included in the list? How many representatives will you need to contact? (**Hint:** Create a layer from the districts in the qualified states.)

9. Using the Major Cities layer, determine how many people in the United States lived in state capitals in 2014. What is the smallest, largest, and average population of the capitals?

10. Which state has the highest percentage of owner-occupied housing units? Which one has the lowest? What are the percentages? Create a map showing the percentages using a standard deviation classification method.

## Challenge Problem: Does Teaching Pay?

Part I. Do an Internet search to find a table, preferably an Excel spreadsheet, of teacher salaries by state. Format the table so that it can be imported into ArcMap and joined to the states feature class in the usdata geodatabase. Export the joined table to a new feature class and remember to create/update the Item Description for it. Create a map layout showing the salaries.

Part II. Search the Internet for the Cost of Living Index by state. Find a way to add this information to your states feature class. Figure out a way to calculate an adjusted salary based on the cost of living in the state. Edit the map layout to show both maps. Include an explanation of how the adjustment was done on the layout, and remember to cite your sources.

# Chapter 7. Basic Editing

## Objectives

▶ Understanding topological errors

▶ Using snapping to ensure topological integrity of features

▶ Adding features to map layers using basic editing functions

▶ Using the sketching tools and context menus to precisely position features

▶ Entering and editing attribute data

# Mastering the Concepts

### GIS Concepts

Editing can update existing feature classes or create new ones. If a housing subdivision is added to a city, the new roads, parcels, sewer lines, and other infrastructure need to be added to the city database to ensure that it is up-to-date. Entirely new feature classes can be created, for example, if the city planning department decides to create a map of garbage collection zones where none existed before. Parcel ownership attributes must be updated when parcels are sold.

**Topology** refers to the spatial relationships between features in terms of adjacency, connectivity, intersection, or overlap. When editing, care must be taken to create and maintain topological integrity between features, so that the relationship in the database matches the relationship in the real world. If two parcels share a common boundary, then the boundaries should match exactly. Line features, such as streets or water lines, should connect to each other in the feature class. Lines that cross each other should intersect at a node. Lines and polygon boundaries should not cross over themselves. These basic rules must be observed to ensure the **logical consistency** of features so that they properly represent the relationships of their real-world counterparts.

Topological data models permit the user to test, locate, and fix topological errors. However, topology is even more important when editing spaghetti models because the user must manage it without help. This chapter introduces two capabilities that aid in creating and maintaining topological integrity: snapping and coincident boundary creation.

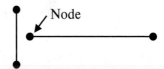

A dangle — the two lines fail to connect.

Correct topology — the horizontal line intersects the vertical one, creating three lines.

Fig. 7.1. Topological relationships between lines

#### Snapping features

When creating line features that connect to each other, such as roads, one must take care that the features connect properly (Fig. 7.1). A line that fails to connect is called a **dangle**. Although you may not be able to see the gap between the lines, the gap will exist unless the endpoints of each line (nodes) have exactly the same coordinate values. Even though the map may look correct, certain functions, such as tracing networks or locating intersections, will not work properly. It is impossible to intersect lines properly by simple digitizing.

**Snapping** ensures that the nodes of lines and the vertices of polygons match. When snapping is turned on, it affects features being added or modified. If you place the cursor within a specified distance of an existing node or vertex (Fig. 7.2), then the new feature is snapped to the existing one—that is, the coordinates are matched at that one point. This distance is called the **snap tolerance**.

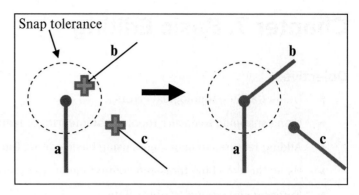

Fig. 7.2. Snapping. Line "b" snapped to "a" because the cursor was inside the snap tolerance. Line "c" did not.

Care must be taken in specifying the snap tolerance. If it is too small, then features won't snap. If it is too large, then features may snap when it isn't appropriate. Snap tolerances are set in either pixels on the screen or meters in the map units. A snap tolerance of 10 pixels indicates that features will be snapped if the cursor moves within 10 screen pixels of an existing feature. When set in pixels, the snap tolerance remains the same as the user zooms in and out, which is convenient. Setting the snap tolerance in map units is useful if the user is trying to maintain a particular level of geometric accuracy independent of zoom scale. Four types of snapping can be used.

**Point snapping** is only used for point feature classes and snaps to an existing point.

**End snapping** only allows a new vertex to be snapped to the endpoints of an existing line. End snapping can ensure that new streams connect to the ends of existing streams, and only to their ends. End snapping only applies to line features.

**Vertex snapping** allows the endpoints of the new line to be snapped to any vertex in an existing line or polygon. It can ensure that adjacent parcels connect only at existing corners.

**Edge snapping** constrains the feature being added to meet the edge of an existing line or polygon feature. In this case, the vertex being added could be placed anywhere on or between vertices. Edge snapping can ensure that a street ends exactly on another street.

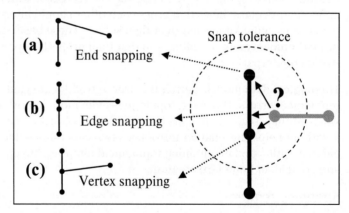

Fig. 7.3. Where will it snap? The snap type dictates where the new horizontal line will end.

Figure 7.3 illustrates how snapping changes depending on the snapping type. The vertical line already exists in the layer, and the horizontal line is being added. The dashed circle shows the snap tolerance. If end snapping is turned on, then the new line will snap to the endpoint of the existing line (Fig. 7.3a). If edge snapping is turned on, then the line can be snapped anywhere between the end and the vertex (Fig. 7.3b). If vertex snapping is set, then the new line will connect to the closest vertex or end (Fig. 7.3c).

If more than one kind of snapping are turned on, then the most inclusive one takes precedence. To ensure snapping only to ends, therefore, both edge and vertex snapping must be turned off.

### Adjacent polygons

Another type of topology to consider is the relationship between two adjacent polygons. A polygon is stored as a series of $x$-$y$ vertices that completely enclose a space. If two polygons share a boundary, then that boundary gets stored twice. If the shared edge contains exactly the same $x$-$y$ pairs for both polygons, it is called a **coincident boundary** (Fig. 7.4). If the pairs do not match exactly, then there will be gaps where the polygons fail to touch, or overlaps where they cross each other, or both.

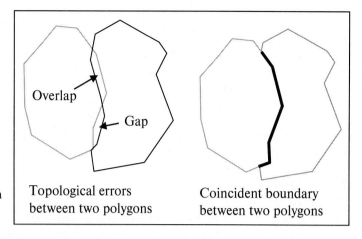

Fig. 7.4. Topological relationships between polygons

Both types of relationships are possible in the real world. If the two polygons represented fires that burned in two different years, then it is possible that the second fire might have overlapped the first, or that some space existed between them; in which case the presence of gaps and overlaps is real and necessary. However, most polygons should have coincident boundaries—soil units, land use areas, school districts, counties, and states all represent features for which gaps and overlaps are errors, usually caused by careless data entry. Imagine the legal headaches that would ensue if the two polygons represent parcels; both overlaps and gaps would likely be in dispute as to who owned what. When editing polygons, it is important to ensure that the right relationship is incorporated into the database, and that adjacent polygons have coincident boundaries, except for the rare instances when gaps or overlaps are justified.

## About ArcGIS

The editing process begins by opening the Editor toolbar, starting an edit session, and identifying which folder or geodatabase will be available for editing. Only one folder or geodatabase (called a workspace) can be edited at a time, but all the feature classes in the folder or geodatabase can be edited together.

### The Editor toolbar

The Editor toolbar provides access and shortcuts to many editing functions (Fig. 7.5). The first button on the left opens the Editor menu, used for turning editing on and off, setting options, and performing certain editing tasks. The Edit tool selects one or more features for editing. It is similar to the Select Features tools on the Standard toolbar but should always be used for editing.

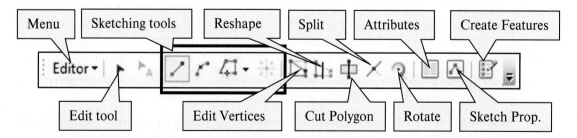

Fig. 7.5. The Editor toolbar contains many editing options and functions.
Source: ArcGIS

Next come the Sketching tools, an assortment of methods to add vertices to a feature. The default is a straight line segment between vertices, but one can also create curves, right angles, midpoints, and other specialized vertices. The Sketching tools provide flexibility in creating features. The next four buttons provide access to common tasks, including editing the vertices of an existing feature to change its shape, splitting a line into two features, cutting a polygon in two, and rotating a feature.

The final three buttons launch editing windows. The Attributes window provides a convenient place for editing feature attributes. The Sketch Properties window is used to examine and, if desired, edit $x$-$y$ coordinates of sketches. The Create Features window contains and manages editing templates, which are used to create new features.

### What can you edit?

ArcGIS offers different levels of editing capability depending on the type of license. ArcGIS Basic can edit shapefiles and personal or file geodatabases. ArcGIS Standard and ArcGIS Advanced can also edit SDE databases, geometric networks, and planar topology. All of these levels are accessed through the same ArcMap interface.

ArcMap can edit several layers at once, as long as they are all in the same folder or geodatabase (workspace). This capability makes it easier to view and edit related layers simultaneously or to copy features from one layer to another.

**TIP:** Coverages cannot be edited in ArcMap. Use the ArcEdit program in Workstation ArcInfo instead, or convert the coverages to shapefiles or geodatabases for editing.

### Editing and coordinate systems

ArcMap can edit layers with different coordinate systems from the data frame. The edits will automatically be converted to the coordinate system of the layer being edited. Consider editing a shapefile of roads stored in a GCS while displaying the roads with a digital orthophotoquad (DOQ) in a UTM projection. The roads coordinates will be converted into decimal degrees before they are put into the shapefile.

### Performance and reliability

Although ArcMap has many useful capabilities, such as editing across coordinate systems and editing multiple files simultaneously, one must be cautious in taking advantage of these benefits. If an editing scenario is demanding extensive system resources by having many files open or by doing many projection transformations during an edit session, system performance and reliability may suffer. Moreover, because spatial data files are complex constructions, the price of an editing glitch may be the loss of data integrity; an ill-timed system glitch might not only lose recent changes but also corrupt the entire data set. Thus, you are encouraged to use the following procedures while editing:

▶ *Always have a backup copy of the file being edited*, stored elsewhere on the disk or on a different medium. This precaution is especially important when working on data sets that would be expensive or impossible to replace should something go wrong.

▶ **Limit the number of open files.** A prolonged editing session should be conducted in its own map document and include only the layers that are needed.

▶ **Avoid extensive editing in map documents** whose primary purpose is cartography or analysis, and in which you have invested hours of effort. The author has found that editing sometimes corrupts map documents. They look fine, but tools will crash when run. The map documents then need to be re-created, potentially wasting hours of effort.

▶ **Save edits frequently in case of a system glitch.** ArcMap does NOT automatically save edits at regular time intervals. However, remember that saving also clears the "Undo" memory of edits. After edits are saved, they can no longer be undone.

### Feature templates

Editing utilizes feature templates. A **feature template** stores the information needed to edit a particular layer in the map document. It includes the feature class in which the edits will be stored, the attributes that are assigned to new features, and the default tool that is used to create them. The template shown in Figure 7.6a is named Commercial. It saves features into a polygon layer called buildings. By default, the plain Polygon construction tool is used. The fields contain default values that are added to each feature, including its designation as a Commercial building.

Fig. 7.6. (a) A feature template's properties; (b) the Create Features window showing the available templates
Source: ArcGIS

Multiple templates may be constructed for each feature class. As shown in Figure 7.6b, the buildings layer has another template called Residential. It is identical to the Commercial template except that the building type field, Bldgtype, contains "Residential." If the user is digitizing a new commercial building, she clicks on the Commercial template. When the building is added, the default attributes "Commercial," "Austin," and "TX" are automatically placed in the attribute fields. If she wishes to add a residential building, she clicks the Residential template, she digitizes the building, and the attributes are entered. Using templates can save a great deal of typing to enter attributes. Only the address field, which is unique to each feature, must be edited manually.

Templates are accessed and managed in the Create Features window (Fig. 7.6b). The upper panel lists each available template, and the user selects the desired template from this panel when ready to create a new feature.

The lower panel shows the available construction tools for that feature type, with the default one defined by the template at the top. Each feature class type has its own set of construction tools used to create various shapes or perform special tasks, such as creating an ellipse or a rectangle. The two buttons at the top of the Create Features window control the grouping and sorting of the templates (left) and the creation and management of templates (right).

Templates should be automatically created for each active layer in a feature class when editing is initiated. However, the user can also create and edit templates as needed. Moreover, templates can be saved and reused, and so they also contain descriptions and tags that can help in searching for stored templates. The Search window provides access to stored templates.

### How editing works

An editing session must be initiated before any changes to a file can be made. This requirement protects the user from accidentally making changes to a file without realizing it. Also, because ArcMap can edit within only one directory or one geodatabase at a time, opening the session establishes the workspace being edited.

Basic editing uses three main components: the Edit tool, the feature templates, and the construction tools. The Edit tool selects existing features when they need to be moved, rotated, deleted, and so on. It is analogous to the Select Features tool on the Standard toolbar, but it should always be used for editing. The templates control the features being added and in which layer they go. The construction tools control the characteristics of the feature being created.

#### Creating sketches

The construction tools are used to create a **sketch**, a provisional feature being entered, analogous to an artist making a light pencil sketch prior to inking in the final lines of a drawing. This sketch is not added to the target layer until it is "finished" by double-clicking the last vertex. The green outline in Figure 7.7 shows a nearly completed polygon sketch, in which green vertices are separated by straight line segments. The latest vertex, just added, is shown in red. New vertices are added to a sketch using one of the Sketching tools on the Feature Construction or Editor toolbar.

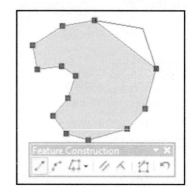

Fig. 7.7. A polygon sketch
Source: ArcGIS

The Straight Segment sketching tool is the default. It appears as the first button on the left on the Editor toolbar. It merely creates a straight line segment between the last and the new vertex. It was used to create the polygon in Figure 7.7. Another tool that is commonly used and placed on the Editor toolbar is the End Point Arc Segment tool, used to create a smooth curve between two endpoints.

Additional Sketching tools are accessed by clicking the drop-down menu to the right of the Straight Segment and End Point Arc Segment tools (Fig. 7.8). The Right Angle tool, for example, forces the sketch corners to be square, making it easier to create buildings. Other sketching tools create curves,

Fig. 7.8. The sketching tools
Source: ArcGIS

make parallel or perpendicular lines, trace along existing features, create midpoints or tangents, and do other specialized tasks. In this chapter, we learn to use two of the most common sketching tools, the default Straight Segment tool and the Right Angle tool.

While the user is sketching, a mini-toolbar known as the Feature Construction menu floats near the cursor on the screen. It provides quick access to different sketching tools and a choice of ways to add the next vertex. If you find the floating Feature Construction menu annoying, you can turn it off from the Editor toolbar using the Editor > Options menu.

### Vertex and Sketch menus

During editing you can access two context menus to aid in certain tasks or change edit settings (Fig. 7.9). The **Vertex menu** appears when you right-click on the sketch (Fig. 7.9a). This menu provides functions for adding, deleting, or moving vertices. The **Sketch menu** appears when you right-click off the sketch (Fig. 7.9b). This menu provides functions for specifying exact angles, lengths, distances, and more.

Fig. 7.9. A sketch and its context menus: (a) Vertex menu and (b) Sketch menu
Source: ArcGIS

### The Snapping toolbar

A small toolbar to control snapping allows easy access to snapping tools in the middle of a sketch. The four basic kinds of snapping (point, end, vertex, and edge from left to right in Figure 7.10) are controlled with the click of a button. A menu extends the snapping types to intersection snapping (where two lines intersect at a common node), midpoint snapping at the midpoint of a line, and tangent snapping to a curve. The snap tolerance can also be controlled here.

Fig. 7.10. The Snapping toolbar
Source: ArcGIS

### Adding adjacent polygons

A special construction tool is used for adding adjacent polygons when they must have coincident boundaries. The Polygon construction tool is first used to enter a complete polygon with no neighbors. To add an adjacent polygon, you must switch to the Auto Complete Polygon tool and digitize only the new part of the polygon. The editor ensures that the polygons share a coincident boundary and are free from topological errors, including gaps and overlaps (Fig. 7.11).

Chapter 7

Another way to add adjacent polygons uses the Cut Polygons construction tool. With this method, an outer polygon is constructed around an area encompassing two or more polygons, and then the internal boundaries are added by cutting the polygons along the intended coincident boundary.

**Editing attributes**

Editing features often includes editing their attributes. The user can modify attributes in two ways. In the first way, edits are typed directly into an attribute table, as discussed in Chapter 6. The second method uses the Attributes window to edit the values of single or multiple features.

The Attributes window is accessed through a button on the Editor toolbar (Fig. 7.12). The upper panel shows the currently selected records; if no features or records are selected, then both areas will be blank. The lower panel shows the attributes of the highlighted feature in the upper panel. When the feature is clicked in the upper panel, the feature flashes on the screen to indicate which one it is, and the attributes in the lower panel may be edited.

The window also allows editing of attributes for multiple features. If the top row containing the feature class (Streets) is clicked, then the attribute values go blank. Further updates to the attributes are applied to all of the selected features.

Draw first polygon with the Polygon tool.

Add new polygon with the Auto Complete Polygon tool.

Two polygons with coincident boundaries

Fig. 7.11. Creating a coincident boundary between two adjacent polygons with the AutoComplete task

Fig. 7.12. The Attributes window can display and edit many records at once.
Source: ArcGIS

212

## Summary

▶ Editing creates and maintains features in a feature class. Care must be taken to ensure topological integrity when editing.

▶ Snapping ensures that features connect properly with each other by automatically matching the coordinates of features within a certain distance of each other (the snap tolerance). Features may be snapped to points, endpoints, vertices, or edges.

▶ When adding adjacent polygons, you must ensure that they have coincident boundaries by using snapping, the Auto Complete Polygon tool, or the Cut Polygons tool.

▶ Editing sessions are initiated and controlled using the Editor toolbar. You may edit multiple layers of data, but only one geodatabase or folder may be edited at a time.

▶ Edits are stored in the same coordinate system as the layer, regardless of the coordinate system of the data frame. However, editing is streamlined when the layer and frame coordinate systems match so that no reprojection need take place.

▶ Basic editing uses the Edit tool, feature templates, and the construction tools. The Edit tool selects features to edit.

▶ Feature templates define all the properties needed to create new features, including the feature class to place them in, the default construction tool to use, and the attributes with any default values in them.

▶ Multiple templates may be created for one feature class, enabling different attributes values to be given to different classes of features.

▶ The construction tools create preliminary figures called sketches. Finishing the sketch creates the new feature in the feature class.

▶ Different Sketching tools provide a variation of functions for placing vertices. Context menus provide additional tools for specialized functions.

### Important Terms

| | | | |
|---|---|---|---|
| coincident boundary | feature template | Sketch menu | Vertex menu |
| dangle | logical consistency | snap tolerance | vertex snapping |
| edge snapping | point snapping | snapping | |
| end snapping | sketch | topology | |

## Chapter Review Questions

1. List the main types of topological errors.

2. Which type of snapping (end, vertex, or edge) would work best in the following situations? (a) digitizing streams, (b) digitizing parcels, (c) digitizing streets, (d) digitizing traffic lights at intersections, (e) digitizing stream guages on streams

3. How are editing functions and options accessed in ArcMap?

4. If a data layer is in UTM coordinates and the data frame is set to State Plane coordinates, in which coordinate system will the edits be stored?

5. True or False? Because you can Undo edits, it is not necessary to make a backup copy of any data set before editing. Explain your answer.

6. When you create a new feature, how do you specify which layer it belongs to?

7. What are dangles and how are they prevented?

8. What is the purpose of the AutoComplete task? What errors does it prevent?

9. How many context menus can you access while sketching? How do you control which one you get?

10. How can feature templates save the user from typing in attribute values?

# Mastering the Skills

## Teaching Tutorial

The following examples provide step-by-step instructions for doing basic tasks and solving basic problems in ArcGIS. The steps you need to do are highlighted with an arrow →; follow them carefully. Click on the video number in the VideoIndex.htm file to view a demonstration of the steps.

→ Start ArcMap and open the map document ex_7a.mxd in the MapDocuments folder.

→ Use Save As to rename the document and remember to save frequently as you work.

### *Creating new feature classes*

You are working on a project for Austin regarding the outcrop of an important aquifer, the Edwards Limestone. You will digitize from a USGS map that is already scanned and georeferenced, using the same coordinate system as the rest of the Austin data. Before starting, you need to create empty feature classes that will contain your new digitized features.

1→ Open the Catalog tab. Right-click the mgisdata\Austin folder and choose New > File Geodatabase. Name it *Edwards*.

 1→ Open ArcToolbox > Data Management Tools > Feature Class > Create Feature Class.

1→ Click the Browse button for Feature Class Location (Fig. 7.13). Navigate to the mgisdata\Austin folder and select the Edwards geodatabase. Click Add.

1→ Name the new feature class *karstfeatures*.

2→ Set the Geometry Type to POINT.

2→ Leave the Template Feature Class blank.

2→ Scroll down, if necessary, to locate the Coordinate System box. Click the Browse button.

 2→ Click the Add Coordinate System drop-down button and choose Import.

2→ Navigate to mgisdata\Austin\Austin geodatabase and select the Environmental feature dataset. Click Add.

**Fig. 7.13. Creating a feature class**
Source: ArcMap

1. We used Import in order to guarantee that the coordinate system of our new feature class matched our other data. What coordinate system is it? _____

 2→ Click the newly added coordinate system and click the Favorites button. Click OK to close the Spatial Reference Properties window.

2→ Click OK to finish creating karstfeatures.

The new feature class is automatically added to the map document. Next we'll create two more feature classes for faults and geology.

3→ Use the Create Feature Class tool again to create a line feature class named faults in the Edwards geodatabase. Make sure that the feature geometry is set to POLYLINE and that you select the right coordinate system from the Favorites folder.

4→ Create a polygon feature class named geology. Make sure that the feature geometry is set to POLYGON and select the right coordinate system from the Favorites folder.

We also need to set up some fields for these feature classes.

5→ Open the Catalog tab, find the Edwards geodatabase, and expand it.

5→ Right-click the faults feature class and choose Properties. Click the Fields tab.

5→ Click inside the first empty Field Name box and type Linetype to name the field.

5→ Click inside the Data Type box for Linetype and choose Text for the Data Type.

5→ Set the Length in the Field Properties box to 10. Click OK.

**TIP:** This method of creating fields is better when multiple fields need to be added to a feature class, instead of using the Table Options method we've used previously.

6→ Use the Catalog tab to open the properties for the Edwards/geology feature class.

6→ Add a text field named Unit with a length of 6. Click OK.

7→ Open the properties for the Edwards/karstfeatures feature class.

7→ Add a text field named Type with a length of 15. Click OK.

### Digitizing points and lines

Now we set up the editing session. We will be digitizing features by tracing them from the scanned and georeferenced image—sometimes called "heads up" digitizing.

 8→ If the Editor toolbar is not visible, click the Editor button.

8→ From the Editor toolbar, choose Editor > Start Editing.

8→ Choose the geodatabase to edit by selecting the Edwards geodatabase from the lower part of the window. Click OK.

8→ If no Create Features window appears automatically, choose Editor > Editing Windows > Create Features.

**TIP:** In Edit mode, some other functions are disabled, such as adding fields, to help prevent damage to data sets. It is wise to be in Edit mode only when you are actually editing.

Examine the Create Features window. It shows a **feature template** for each feature class. A template is a collection of edit settings for a layer. Below the template panel lies a Construction Tools panel, showing the tools available for that template.

8→ Click on the karstfeatures template. Notice that two construction tools are available, *Point* and *Point at end of line*, and that *Point* is currently active.

8→ Click on the geology template. A longer list of different tools appears.

**TIP**: If you cannot see the entire list of tools, place the cursor on the line between the upper and lower parts of the window and move it up to enlarge the lower window.

First we are going to digitize some springs, shown on the geological map as blue dots. Look for a group of six springs near the SW corner of the map (Fig. 7.14).

9→ Zoom in to the group of springs.

9→ Click the karstfeatures template again.

9→ Click inside one of the blue dots to add a point.

9→ Continue adding points until you have digitized all of the springs in view.

Notice that the last one clicked is highlighted, indicating that it is selected.

Fig. 7.14. Adding the springs
Source: ArcMap

9→ Click the Delete key. It will delete the highlighted point, the last one entered.

9→ Click on the spring again to reenter the point.

We need to enter attributes for the springs as well.

10→ Right-click the karstfeatures <u>layer</u> in the Table of Contents and choose Open Attribute Table.

10→ Click inside the highlighted box for the Type field and type *spring*.

10→ Click the next box up and type *spring* in it also.

10→ To select a feature from the table, click on the gray area to the left of a row with an empty Type box to select the feature.

The Attributes window is a handier way to enter attributes while editing.

 11→ Click the Attributes button on the Editor toolbar.

**TIP:** If the Attributes window does not appear, look at the bottom of the Create Features window for an additional tab. To open it, click the tab and drag it away from the Create Features window.

11→ The top panel has one feature, the currently selected one, labeled with its ObjectID.

11→ In the lower panel, find the Type field. Click in the box that currently says <Null> and type *spring*. Click Tab to enter it. Watch it update in the Table window.

It would be tedious to type "spring" for every single feature. There's a better way.

12→ Dock the Attributes window under the Table of Contents.

12→ Adjust the boundaries between the Table of Contents and the two Attributes panels so that you can see several rows of space in the upper panel and the two fields in the lower panel (Fig. 7.15).

12→ Resize and position the Table window to see it.

13→ Choose the Edit tool from the Editor toolbar.

13→ Click and drag a rectangle around all six springs to select them. The Attributes window shows all of them (Fig. 7.15; only four are visible in the figure).

13→ Click the karstfeatures heading at the top of the list of selected features.

13→ Type *spring* in the Type field at the bottom and click Tab. The value is entered in all rows of the table.

Fig. 7.15. The Attributes window
Source: ArcMap

**TIP:** Highlight the feature class name at the top of the Attributes window to make the field edits apply to all of the selected records.

14→ Right-click one of the features in the top panel of the Attributes window and examine the entries of the context menu.

14→ Close the Table window.

14→ On the Editor toolbar, choose Editor > Save Edits.

14→ Clear the selected features using the usual button on the Standard toolbar.

One of the advantages of templates is that you can easily switch between layers.

15→ Find the fault that travels diagonally across the map through one of the springs. Pan/Zoom so that you can see the entire fault, both solid and dashed ends.

15→ Click the faults <u>template</u>. Notice that *Line* is the default construction tool.

15→ Click to place a vertex at the lower end of the <u>solid</u> part of the line.

15→ Click another point on the line some distance from the first.

15→ Keep adding vertices until you reach the end of the solid part. Double-click there to complete the sketch. A highlighted line feature appears.

**TIP:** If you are not satisfied with the fault, click Delete (while it is still highlighted) to get rid of it and try again. If it is no longer highlighted, use the Edit tool to select it.

16→ In the Attributes window, click in the Linetype box and enter *solid*. Click Tab.

16→ Move the cursor toward the lower end of the fault just added, as if you were creating the next fault, but don't click yet.

Notice that, as you get close to the end of the line, some text and a symbol appear, called Snap Tips. They show what the next vertex will snap to.

16→ Move the cursor along the fault and watch as the Snap Tips change.

The text "Faults: Endpoint" appears at the ends of the existing fault. The text "Faults: Vertex" appears when you run over a vertex. The text "Faults: Edge" appears in between vertices.

16→ Place the cursor on the lower end of the fault, making sure that "Faults: Endpoint" is being displayed. Click to enter the first vertex for the new fault.

16→ Add several vertices on the dashed line. Double-click at the end to finish the sketch.

16→ Enter *dashed* in the Linetype box of the Attributes window and click Tab or Enter.

You probably can't see the first fault now because its thin symbol does not show up well.

17→ In the Table of Contents, change the symbol for faults to a 2-pt. green line.

17→ Add the upper dashed fault, taking care to snap to the end of the solid fault.

17→ Enter *dashed* in the Attributes window.

**TIP:** The floating mini-toolbar can be annoying. To turn it off, choose Editor > Options, click the General tab, and uncheck the box to *Show feature construction toolbar*.

18→ Zoom to the full extent of the map. Find some black dots east of the six springs and zoom to them. These are more karst features—caves or sinkholes.

18→ Enter the black points using the karstfeatures template.

18→ Click the Edit tool and draw a rectangle to select them.

18→ Highlight the top karstfeatures entry in the Attributes window and enter the text *cave/sink* into the Type box in the Attributes window.

Source: ArcMap

**TIP:** To change the text used to label the records in the Attributes window, change the Display Expression on the Display tab of the layer properties.

19→ Just for a little more practice, click on the faults template.

19→ Digitize two or three of the short dashed faults in this view. Remember to assign their attributes as *dashed*.

19→ Choose Editor > Save Edits from the Editor toolbar.

19→ Click the Save button on the Standard toolbar to save the map document.

**TIP:** Saving the map document does NOT save the edits, and vice versa.

### Working with snapping

Snapping ensures that all the features that are supposed to touch actually do. It helps prevent topological errors between features. Let's see how it works.

20→ Choose Editor > Snapping > Options.

20→ Make sure that the snap tolerance is set to the default 10 pixels.

20→ Examine the Snap Tips check boxes, which control the snapping guides. Click OK.

**TIP:** Use the Show Tips check box to control whether Snap Tips appear and what they include.

20→ Choose Editor > Snapping > Snapping Toolbar. (If it does not appear, try choosing it again from Customize on the main toolbar.)

20→ Choose Snapping from the Snapping toolbar and examine the menu choices.

The Snapping toolbar is a quick way to turn each type of snapping on or off.

21→ Dock the Snapping toolbar in an accessible place.

21→ Look above the Kprd text in the blue polygon and find the fault that starts dashed at the upper end and changes to solid at the lower end (A in Fig. 7.16).

21→ Click the faults template and digitize the dashed upper end, moving from NE to SW. Stop where the dash ends and double-click to finish the sketch.

21→ Enter *dashed* in the Attributes window.

Fig. 7.16. Digitizing faults
Source: ArcMap

In the next section, you want to snap only to the end of the existing fault—not to its edge.

 22→ On the Snapping toolbar, click both the vertex and edge snapping symbols to the off position.

22→ Begin the next solid section of the fault. As you get to the cave symbol, however, the point snapping is a problem.

 22→ Turn off Point snapping on the Snapping toolbar (even in the middle of a sketch!).

22→ Continue digitizing the fault. Snap to the ends of any faults that touch it, if they have been digitized already.

22→ When you run against the edge of the map, don't finish the sketch yet.

22→ Switch to the Pan tool and move the view to show the rest of the fault.

22→ Click on the faults template again, and your sketch resumes where you left off.

22→ Continue digitizing, and finish the sketch when you get to the end of the fault.

22→ Enter *solid* for the attribute.

23→ Click on the edge of the image to start the next fault below the first.

23→ Alternate between sketching and panning to digitize the entire solid part of this fault, ending where it appears to split into two segments by the caves. Enter *solid*.

23→ Start another fault on the end, taking the upper branch and digitizing up to where the fault splits again near the text "Buda." Enter *solid*.

24→ North and northeast of "Buda" you will see additional caves and springs.

24→ Change to the karstfeatures template and add the springs (blue circles).

24→ Use the Edit tool to select them all and assign them the *spring* attribute.

25→ Click the karstfeatures template and add the caves (black circles).

25→ Change to the Edit tool, select them all, and assign them the *cave/sink* attribute.

25→ Choose Editor > Save Edits. Clear any selected features.

2.  Use the Measure tool to find the diameter of the spring point markers in ground units _____. What is the source scale of the EdwardsCrop raster? _____

It is not easy to get the points in the exact center of the circles, and a circle occupies a large area on the ground. How is the source scale of this map impacting the geometric accuracy of the springs? If you put the points on a GPS and tried to find the springs, would it be easy or difficult?

### *Using the sketch context menus*

Let's look at ways in which the Sketch menu enhances editing, such as correcting mistakes.

26→ Zoom to the full extent of the map. Find a clear area and zoom to it. The next editing is just practice.

26→ Click on the faults template.

26→ Enter some vertices along a random line. Notice that the last vertex is red and the other vertices are green.

26→ Right-click on top of the last red vertex to open the Vertex menu.

26→ Choose Finish Sketch.

Choosing Finish Sketch does the same thing as double-clicking at the end of the sketch.

27→ Start another line and add some vertices.

27→ Right-click the last red vertex to open the Vertex menu and choose Delete Vertex.

The last vertex was deleted and the second-to-last one is now red. This menu can delete any of the previous vertices.

27→ Right-click on any of the green vertices and choose Delete Vertex.

27→ Right-click another vertex and examine the menu again.

27→ Click on the window bar at the top to close the menu without choosing anything.

Notice that all this time we are still in the middle of creating this sketch.

28→ Right-click on the sketch *between* two vertices.

28→ Choose Insert Vertex from the context menu.

Remember, clicking ON the sketch gives the Vertex menu. A different menu appears if you click OFF the sketch.

28→ Right-click OFF the sketch, somewhere on the map. The Sketch menu appears.

28→ Choose Delete Sketch from the menu.

Some of the entries, such as Finish Sketch and Delete Sketch, are found in both menus. If you make a mistake as we continue, you can delete the sketch and start over at any time.

29→ Start a new sketch with several vertices.

29→ Right-click off the sketch to open the Sketch menu.

29→ Choose the second entry, Direction.

29→ Type 90 in the Direction box and press Enter.

As the mouse moves, the cursor is constrained along a vertical line, forcing the next segment to follow the specified angle. Angles are measured counterclockwise from the east (Fig. 7.17).

Fig. 7.17. Angles are measured in degrees from horizontal.

      29→ Click along the line to enter a vertex. The new segment is exactly vertical.

      29→ Double-click to finish the sketch.

3.   What angle would you enter to create a line going S25°W?

      _____

Next, let's create a segment of a specific length.

      30→ Start a sketch and add several vertices.

      30→ Right-click off the sketch and choose Length from the Sketch menu.

      30→ Type 5000 into the box and press Enter.

4.   What do you think are the units of these distance values? _____

      30→ Move the cursor around, noting how the last segment remains the same length.

      30→ Click somewhere to enter the vertex.

**TIP:** If the last segment goes off the page or is too small to see, try entering a different length. Simply right-click off the sketch again and type in a different length value.

The Sketch menu also combines the length and direction functions to enter a line with a particular bearing and direction. We will use this feature to enter the next segment of the same sketch.

      31→ Right-click off the sketch and choose Direction/Length.

      31→ Type 45 for the direction and 5000 for the length (you can use different values, if necessary, to keep the sketch on the screen). Click Enter when both boxes are filled.

      31→ Note that you didn't have to click to enter the vertex. Double-click to finish.

**SKILL TIP:** Learn to use the other functions on the Vertex and Sketch menus (Editing).

### Digitizing polygons

Now we will learn how to create polygons. First we will practice with the construction tools.

      32→ Use the Edit tool to select the practice lines and delete them.

      32→ Click on the geology template. The Polygon construction tool is the default.

      32→ Start a polygon by adding several vertices. Notice how the shape remains closed, with the last vertex always connected to the first.

      32→ Double-click the last vertex to finish the sketch.

      32→ Try adding several more polygons that don't touch each other, for practice.

Next we will create some adjacent polygons. It is important to use the Auto Complete Polygon tool so that the polygons have a coincident boundary.

 33→ Turn on Vertex snapping by clicking the toggle on the Snapping toolbar.

33→ Change the construction tool to Auto Complete Polygon.

33→ Click INSIDE an existing polygon to start the sketch (Fig. 7.18).

33→ Snap to a vertex of the existing polygon on the way out, for neater polygons.

33→ Add more vertices to define the new polygon.

33→ Add the last vertex INSIDE the existing polygon and double-click to finish.

Fig. 7.18. Adding an adjacent polygon
Source: ArcMap

**TIP:** The Auto Complete tool will not construct a polygon unless the space defined by the existing polygon(s) and the sketch is completely enclosed.

34→ Begin a new sketch inside one of the polygons, digitize the boundary of a new polygon, and end inside the other polygon (Fig. 7.19a). Finish the sketch.

Splitting provides another way to create adjacent polygons with coincident boundaries.

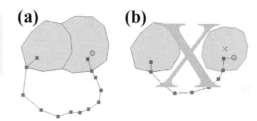

Fig. 7.19. Creating adjacent polygons: (a) works; (b) does not work because the area is not closed.
Source: ArcMap

 35→ Select one of the polygons with the Edit tool (Fig. 7.20).

35→ Click the Cut Polygons tool on the Editor toolbar.

35→ Start a sketch OUTSIDE the selected polygon and add a line that goes through it.

35→ Add the last vertex OUTSIDE the polygon and double-click to finish the sketch.

Cutting works on multiple polygons, too.

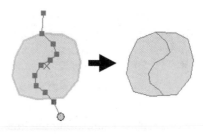

Fig. 7.20. Cutting a polygon
Source: ArcMap

35→ Check that the two polygons are still selected. If not, select them (Fig. 7.21).

35→ Sketch another line through both and finish the sketch.

Now you have the basic skills to do some real polygon editing. First, let's clean up.

Fig. 7.21. Cutting multiple polygons
Source: ArcMap

36→ Use the Edit tool to select and delete the practice polygons.

36→ Zoom to the full extent of the map.

36→ Pan and zoom until you can see the
polygons shown in Figure 7.22.

**TIP:** Sometimes features do not disappear when you
press Delete. If this happens, click the Edit tool on the
Editor toolbar and press Delete again. The Delete key
does not work with some of the cursor tools, but it
always works if the Edit tool is active.

Fig. 7.22. The first polygons
Source: ArcMap

Now we will try digitizing some geology polygons on the
SW end of the outcrop area. Treat the faults as geological
contacts that define boundaries of the polygons.

37→ Click the geology template.

37→ Start digitizing polygon A in Figure 7.22.
Take care, but it doesn't have to be perfect
for this exercise. Double-click to finish.

**TIP:** For best results, place vertices at sharp corners. Use fewer vertices on straight lines and
more on curves.

38→ Type *Kkd* into the Unit field in the Attributes window.

38→ In the Table of Contents, change the geology symbol to 10% Cross Hatch.

38→ Open the geology layer properties and click the Labels tab. Turn the labels on and
change the Label Field to Unit. Set the font to Arial 12 pt. bold. Click OK.

Polygon B is adjacent to polygon A. To ensure that they have a common boundary, we must use
the Auto Complete Polygon tool.

39→ Turn on Edge snapping by clicking the toggle button in the Snapping toolbar.

39→ Click on the Auto Complete Polygon tool in the Construction Tools panel.

39→ Start by snapping to the upper-right vertex of polygon A.

39→ Click along the fault until the purple area ends; then follow the geological contact
between the purple and brown polygons to the lower fault.

39→ Click along the lower fault. End by snapping to the vertex of polygon A.
Double-click to finish the polygon.

39→ Click in the Unit box in the Attributes window and type *Kkbn*.

Notice how the Auto Complete Polygon tool used the boundary from polygon A, ensuring that the
two polygons shared an identical boundary and that you did not have to digitize it twice.

40→ Select the Auto Complete Polygon tool again.

40→ Start at the lower-left corner of the Kkd polygon A and snap to the vertex.

40→ Digitize counterclockwise around pink polygon C.

40→ At the end, snap to an edge or a vertex of polygon B. Double-click to finish.

40→ Enter the unit name, *Kkd*, in the Attributes window.

41→ Start polygon D in the lower-middle part of polygon C, snapping to its vertex.

41→ Start trying to digitize this narrow slice of polygon D.

Polygon D is close to polygon C in places, and you may find that it is snapping when you don't want to. Two things might help: zooming in and reducing the snap tolerance.

41→ Right-click near the sketch and choose Delete Sketch from the context menu.

41→ Zoom in so that polygon D fills the screen.

41→ From the Editor toolbar, choose Editor > Snapping > Options. Change the Tolerance from 10 pixels to 5 pixels and click OK.

42→ Click the Auto Complete Polygon tool and start digitizing polygon D again, snapping to the vertex of polygon C as before.

42→ Don't let the sketch cross into polygon C—keep the unit narrow and thin along the existing polygon.

42→ When ending, be sure to snap to an edge or a vertex of polygon B.

42→ Enter the Unit name Kkbn in the Attributes window.

Cutting polygons is sometimes easier than using Auto Complete.

43→ Pan so that you can see the long, thin slice above the fault.

43→ Click the Auto Complete Polygon tool. Snap to the upper-right corner of Polygon B, and enter vertices along the fault, then up and around to enclose the entire slice. End by snapping to the edge or vertex of Polygon A.

Source: ArcMap

44→ While the slice is highlighted, click the Cut Polygons tool from the Editor toolbar.

44→ Digitize across the slice along the fault, starting and ending outside the slice. The slice is cut in two.

44→ Digitize across the still-selected slice again, following the geological contact between the purple and brown units.

45→ Click the Edit tool and select the two brown parts of the slice (hold the Shift key down to select both).

45→ Click the top geology entry in the Attributes window and enter the unit for both polygons, Kgru.

45→ Select the middle purple unit, and enter its attribute, Kkbn.

46→ Switch to the Auto Complete Polygon tool and digitize the purple polygon below the slice, starting and ending along the slice edge.

46→ Enter its attribute, Kkbn.

47→ Pan or Zoom, if needed, to see the entire large Kgru brown polygon.

47→ Use Auto Complete Polygon to enter it, being sure to snap it to existing polygon vertices at both ends. Enter its attribute, Kgru.

225

Now we have a problem. The new polygon has completely enclosed the purple Kkbn polygon, and now they overlap each other. We should fix it.

Source: ArcMap

48→ Click the Edit tool and select the purple Kkbn polygon.

48→ Chose Editor > Clip from the Editor toolbar.

48→ Fill the button to *Discard the area that intersects* and click OK.

49→ Click the Auto Complete Polygon tool and digitize the small purple polygon that remains, snapping it to the polygon above it. Enter *Kkbn*.

49→ Close off the last section of brown polygon along the dashed fault. Enter *Kgru*.

50→ Zoom to the new polygons (Fig. 7.23).

50→ Choose Editor > Stop Editing. You will be prompted to save your edits. Say yes!

50→ Save the map document.

Fig. 7.23. The first polygons
Source: ArcMap

When digitizing adjacent polygons, the order is important. New polygons must enclose a space completely using a single line. In this exercise, you were instructed to create them in a suitable order, but you must learn to look ahead and plan.

### Adding buildings

Now that you have some experience creating features, we will investigate some additional tools for city editing.

→ Open the ex_7b.mxd map document and use Save As to save it with a new name.

Let's look at another way to create a new, empty feature class for building footprints.

51→ Open the Catalog tab and navigate to mgisdata\Austin.

51→ Right-click the Crestview geodatabase and choose New > Feature Class.

51→ Enter *buildings* for the name of the feature class. No alias is needed.

51→ Leave the Type set to Polygon Features and click Next.

51→ For the coordinate system, click the NAD 1983 State Plane Texas Central coordinate system from the Favorites folder. Click Next.

51→ Accept the default XY Tolerance and resolution, and click Next.

52→ Click in the first available blank field box and type *Bldgtype*. Make it a text field with a length of *25* characters.

52→ Click in the next empty field box and create a text field called *Address* with *50* characters.

53→ Click in the next empty field box to create a text field called *City* with *25* characters.

53→ Click in the Default Value box above the Length box properties and enter *Austin*.

53→ Click in the next available field box to create a text field called State. Give it a length of 2 characters and a default of TX.

53→ Click Finish.

You will see the use of the Default Values in a moment.

54→ Zoom to the area of the green box (Fig. 7.24).

54→ Choose Editor > Start Editing.

Only one geodatabase was used in this map document, so you don't have to choose one to edit.

**SKILL TIP:** An editing template should be created for every layer in the editing workspace. If one is missing, you can create it yourself (Data Management: Editing).

Fig. 7.24. Zoom here.
Source: ArcMap

55→ Right-click the buildings template in the Create Features window and choose Properties.

55→ Examine the template settings (Fig. 7.25).

The City and State fields already have values, Austin and TX, because you set default values when you created the fields.

55→ Click Cancel to exit without making any changes.

56→ Open the Attributes window and dock it below the Table of Contents if necessary.

56→ Click the buildings template. Start with a corner of the first large building on the left, and digitize polygon vertices around it. Double-click to finish.

| General | |
|---|---|
| Name: | buildings |
| Description: | |
| Tags: | Polygon |
| Default Tool: | Polygon |
| Target Layer: | buildings |

Drawing Symbol

View ▼

| Bldgtype | <Null> |
|---|---|
| Address | <Null> |
| City | Austin |
| State | TX |

Fig. 7.25. Editing template properties
Source: ArcMap

56→ In the Attributes window, enter Commercial for the Bldgtype attribute. Notice that Austin and TX are already entered!

56→ Digitize the next building to the right, and enter its attribute as Commercial.

Default values are nice, but templates can make it even easier.

57→ Right-click the buildings template and choose Properties.

57→ Change the template name to Commercial.

57→ Enter the text Commercial in the Bldgtype field. Click OK.

57→ Digitize the next building to the right. When you finish it, look in the Attributes window. "Commercial" is already entered!

58→ Right-click the Commercial template and choose Copy.

58→ Right-click Copy of Commercial and choose Properties.

58→ Change the template name to *Residential*.

58→ Enter *Residential* in the Bldgtype field and click OK.

59→ Click the Residential <u>template</u> and digitize one of the smaller residential buildings above the commercial buildings.

59→ Open the symbol properties for the buildings <u>layer</u> from the Table of Contents.

59→ Create a unique values map for buildings based on the Bldgtype field, so that the commercial and residential buildings are shown with different colors.

Note that the template colors now reflect the map colors. You can switch templates to enter the appropriate kind of building and save a lot of typing.

59→ Save the map document so it will remember the templates.

Buildings have square corners. The Right Angle tool makes them square by helping the next vertex form a right angle from the previous segment.

60→ Zoom/pan, if necessary, to put several undigitized commercial buildings in view.

60→ Click the Commercial template.

60→ Enter two vertices defining one edge of the building (Fig. 7.26).

60→ Click the first drop-down button on the Editor toolbar. A set of alternate tools appears.

 60→ Find the Right Angle tool and click it.

60→ Enter the vertex on the third corner of the building.

Fig. 7.26. Using the Right Angle sketch tool
Source: ArcMap

60→ Add the remaining vertices on the building corners until you reach the last one. Instead of adding it, right-click the spot and choose Square and Finish from the menu.

**TIP:** For best results, use the longest, clearest side of the building for the first edge.

61→ Continue digitizing the commercial buildings until you reach the end of the block.

**TIP:** If you click on a zoom tool, or on another template, it defaults back to the Straight Segment tool, and you will need to select the Right Angle tool again from the Editor toolbar.

62→ Zoom in a little more, if needed, and switch to the Residential template.

62→ Select the Right Angle tool from the Editor toolbar. Digitize a house and finish it.

62→ Finish digitizing the rest of the houses in this block (south of the street only).

62→ Choose Editor > Save Edits.

Now we must assign the addresses to the buildings. The Attributes window is useful here.

63→ Turn on the Addresspts layer and open its label properties.

63→ Check to label the features, use the FULL_NAME field, and make the symbols 10-pt. text. Click Symbol and Edit Symbol and the Mask tab to set a halo.

63→Right-click the buildings layer and choose Selection > Make This The Only Selectable Layer.

64→ Click the Edit tool and select the two houses (and one outbuilding) with addresses on Isabelle Dr.

64→ In the Attributes window (Fig. 7.27), click the buildings entry at the top so that edits are applied to all three selected buildings.

64→ Click the Caps Lock key. Type XX ISABELLE DR in the Address field.

Fig. 7.27. Adding the addresses
Source: ArcMap

That put the street in all three attributes. Now all we need to do is edit the house number. The XX helps select it.

65→ Click on the first record in the top of the Attributes window. The building it represents will flash on the screen.

65→ Click several times, if need be, until you see which house number it needs.

65→ Double-click the XX in the Address field and add the house number.

65→ Continue the same steps with the other two addresses until all three have numbers.

66→ Use the Edit tool to select another group of residential buildings on ESTHER DR.

66→ Use the same technique to enter the addresses.

67→ Open the attribute table for buildings and examine the fields. Close the table.

67→ Save your edits, and choose Stop Editing from the Editor toolbar.

67→ Save the map document.

This is the end of the tutorial.

→ Close ArcMap.

## Exercises

1. Digitize all of the remaining caves/sinks and springs in the Edwards outcrop area map, placing them in the karstfeatures feature class. Remember to assign the attributes.

2. The Travis-Hays County line runs NW-SE across the Edwards outcrop area map, cutting it roughly in half. Digitize all of the faults SW of the county line, placing them in the faults feature class. Remember to assign the attributes.

3. Create a map showing only the Edwards (State Map) layer outline, the karst features, and the faults. Assign different symbols to cave/sink and spring features, and to the dashed and solid fault lines. Capture the map.

4. Digitize the geological units shown inside the yellow line in Figure 7.28, putting them in the same feature class as the polygons you entered in the tutorial. The area is just NE of the polygons you digitized in the tutorial. Assign each polygon the appropriate attribute. Be sure to use the Auto Complete Polygon tool or the Cut Polygons tool, and plan ahead.

5. Create a unique values map of your geology polygons based on the Unit field. Capture a map showing only these polygons.

6. Digitize the buildings in the full block north of Esther Dr. Also include the houses on that section of Isabelle Dr., as well as the very large building behind them. Finish entering the addresses for all of the buildings.

Fig. 7.28. Digitize this area.
Source: ArcMap

7. Create a map showing all of your digitized buildings, labeled with their addresses and symbolized by type. Capture the map.

## Challenge Problem: Where Does the Space Go?

A college campus has many types of space, but how is it all used? You will estimate the total area dedicated to different activities on your campus and present it using a map layout.

Create a file geodatabase containing a feature class for building footprints, using an appropriate coordinate system. Include fields for the building names, the number of floors, and the <u>primary</u> use of the building (e.g., residential, academic, administration). Use whatever categories you think appropriate, but keep it to eight or fewer.

Use the World Imagery layer in ArcGIS Online to digitize the buildings and enter the data in the attribute tables. Calculate the total area for each building, including all floors. Summarize the area by use category, both as total area in each use type and the percent of all campus floor space devoted to that use. Create an Item Description for the feature class.

Create an informative map layout showing the imagery, the buildings, and the usage table. Use all you have learned to tell a good story. Remember to create a balanced layout with a clear visual hierarchy. Include a citation for the image data source.

Compare your map with those of other students. How similar are your estimates? What factors contribute to the differences you see?

# Chapter 8. Queries

## Objectives

▶ Understanding queries and how they are used

▶ Selecting features based on attributes using SQL and Boolean operators

▶ Selecting features based on their spatial location with respect to other features

▶ Applying selection options, including the selectable layers and the selection method

## Mastering the Concepts

### GIS Concepts

#### *About queries*

Queries are a fundamental tool in GIS, used in nearly every analysis. They serve a critical function in data exploration or in picking one's way through a data set looking for patterns. They can also serve as the preparatory step of extracting certain features prior to performing another function. Some common applications involving queries include the following:

- *Selecting features of interest.* Queries can be used to search a table and to find which features meet certain criteria. How many houses currently for sale fall into my price range? Combined with the statistics function, this tool becomes powerful. What is the average cost of three-bedroom homes in this town?

- *Exploring spatial patterns.* Creating a map from selected features and examining their spatial distribution can be illuminating. Where do the wells with high values of contaminants occur? Do they show a pattern relative to known point source pollution emissions? Are they widely spaced or clustered together?

- *Isolating features for more analysis.* Select the largest earthquakes and find out how many of them lie close to volcanoes. Choose the states that have lost population and subject them to a multivariate statistical analysis of demographic data to find out why.

- *Exploring spatial relationships.* Which parcels lie in the floodplain? Which cities are close to a volcano? Where do roads cross unstable shale units?

- *Creating raster queries.* Although we have only discussed vector queries, cells of a raster can be queried also. Which cells have a slope greater than 10 degrees? Which cells show an increase in vegetative greenness as measured by satellite images taken in 1998 and 1999? Where did land use change between 1970 and 2000?

A query extracts features or records from a feature class or from a table and isolates them for further use, such as printing them, calculating statistics about them, editing them, graphing them, creating new files from them, or doing more queries on them. In the simplest kind of query, you look at a map or a table and use the mouse to select the desired record(s). An **attribute query** uses records in the attribute table to test a condition, such as finding all cities with population greater than 1 million people. Attribute queries are sometimes called aspatial queries because they do not require any information about location. A **spatial query** uses information about how features from two different layers are located with respect to each other. A spatial query might choose cities within a certain county, rivers that are within a state, or hospitals within 20 miles of an airport. Attribute queries can be performed on either attribute tables or standalone tables. A spatial query requires a spatial data layer. Spatial and attribute queries may be combined, such as selecting cities with more than 50,000 people who are within 50 miles of an airport. The spatial and attribute queries must be performed separately but can be done in any order.

### Attribute queries

In this method, the user specifies a certain condition based on fields in the attribute table and selects the records that meet the criteria, such as choosing all states with a population greater than 5 million. We already introduced selecting by attributes from tables in Chapter 4. In this section, we extend selection to features and discuss more advanced selection options.

Attribute queries test conditions between two or more attributes using **logical operators**, which include the familiar <, >, and = operators. The condition is written as an **expression** in a format called **Structured Query Language** (SQL), which is used by most major databases to perform selections. SQL expressions can be complex and sophisticated, or fairly simple.

SELECT *FROM cities WHERE [POP1990] >= 500000

SELECT *FROM counties WHERE [BEEFCOW_92] < [BEEFCOW_87]

The SELECT *FROM and WHERE key words must be present as dictated by the syntax rules of SQL. The lowercase words (e.g., *cities*) give the name of the table from which to select, and the terms in double quotes or brackets are the field names. This syntax would be used in any SQL-compliant database, so SQL is a transportable language. However, minor differences between database SQL varieties can be found, such as whether field names are enclosed in brackets or quotes, or nothing.

Care must be taken when performing queries on numeric values depending on the type of field used to store the data. The = operator can only be used on integers, because a floating-point value is never exactly equal to an integer or specific decimal value. The expression NUM = 3 will not work on a floating-point field, and the expression NUM = 3.6 would never work at all. Queries on floating-point fields must use only the < or > operators.

### Boolean operators

When a query includes more than one condition, the additional operators AND, OR, XOR, and NOT are needed. Consider the following queries with two conditions:

SELECT *FROM accounts WHERE Cust = 'COM' AND Balance > 500

SELECT *FROM accounts WHERE Cust = 'COM' OR Cust = 'GOV'

SELECT *FROM accounts WHERE Balance > 500 AND Balance < 1500

The first expression selects records from a customer accounts table that are held by commercial customers and have a balance over $500. The second expression finds the accounts held by commercial and governmental customers. The third expression finds accounts with balances between $500 and $1500.

**TIP:** SQL requires that a field name appear in every condition of the expression, even if it is the same field. SQL will not correctly evaluate the expression Cust = 'COM' AND 'GOV'.

AND, OR, XOR, and NOT are called **Boolean operators**, and they evaluate two input conditions, which may be either true or false, and return a set of records meeting those conditions. Figure 8.1 represents these operators graphically using **Venn diagrams**. The large ellipse labeled T represents all the records in the table. Circle A represents the subset of records meeting condition A, circle B represents the subset meeting condition B, and the blue region indicates the records selected by the different Boolean operators.

Examine the first expression in the preceding examples of Boolean operators. The first condition, Cust = 'COM', constitutes condition A; condition B would be Balance > 500. A AND B finds the accounts for which both input conditions are true (commercial customers and balance over $500). If either condition is false, the record is not selected.

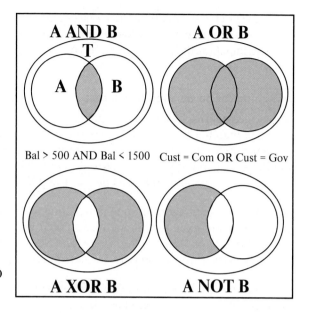

Fig. 8.1. Boolean operators

The expression A OR B selects the records for which either condition is true, finding all the commercial customers and all the accounts with balances over $500 (commercial or not). A NOT B finds all the commercial accounts but excludes the ones with balances over $500, instead finding the commercial customers with balances under $500. The "exclusive or" operator XOR is not available for attribute queries in ArcGIS but is used in raster analysis. XOR finds instances when one or the other condition is true, but not both.

The AND and OR operators are sometimes confused. In English we would say, "select the commercial and the governmental accounts" if we want both, and it seems natural to use the expression Cust = 'COM' AND Cust = 'GOV'. However, this expression yields no records because a customer account cannot be both commercial and governmental at the same time. The OR operator should be used.

The AND operator is used to find records within a numeric range. Consider the third expression, Balance > 500 AND Balance < 1500. If a record with Balance = 1000 is being considered, it would test true for both conditions and be selected. Using OR is incorrect; it would return ALL of the values instead of those inside the range. A record with Balance = 300 would test true for the second condition and so be included. A record with Balance = 3000 would test true for the first condition and be included.

SQL can accept three or more Boolean conditions. Boolean operators have no order of preference, and so expressions are evaluated from left to right, unless parentheses are used to change the order. The order does matter. Compare these two expressions for selecting parcels:

(LUCODE = 42 AND VALUE > 50000) OR SIZE > 50

LUCODE = 42 AND (VALUE > 50000 OR SIZE > 50)

In the first expression, all parcels > 50 hectares are chosen regardless of their zoning or value, and additional parcels are added if they are code = 42 and high in value. In the second expression, all of the parcels must have a zoning of 42 and can be either high in value or large in size.

## Searching for partial matches in text

Sometimes queries based on text fields need to look for partial matches to a condition rather than a perfect match. Consider a table containing customer names. A user might wish to find all customers with the last name *Smith*. If the first and last names were stored in separate fields it would be easy, but if they were stored together in a single field it would be necessary to search for names that contain *Smith* as part of the full name.

Also, when selecting strings, one must allow for extraneous spaces or inconsistencies in how data are entered. For example, the string *Maple St* with a space following *St* would not match the string *Maple St* or *Maple St.* or *Maple Rd*. Often databases have inconsistencies like this, because it is difficult to enforce standard formats perfectly during data entry.

To solve such problems, most GIS systems (and databases) provide an operator, such as LIKE or CONTAINS, that allows searches for a particular substring within another string. This operator is helpful in searching for *Smith* within a field containing both first and last names or *Maple* within a field of street addresses. LIKE searches for the specified set of characters within the field and returns any record that contains those characters. When using LIKE, a wild card character (% for shapefiles; * for geodatabases) is used to stand for other letters that may appear. Consider selecting cities based on their NAME field. The two queries

NAME LIKE 'New %' *(shapefile)*

NAME LIKE 'New *' *(geodatabase)*

would select all cities beginning with the word *New*, such as New London and New Haven, but would not include Newcastle because there is no space in the latter. The expression

NAME LIKE '%Smith%'

would return the customers with the name *Smith*. Notice the second % sign used in the expression. One might expect it to be superfluous, and often it would be, but it does ensure that an extraneous space character at the end of the name would not prevent it from being selected.

## *Spatial queries*

Spatial queries are a powerful tool unique to GIS because they select based on spatial relationships, such as finding wells within five miles of a river or finding parcels inside a floodplain. A spatial selection uses two layers and one spatial condition. The features of the layer being selected are compared spatially with the features of the second layer to see which ones meet the criteria, and features meeting the criteria are selected.

Panning and zooming techniques, already familiar to the reader, are actually a type of spatial query. Pans and zooms define a rectangle of interest based on $x$-$y$ coordinates and ask that the software return all features that fall inside it.

Because feature classes vary in precision and geometric accuracy, it can often happen that two objects that coincide in the real world will not match when their GIS coordinates are compared. It is helpful to be able to specify a search radius when evaluating a spatial condition so that the features need not exactly match.

Consider the problem of selecting cities intersecting rivers. A city at a national scale is represented by a point, and a river by a generalized line. Even if the real-life city intersects a river, it would be only luck if the point occurred on the line feature representing the river (Fig. 8.2). Using the search radius provides some slack. If one assumes that a typical large city might be represented by a square 10 kilometers on a side, then setting the search radius to 5 kilometers might produce an acceptable result. Even so, one would probably encounter some false positive or false negative results.

Fig. 8.2. Features for Minneapolis–St. Paul and the Mississippi River
Source: ArcGIS

Users must always remember the difference between real-world entities and how they are represented by a GIS, and be prepared to mitigate problems that arise because of scale or accuracy issues. In this case, for example, the user might go ahead with the query, but then check each selected city using the online imagery to ensure that the correct relationship exists.

Attribute queries test relationships between attributes using logical operators, such as <, >, or =. Spatial queries test for relationships using **spatial operators**, which include **containment**, **intersection**, and **proximity**. Consider two feature classes, A and B. The spatial operators shown in Figure 8.3 test for the relationship of each feature in A to each feature in B and return the features from A that meet the criteria with respect to B. A complete description of each operator can be found in the ArcGIS Help.

```
intersect
intersect (3d)
are within a distance of
are within a distance of (3d)
contain
completely contain
contain (Clementini)
are within
are completely within
are within (Clementini)
are identical to
touch the boundary of
share a line segment with
are crossed by the outline of
have their centroid in
```

Fig. 8.3. Spatial operators
Source: ArcGIS

**Containment** tests whether one feature includes another. The test A *completely contains* B returns all features in A that fully surround the features in B. A *contains* B if B is inside A, even if they share a boundary. For example, both the yellow and blue counties in Figure 8.4 are contained by the state of Oregon, but only the blue counties are completely contained by the state. The *are within* and *are completely within* operators are the inverse of *contain* and *completely contain*. The *have their centroid in* operator tests whether the center of a feature in A lies inside a feature in B.

Fig. 8.4. The blue counties are completely contained by Oregon.
Source: ArcGIS

**Intersection** is the most generic operator and returns any feature in A that touches, crosses, or overlaps any part of a feature in B. The *are crossed by the boundary of* operator is a special case of intersection that returns features in A that are crossed only by the boundary of features in B, such as the parcels that are crossed by the city limits. The most stringent condition is that the feature from A equals the feature from B exactly (i.e., they have precisely the same geometry). This corresponds to the operator *are identical to*.

**Proximity** tests how close features in A are to features in B. The most generic test, *are within a distance of*, selects features in A that are within a certain distance of B, such as returning all parcels within two miles of a school. Adjacency is a special case of proximity where the distance goes to zero and the boundaries of the features actually touch each other. These conditions are covered by the *share a line segment with* and *touch the boundary of* operators (which might also be considered to be intersection operators because the features touch). Figure 8.5 shows several examples of applying various operators to two different feature classes.

**TIP:** A complete description of the different spatial operators with graphic examples can be found in the ArcGIS Help by searching in the Index for *Select By Location, described*.

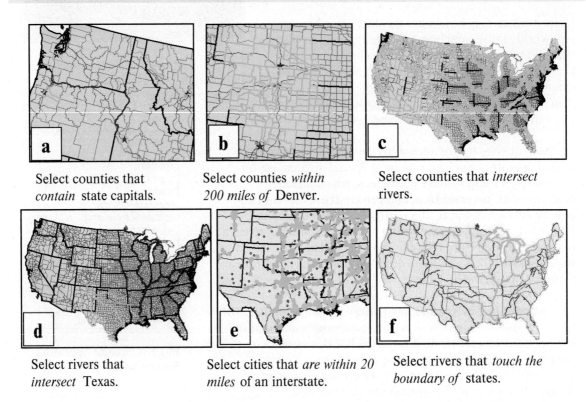

|  |  |  |
|---|---|---|
| Select counties that *contain* state capitals. | Select counties *within 200 miles of* Denver. | Select counties that *intersect* rivers. |
| Select rivers that *intersect* Texas. | Select cities that *are within 20 miles* of an interstate. | Select rivers that *touch the boundary of* states. |

Fig. 8.5. Examples of selecting a layer based on its spatial relationship to another layer
Source: ArcGIS

Sometimes a problem calls for testing spatial relationships using a *subset* of features rather than the entire layer. For example, a user might want to select cities within 50 miles of Interstate 80, rather than cities within 50 miles of any interstate. This procedure begins with an attribute query to isolate the desired interstate, I-80, and then a spatial query to select the cities.

It must be kept in mind that spatial queries can only select or not select an entire feature. Consider Figure 8.5d, which shows the rivers that intersect Texas. Most of the rivers extend beyond the Texas border. If the within operator is used, then only rivers inside the state are selected (in this case, none of them). A different type of function is needed to extract only the *portions* of rivers that fall inside Texas.

### Extraction functions

**Extraction** functions separate features of interest from a larger group. Some consider queries as one type of extraction. The functions **clip** and **erase** are used to extract whole features and portions of features from one feature class based on a spatial boundary supplied by another, saving the extracted features in a new feature class. Clip and erase have the ability to truncate features when they cross a boundary, for example, a road crossing a state line. This ability differs from the Select By Location function, which must select (or not select) the entire feature.

A **clip** works like a cookie cutter to truncate the features of one file based on the outline of another. In Figure 8.6, the county roads layer (gray lines) has been clipped by the land use layer (beige) to produce a shapefile of the roads that fall inside the city planning boundaries (red). The features being extracted come from the input layer. The feature class providing the boundary is called the clip layer. The clip layer must always be a polygon feature class, but the input layer may be points, lines, or polygons. Only the outside boundary of the clip layer is used for clipping; internal boundaries, if present, have no effect on the output.

> **SKILL TIP:** To temporarily clip features for display only, use the Data Frame tab of the Data Frame Properties menu (Layouts and Data Frames).

**Erase** works in the opposite sense of a clip, keeping features that fall outside the erase layer and eliminating those inside. Given the same input and clip layers in Figure 8.6, an erase would keep the roads outside the city (gray) and eliminate the roads inside the city (red). Erase is only available to users with an ArcGIS Advanced license.

Fig. 8.6. A clip truncates the features of a layer by the boundary of another.
Source: ArcGIS

Attributes from both truncated and nontruncated features in the input layer are preserved intact during a clip or an erase operation. However, attributes from the clip or erase layer are ignored. The result of a clip or erase is a copy of the input layer that now occupies a smaller region of interest.

In clip and erase, any features crossing the boundary will be truncated, potentially changing areas, perimeters, or lengths of features. The geodatabase fields Shape_Area and Shape_Length are automatically updated, but user-defined area or length fields are not. A field named MILES or ACRES containing highway miles or tract acres, for example, would contain an incorrect value for any truncated feature. The lengths and areas of user-defined fields must be updated manually using the Calculate Geometry function.

Rasters may also be clipped to a smaller area of interest defined by a feature class or another raster. The tools used for raster extraction are different from those designed for feature data, but they work in much the same fashion.

## About ArcGIS

ArcMap offers three ways to select features. **Interactive selection** uses a pointer to select features on the screen. Attribute queries are performed using **Select By Attributes**. Spatial queries are executed using **Select By Location**. After a query, the selected features are highlighted in the map and in the table.

### Processing layers with selections

ArcGIS follows an important rule when processing features or records in any of its functions or tools: ***after a query is performed and a subset of features or records is selected, any*** *subsequent operation on that layer or table honors the selected set.* You have already seen this rule in action. If you were to select the New England states and calculate statistics, only the New England states would be included. If you selected parcels in a floodplain and then exported the features to a new feature class, only the parcels in the floodplain would be exported. Once a query has been executed on a layer, the layer behaves as if the unselected features do not exist.

By default, a layer has no query applied. In this state, all of the features are available for processing by a command. If a Statistics command were requested in this state, all of the features would be included in the calculations. Executing a query on a layer pushes it into a new state. The layer has a selection, and any function applied in ArcMap or in ArcToolbox will restrict the operation of the function to the selected features or records.

Sometimes a user may execute a query that has proper syntax but yields no matching records. For example, it is possible to query for cities above 10,000 feet in elevation, but no cities actually meet the condition, and no features would be returned. When this happens, ArcGIS removes the query and sets the layer back to its default state with all features available. Essentially, it does not permit the user to execute a query that returns no records or features.

### Interactive selection

In an interactive selection, you use the mouse to click on one or more objects in the map or table until you have all the desired records. For example, to get all the states beginning with A, you could sort the table by state name and then find and click on each record containing states starting with A. Or, if you wanted the states that have a Pacific Ocean coastline, you could look on the map, hold down the Shift key, and choose California, Oregon, Washington, Alaska, and Hawaii.

Fig. 8.7. Selecting usning shapes
Source: ArcGIS

You can use graphics to select features, such as a box, a polygon, a circle, or a line. The Standard toolbar has several tools that select features by constructing a shape on the fly (Fig. 8.7). You can also select using an existing graphic, as shown in Figure 8.8. When selecting using a shape or graphic object, features only need to touch the object. Notice that three states are selected in Figure 8.8b, even though the rectangle does not completely enclose any of them. This default setting can be changed in the Selection options menu, however.

The user can also control which layers may be selected using an interactive query. By default, all layers are selectable, so that the rectangle in Figure 8.8a has selected states, counties, rivers, roads, and more. The Selectable Layers option can turn off selection for one or more layers. In Figure 8.8b, the same rectangle has been used as before, but the user has set the selectable layers to states only.

Fig. 8.8. (a) Selecting with all layers selectable; (b) selecting with only states selectable
Source: ArcGIS

**TIP:** The Selectable Layers option applies only to interactive selection with the mouse. It does not affect selecting by attributes or selecting by location using the menus.

### Selecting by attributes

SQL expressions in ArcMap are constructed using the fields and the buttons in the Select By Attributes window (Fig. 8.9). The window automatically enters the first part of the expression (SELECT *FROM *layer* WHERE), so in subsequent examples we will drop this part and focus on the text to be entered. Notice a few additional options that can be used in this window.

The check box allows the user to restrict the available layers to include only the ones set as selectable. By default this box is unchecked.

The ***Method*** option controls what happens to previous selections when this one is executed. By default, a previous selection is discarded before the new one is applied. For information about the other options, see the later section "Choosing the selection method."

Fig. 8.9. The Select By Attributes window
Source: ArcGIS

The ***Get Unique Values*** button adds values from the table to the window where the user may select them instead of typing them. This button is useful for text and categorical data when you don't know the values to use.

The ***Verify*** button checks an SQL expression before executing it.

The ***Save and Load*** buttons store queries for use again later or to record what criteria were used, especially for long, complex queries with many conditions. Queries are saved in text files with the extension .exp and can be edited with a simple text editor.

The ***Clear*** button erases old or invalid queries and starts again.

**TIP:** Field names may have quotation marks, brackets, or nothing around them, depending on the type of data format and which version of ArcGIS is in use. The Select By Attributes window will always contain the proper syntax when a query is performed.

## Selecting by location

The Select By Location window provides the means to set up a spatial query (Fig. 8.10). The target layer refers to the layer being selected, in this case Interstates. They will be the features highlighted after the query. The source layer is the one against which the target layer is being compared. The spatial selection method identifies the spatial operator used to compare the layers. This window is set up to select the interstates that intersect rivers.

If the source layer has a query applied, you must choose whether to honor the selection. For example, if the user previously selected only the Mississippi River and checked the box to use the selected features, then only interstates crossing the Mississippi River would be selected.

The **search distance** option provides a way to manage potential data accuracy issues.

## Choosing the selection method

ArcMap has four methods for selecting features (Fig. 8.11). The methods may be specified in two places, from the main Selection menu when doing interactive selection or within the Select By Attributes or Select By Location window.

Fig. 8.10. The Select By Location window
Source: ArcGIS

The **selection method** applies to all three types of selections (interactive, selecting by attribute, and selecting by location). These methods give greater flexibility in defining selection sets and enable selection using multiple steps. When performed sequentially, complex expressions involving many criteria are often less confusing than when executed by entering a long expression full of ANDs and ORs and trying to figure out where to put the parentheses.

Fig. 8.11. Selection methods
Source: ArcGIS

**Create New Selection** selects the specified records regardless of what else is already selected, and it replaces any previous selection with the new selected records. For example, if you have selected green jelly beans and then choose to select red jelly beans, then you will only have red jelly beans selected. This method is the default.

**Add to Current Selection** yields the records that fit the query in addition to any that are currently selected. Thus, if you have green jelly beans already selected and then select the red ones, then you will have both red and green jelly beans selected. Using this method is equivalent to doing a single selection step with a double expression using the OR operator, as when selecting beans where "color" = red OR "color" = green.

**Remove From Current Selection** takes away the records that meet the selection criteria. If you already have red, green, and yellow jelly beans selected and you "select" green jelly beans, then you will have red and yellow jelly beans left. In this option, what you choose gets deleted from your selected set, so you must think backward on this one. It is equivalent to the NOT operator.

***Select From Current Selection*** chooses the records that fit the criteria from the already selected set. If you have red, green, and yellow jelly beans selected and choose to select the green ones, then you will have the green ones selected. If you choose to select orange ones, then you will end up with *nothing* because orange jelly beans were not in the original selection. Performing several selections in a row with this method is similar to performing a series of AND expressions.

Imagine selecting earthquakes larger than 5.0 magnitude in Mineral County, Nevada (Fig. 8.12). This problem involves selecting by attribute *and* by location, so it occurs in two steps. First, use Select By Location to select the earthquakes within the county, using *Create New Selection* as the method (the green earthquakes in Figure 8.12). Then use Select By Attributes to choose earthquakes with MAG > 5. However, using *Create New Selection* again in the second step would yield ALL earthquakes with MAG > 5, not just those in Mineral County. You need to specify *Select From Current Selection* in order to obtain the right earthquakes (circled ones).

One other selection option, ***Switch Selection,*** replaces the selected set with the unselected records. In other words, if you have red, yellow, and green jelly beans available and the red ones are selected, and if you do a switch selection, then you will have green and yellow ones selected.

Fig. 8.12. Using selection methods
Source: ArcGIS

### Managing results from queries

After a query, it is often useful to preserve the selected features for future use. Several options are available, both temporary and permanent.

#### Creating a selection layer

In Figure 8.13, the user selected the southeastern states and created a layer from the selected features, resulting in the new states selection layer shown above the original states layer.

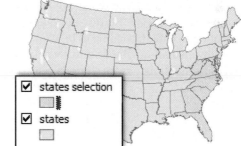

Creating a selection layer has several advantages. The layer can be given its own symbols and can be displayed separately from the original layer. The layer preserves the selected features for future reference and eliminates the risk of accidentally clearing the selection and having to redo it. Selection layers can

Fig. 8.13. The South Atlantic states were selected and used to create a layer.
Source: ArcGIS

be used to input the same set of features to several tools. They are helpful for viewing and recording intermediate results in a complex series of queries. The layer can also be saved as a layer file for use in other map documents.

Creating a layer is the best solution when a selected set is intended for use within a single map document (although it can also be saved as a layer file for use in other map documents). It does not create multiple copies of the data, and edits to the source data will be updated automatically. It is the best solution when you need a temporary duplicate of features from your own data or from an organizational database shared by multiple users.

### Definition queries

A **definition query** is a property of a layer. It is similar to a selection layer, except that instead of being created from a selected set, the query is performed in the layer properties window. Like selection layers, definition queries refer back to the original database rather than creating a copy of it, and they share the same advantages and disadvantages.

### Exporting data

The **export** function saves a new feature class that contains only the selected features, and it is used when a permanent copy of the selection is desired. Because it creates a physical copy of the data, the link to the original feature class is broken. This can be a disadvantage, such as when an organizational database with frequent updates is the source. Changes made to the database will not be reflected in the exported file, which may eventually become out of date. On the other hand, it can be an advantage if the user needs to make changes but does not have write access to the original data. Exports may also be used to develop a new geodatabase based on a subset of existing data, such as developing a New Jersey geodatabase using feature classes from a US geodatabase, as we did in Chapter 2.

### *The Selection window*

The Table of Contents has a viewing window that facilitates working with queries, the List By Selection window (Fig. 8.14). It reports which layers currently have selections applied and how many features are selected (Counties and Capitals), which layers are selectable but have no current selection applied (Rivers and Cities), and which layers are not available to interactive selection (States and Interstates). The dim layers are not currently turned on to be visible on the map. This window also allows the user to set selection options and access context menus for various selection options and functions.

Fig. 8.14. The List By Selection view
Source: ArcGIS

## Summary

- Queries extract a subset of records or features from a data set based on a set of one or more conditions.

- Extraction functions, such as clip and erase, separate features of interest based on a "cookie cutter" provided by a polygon layer and can truncate features at the cutter boundaries. Attributes are carried through without change.

- Attribute queries extract features based on conditions from fields in the attribute table. They are written in SQL and use operators such as =, >, <, AND, and OR. Partial matches to text strings are evaluated with the LIKE operator.

- Spatial queries extract features based on criteria of how two layers are spatially related, using spatial operators to test containment, intersection, or proximity/adjacency.

- Data accuracy and scale issues may affect spatial queries and need to be considered in evaluating the methods and results.

- ArcGIS uses only the selected features in a layer when a function, command, or tool is applied to the layer. If no query has been applied, all features are used. If a query has been applied, only the selected features are used.

- An interactive query uses the Select Features tool and lets the user pick certain features by finding them on the screen and clicking them. Attribute queries are implemented using Select By Attributes, and spatial queries use Select By Location.

- The Selectable Layers option restricts the operation of the Select Features tool to the specified layers. By default all layers are selectable.

- Four selection methods add greater flexibility to queries. These methods include creating a new selection, adding to the current selection, removing from the current selection, and selecting from the current selection.

- Creating a new layer from a set of selected features allows the selection to be stored, displayed, and passed on to tools or commands. Exporting creates a permanent copy.

## Important Terms

| | | | |
|---|---|---|---|
| attribute query | expression | Select By Attributes | Structured Query |
| Boolean operators | extraction | Select By Location | Language (SQL) |
| clip | interactive selection | selection method | Venn diagram |
| containment | intersection | spatial operators | |
| definition query | logical operators | spatial query | |
| erase | proximity | | |
| export | | | |

## Chapter Review Questions

1. How does a clip function differ from Select By Location?

2. Write a valid SQL expression to select cities between 1000 and 10,000 people using a field called POP2000.

3. Write a valid SQL expression to select all counties whose names begin with the letter Q.

4. Let T be a table containing all students attending a community college in New York. Let A be the subset of students living in New Jersey. Let B be the students with a GPA greater than 3.0. The query A AND B yields 200 records. The query A OR B yields 1100 records. The query A NOT B yields 400 records. Construct a Venn diagram for the sets, labeling each section with the number of students. How many students live in New Jersey? How many students have a GPA greater than 3.0?

5. From the information in Question 4, can you determine the number of students attending the community college? If yes, state how many. If not, explain why.

6. What does it mean to set the selectable layers? What is the default setting?

7. Imagine that you have some trail mix composed of peanuts, raisins, almonds, cashews, dried cranberries, and chocolate candies colored red, green, yellow, and orange. Imagine that you apply the following set of "queries" to the trail mix:

   Create new selection all candies
   Add to selection cashews
   Remove from selection red and green candies
   Select from selection all nuts and candies

   What do you have selected now? _____

8. For each of the following queries, state whether the syntax is correct or incorrect. If incorrect, explain why.

   a. ZONE = 'COM' AND ZONE = 'RES'

   b. COVTYPE = 'SPRUCE' AND CROWNCOV > 50

   c. POP2000 > 2000 OR POP2000 < 9000

   d. INCOME < 100000 AND INCOME > 50000

9. What is an operator? Describe and give examples of each of the following: arithmetic operators, logical operators, spatial operators, and Boolean operators.

10. List some advantages of creating a new layer from the selected features.

# Mastering the Skills

## Teaching Tutorial

The following examples provide step-by-step instructions for doing basic tasks and solving basic problems in ArcGIS. The steps you need to do are highlighted with an arrow →; follow them carefully. Click on the video number in the VideoIndex.htm file to view a demonstration of the steps.

> → Start ArcMap and open the document ex_8.mxd in the MapDocuments folder.
>
> → Use Save As to rename the document and remember to save frequently as you work.

### *Using interactive selection*

First, let's experiment with selecting features with the mouse.

> 1→ Choose Bookmarks > Texas from the main menu bar.

> 1→ Click on the Select Features by Rectangle tool (on the Tools toolbar), and click on the star that represents Austin, Texas.

**SKILL TIP:** Learn how to set and use bookmarks with map documents (ArcMap Basics).

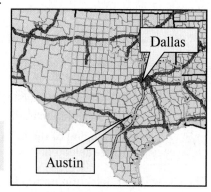

Fig. 8.15. Interactive selection with all layers selectable
Source: ArcMap

Your results may differ slightly, but you should now have several features selected, including Texas, one or more counties near the capital, and an interstate highway (Fig. 8.15). Any feature near the clicked point was added to the selected set regardless of which layer it came from.

> 1→ Try clicking several other spots to see what gets selected.
>
> 1→ Click and drag a box around the Dallas area and notice that anything in the box or anything that goes through the box is selected.

You can control which layers are selectable by using the List By Selection window in the Table of Contents.

> 2→ Click the List By Selection icon at the top of the Table of Contents.

The upper part of the window shows the layers that currently have selected features, and how many features are selected in each (Fig. 8.16). The lower part shows selectable layers that don't currently have a selection. You can control whether a layer is selectable using the icons in this window.

> 2→ Click the Clear Layer Selection button in the Counties row to clear the counties selection. It moves to the lower panel.

Fig. 8.16. The List By Selection window
Source: ArcMap

2→ Click the Clear Layer Selection button for the other layers that have selections also.

 2→ Click the Toggle Selectable button to make the Interstates unselectable. Click it again to turn selection for the layer on again.

3→ Right-click the Counties layer to open a context menu. Examine the choices, and then choose Make This The Only Selectable Layer.

3→ Use the Select Features by Rectangle tool again to draw a rectangle around the Dallas area. This time, only counties are selected.

3→ Make Interstates the only selectable layer, and draw a box around Dallas again.

Notice that when a new selection is made, the previous selection disappears. This behavior is called the selection method. By default, when a new selection is made, the previous one is discarded. The selection method can be changed.

4→ Choose Selection > Interactive Selection Method > Add to Current Selection from the main menu bar.

4→ Click the toggle switch to make the States layer selectable.

4→ Click somewhere inside the state of Texas. It is added to the selected set.

4→ Click on Oklahoma, the state north of Texas. It is added as well.

4→ Click on one of the interstates running west out of Texas. Both the interstate and the state are selected.

The Clear Layer Selection button in the Selection window clears only that layer. You can also clear the selection for all layers.

 4→ Click the Clear Selected Features button on the Tools toolbar, next to the Select Features by Rectangle Tool.

4→ Choose Selection > Interactive Selection Method > Create New Selection from the main menu bar (returns to the default method).

It's time to experiment some more with interactive selection.

 5→ Choose Bookmarks > Michigan from the main menu.

5→ Click the Visibility icon next to Interstates to turn them off.

5→ Make Counties the only selectable layer.

6→ Click one county in Michigan with the Select Features by Rectangle tool.

6→ Hold down the Shift key and click another county. It is added to the selected set.

6→ Add three more counties using the Shift-click method.

6→ Now hold down Shift and click one of the counties that are *already selected*. It will be *removed* from the selected set.

Imagine that you want to efficiently select all the counties in the northern portion of Michigan that lie between Lake Superior and Lake Michigan.

7→ Click and drag a box that goes through as many counties as possible in northern Michigan (Fig. 8.17). Do it several times, if necessary, to get a good result. You will probably not be able to get all the counties, however.

The Interactive Selection Method makes it easier to add to an already selected group.

> 7→ Choose Selection > Interactive Selection Method > Add to Current Selection.
>
> 7→ Click on each of the remaining northern Michigan counties. Each one clicked becomes selected with the previous ones, without needing to hold down the Shift key.
>
> 7→ Continue until all of the northern Michigan counties are selected.

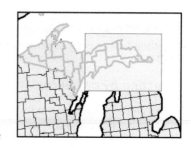

> 8→ Now add a few southern Michigan counties by drawing a box around them.
>
> 8→ Choose Selection > Interactive Selection Method > Remove from Current Selection.

Fig. 8.17. Counties of northern Michigan
Source: ArcMap

> 8→ Click on or draw a box around the southern Michigan counties to remove them from the selected set.

At this point, you should have all the northern counties selected and none of the southern ones (Fig. 8.17). Recall that the Statistics command shows information about the selected features.

> 8→ Right-click the Counties layer and choose Open Attribute Table.
>
> 8→ Right-click the POP2014 field and choose Statistics.

1.  How many counties are selected and what is the total number of people in them (in 2014)?

_____

> 8→ Close the Statistics box and the attribute table.

Now let's get back to testing the different selection methods.

> 9→ Choose Selection > Interactive Selection Method > Select from Current Selection.
>
> 9→ Use the Select Features by Rectangle tool to draw a box that encloses the northern peninsula, covering the area shown in Figure 8.18. Be sure to include some of the southern counties in the box.

Fig. 8.18. The box
Source: ArcMap

Notice how the northern counties touching the box were selected, but the southern counties were not because they were not already part of the selected set. Let's use some other selection tools.

> 9→ Set the Interactive Selection method to the default, Create New Selection.
>
> 9→ Click the tool to Clear Selected features on the Tools toolbar.

> 10→ Click the Down arrow next to the Select Features by Rectangle tool and examine the other tools (Fig. 8.19).
>
> 10→ Choose the Select by Circle tool.

10→ Click on the state capital, Lansing, and drag to make a circle (Fig. 8.19). Watch the radius measurement in the lower-left corner of the window and get it close to 100,000 meters. Release the mouse button when you are satisfied with the circle.

Click here

Select by Rectangle

Select by Polygon

Select by Lasso

Select by Circle

Choose this tool

Fig. 8.19. Selecting using a circle
Source: ArcGIS

As you drew the circle, you may have noticed that the cursor snapped to Lansing and various county corners or edges. As in editing, the snapping options are controlled by the Snapping toolbar. You can specify a circle of exact radius if you wish.

11→ Click on Lansing again and drag to start the circle.

11→ With the mouse button still down, press the letter *r* (for *radius*) on the keyboard. When the box appears you can let go of the mouse button.

11→ Type the desired radius in the box, *100000* meters, and press Enter.

12→ Choose the Select by Polygon tool.

12→ Click around the lower peninsula to add the vertices of a polygon that touches all of the southern Michigan counties without including those from other states or the northern peninsula. Double-click to finish the polygon and make the selection.

→ Experiment with the other selection tools. When you are satisfied that you know how to use them, clear any selected features and go on with the tutorial.

**SKILL TIP**: Learn to change the Selection Options, such as the color of highlighted features and whether features must pass through the rectangle/circle or be inside it (Queries).

### Selecting by attributes

In previous chapters, you have done queries using the Options button in the table. Another method does not require opening the table.

13→ Choose Bookmarks > USA from the main menu.

13→ Right-click the Cities layer and choose Properties. Click the General tab.

13→ Set the scale range to *Show layer at all scales*, and check the *Visible* box to make sure it is visible. Click OK.

13→ Choose Selection > Select By Attributes from the main menu.

The Select By Attributes window is identical to the one accessed by the Table Options menu, except that you must set the layer to select in the first drop-down box.

14→ Use the drop-down box to choose Cities as the layer to select.

14→ Confirm that the selection method is *Create a new selection*.

14→ Enter the expression POP2014 > 1000000 (1 million).

14→ Use the Verify button to make sure the expression is correct.

**TIP:** Most users will find it faster to use the buttons and the windows of the query window rather than trying to type the expressions.

14→ Click Apply and move the Select By Attributes window to show the Table of Contents, the map, and the window at the same time.

2. How many cities in the United States had more than 1 million people in 2014? _____

One advantage of using this window is being able to query another layer from the same window.

15→ Click the Select By Attributes window again to activate it.

15→ Change the layer being selected to States. Keep the Create New Selection method.

15→ Enter the expression to POP2014 > 5000000 (5 million). Click Apply.

Notice that both cities and states are now selected. States and Cities are separate layers and are treated separately in this query window. Also note that the states were selected even though they were not a selectable layer. The selectable layer applies only to interactive selection.

Now let's try a selection with double criteria, all counties that lost population between 1990 and 2000 and have more males than females.

16→ Clear the selection using the Clear Selected Features button on the Tools toolbar.

16→ Click the Select By Attributes window to activate it, if necessary.

16→ Set the layer to Counties.

16→ Enter the expression POP2014 < POP2010 AND MALES > FEMALES in the query box. Click Apply.

3. How many counties with more males than females lost population between 2010 and 2014?

Now from these we will select those counties that also have fewer than 10,000 people. Be careful to ensure that the new set of counties is pulled only from the currently selected ones.

17→ In the Select By Attributes window, change the selection method to *Select from current selection.*

17→ Click the Clear button to erase the old expression.

17→ Enter the expression POP2014 < 10000. Click Apply.

Watch as a few of the counties disappear from the selected set.

4. How many counties remain selected? _____

Springfield seems to be a popular name for cities. Let's see how many there are. Text queries must have quotes around the value being selected.

18→ Clear the selection for the Counties layer.

18→ In the Select By Attributes window, change the layer to Cities.

18→ Change the Selection method back to *Create a new selection.*

18→ Enter the expression NAME = 'Springfield'. Click Apply.

When selecting text, using the Get Unique Values field may be valuable and save some typing, but it tends to work best with categorical data. With nominal data like city names, there are so many different choices to scroll through that it is usually faster just to type the desired name.

We might be curious how many cities end with the word *City*, such as *Oklahoma City*. The LIKE operator is used to search for text within a string instead of an exact match.

19→ Clear the expression box and enter NAME LIKE '%City'. Click Apply.

19→ Right-click the Cities layer in the Table of Contents and choose Open Table Showing Selected Features to check the results. Did you get what you expected?

19→ Close the Table window when finished looking at it.

5. How many city names begin with the word *New*? Be sure to include cities such as New Brunswick, but exclude cities like Newcastle. Check the names in the table. _____

**TIP:** The Clear Selected Features tool on the Tools toolbar and in the Selection menu clears the selection from ALL layers. To clear the selection for a single layer, use one of the context menus.

19→ Clear all selections and close the Select By Attributes window.

### Selecting by location

Next we will try selecting by location. First we will select all the counties that are adjacent to rivers. The features you want to select (Counties) are the target layer.

 20→ Click the Visibility icon to turn off the Cities layer.

20→ Choose Selection > Select By Location, and adjust the window to show both the Selection window and the map.

20→ In the first drop-down box, choose the selection method to *select features from* (Fig. 8.20).

20→ Check Counties as the target layer.

20→ Choose Rivers as the source layer.

20→ Set the spatial selection method to *intersect the source layer feature*.

20→ Click Apply and view the map.

Selection method:

select features from

Target layer(s):
- ☐ Capitals
- ☐ Cities
- ☐ Interstates
- ☐ Rivers
- ☐ States
- ☑ Counties

☐ Only show selectable layers in this list

Source layer:
Rivers

☐ Use selected features    (0 features selected)

Spatial selection method for target layer feature(s):
intersect the source layer feature

☐ Apply a search distance
400000.000000    Meters

Fig. 8.20. The Select By Location window
Source: ArcMap

6. How many counties in the United States are intersected by rivers? _____

Next let's select the counties that contain state capitals.

    21→ Change the source layer to Capitals and the operator to *contain the source layer feature*. Click Apply.

7. Which of these counties containing capitals has the smallest 2014 population? _____
    What is the capital and which state is it in? _____

Now let's do an example with points and lines by selecting the cities close to interstate highways.

    21→ Click the Visibility icons (now gray) to turn on the Cities and the Interstates.

    22→ In the Select By Location window, change the target layer to Cities (be sure to uncheck Counties as well) and set the Source layer to Interstates.

    22→ Change the spatial selection to *are within a distance of the source layer feature*.

    22→ Set the search distance to 20 miles. Click Apply and view the results.

**TIP:** Layers from a previous query will remain checked in this window until they are unchecked, so get in the habit of making sure only the desired layers are checked.

8. How many US cities are within 20 miles of an interstate highway? _____

You might wonder what percentage of cities this is, because the map looks full of selected cities. To find out, you need to know how many cities are in the whole layer.

    23→ Right-click the Cities layer and open the attribute table.

    23→ Notice that the bottom of the table says (3641 out of 3905 Selected).

    23→ Use a handheld calculator, or the one on your computer, to calculate the percentage.

9. What is the percentage of cities that lie within 20 miles of an interstate? _____

**TIP:** Look carefully at less populous states, like Wyoming and Nevada. Sort the cities by ascending population and examine the CLASS field. Do you think this data set contains *all* cities in the United States? What impact might the completeness of this data set have on your analysis?

In the previous example, we compared *all* the features in one layer to *all* the features in the other. You can also select features based on a subset, such as selecting the rivers that intersect Texas. We must select Texas first and then perform the Select By Location for the rivers.

    24→ Clear all selected features and close the Table window.

    24→ Turn off the visibility for Cities, Capitals, and Interstates.

    24→ Choose Selection > Interactive Selection Method > Create New Selection on the main menu bar.

    24→ Make States the only selectable layer.

    24→ Click the Select Features by Rectangle tool and select the state of Texas.

25→ Open the Select By Location window, if necessary.

25→ Set the target layer to Rivers and the source layer to States.

25→ To use only Texas, you must check the box to *Use selected features.*

25→ Set the spatial selection method to *intersect the source layer feature.* Click OK.

10. Which rivers intersect Texas? _____

Spatial queries may also be performed within a single layer, using it both as the target and as the criterion layer. The features comprising the criteria must be selected first. For example, select all cities within 400 miles of Kansas City.

26→ Clear all selected features.

26→ Turn off the Rivers and turn on the Cities in the Table of Contents.

26→ Open the Select By Attributes window and make sure the selection method is set to *Create a new selection.*

26→ Make sure that the box to *Only show selectable layers* is unchecked.

26→ Select Cities as the layer to select.

26→ Clear the previous expression and enter NAME = 'Kansas City'. Click OK.

27→ Open the Select By Location window.

27→ Set both the target and the source layer to Cities.

27→ Because the target and source layers are the same, you <u>must</u> use the selected set. The box is checked and dimmed, so you cannot change it.

27→ Change the spatial selection method to *are within a distance of...* and set the distance to 400 miles. Click OK.

Attribute and location queries gain power when combined sequentially. Let's select all the rivers that intersect states in the Mountain subregion. We do this by first selecting the mountain states using an attribute query and then selecting the rivers that intersect these states (Fig. 8.21).

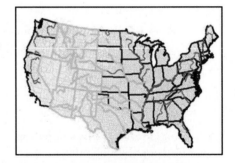

28→ Clear all selected features. Turn off Cities and turn on Rivers.

28→ Choose Selection > Select By Attributes.

28→ Set the layer to States, and enter the expression SUB_REGION = 'Mountain'. (This is categorical data, so use the Get Unique Values button.) Click OK.

Fig. 8.21. Rivers that cross the Mountain subregion
Source: ArcMap

29→ Open the Select By Location window.

29→ Change the target layer to Rivers, the source layer to States, and the spatial selection method to *intersect...*

29→ Check the *Use selected features box.*

29→ Make sure that the *Apply a search distance* box is NOT checked. Click Apply.

Notice that many of the rivers extend far to the east to join the Mississippi (Fig. 8.21). Because they do pass through the mountain states, though, they are included. You could try selecting only the rivers that are within the mountain states instead.

29→ Change the selection method to *are within the source layer feature* and click Apply.

11. Now which rivers are selected? _____

Now hardly any rivers are selected. Ideally, you might like to select the *portions* of rivers that fall inside the mountain states. However, Select By Location only works on whole features. We will learn to extract portions of features later in the tutorial.

Combining attribute and location queries can create sophisticated selections. Imagine that a meatpacking company is considering adding the capacity to slaughter bison and wants to survey customers between Pierre and Denver to see if the demand for bison meat justifies the added expense of handling these more difficult animals. The company decides to target the survey to counties that are within 300 miles of both Denver and Pierre and have more than 10,000 people.

30→ Clear all the selected features and turn on the Capitals layer.

30→ Right-click the Capitals layer in the Table of Contents and choose Properties. Click the Display tab.

30→ Check the box to *Show MapTips using the display expression.* Click OK.

30→ Right-click Capitals in the Table of Contents and choose Make This The Only Selectable Layer.

30→ Click on the Select Features by Rectangle tool and select Denver in Colorado.

30→ Open the Select By Location window and set it to select all Counties within 300 miles of the selected features of Capitals (Denver). Click Apply.

31→ Clear the selection on the Capitals layer <u>only</u> by right-clicking Capitals and choosing Clear Selected Features.

31→ Choose the Select Features by Rectangle tool, hold down the Shift key, and select Pierre in South Dakota. If you forget to Shift, the county selection will be cleared.

31→ In the Select By Location window, change the selection method to *Select from the currently selected features in.* Check the *Use selected features* box and click Apply.

31→ Close the Select By Location window.

**TIP:** You may encounter a bug in redrawing the selected features. If your selected set does not look like Figure 8.22, try using the Refresh button in the lower-left corner of the map window.

At this point, a patch of counties between Denver and Pierre should show as selected (Fig. 8.22). Now we select from this set the counties with more than 10,000 people.

Fig. 8.22. Counties within 300 miles of Pierre and Denver Source: ArcMap

32→ Open the Select By Attributes window.

32→ Set the layer to select to Counties.

32→ Set the selection method to *Select from current selection.*

32→ Enter the expression POP2014 > 10000 and click OK.

32→ Right-click Capitals and choose Clear Selected Features. The map should look similar to Figure 8.23.

33→ Right-click the Counties layer and choose Create Layer from Selected Features.

33→ Change the Table of Contents view to List by Drawing Order.

33→ Name the new layer *Packing Survey*. Change the symbol, if necessary, to show up well against the background.

33→ Clear all selected features.

Fig. 8.23. Target counties for the packing survey
Source: ArcMap

12. How many counties will be in the survey? _____ What is the total number of people who live in these counties? _____

## Definition queries

A definition query sets a property for a layer that temporarily restricts the layer to features meeting the condition. Imagine that you want to create a map showing only your home state.

34→ Turn off all layers except States and Counties and zoom to your home state.

34→ Open the layer properties for Counties and click the Definition Query tab.

34→ Click the Query Builder button and enter the expression STATE_NAME = '*your state name*'.

34→ After using Verify on the expression, click OK and OK.

**TIP:** Be sure to Verify when doing definition queries. An error may cause the layer to disappear entirely, which may confuse you.

35→ Open the properties for the States layer, click the Definition Query tab, and build a query to show only your home state.

The state may appear rotated, especially if it is near the east or west coast. The coordinate system for the data frame is set to US Equidistant Conic, optimized for displaying the entire United States. For a state map, you should choose a coordinate system that is optimized for the state.

36→ Right-click the Layers data frame name in the Table of Contents and choose Properties. Click the Coordinate System tab.

36→ Expand the folders for Projected Coordinate Systems > State Plane > NAD 1983 (Meters). Be sure to use the plain NAD 1983 datum and not one of the special ones.

36→ Choose one of the State Plane zones for your state. If a Central zone is listed, it will likely have the least distortion. However, any of them will work for now.

A definition query works for a quick map, but for an extended project it is more convenient to create a copy of the data with only the selected feature(s) in it. We'll try that next, but first remove the definition queries on these layers.

37→ Open the layer properties for States and click on the Definition Query tab. Delete the query expression and click OK.

37→ Open the layer properties for Counties and delete its definition query also.

## Exporting data

A layer with a definition query still refers back to the original feature class. It is a temporary exclusion of the features you don't want. Export creates a permanent subset of a feature class that is stored in its own file. You could use it to build a geodatabase for your own state. You did this already in Chapter 2, but let's do it again as a quick reminder of how it works.

Note: If you still have your home state geodatabase from Chapter 2, you can store the new data in it, so skip creating another geodatabase and proceed to Step 39.

38→ Open the Catalog tab, if necessary, and expand the gisclass folder.

38→ Right-click the gisclass folder and choose New > Folder. Give the folder the name MyHome2 and click Enter.

38→ Right-click the MyHome2 folder and choose New > File Geodatabase. Give the geodatabase the name of your state (omitting spaces, if any) and click Enter.

39→ Use Select By Attributes to select the counties in your state. (Make sure that the selection method is reset to *Create a new selection*.)

40→ Right-click the Counties layer and choose Data > Export Data.

40→ Choose to export the *Selected features* (Fig. 8.24).

40→ Fill the button to *Use the same coordinate system as the data frame*, so that the new file is stored in State Plane.

40→ Click the Browse button. Make sure the *Save as type is set to File and Personal Geodatabase feature classes*.

Fig. 8.24. Exporting features
Source: ArcMap

40→ Navigate inside the new geodatabase and name the feature class stcounties. Click Save and OK.

40→ Click Yes to add the new feature class to the map.

40→ Clear the selected features.

## Clipping data

In previous chapters, you learned to assemble data for one state based on an attribute query, but we found that certain types of features, such as major roads, cannot be extracted that way because they do not belong to a single state. The Clip tool solves this problem. Let's extract the Interstates in your home state, being careful to save it in a State Plane coordinate system.

41→ Turn on the Interstates layer in the Table of Contents.

41→ Open the ArcToolbox > Analysis Tools > Extract > Clip tool.

41→ Set the Input Features to Interstates and the Clip Features to stcounties.

41→ Save the output feature class in the home state geodatabase as *stinterstates*.

41→ Click the Environments button on the tool. Click on the Output Coordinates. Set it to Same as Layer 'stcounties', so it will use the State Plane coordinate system.

41→ Click OK to finish setting the Environments and OK to start the clip.

42→ Symbolize the stinterstates layer with a more visible symbol that is different from the Interstates layer symbol.

42→ Open the stinterstates attribute table.

There are two user-defined length fields in this table: DIST_MILES and DIST_KM. The lengths of some are likely to be incorrect because the features were truncated during the clip. We will recalculate them to be sure they are correct. The data frame is set to the State Plane coordinate system, which should preserve distance, so we can use it in the calculation.

43→ Right-click the DIST_MILES field and choose Calculate Geometry.

43→ Use the data frame coordinate system and set the units to miles. Click OK.

43→ Recalculate the DIST_KM field as well, setting the units to kilometers.

Clipping can also be used on rasters, but the process is a little more involved. We will obtain a coarse resolution DEM for our state geodatabase.

44→ Search ArcGIS Online for *gtopo30* and find the 30 arc-second DEM of North America (layer package). Add it to the map.

44→ Change the data frame coordinate system to match the namer_dem layer.

45→ Turn off the Counties layer and symbolize the stcounties layer with the Hollow symbol.

The DEM is in the GCS WGS 1984 coordinate system. Because clipping and projecting rasters is complex and computer-intensive, we will do it in two steps: clip first, then project. We set the data frame to match the raster to ensure that only the clip will be done in the first step.

 46→ Open the Search window, if necessary, and dock the Search window on the ArcMap window.

46→ Type *clip raster* in the search box and click the Search button.

46→ Click the Clip (Data Management) tool to open it and verify that it is the tool for rasters (Fig. 8.25).

Fig. 8.25. Clipping the DEM
Source: ArcMap

47→ Set the input raster to namer_dem.

47→ Set the output extent to stcounties.

47→ Check the box to *Use Input Features for Clipping Geometry*, so that the DEM will be clipped at the state boundary instead of the bounding rectangle.

47→ Save the output raster in your MyHome2 geodatabase as statedem.

47→ Uncheck the box to *Maintain Clipping Extent* to avoid resampling the raster now. Click OK.

Recall that resampling impacts the accuracy of a raster. We will have to resample the raster when projecting it, and we don't want to resample it twice if we don't have to.

48→ Remove the original namer_dem layer package and set the data frame coordinate system back to State Plane.

49→ Type project raster in the Search window and open the Project Raster (Data Management) tool.

49→ Enter statedem as the input raster.

49→ Save the output raster in MyHome2 as *dem900m*.

49→ Set the output coordinate system (use the Layers folder) to your State Plane coordinate system. A datum transformation will be entered automatically.

49→ Set the resampling technique to bilinear, since elevation is continuous data.

49→ Enter 900 as the new X and Y output cell sizes. Click OK.

50→ Remove all the layers except the three for your home state.

50→ Symbolize dem900m nicely and display it with the counties and interstates (Fig. 8.26).

50→ Use Save As to save this map as a new map document in the MyHome2 folder.

Fig. 8.26. Interstates, counties, and DEM for Oklahoma Source: ArcMap

51→ Update the Item Description for dem900m, giving an original source citation and describing the two operations performed on it (clip and project).

This is the end of the tutorial.

→ Exit ArcMap and save your map document.

## Exercises

Use the files in ex_8.mxd to answer the following questions. Use the most recent population fields available in the table unless directed otherwise.

1. How many states have counties named for Thomas Jefferson (i.e., how many Jefferson Counties are there)? Which state has the Jefferson County with the most people in the year 2014?

2. How many counties in the United States have more men than women? What percentage of the counties do they represent?

3. How many cities in the United States have more Hispanics than African Americans, a median age greater than 40, and a population between 50,000 and 100,000? List the names of the three largest. (**Hint:** Do your queries in more than one step.)

4. What is the percentage of United States counties that have a river intersecting them?

5. How many counties have more than 1 million people and contain a state capital? List the states these counties are in.

6. How many cities are within 50 miles of a volcano? (Add the volcano feature class from the Usa\usdata geodatabase.) What is the total number of people living in those cities?

7. How many other volcanoes are there within 300 miles of Crater Lake, a volcano in Oregon? How many of these volcanoes are also within 50 miles of an interstate?

8. How many cities in the West South Central subregion of the United States are less than 200 miles from Oklahoma City? Capture a map showing your selected cities.

9. How many capitals are more than 50 miles from a river? Which one has the most people?

10. Congress has awarded FEMA $10 million to help large cities prepare for earthquakes. The cities that qualify for the funding must have more than 500,000 people and be less than 50 miles from one or more earthquakes exceeding 6.0 in magnitude. The bill stipulates that the funding is to be divided among the qualified cities in proportion to their population. Create a feature class that contains only the qualified cities and has a table field listing the amount of funding to be given to each. Capture a view of the table showing the city, population, and funding amount.

## Challenge Problem: My Ol' Home (Reprise)

Use the clip tools to add more data to your home state geodatabase. Search the Internet and/or ArcGIS Online to find at least two more line feature classes and one raster at a suitable scale for your home state. Keep in mind that rasters can get very large, so try to find something with a resolution of about 500–1000 m or 30 arc seconds. Make sure that the final data sets are in your home state coordinate system, and be sure to create an Item Description with a citation for each one. Create a map layout showing your new data, including your citations.

# Chapter 9. Spatial Joins

## Objectives

▶ Learning the purpose and capabilities of spatial joins

▶ Correctly setting up spatial joins based on cardinality and feature type

▶ Learning to solve problems with spatial joins

## Mastering the Concepts

### GIS Concepts

#### *What is a spatial join?*

Chapter 6 presented attribute joins performed on tables—for example, joining information in an earthquake statistics table to a map of states to create a graduated color map of damages. This join is based on a common field, the state name, and has a **cardinality** of one-to-one. The join results in combining the two tables as if they were one table. The destination table receives the information from the source table.

A **spatial join** is similar to an attribute join, except that, instead of using a common field to decide which rows in the table match, the *locations* of the spatial features are used. The spatial join uses either a containment criterion (one feature inside the other) or a proximity criterion (one feature close to another).

Like attribute joins, spatial joins designate a source feature class and a destination feature class. Unlike an attribute join, which appends the source attributes to the existing destination table, a spatial join creates a new feature class. It retains the features from the destination layer and appends the attribute information from the source layer (Fig. 9.1). The two original feature classes are unaffected. The destination feature class determines the type of features in the output feature class. If an airports feature class is joined to a cities feature class with cities as the destination, then the output feature class contains cities.

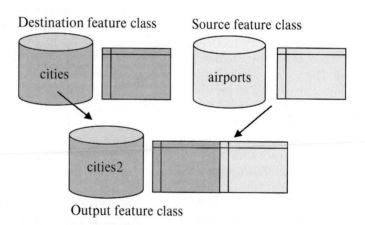

Fig. 9.1. A spatial join keeps features from the destination and appends attributes from the source.

#### Distance joins

A **distance join** uses a proximity criterion to link one feature and its attributes to another based on whether one feature is closest to another. Figure 9.2a shows the details of the join between airports and cities diagrammed in Figure 9.1. The source feature class, airports, has been joined to the destination feature class, cities. The output feature class contains cities. Each city has been given the attribute information from the airport that lies closest to it, and a new field has been

**(a)**

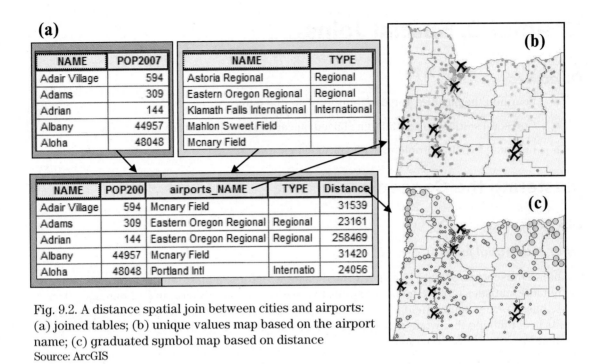

Fig. 9.2. A distance spatial join between cities and airports:
(a) joined tables; (b) unique values map based on the airport
name; (c) graduated symbol map based on distance
Source: ArcGIS

added to record the distance. The attribute table contains two parts, the original data from cities
and the joined data from airports. So McNary Field is the closest airport to Adair Village, and
Eastern Oregon Regional is the closest airport to Adams.

Two maps have been created from the new joined layer. Figure 9.2b is a unique values map based on
the airport name, so each dot gets a color based on the closest airport. The colors indicate which cities
are served by each airport, assuming
that people will drive to the closest
one. The second map is a graduated
symbol map based on the distance
field, with the larger circles
indicating greater distance from the
airport (Fig. 9.2c). The units of the
distance field are always given in
the stored coordinate system units.
These data are in the Oregon
Statewide Lambert coordinate
system, and the units are meters.

### Inside joins

In an **inside join**, the records of
the feature classes are joined
based on whether one feature is
inside another (wholly or partly).
Figure 9.3 shows a point layer
containing septic system locations
and a polygon layer showing
geological units. Imagine that a
community has a porous geological
aquifer that provides the city water
supply.

Fig. 9.3. A spatial join gives each septic system the
attributes of the geology polygon within which it lies.
Source: ArcGIS

Extensive development outside city limits has caused the city concern about potential contamination of the aquifer by faulty septic systems. Assessing the threat requires identifying the number of septic systems that occur in the outcrop area of the aquifer.

A spatial join solves this problem perfectly. Septics is the destination layer and becomes the output feature class. The geology attributes of the polygon containing the septic system are appended to the output septics table. As shown in Figure 9.3, septic system 800 falls inside the Terrace Deposits (Qt) polygon in the geology layer, and septic systems 836 and 839 fall inside the Navarro Group (Knb) unit. The output table will be helpful in assessing how many septics fall on sensitive geological units.

Spatial joins may be performed on any two spatial data layers. A user can join points to points, polygons to polygons, lines to points, and nearly any combination of the three types of data. The output layer will always have the same feature type as the destination layer.

## Cardinality

**Cardinality** is an important issue for spatial joins, just as it is for attribute joins. Because records in tables are being matched together, the Rule of Joining must be fulfilled in spatial joins also. Each feature record in the destination table must have one and only one matching record in the source table. This condition is met if the cardinality of destination to source is one-to-one or many-to-one. In attribute joins, we had to use a relate if the Rule of Joining was not fulfilled. With spatial joins, we must use a **summarized join** if we encounter a one-to-many relationship.

Recall that the Summarize function calculates statistics for groups of records in a table. It uses one field to divide the records into groups, and then it calculates statistics for other fields for each group. In a summarized join, each feature in the destination layer is matched to many features in the source. Statistics are calculated for that group of features, and the result is appended to the feature record.

| NAME | TYPE | Count | Sum_POP |
|------|------|-------|---------|
| Rogue Valley Intern | Interna | 22 | 192645 |
| Astoria Regional | Region | 14 | 34898 |
| Mahlon Sweet Field | | 23 | 266623 |
| Klamath Falls Intern | Interna | 8 | 46110 |
| North Bend Muni | Munici | 21 | 99896 |

In Figure 9.2, imagine reversing the join so that airports is the destination layer and cities is the source layer. Each airport has many cities that are closer to it than to any other airport. Instead of attaching the single closest city, a summarized join finds all of the cities closer to the airport and calculates one or more statistics, for example, the sum of the city populations. Then for each airport we would know the total number of people being served by that airport (i.e., the sum of the populations of the cities that are closer to that airport than to any other). Figure 9.4a shows the output table of the joined layer with a Count field representing the number of cities and the Sum_POP field representing the total people in those cities. So Astoria Regional has 14 cities close to it with a total of 34,898 people. Figure 9.4b shows a proportional symbol map based on the Sum_POP field to represent the total potential population served by the airport.

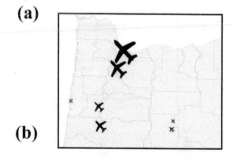

**(a)**

**(b)**

Fig. 9.4. A summarized spatial join: (a) output table; (b) map based on sum of population served
a: Source: ArcGIS

This type of problem, estimating usage of airports by surrounding populations, is known as an allocation problem, in which provision of services is being assessed. In truth, spatial joins are not the best approach—people select airports based on more than distance: cost, schedules, and so on. More sophisticated techniques for solving allocation and logistics problems are available but require the purchase of the Network Analyst extension.

Analysis problems typically involve making assumptions about the forces at work, and nearly always these assumptions are simplifications of the actual situation. We made an assumption that people choose the closest airport based on distance, and although it is true in a general way, it is not the whole story. That does not make the analysis wrong—airports close to many people are going to be serving many more passengers than airports in a lower population area—but it is important that we are cognizant of the assumptions being made in any analysis and remain sensitive to how they may impact the results. In this case, we have ignored tourist/business air traffic as well as the fact that many people will drive further to get a better deal. Thus we are probably overestimating use at the smaller airports and underestimating use at the large ones.

### *Types of spatial joins*

Spatial joins fall into four main types according to the cardinality of the relationship between the joined layers (simple or summarized) and the choice of spatial criteria (inside or distance). Figure 9.5 shows a matrix of the four possible combinations. A **simple join** may be used whenever the cardinality is one-to-one or many-to-one so that the Rule of Joining is maintained. In a one-to-many relationship, a summarized join must be employed.

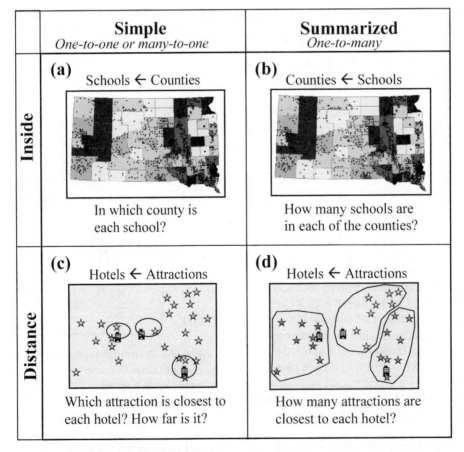

Fig. 9.5. Matrix of spatial joins resulting from different choices of destination table, spatial condition (inside or distance), and cardinality
Source: ArcGIS

Consider the task of finding the county for a particular school, which requires an alphabetized table of schools with a county field. This table can be created by joining the counties to the schools, with schools as the destination, as shown in Figure 9.5a. The output layer contains schools, and each school will have the attributes of the county it falls inside. This example is a *simple inside join*. The previous example of the septic systems is also a simple inside join.

Now consider a different question: How many schools does each county have? This time we need a table of counties with a field for the number of schools, so counties must be the destination. The cardinality of counties to schools is one-to-many, and a simple join is not possible. A summarized join may be employed in this case. The summarized join groups the schools together based on which county they are in and generates a single record of statistics for the group. This single summarized school record can then be appended to the matching county in the destination table. A Count field is always generated, containing the total number of schools in each county. Figure 9.5b shows this as a *summarized inside join*.

Distance joins also come in simple or summarized varieties. Consider evaluating the desirability of several hotels based on their distance to local tourist attractions, using a layer of hotels and a layer of attractions with the yearly number of visitors in each one. What is the cardinality of this relationship? This question seems confusing because there are several hotel sites and many attractions. In a distance join, it is the question that dictates the cardinality. If we ask "Which attraction is closest to the hotel?" in order to find out whether that attraction has a large visitor pool, we have specified a one-to-one criterion because only one attraction can be *closest* to each hotel. A *simple distance join* suffices, and each hotel appears in the output table along with the attributes of the attraction closest to it and the distance between them. In Figure 9.5c, the circles connect each hotel site with its closest attraction.

A different question may be asked: "How many tourist attractions are closer to this hotel site than they are to any other?" In this case, we are interested in evaluating the richness of choice of attractions for each hotel or perhaps the total combined visitor pool from all the closest attractions. In this case, each hotel is connected with many attractions, as shown in Figure 9.5d, and summary statistics for the group, such as the number of attractions and the sum of the visitor pools, may be added to the record for each site. This would be a *summarized distance join*.

### Feature geometry and spatial joins

The available join types will depend in part on the geometry of the features being joined. When joining points to points, for example, an inside join type does not apply. Thus, only two options are available when joining points to points, simple distance or summarized distance. Table 9.1 lists all of the possible geometry combinations and the join types that can be applied to each. (In this table we break the convention of putting the destination first in order to match the descriptions in the spatial join window in ArcGIS.)

Let us examine some other combinations of geometry types. Consider this problem: Many counties in South Dakota do not have a hospital. The state emergency planning office wishes to know the hospital closest to each county and how far away it is. They require a list of counties, each with the closest hospital attached to it. Thus, counties is the destination layer, and hospitals is the source layer. According to Table 9.1, joining points to polygons requires either a summarized inside or a simple distance

cntyhosp

| NAME | hospitals_NAME | Distance |
|------|----------------|----------|
| Shannon | Battle Mountain National Sanitar | 38424 |
| Fall River | Battle Mountain National Sanitar | 0 |
| Bennett | Battle Mountain National Sanitar | 110128 |
| Pennington | Bennett Clarkson Hospital | 0 |
| Lincoln | Canton-Inwood Hospital | 0 |
| Yankton | Canton-Inwood Hospital | 51493 |

Fig. 9.6. This table resulted from a distance join of hospitals to counties.
Source: ArcGIS

join. In this case, the simple distance join is the correct choice. After the join, each county has a field indicating the name of the closest hospital (Fig. 9.6) as well as the distance from the county to the hospital. When the hospital is inside the county, the distance of zero is assigned.

Table 9.1. Join types are available for each combination of feature geometries in a spatial join. The second feature type is the destination layer in each case.

| Geometry Type | Join Type | Example |
|---|---|---|
| Points to points | Simple distance | Find the hospital closest to each town. |
| | Summarized distance | Find all the towns closer to one hospital than to any other hospital. |
| Lines to points | Simple distance | Find the water main closest to the proposed building site. |
| | Summarized inside | Find the total voltage of all electric lines meeting at a substation. |
| Polygons to points | Simple inside | Find the soil type that underlies each gas station. |
| | Simple distance | Find the lake that is closest to each campground. |
| Points to lines | Simple distance | Find the elementary school that is closest to each residential street. |
| | Summarized distance | Find the total number of septic systems closer to a particular stream than to any other stream. |
| Lines to lines | Summarized inside | Find the number of roads that cross each river. |
| | Simple inside | Give a section of hiking trail the attributes of the road it follows for a short distance. |
| Polygons to lines | Summarized inside | Give a stream the average erosion index of the soil types it crosses. |
| | Simple distance | Find the lake closest to a hiking trail or the national park within which a road lies. |
| Points to polygons | Summarized inside | Find the total number of schools and students in a county. |
| | Simple distance | Find the town that is closest to a lake. A point inside a polygon is given a distance of zero. |
| Lines to polygons | Summarized inside | Find the total number of rivers crossing a state. |
| | Simple distance | Find the carrying capacity of the closest power lines to an industrial site. |
| Polygons to polygons | Summarized inside | Find the total population of all counties that intersect part of a watershed. |
| | Simple inside | Find the county within which a lake falls completely. |

In Figure 9.7, a unique values map was created based on the hospital name. In the northeast part of the state, the purple counties are closest to the hospital in the purple area, the green counties to the hospital in the green area, and so on. The long, pink, hatchet-shaped county in western South Dakota (Pennington County) has several hospitals. Since all three lie inside Pennington, they all have a distance of zero, and Pennington's closest hospital was arbitrarily assigned based on which hospital was found in the table first. The lighter pink counties in the central part of the state were assigned to a different hospital in Pennington County and appear in a different color. The map in Figure 9.8 shows each county displayed according to its distance to the closest hospital.

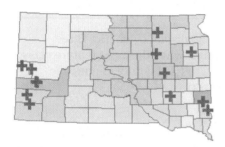

Fig. 9.7. Counties with the same color are closest to the same hospital.
Source: ArcGIS

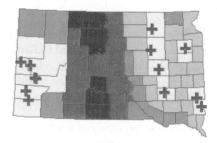

Fig. 9.8. Counties colored according to distance from the closest hospital
Source: ArcGIS

Figure 9.9 demonstrates an example of joining points to lines to predict the susceptibility of streams to contamination from septic systems. In this analysis, we make the assumption that the more septic systems that are close to the stream, the greater the stream's susceptibility to contamination. The point locations represent the centers of one-mile by one-mile sections, and the size of the symbol indicates the number of septic systems in the section. We need to join each stream line to the closest septic systems, so streams is the destination layer. Joining points to lines requires either a simple distance join or a summarized distance join. In this case, a summarized distance join, which sums the total septic systems that are closest to each stream, is correct. The map shows the results: the thicker the line symbol of the stream, the more septic systems are closest to that stream and the higher the susceptibility of the stream to contamination.

Fig. 9.9. Joining septic systems to streams to evaluate stream susceptibility to contamination
Source: ArcGIS

### Coordinate systems and distance joins

Distance joins should always be performed on layers with projected coordinate systems that preserve distance. If a join is performed on a layer with a geographic coordinate system (GCS) and units of decimal degrees, two problems arise. First, the distances reported in the table will be in decimal degrees. Degrees cannot be easily converted into miles or kilometers because the conversion factor changes with latitude. Second, the result could be invalid. The distance algorithm relies on the relative $x$-$y$ coordinates of the features to calculate distances and assumes a Cartesian coordinate system to do so. Degrees of latitude and longitude are spherical coordinates, not Cartesian, and the relative distances calculated may be incorrect.

Fig. 9.10. The effect of the coordinate system on a spatial distance join. (a) Three attractions are incorrectly joined to Hotel C (purple stars) when the input feature classes have a GCS. (b) When the feature classes are projected to UTM before joining, the attractions are correctly assigned to Hotel B (green stars).
Source: ArcGIS

Figure 9.10 shows two cases of a summarized distance join of tourist attractions to hotels. In Figure 9.10a, the join was done with both layers in a GCS. Notice that the three attractions in the northeast corner were assigned to Hotel C based on their distance from it in decimal degrees. In Figure 9.10b, both layers were projected to UTM prior to joining, and the attractions were assigned instead to Hotel B. Since a UTM projection preserves distance within the zone, we know that the second example gives the correct spatial distances between the attractions and the hotels and is the valid result. Notice that the spatial distribution of attractions is elongated in Figure 9.10a, as a result of being displayed in a GCS.

It is not sufficient to set the data frame to a projected coordinate system; the layers themselves must be projected. Furthermore, the projected coordinate system should be one that preserves distance over the region of analysis, such as UTM, State Plane, Equidistant Conic, and so on. Otherwise, the distance analysis may again be incorrect.

## About ArcGIS

Spatial joins are performed on two input feature classes and result in a new feature class that may be stored as a shapefile or geodatabase feature class. The original inputs are unchanged. Spatial joins are initiated using the same method as attribute joins.

### Choosing the join type

As in an attribute join, the process begins by deciding which layer is the destination. The user right-clicks the destination layer, chooses to join the data based on spatial location (Fig. 9.11), and enters the desired source layer.

The join menu always offers two ways to decide how to match the fields, taken from the four types shown in Figure 9.5. Consider joining schools to counties, with counties as the destination layer as described in Figure 9.5a. Each county can have more than one school, so it is a one-to-many join. There are two options in the window based on the choices listed in Table 9.1. The user must choose between them based on the desired result (Fig. 9.11).

For this example, Option 1 offers an *inside summarized join*, which gives each county a statistical summary of the attributes of all the schools inside it, such as the total number of schools or the sum of the students. Option 2 specifies that a *simple distance join* should be performed, which gives each county the attributes of the closest school. In this case, the second option is nonsensical. All of the schools are inside the county and would be assigned a distance of zero, so it avails little to find the "closest" one. In this example, the summarized join is the appropriate choice.

Figure 9.12 shows the result of the join. The resulting layer is a map of counties with the original county attributes plus a field called Count_ that contains the number of schools in the county. From this field, the graduated color map was created, showing the number of schools in each county. When summarizing, the user can choose from several statistics—minimum, maximum, average, and so on. All numeric attributes in the source table are summarized using the chosen statistics and are placed in the output table. String fields cannot be summed or averaged, so they are not included in the output table.

Fig. 9.11. A one-to-many cardinality is handled by either a summarized join or a distance join.
Source: ArcGIS

Fig. 9.12. Using a spatial join of schools to counties, one can create a graduated color map showing the number of schools in each county.
Source: ArcGIS

### Setting up a spatial join

Performing the spatial join itself is a simple process. However, determining that a spatial join is required and identifying the destination table and the type of join can challenge beginners. This section presents a series of questions to be answered when setting up a spatial join to help produce the correct result. Too many novices simply try random combinations until they hit on the right one—this process is designed to give the right answer the first time.

When faced with a suspected spatial join problem, first make a simple sketch of the relationships between the layers to be joined. Then answer this series of questions, designed to lead to the correct solution:

> What should the final output layer or table look like?
>
> Which is the destination layer?
>
> Should a distance join or an inside join be used?
>
> What is the cardinality of the join?
>
> Should a simple join or a summarized join be used?

Let us demonstrate this process by using the problem of estimating stream susceptibility to contamination from septic systems, as shown in Figure 9.9. Recall that each point represents a section and stores the total number of septic systems in the section. First sketch the problem and then go on to the questions.

**What should the final output layer/table look like?** The desired result is a layer of streams. In the attribute table, each stream must be assigned the total number of septic systems that are closer to it than to any other stream. Imagine the fields in the table, with the stream followed by the sum of the number of septic systems. Once the output is envisioned, it is usually easy to see which feature class is the destination layer.

**Which is the destination layer?** The features in the output layer are always the same features as those in the destination layer. If streams is chosen as the destination, then the output layer will contain streams. If the septics layer is chosen, then the output layer will contain septic points. Imagining our output table again, we see that what we really want is streams, each of which has septic information assigned to it. Thus, streams is the destination layer.

**Should a distance join or an inside join be used?** Since we're looking for septic systems closest to each stream, this is clearly a distance join.

**What is the cardinality of the join?** Assignment of cardinality depends on the destination layer, in this case streams, so consider the relationship between streams and the septic systems. Since one stream can have multiple septic points closer to it than to any other stream, this is a one-to-many cardinality.

**Should a simple join or a summarized join be used?** Since the cardinality is one-to-many, a simple join cannot work. Because the goal is to sum all the septic systems closest to the streams, rather than simply finding the single closest septic to the stream, a summarized join is indicated, with Sum as the statistic.

Now that the questions are answered, the problem setup becomes clear: We need to do a summarized distance join with streams as the destination layer and septics as the source layer. Even experienced users may find that following this suggested procedure when setting up a join makes it easier to find the right approach. In the next few examples, we will apply this process to set up and solve three different spatial join problems.

### Problem 1
### Number of earthquake deaths in congressional districts

Imagine that a representative from one of the congressional districts in California is sponsoring legislation to provide earthquake emergency planning funds and is looking for support for the bill. She is planning to throw a big party and invite all of the reps from districts with a significant number of earthquake deaths. She asks one staffer, who is a GIS specialist, to draft a list of names for the invitation list. The staffer might approach the problem as follows, using the data in the mgisdata folder.

STOP! Write the answers to the questions and then read on to see if you analyzed the problem correctly.

What should the output layer/table look like? _____

Which is the destination layer? _____

Should a distance join or an inside join be used? _____

What is the cardinality of this join? _____

Should a simple join or a summarized join be used? _____

The earthquake table has a state attribute but none for congressional district. The only way to associate the number of earthquakes with districts is to perform a spatial join. The goal is a table of districts with a field containing the number of earthquake deaths for each one. Thus, the districts will be the destination layer. The relationship between districts and earthquakes is potentially one-to-many, so a simple join is out of the question. Since the staffer wants to know the total deaths from quakes in each district, the summarize option will be the best, and the proper statistic is Sum. We will do a summarized inside join, with districts as the destination layer.

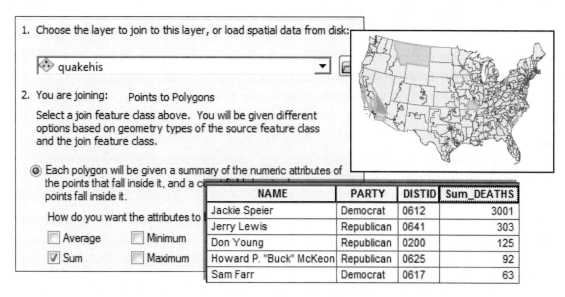

| NAME | PARTY | DISTID | Sum_DEATHS |
|---|---|---|---|
| Jackie Speier | Democrat | 0612 | 3001 |
| Jerry Lewis | Republican | 0641 | 303 |
| Don Young | Republican | 0200 | 125 |
| Howard P. "Buck" McKeon | Republican | 0625 | 92 |
| Sam Farr | Democrat | 0617 | 63 |

Fig. 9.13. Using a summarized spatial join to find the congressional districts that have had more than 10 earthquake-related deaths
Source: ArcGIS

Figure 9.13 shows the spatial join window filled out for this problem, as well as the resulting table. After performing the join, the staffer uses an attribute query to select all of the districts that have had 10 or more earthquake-related deaths. The table and the map show the selected records from the query—many of the districts are in California, as one might expect. Finding only 12 districts on the list and knowing that the boss wants a BIG party, the staffer then uses Select By Attributes to find all the districts that have had ANY earthquake deaths. This brings the total up to 29 districts. The guest list should be even bigger, so the staffer decides to use Select By Location to also select any districts that touch the districts with earthquake deaths (Fig. 9.14). This brings the guest list to 91, and the staffer can make up the invitations.

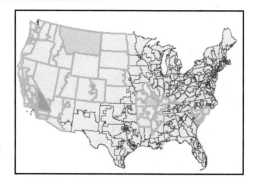

Fig. 9.14. Congressional districts with earthquake deaths or that touch a district with earthquake deaths
Source: ArcGIS

**Problem 2**

**Pollution risk of rivers based on county populations**

Imagine that Lindsey must do an analysis estimating the risk of pollution to rivers in the United States. She has no detailed information of the sources and types of pollutants, but she can make use of the observation that, generally speaking, large numbers of people are generally correlated with high risks of pollution. She decides to use the population of counties as a proxy for the pollution risk. For each river, she wants to find the total number of people living in counties that intersect the river.

STOP! Write the answers to the questions and then read on to see if you analyzed the problem correctly.

What should the output layer/table look like? _____

Which is the destination layer? _____

Should a distance join or an inside join be used? _____

What is the cardinality of this join? _____

Should a simple join or a summarized join be used? _____

The final output layer should contain rivers, with a table listing each river and the total population of counties adjacent to that river. Thus, rivers must be the destination layer. Since the counties must actually touch the river, this is an inside join rather than a distance join. The cardinality is one-to-many, since one river can have many counties touching it. An inside summarized join must be used with rivers as the destination layer, and the Sum statistic should be requested.

Figure 9.15 shows the spatial join window filled out for the join. The output table shows the river name, the number of counties adjacent to it, and the sum of each numeric field, including the POP2000 and POP2010 fields. The figure also shows a graduated symbol map based on the Sum_POP2010 field, in which the width of the river represents the total county population living adjacent to it. Notice that ALL of the numeric fields are summed during the join, not just the POP2000 field. One drawback to summarized spatial joins is that the chosen statistic is applied to all the fields, yielding potentially large attribute tables.

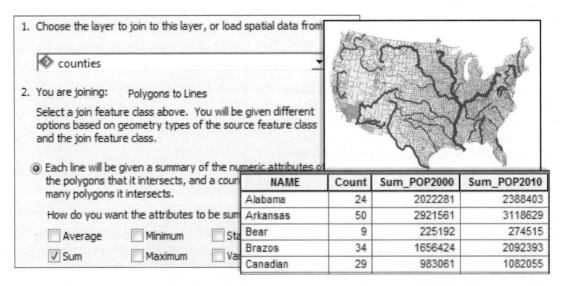

Fig. 9.15. Estimating pollution risk for rivers based on adjacent county populations
Source: ArcGIS

Also note that the field names appear truncated. Because field names are limited to 13 characters, prefixing each with "Sum_" may crop other characters from the end. Notice that the POP2000 field may become Sum_POP200, the POP00_SQMI field may become Sum_POP00_, and so on. The shorter names may cause confusion, and it may be necessary to go back to the original table of the source layer to find out which field is which. The fields will be in the same order as the original. An alias can then be created, if desired, to avoid further confusion.

## Problem 3

### Closest volcano to a city

In this final example, imagine that a professor who specializes in volcanoes has built a web site about them. He would like to put a table on his web site so that school children all over the United States can enter the name of the city they live in and get an information page about the volcano that is closest to them. Since the professor knows nothing about GIS, his graduate student, Cody, gets to make the table.

STOP! Write the answers to the questions and then read on to see if you analyzed the problem correctly.

What should the output layer/table look like? _____

Which is the destination layer? _____

Should a distance join or an inside join be used? _____

What is the cardinality of this join? _____

Should a simple join or a summarized join be used? _____

The final goal is a cities layer with fields indicating the closest volcano and the distance to it. Thus, Cody realizes, Cities is the destination table, and a distance join to the volcanoes layer will provide the necessary information. Because the closest volcano is the target, this is a simple distance join (only one volcano can be the closest one).

Figure 9.16 shows the spatial join window settings to produce this analysis and the result of the spatial join. The table shows the city name, the state abbreviation, the volcano name, the place where the volcano is located, the volcano elevation and type, and the distance to the volcano.

1. Choose the layer to join to this layer, or load spatial data from disk:

| NAME | ST | volc_NAME | TYPE | Distance |
|------|-----|-----------|------|----------|
| Algonquin | IL | Dotsero | Maar | 1,505,644 |
| Alhambra | CA | Lavic Lake | Volcanic field | 151,255 |
| Alice | TX | Durango Volc | Cinder cones | 738,157 |
| Aliquippa | PA | Dotsero | Maar | 2,139,036 |
| Aliso Viejo | CA | Lavic Lake | Volcanic field | 162,176 |

volcanos

2. You are joining:    Points to Points

Select a join feature class above. You will options based on geometry types of the s and the join feature class.

◉ Each point will be given all the attributes of the point in the laye being joined that is closest to it, and a distance field showing how close that point is (in the units of the target layer).

Fig. 9.16. Table showing each city and its closest volcano, with the distance to the volcano in meters: graduated color map is based on the Distance field.
Source: ArcGIS

Notice the field name volcanoes_NAME in the output table. It happens that the Cities layer has a NAME field, and so does the volcanoes layer. The output table cannot have two fields with the same name, so during the join the repeated names are automatically converted into unique names. The Cities name field remains NAME, but the Volcanoes name field becomes NAME_1. In addition, an alias is created, volcanoes_NAME. It is the alias that is being displayed in the table, but if the table is set to show actual field names, then the NAME_1 field will be shown instead.

Also notice that the distance from Boone, Iowa, to the Dotsero volcano in Colorado is over one million. These units are clearly not miles or kilometers. In fact, they are meters, indicating that the coordinate system of the original data is stored with units of meters.

**TIP:** Always check the original coordinate system after a join to be sure you know the units for the distance values. Also, remember that distance joins should be performed only on projected coordinate systems that preserve distance, or the results will be invalid.

## Summary

▶ A spatial join combines the records of two feature tables based on the location of the features. A new feature class is created by a spatial join.

▶ An inside join uses the criterion that one feature falls inside another or, in the case of points and lines, on top of each other. A distance join matches the destination layer feature to the record of the closest feature in the source layer. A distance field reporting the distance between the joined features is added to the table.

▶ A simple join may be used whenever the cardinality of the layers is one-to-one or many-to-one. A summarized join generates summary statistics for all the source features matching the destination features and is used when the cardinality is one-to-many or many-to-many.

▶ Four types of spatial joins exist based on the combination of the criterion and the cardinality of the relationship: simple inside joins, simple distance joins, summarized inside joins, and summarized distance joins.

▶ Distance measurements are reported in map units of the input layers.

▶ Distance joins should only be performed with layers having projected coordinate systems that do not distort distances. Using layers with a geographic coordinate system may yield incorrect results.

▶ In the joined table, fields with identical names will be renamed in the output file so that all field names are unique, such as NAME to NAME_1. Usually, the source field is renamed.

▶ Use the following series of questions to help you set up a spatial join properly:

What should the output layer/table look like? _____

Which is the destination layer? _____

Should a distance join or an inside join be used? _____

What is the cardinality of this join? _____

Should a simple join or a summarized join be used? _____

### Important Terms

| | | | |
|---|---|---|---|
| cardinality | inside join | simple join | summarized join |
| distance join | logical consistency | spatial join | |

## Chapter Review Questions

1. What primary characteristic distinguishes a spatial join from an attribute join?

2. What two options may be used to handle one-to-many relationships in a spatial join?

3. If a polygon feature type is joined to a line layer, with the lines as the destination table, what will the feature type of the output layer be?

4. How many output fields will result if a summarized join is specified and a single statistic (e.g., Sum) is selected?

5. Why should distance joins always be performed on layers with a projected coordinate system? What kind of projection should be used?

6. What happens if the two input layers in a join each have a field with the same name?

For the following spatial join problems, answer the series of questions in the text and then state the type of join that should be used: simple inside, simple distance, summarized inside, or summarized distance.

7. Determine the number of parcels within each of Austin's watersheds.

8. Find the closest school for each house in a realtor's database.

9. Find the land use zoning type associated with each well in Atlanta.

10. Determine the number of counties and the total number of people served by each airport in the United States.

# Mastering the Skills

## Teaching Tutorial

The following examples provide step-by-step instructions for doing basic tasks and solving basic problems in ArcGIS. The steps you need to do are highlighted with an arrow →; follow them carefully. Click on the video number in the VideoIndex.htm file to view a demonstration of the steps.

### *Simple inside joins*

→ Start ArcMap and open the map document ex_9.mxd.

→ Use Save As to rename the document and remember to save frequently as you work.

Spatial joins produce new feature classes. We only need these for practice and don't want them to become mixed with the data in our permanent Austin geodatabase. So, for this lesson, we'll create a new file geodatabase to store the outputs.

1→ Click the Catalog button or go to the Catalog tab if it is already docked in ArcMap.

1→ Expand the Folder Connections entry and navigate to the mgisdata\Austin folder.

1→ Right-click the Austin folder and choose New > File Geodatabase.

1→ While it is highlighted, name the new geodatabase chap9results and click Enter.

**TIP:** When you see the word **STOP**, pause a moment and set up the problem using the questions presented in the Concepts section. Then read on to see if you analyzed the problem correctly.

▶ What should the output layer/table look like?

▶ Which is the destination layer?

▶ Should a distance join or an inside join be used?

▶ What is the cardinality of this join?

▶ Should a simple join or a summarized join be used?

The map currently shows the geology and water wells for Austin, TX. For each well, we would like to know the geological unit associated with it at the surface (i.e., for each well, we want to know the geological unit that the point falls inside). **STOP** and think it through.

The desired output is a table of wells containing a field with the geological unit. Therefore, the Wells layer is the destination. The geology polygon might have many wells, but each well can only fall in one polygon, making it a many-to-one cardinality. Therefore, a simple inside spatial join is indicated.

2→ Right-click the Wells layer and choose Joins and Relates > Join from the context menu.

2→ In the top drop-down box, choose to *Join data from another layer based on spatial location* (Fig. 9.17).

2→ Choose Geology as the layer to join.

2→ Choose to join the point to the polygon that *it falls inside*.

2→ Click the Browse button and change the *Save as type* to *File and Personal Geodatabase feature classes*.

2→ Navigate inside the Austin\chap9results geodatabase and enter *wellgeology* as the name of the feature class to be created. Click Save and click OK.

The new feature class appears at the top of the Table of Contents.

3→ Turn off the original Wells layer.

3→ Right-click the wellgeology layer and choose Open Attribute Table.

3→ Scroll to the right to find the fields from the Geology table.

**TIP:** The first field from the Geology feature class is named geology_OBJECTID. All fields to the right of it are from the Geology table. Look for the second OBJECTID after any spatial join to find where the appended information begins.

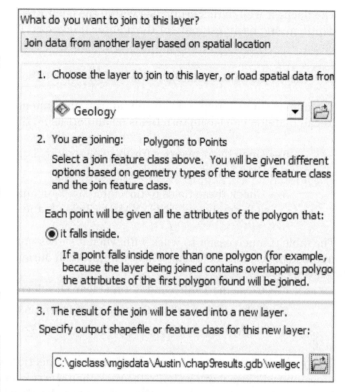

Fig. 9.17. The Join Data window for spatial joins
Source: ArcMap

Locate the UNIT_NAME field at the very end of the new table, containing the geological unit. Let's find out how many of the wells are on limestone units.

3→ Click the Select By Attributes button in the Table window and select the wells based on the expression UNIT_NAME LIKE '%Limestone%'. Close the selection window.

1. How many wells are situated on limestone? _____

4→ Right-click the wellgeology layer and choose Zoom to Layer.

4→ Expand the legend for the Geology layer and find the Glen Rose Limestone. Where does it outcrop on the map? (Try changing it color to one that stands out.)

4→ Find the aquifer_code field from the wells part of the table and examine the entries. Look for codes containing "GLR."

The aquifer code field indicates the formation that supplies water to the well at an opening some depth underground. The GLR aquifer codes refer to the Glen Rose Limestone, which is a prominent aquifer (water-bearing rock) in this part of the country. Wells often get water from a different rock formation than is exposed at the surface. Let's find out how many of the wells ON the Glen Rose Limestone actually get water from the Glen Rose Limestone.

4→ Open the Select By Attributes window.

4→ Enter the new expression

UNIT_NAME = 'Glen Rose Limestone' AND aquifer_code LIKE '%GLR%'

The deepest well in this group lies within the Glen Rose Limestone both at the surface and at depth, giving us an estimate of the aquifer thickness. The well_depth field gives the depth in feet.

2.   What is the approximate thickness of the Glen Rose Limestone near Austin? _____

The number of different fields resulting from this join makes the table unwieldy to navigate. Recall that you can easily turn fields on and off using the layer properties.

   5→ Open the properties for the wellgeology layer and click the Fields tab.

   5→ Click the *Turn all fields off* button.

   5→ Check these fields to turn them back on again: state_well_number, aquifer_code, well_depth, and UNIT_NAME. Click OK and examine the table.

The table is much easier to work with when it shows only the fields in which we are currently interested. Use this technique whenever you wish during this tutorial and afterwards.

   6→ Close the Table window and clear all selected features.

   6→ Turn off the wellgeology layer and turn on the Creeks layer.

   6→ Right-click the Creeks layer and choose Zoom to Layer.

The geological substrate of a creek affects its connection with groundwater. A creek on porous limestone will lose more water to the rock than a creek on clay. We assigned geology information to wells using a spatial join; we can also do it for creeks. **STOP** and think through the problem.

We want a table containing creeks with a field indicating the geological unit that the creek is on. Thus, Creeks is the destination feature class. A simple inside join is called for.

   7→ Right-click the Creeks layer and choose Joins and Relates > Join.

   7→ Set the source layer to Geology.

   7→ Use the simple join option. (*Each line will be given the attributes of the polygon that it falls completely inside.*)

   7→ Store the output feature class in the chap9results geodatabase and name it creekgeology. Click OK.

   8→ Collapse the Geology layer legend and turn it off.

   8→ Open the properties of the new creekgeology layer and choose a Categories: Unique values map based on the UNIT_NAME field.

   8→ Click the Symbol heading to change properties for all symbols to a 2-pt.-thick line.

   8→ Choose a color scheme with dark, bold colors that will show up well. Click OK.

In the Table of Contents, notice that one of the symbols has a blank, or <Null>, value instead of a unit name.

   9→ Right-click the symbol with the blank and set it to a light gray color (Fig. 9.18).

   9→ Turn on the Geology layer again.

The gray creeks cross geological contacts. We assumed that each creek was within a single geological unit, but we can see that it is not true in all cases. When a creek crossed a boundary, no unit was joined to it, and it received a <Null> value in the attribute table for the geology attributes.

9→ Open the creekgeology attribute table. Examine the fields, particularly the UNIT_NAME field containing the geology.

9→ Right-click the UNIT_NAME field and choose Sort Ascending to see the <Null> values. Then close the Table window.

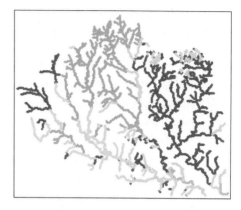

Fig. 9.18. Creeks displayed by geological unit
Source: ArcMap

One weakness of inside spatial joins is that they join only when features fall completely inside other features. In Chapter 10, we will learn about the Intersect tool, which is similar to a join but can split the features when they cross. Let's take a sneak peek ahead.

10→ Open the ArcToolbox > Analysis Tools > Overlay > Intersect tool.

10→ Click the dropdown button for input features and select Creeks.

10→ Click the dropdown button again and select Geology.

10→ Name the output feature class *creekgeolint* and save it in chap9results.

11→ Symbolize the creekgeolint layer as you did creekgeology. Compare the two layers and also examine the tables.

The creekgeolint layer has no <Null> values in its table. Creeks that crossed a geologic unit boundary have been split into pieces, and the geologic unit has been assigned to each piece.

### A summarized inside join

A simple join is appropriate for a one-to-one or a many-to-one cardinality. A summarized join is needed for a one-to-many cardinality. Think about a watershed, the collection area for surface water. Runoff that comes from a watershed with many people is likely to have more pollutants than runoff from a watershed with fewer people. Let's analyze which watersheds in the Austin area are at greatest risk for polluted runoff.

12→ Remove the creekgeology, creekgeolint, and the wellgeology layers from the map. They are still saved in the chap9results geodatabase if you want them later.

12→ Turn off the Geology and Creeks layers.

12→ Turn on the Watersheds and Block Population layers.

12→ Right-click the Block Population layer and choose Zoom to Layer.

12→ Open the table for the Block Population layer.

Block Population contains points representing the centroids of block groups, the smallest unit for which the Census Bureau summarizes population data. Each point represents numbers of people and households. We will use these points to determine the number of people in each watershed. **STOP** and think it through.

This time we want to have a list of watersheds with the total number of people in it. Watersheds is the destination layer. One watershed can have many block points in it, so the relationship is one-to-many. A summarized inside join must be used. Since we want to know the total number of people, we need to request the Sum statistic.

13→ Close the Table window.

13→ Right-click the Watersheds layer and choose Joins and Relates > Join.

13→ Set Block Population as the layer to join to.

13→ Choose the option to summarize the points inside the polygons and check the box for the Sum statistic.

13→ Name the output feature class watershedpop and put it in the chap9results geodatabase. Click OK.

14→ Drag the watershedpop layer just below the Block Population layer.

14→ Open the watershedpop table and examine the fields on the far right.

14→ Close the Table window.

The Count field records how many block points were found in the watershed. The fields prefixed with "Sum_" show the totals for each of the numeric fields in the source table. Sum_POP2010 tells us the total number of people living in each watershed.

15→ Open the Symbology properties for the watershedpop layer and create a Quantities: Graduated color map based on the Sum_POP2010 field and the Jenks classification.

15→ The population values are influenced by polygon area, so normalize the classification using the Shape_Area field.

15→ Symbolize it with a monochromatic color ramp and click OK.

15→ Name the layer Watershed Hazard.

Fig. 9.19. Watershed pollution risk from population
Source: ArcMap

The map should look similar to Figure 9.19. Note that the watersheds extend beyond the block population data. If this were a real project, you would need to address this issue by dropping polygons with incomplete data or expanding the block group data to the full extent of the watersheds.

16→ Turn off the Watershed Hazard, Watersheds, and Block Population layers.

16→ Turn on the Streets layer. Add the police_districts feature class from the mgisdata\Austin\Austin\Administrative feature dataset.

Imagine that the city has been reviewing its police department staffing levels. To help assess needs, it would be helpful to know the total length of streets for which each district is responsible. You have been given the task of finding this information. **STOP** and think it through.

You want to produce a list of the police districts, each of which has the total street length. Thus, the districts are the destination layer. Each district contains many streets, so the cardinality is one-to-many. A summarized inside join must be used, with Sum as the statistic. First, notice that many streets do not fall in any of the districts. You can reduce the processing time for the spatial join by first selecting only the streets that are within the districts.

17→ Open the Select By Location window.

17→ Set the target layer to Streets and the source layer to police_districts.

17→ Set the spatial selection method to *are within the source layer feature*. Click OK.

Now when you do the spatial join, only the selected streets will be used, saving time. You will also have fewer streets with unassigned districts.

18→ Right-click the police_districts layer and choose Joins and Relates > Join.

18→ Set the source layer to Streets and choose the summarized option. Check the Sum statistic.

18→ Place the result in chap9results and name it *policestreets*.

18→ Click OK and wait. This join may take several minutes.

19→ Clear the selection and turn off the Streets layer.

19→ Open the policestreets table. The original Streets layer contained a field called MILES with the length of the street. Find the Sum_MILES field in the joined table.

19→ Close the Table window.

20→ Create a graduated color map of the policestreets layer based on the Sum_MILES field. For this map, it is best to classify the distances using a defined interval of 50 miles.

The downtown districts are smaller and have fewer miles of road than the outlying districts (Fig. 9.20). The next step might be to analyze the block population for each district to see if they are balanced in that way instead. We will leave that as an exercise, and move on.

21→ Turn off the police_districts layer.

21→ Remove the policestreets layer from the map document.

Fig. 9.20. Total road mileage each police district covers
Source: ArcMap

### Simple distance joins

Distance joins combine records for features that are closest to each other. Consider the problem of post office usage. One might assume that, generally, people will go to the closest post office. For each street, it would be nice to know which post office to go to. **STOP** and think it through.

We want a list of streets with a field indicating the closest post office, so Streets is the destination layer. Since only one post office can be "closest," this is a one-to-one relationship. A simple distance join is appropriate.

22→ Turn on the Streets and Post Office layers.

22→ Right-click the Streets layer and choose Joins and Relates > Join.

22→ Set Post Office as the layer to be joined.

22→ Choose the simple join option, where *Each line will be given the attributes of the point that is closest to it.*

22→ Name the output file *streetpost* and save it in chap9results. Click OK.

23→ Open the streetpost attribute table and examine all of the fields. Find the field that contains the names of the different post offices.

23→ Close the Table window.

3. Which field contains the post office names? _____

23→ Create a unique values map of the streetpost layer based on the FACILITY_N field, so that each street is symbolized by the post office it is closest to. Use thick lines to see the colors better (Fig. 9.21).

A distance join also creates a distance field. In this case it shows how far each street lies from the nearest post office (as the bird flies, not driving distance). The units will match whatever the storage units are for the feature class coordinate system.

4. What are the coordinate system and units for the streetpost feature class? _____

5. How many streets are more than two miles from a post office (as in Fig. 9.22)? _____

24→ Clear the selected features.

24→ Create a graduated color map of the streetpost layer based on the Distance field, using a thicker line and a monochromatic color ramp (Fig. 9.23).

25→ Remove the streetpost layer and turn off the Post Office layer.

25→ Add the trail_heads and restrooms feature classes from the Parks feature dataset in the Austin geodatabase.

Before people start a hike, or after they finish, they like to find a restroom. Imagine that a hiking club wants to analyze the proximity of trailheads to restrooms and make recommendations for adding more restrooms to serve hikers. **STOP** and think through the problem.

The club needs a list of trailheads that includes the distance of each trailhead to the closest restroom. Thus, trail_heads

Fig. 9.21. Post office service areas
Source: ArcMap

Fig. 9.22. Streets more than two miles from a post office
Source: ArcMap

Fig. 9.23. Distance of streets from nearest post office
Source: ArcMap

is the destination layer. There can be only one closest restroom to a trailhead, so the cardinality is one-to-one, and a simple distance join will suffice.

26→ Right-click the trail_heads layer and choose Joins and Relates > Join.

26→ Set restrooms as the layer to join.

26→ Choose the simple join option to give each point the attributes of the closest point.

26→ Name the output feature class *trailrest* and save it in chap9results. Click OK.

The club is most concerned about the trailheads that are more than 1000 feet from a restroom. You will create a map that will help the club identify optimum locations for new restrooms.

27→ Open the trailrest table and examine the fields. Find the Distance field.

27→ Use Select By Attributes to select the trailheads that are more than 1000 feet from a restroom. Close the Table window.

27→ Right-click the trailrest layer and choose Selection > Create Layer From Selected Features. Name it *Problem Trails*.

27→ Clear the selected features.

28→ Add the parks and the trails feature classes from the Parks feature dataset.

28→ Symbolize the parks in a light green color and the trails as a dark green thick line.

29→ Click on the restrooms layer symbol to open the Symbol Selector.

29→ Type *restroom* in the search box at the top and click the Search button. Choose the symbol you prefer and make it about 16 pt.

29→ Open the restrooms layer properties, click the General tab, and set the minimum scale to 1:60,000.

30→ Remove the trailrest and trail_heads layers.

30→ Create a graduated color map of the Problem Trails layer based on distance from the restrooms. The values are skewed, so use the Jenks classification method. Make the symbol slightly larger so you can better see the colors (Fig. 9.24a).

30→ Zoom in to an area with clusters of Problem Trails to test the map (Fig. 9.24b).

Let's save this work as a group layer file for future use by the club.

31→ Click on the Problem Trails layer name to highlight it. Hold down the Ctrl key and click the restrooms, parks, and trails layers.

31→ Right-click one of the highlighted layers and choose Group.

31→ Name the new group layer *Restroom Map*.

31→ Right-click the Restroom Map group layer and choose Save As Layer File. Save it in the Austin folder as *Restroom Map.lyr*.

Fig. 9.24. Map showing trailheads more than 1000 feet from a restroom
Source: ArcMap

32→ Turn off the Restroom Map group layer and collapse it.

32→ Right-click the City Limit layer and zoom to it.

### Summarized distance joins

A simple distance join asks "Which feature is closest?" A summarized distance join is needed to address the question "How many features are closer to one location than they are to another?" Imagine that the Department of Recreation is applying for a grant to renovate its recreation centers. For guidance, they would like to know the potential population served by each center, making the assumption that most people will go to the closest center. The Block population data will provide the base data set for the analysis. **STOP** and think it through.

We want a list of the recreation centers with a field containing the total number of people who are closest to it. Centers is the destination layer, and the cardinality is one-to-many. A summarized distance join will be used.

33→ Add the facilities feature class from the Facilities feature dataset in the Austin geodatabase.

33→ Use Select By Attributes to select recreation centers using the expression FACILITY_T = 'RECREATION CENTER'.

33→ Right-click the facilities layer and choose Selection > Create Layer from Selected Features. Name the layer *Rec Centers*.

33→ Remove the facilities layer.

**TIP:** Creating intermediate selection layers is often a smart thing to do. It provides a visual test of your query, and it saves the selection in case a later mistake requires redoing the analysis. Give each selection layer a descriptive name to avoid confusion later.

34→ Right-click the Rec Centers layer and choose Joins and Relates > Join.

34→ Set Block Population as the layer to join.

34→ Choose the summarized option. You want to know the total number of people, so check the Sum statistic.

34→ The default save folder was changed to Austin when you saved the Restroom Map layer file. Click the folder Browse button to change it back.

34→ Change the Save as type to *File and Personal Geodatabase feature classes.*

34→ Navigate inside the chap9results geodatabase. Name the output recpop and click Save and OK.

35→ Open the recpop table and examine the Sum fields.

35→ Examine the statistics for the Sum_POP2010 field.

6. What are the minimum, maximum, and average numbers of people served by the recreation centers? _____

Clearly there are large discrepancies in the potential usage of each center. What is the spatial distribution of this potential usage?

35→ Close the Statistics and Table windows.

35→ Create a graduated symbol map of the recpop layer based on the Sum_POP2010 field. Use Jenks Natural Breaks.

Fig. 9.25. Potential usage for recreation centers
Source: ArcMap

The map is interesting (Fig. 9.25). Small centers clustered in the downtown area appear to compete with one another, while the outlying centers draw large groups over larger distances. The department decides to adopt a strategy to turn the downtown cluster into special-interest centers that may draw more people, while expanding standard activities in the outlying centers. Of course, the final plan would take into account many more issues, such as parking, public transportation, and current facilities and usage.

Meanwhile, another group is concerned about the effects of population on stream pollution. The number of people living close to a stream impacts the likelihood for the stream to be affected by pesticides and other pollutants. This group wants to characterize the basic threat to each stream by summing the population closer to that stream than to any other. **STOP** and think it through.

We want a feature class of streams that contains a field with the total population close to it, so streams are the destination layer. The cardinality is one-to-many, so a summarized distance join with the Sum statistic is needed.

36→ Add the named_creeks feature class from the Environmental feature dataset in the Austin geodatabase.

36→ Turn off all layers except the City Limit and named_creeks layers.

36→ Turn on the Block Population layer, but change the color to Gray 10%.

37→ Right-click the named_creeks layer and choose Joins and Relates > Join.

37→ Set the source layer to Block Population.

37→ Choose the summarized join option. Fill the button to use the closest features and check the Sum statistic.

37→ Name the output creekpop and save it in chap9results. Click OK.

38→ Open the creekpop attribute table and examine the Sum_ fields.

38→ Sort the Sum_POP2010 field in descending order.

7.  Which three creeks have the highest risk based on total population?

_____

38→ Rename the creekpop layer Population Load.

38→ Create a graduated color map for the Population Load layer based on the Sum_POP2010 field. The data are skewed, so use Jenks Natural Breaks.

38→ Use a monchromatic color ramp and thick lines so the colors show (Fig. 9.26a).

It may occur to you that the longer creeks are naturally exposed to more people. You decide to create a hazard index based on the number of people <u>per unit length</u> of stream.

Fig. 9.26. Maps showing pollution hazard of Austin creeks based on (a) population load and (b) hazard index
Source: ArcMap

39→ Click the Table Options menu and choose Add Field.

39→ Name the field HazIndex and set the field type to Float. Click OK.

39→ Right-click the <u>empty</u> HazIndex field and choose Field Calculator. Enter the formula Sum_POP2010 / Shape_Length and click OK.

39→ Examine the HazIndex field and sort in descending order.

8.  Which three creeks have the highest risk based on the index?

_____

Notice that most of the creeks have an index near or below 1, but the top three are in the range of 10–14. Notice the Shape_Length field also. These high-index creeks have very short lengths, and they are probably scraps of creek remaining after they were clipped to the city boundary. These high-index values are probably a fluke and should be eliminated from consideration. To avoid having them skew the map, you will exclude short creeks from the map.

40→ Close the Table window.

40→ Right-click the Population Load layer and choose Copy.

40→ Right-click the Layers data frame name and choose Paste Layer(s). Rename the new layer Hazard Index.

41→ Open the layer properties for the Hazard Index layer.

41→ Click the Definition Query tab and enter the expression Shape_Length > 1000. Click Apply so that it takes effect before you set the symbols.

41→ Click the Symbology tab and create a graduated color map based on the HazIndex field using the same symbols as the Population Load layer (Fig. 9.26b). Click OK.

As you can see, there are some differences between the two approaches for evaluating pollution hazards. (Some color differences may also be due to differences in the Jenks breakout of the different values.) Which model do you think might be a more realistic approach?

A hydrologist who specializes in surface water quality would probably question both models, because she knows that surface water runoff is controlled more by elevation than by distance, and that a watershed provides a better unit for summing the population impacts. You already did this using a summarized inside join.

42→ Turn on the Watershed Hazard layer and open its properties.

42→ On the Display tab, set the transparency to 50%.

42→ Turn off the Block Population layer.

42→ Compare the Population Load and Hazard Index maps visually against the Watershed Hazard map by turning the upper one on and off several times.

Which of the two creek-based methods appears to agree better with the watershed method? Overall, do the three methods agree on which watersheds and creeks bear closer scrutiny?

This analysis brings up an important consideration in GIS analysis. A problem may be approached several ways, using different models based on different assumptions. All three approaches had potential problems. In the watershed-based model, the population data set did not cover the entire watershed area. In the creek-based analysis, the fundamental assumption that distance was an appropriate predictor of influence is flawed, and short creeks resulting from data problems could have impacted the results. Each approach yielded similar, but not the same, results. Three important lessons should be gleaned from this example.

First, ***data issues and problems affect nearly all analysis procedures.*** You cannot mitigate all issues, but it is important to be aware of what they are, as well as what impacts they might have.

Second, ***the fundamental assumptions behind a model should be the best available.*** The watershed model had the most realistic assumption about water flow and is the best of the three models (or would be if the population data covered all the watersheds).

Third, ***do not push the results of a model beyond its limits.*** Even the watershed model neglects important variables. For example, industrial areas produce different amounts and types of pollutants than residential or farming areas do. Stream flow impacts the ability of a stream to dilute pollutants. Although we mapped five hazard levels, whether the model can realistically separate the hazard into many categories with distinct boundaries is questionable.

In the long run, the best you can say about this analysis is that it highlights potential problem areas that would benefit from further scrutiny. Toward that goal, all three models performed equally well. This observation yields a fourth lesson, that ***the most detailed and accurate model is not always necessary to achieve the objective.***

### *Data quality issues with spatial joins*

Now we will do one more polygon-to-polygon join to demonstrate some issues to consider when joining. We have a census tracts layer for Austin that has a county FIPS code, but no county name. We would like to do a spatial join of counties to the tracts in order to provide a field with the county name. This is a simple inside join with a many-to-one cardinality.

43→ Collapse the legends of all of the layers in the data frame and turn them off.

43→ Add the tracts and the counties feature classes from the Administrative feature dataset of the Austin geodatabase.

43→ Put tracts above counties in the Table of Contents and zoom to the full extent.

44→ Right-click the tracts layer and choose Joins and Relates > Join.

44→ Set the join layer to counties.

44→ Choose the simple join option.

44→ Name the output *tractcounty* and place it in the chap9results geodatabase.

45→ Change the counties symbol to a hollow shade with 2-pt. purple borders, and place it above the new tractcounty layer in the Table of Contents.

45→ Create a unique values map for the tractcounty layer, using the joined COUNTY field containing the county names (it is near the end of the table).

Notice that the unique values legend has a <Null> entry, and many of the tracts are symbolized with this value. What happened?

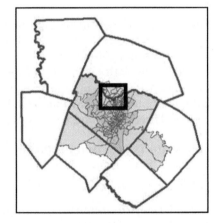

46→ Open the tractcounty table and scroll far to the right where the COUNTY field is.

46→ Scroll down to examine all the rows in the table.

You can see many <Null> fields in the table. These tracts were not matched to a county at all. This fact is puzzling, because, from the map, the tracts appear inside the counties and should have been matched with one. However, the map shows that the affected tracts all share a boundary with the county (Fig. 9.27).

Fig. 9.27. The orange tracts received no match during the join. Source: ArcMap

46→ Close the Table window and zoom in to the area shown in the black box in Figure 9.27.

46→ Switch to the List by Selection view in the Table of Contents and make tractcounty the only selectable layer.

46→ Click the Select Features by Rectangle tool and click on one of the tracts that share the county border. Click several more.

It becomes apparent that the tract boundary and the county boundary are not exactly the same (Fig. 9.28). In the real world, a tract always falls inside a county. In this GIS data set, the tracts and counties came from two different sources, and as a result their boundaries do not coincide. To be joined, the tract must fall within the county. Even the small discrepancies seen here are sufficient to prevent the county from being matched to the tract during the join.

How well the GIS data features reflect relationships in the real world is called **logical consistency**. In this case, the data sets are not logically consistent because the data features with mismatched boundaries do not reflect the real-world boundaries, which are identical.

Fig. 9.28. Part of this tract lies outside the county.

**TIP:** Always keep in mind that data accuracy and logical consistency issues may have an impact on an analysis. Be alert for potential problems.

This is the end of the tutorial.

→ Close ArcMap. You can save your changes.

## Exercises

Use the data in the mgisdata\Austin\Austin geodatabase to answer the following questions.

1. Give the name of the watershed in Austin that contains the most wells. How many wells does it contain? Which watersheds have the deepest and shallowest average well depths?

2. How many wells are there in the WATERFRONT zoning code in Austin? Use the O_NAME field from the zoneoverlays feature class. How many of the wells have depths less than 100 feet?

3. Which zoning category (O_NAME) in Austin contains the most wells?

4. What are the minimum, maximum, and average distances from barbeque (bbq) pits to the closest restroom?

5. An elementary school pool fun day is planned. If each school goes to the closest pool to it, which pool will have the most schools attending? Create a map that would be helpful to planners in reassigning schools to less crowded pools. (**Note:** You must export the elementary school selection to a new feature class before joining, because the distance join does not honor the selected set. This may be a bug.)

6. Examine the table from Exercise 5 closely, and you will find a problem with your initial analysis. What is it? What would you need to change to get a better result?

7. Which post office potentially serves the greatest number of people, based on the block group population data? Which one serves the fewest? (**Note:** The post office locations are one of the types of information in the facilities feature class.)

Use the data in the mgisdata\Oregon\oregon geodatabase to answer the following questions.

8. Which city in Oregon is farthest from an airport? What is the distance in *kilometers*?

9. Assuming that an airport's service area includes all of the cities that are closer to it than to any other airport, determine how many cities each airport serves. Which airport serves the most cities? How many cities? Which serves the most people? How many people? Sort and examine the POP2000 field in the cities feature class. What do you need to do to ensure that you have the correct answer for the number of people served?

10. A boating club would like to know which parks in Oregon have the best access to lakes. Find the number of lakes and total lake area in each park. Which park(s) have the most lakes and how many? Which park has the greatest area of lakes? Examine parks with only one lake. The sum of the area fields for these lakes keeps repeating the same few numbers. Explain why. How is the Columbia River affecting this analysis?

## Challenge Problem

Imagine that a national body recommends the minimal staffing levels for a police district as one officer for every 1000 people plus one officer for every 25 miles of road. The police department has asked you to calculate the minimum recommended staffing levels for each district. Create a table showing the largest 15 to 20 districts with the district name, the population, the miles of road, and the number of officers. Place the table on a supporting map layout (use the Table Options menu), along with a graduated color map of population and labels for the staff numbers.

# Chapter 10. Map Overlay and Geoprocessing

## Objectives

▶ Learning about spatial analysis functions, including overlay, clipping, and buffering

▶ Using map overlay to analyze multiple spatial criteria

▶ Understanding differences between spatial joins and overlays

▶ Geoprocessing with menus, ArcToolbox, and Model Builder

## Mastering the Concepts

### GIS Concepts

Over the years many procedures have accumulated for characterizing spatial relationships within and between spatial features. Functions can find areas shared by two or more conditions, evaluate distances, extract or erase areas of interest, merge similar features together, examine distance relationships between points, and more. The entire enterprise of GIS, including gathering the data, putting them in digital form, designing the geodatabase, managing the data, and creating maps, is justified by the ability to apply these tools to extract information that might be difficult to obtain any other way. GIS shares many capabilities with CAD systems and database software, but spatial analysis gives GIS unique power to get the most out of map and attribute data.

Chapter 9 discussed spatial joins as one way to analyze spatial relationships. In this chapter, we'll learn more tools to solve spatial problems. Often two or more functions will be strung together to solve a specific problem, a practice called **geoprocessing**. We'll begin with a general discussion of spatial functions and then investigate how these functions are implemented within ArcGIS.

### *Map overlay*

**Map overlay** combines two layers to create a new output feature class containing information from both of the inputs. In Figure 10.1, the polygons from a geology layer and from a land ownership layer have been combined to produce a new set of polygons. Each new polygon has

geology

ownership

overlay of geology
and ownership

Fig. 10.1. Map overlay combines two feature classes to create a new feature class containing information from both inputs. Both features and attributes may be combined. Here, each new polygon in the output contains both geology and ownership attributes.
Source: ArcGIS

been given the attributes of the originals, and each contains information about both geology and ownership for the same feature. The output layer might be useful for selecting areas containing certain gem-bearing rock formations on public land that would be suitable for mineral hunting.

Overlay functions fall into two categories, those that do not combine attributes and those that do. Functions that do not combine attributes, also called extraction functions, include **clip** and **erase,** which were discussed in Chapter 8. In extraction, the input features retain their attributes, but the attributes of the clip or erase boundary layer are ignored. The two functions that combine attributes include **intersect** and **union**. Union preserves all of the features from both inputs; intersect only keeps features common to both inputs.

### Overlay with attributes

The intersect and union functions are related to spatial joins in that they correlate features based on their spatial relationships to each other. However, spatial joins fail when spatial features do not overlap exactly (recall the creeks and geology example from Chapter 9). Consider the map of roads and land use shown in Figure 10.2. The state government has requested a report of the total miles of road falling into each land use category. At first glance, a spatial join might be considered to generate this information; by joining the land use polygons to the roads, one might expect to get a field showing the land use type each road crosses.

Fig. 10.2. Splitting a road to assign attributes to each piece
Source: ArcGIS

However, the single selected road, highlighted in Figure 10.2, crosses three different land use classes. How would a land use type be assigned in this case? If one used the *completely within* join option, no land use class would be assigned for the road because it does not lie entirely inside a polygon. A *summarized join* would do no good because a nominal data type such as land use class can't be averaged. The ideal solution would split the road into three sections and assign the land use to each section, and this is what overlay functions do, thereby enforcing a one-to-one relationship between features and enabling a perfect correspondence when joining the tables. In Figure 10.2, the original single road is now three sections, and each segment in the output table (Fig. 10.3) is assigned the attributes of the land use polygon containing it.

Map overlay with attributes occurs in two forms. **Union** combines two polygon layers, keeping all areas and merging the attributes for both layers. **Intersect** also merges the attributes but retains only the areas common to both layers, and may be performed with points, lines, or polygons.

Fig. 10.3. During overlay, each new road segment retains its original attributes and receives the land use code and other attributes of the polygon in which it falls.
Source: ArcGIS

| roadlu | | | | |
|---|---|---|---|---|
| OBJ_ID | ROADNAME | SUFFIX | LU-ID | LU_CODE |
| 3033 | 2 ST | ST | 153 | High Density Residential |
| 3034 | QUINCY ST | ST | 153 | High Density Residential |
| 3035 | RANGE RD | RD | 72 | Office/Commercial |
| 3036 | RANGE RD | RD | 305 | Public |
| 3037 | RANGE RD | RD | 70 | Medium Density Residential |
| 3038 | SOO SAN DR | DR | 72 | Office/Commercial |

A **union** creates all possible polygons from the combination of features in two polygon layers. Figure 10.4 shows a union performed on geology and slope class layers to create a landslide hazard map. Landslide risk depends primarily on two factors, the strength of the geological units (sandstones are less susceptible to sliding than shales) and the slope. The lighter areas in the slope class map have lower slopes, and the dark areas have higher slopes. A union produces a new feature class with new polygons, each of which possesses the original attributes of its parents. The new feature class can then be evaluated for the various combinations of slope and geology that may constitute a landslide hazard.

Fig. 10.4. A union creates every possible new polygon from the combined input layers. Each new polygon assumes the attributes from both layers; the new polygon marked with the circle contains attributes from both the geology and slope class layers.
Source: ArcGIS

The **intersect** operation with two polygon layers is similar to a union, except that it only keeps the polygon areas that are shared by both input layers. This function provides a way to find out where two or more conditions hold simultaneously, the foundation of a class of problems known as **suitability analysis**—evaluating a landscape to find which areas best serve a given purpose based on a set of factors. For example, one could find the potential habitat for a species with specific environmental requirements.

Imagine that a rare species of snail inhabits the Black Hills. This snail prefers limey soils in cool and dense coniferous forest and is rarely found above an elevation of 1600 meters or below 1200 meters. Each data set (elevation, geology, vegetation) is queried to find the areas meeting the specified condition, and the resulting three layers are intersected to find where all three conditions hold (Fig. 10.5). The habitat polygons in the output contain the full attributes of all three layers,

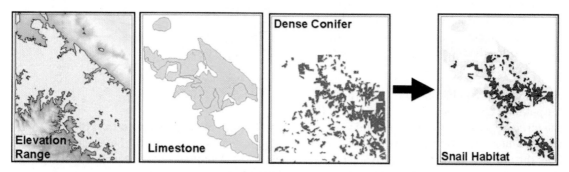

Fig. 10.5. Intersecting elevation, limestone areas, and dense conifer vegetation can help identify areas of potential snail habitat.
Source: ArcGIS

should that information be needed. Such a map would help biologists create sampling strategies for counting populations, analyze whether the habitats are interconnected or widely separated, and make decisions about forest management to protect the snails.

The snail habitat demonstrates polygon-on-polygon intersection. Intersection can assign attributes from polygons to lines or points within them. The roads and land use problem described in Figure 10.2 is an example of line-in-polygon intersection, assigning land use categories to the

roads. In Figure 10.6a, line-in-polygon intersection is used to assign a geologic unit to each stream reach (as in Chapter 9), useful for estimating stream loss to groundwater. One can also intersect points with polygons, or point-in-polygon intersection. A realtor might have a point feature class of houses for sale and intersect it with polygons representing school districts in order to be able to list the school for each house (Fig. 10.6b). Note that a point-in-polygon intersection is equivalent to a simple inside join with the point layer as the destination, even though they use different tools to do it.

Fig 10.6. (a) Assign geologic units to streams with line-in-polygon intersection; (b) assign schools to houses for sale with point-in-polygon intersection.
Source: ArcGIS

Whereas the union function requires two polygon input layers, the intersect command is more versatile. Both input and overlay layers may contain points, lines, or polygons. The output geometry may vary, but it cannot exceed the dimensionality of the lowest input.

If the inputs are both polygons, the output may be points, lines, or polygons. In Figure 10.7a, the two circles might represent the drug-free zones around two schools. The polygon output shows the area of overlap; the line output could become a boundary on a map.

If the inputs are both line feature classes, the output may be lines or points (Fig. 10.7b). These lines might represent hiking trails, in which case the line output would find locations where two trails travel together. The point output could find trail intersections that need to have signposts placed on them.

If the inputs are polygons and lines, the output may be points or lines (Fig. 10.7c). If the circle represents the area around a school where sex offenders are forbidden to live, the line output could provide a list of the streets and address ranges that are

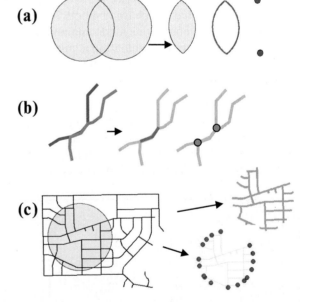

Fig. 10.7. Intersect geometries. (a) Polygons yield polygons, lines, or points. (b) Lines yield lines or points. (c) Polygons and lines yield lines or points.
Source: ArcGIS

off limits. If the city intends to put a sign on every street entering the zone, the points output would estimate the number of signs needed and where they should go.

Intersecting two sets of point data requires the output to be points, but is generally not a common analysis, unless someone is looking for duplicate points in two different data sets.

### Comparing overlay functions

Figure 10.8 summarizes the different overlay operations that can be performed using two polygon input layers, including the extraction functions clip and erase, which do not combine attributes, and the intersect and union functions, which do. Clip and erase can also accept point and line feature classes for one input, but the boundary input must contain polygons. The output feature classes will have the same geometry as the input classes.

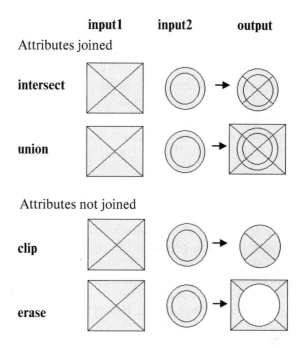

Fig. 10.8. Summary of polygon overlay operations and results
Source: ArcGIS

Unions and polygon-on-polygon intersections are similar, so a word about when to use each may be helpful. The primary use of intersect is to find areas where certain conditions overlap. Typically the input layers occupy different regions, as do the elevation range, limestone, and conifer layers shown in Figure 10.5. The goal is to find the areas common to the inputs. Although the attributes from each input are combined in the final table, the attributes themselves are sometimes of little interest. In the snail habitat problem, the pertinent attributes were already preselected using queries. The main goal was developing a map of the overlaps.

In a union, the primary goal is to combine the tables. Generally the input layers all fully occupy the same region, as the geology and slope class layers do in Figure 10.4. It is expected that the entire map area will remain, but that the new features will contain the attributes of both inputs. After a union, the usual next step is to perform attribute queries to find areas with specific combinations of attributes, or to symbolize the combinations to show areas of interest. In the landslide hazard example, we might have extracted the shale and high slope areas and done an intersect to show where the high-hazard areas exist. The hazard is either there or not there. With a union, though, we are able to analyze different combinations of geology and slope to assess potential impacts on slope stability.

Fig. 10.9. Overlay may produce small sliver polygons.
Source: ArcGIS

### Slivers and tolerances

Overlaying layers often produces small extraneous polygons or lines called **slivers** (Fig. 10.9). Sometimes these slivers represent real features. However, more often they arise when overlaying layers share some boundaries, such as voting districts and counties. In theory the shared boundaries should match exactly, but in practice few data sets have been corrected to this level of integrity. These slivers do not represent real combinations of values and are a nuisance when calculating statistics or performing subsequent operations. It is often desirable to prevent slivers when overlaying.

An **XY tolerance** can be set during an overlay operation. This tolerance specifies the minimum distance between vertices, and it will combine vertices that fall close together. Setting an appropriate tolerance during an overlay will eliminate many slivers. However, the tolerance applies to *all* vertices, and using a tolerance may degrade the accuracy of the data set. Caution should be used in setting tolerances that are large enough to correct slivers without shifting other vertices an unacceptable amount. It must also be remembered that outputs from an overlay that uses a tolerance generally have a lower geometric accuracy than the original layers.

Slivers are especially troublesome when multiple overlay or analysis operations are planned, because the errors propagate and multiply with each successive step. If XY tolerances are being used, the generalization of the boundaries becomes progressively more severe and may significantly affect the geometric accuracy of the results. It is impossible to avoid these impacts, so users should be aware of what is happening and have realistic expectations for the spatial accuracy of the final output. If an XY tolerance is used during processing, its value must be included in metadata for the final output.

### Other spatial analysis functions

Overlay is only one type of spatial analysis available in GIS systems. The remainder of this chapter presents additional commonly used functions, including dissolving, buffering, appending, and merging. All of these analysis functions, and many others, are found in ArcToolbox.

#### Dissolve

A **dissolve** is used to group features together based on whether they share the same value of an attribute field. For example, the road segments in Figure 10.10 have the same street name (Main St), but they are separate features. A dissolve based on the street name field would yield a new file in which all streets having that name were one feature. A dissolve can also be used to remove lines between polygon features that share the same value. For example, the map in Figure 10.11 shows ponderosa pine stands of different ages. Dissolving on the cover type field removes the age boundaries and yields polygons based on the cover type (ponderosa) only.

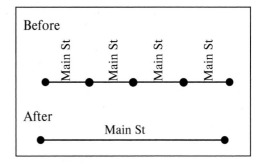

Fig. 10.10. Separate road segments were combined using the dissolve function.
Source: ArcGIS

The output from a dissolve operation may produce either single features or multifeatures. A multifeature exists when multiple unconnected areas constitute a single feature, as when the seven Hawaiian islands are combined to produce

Fig. 10.11. A dissolve removes boundaries between polygons with the same attribute value, in this case, tree species.
Source: ArcGIS

one state feature. In Figure 10.11, the green polygons constitute many individual features. After the dissolve, if the multifeature option was selected, than all of the orange polygons will belong to one feature. If the multifeature option was declined, then the data set will have many orange features.

When features are dissolved, the output layer is a new file containing the single attribute on which the dissolve was based. However, the user can specify additional fields to summarize the information from the original features. For example, in the dissolve shown in Figure 10.11, one might request the average canopy percentage of each output polygon, based on averaging the canopy percentages of the polygons before the dissolve.

### Buffer

A **buffer** is constructed to delineate areas that fall within a certain distance of a set of features. Buffers can be created for points, lines, or polygons (Fig. 10.12). They could be used to find 300-yard drug-free zones around schools or sensitive protected areas within 100 meters of a stream. Negative buffers can be applied to determine setback limits from the edge of a piece of property. Buffers can be created as simple rings or as multiple rings. An attribute can even supply different sizes of buffers for different features—for example, buffering primary roads by 200 meters and secondary roads by 100 meters.

Fig. 10.12. Examples of buffers around (a) wells and (b) roads
Source: ArcGIS

Buffers are created for each individual feature and may overlap. This option might be appropriate in analyzing the distribution of soils in buffers around wells. In many cases, however, it is best to dissolve the buffers, which gets rid of the boundaries between them and removes the overlapping areas to create a single region. Figure 10.13 shows the difference in results obtained when the buffers around roads are not dissolved versus when they are. If the areas of the buffers are of interest, then dissolving is always the correct action. If areas were calculated from nondissolved buffers, they would be grossly overestimated, because the overlapping areas would be counted multiple times.

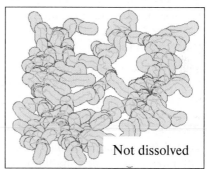

Not dissolved

Finally, buffering is an intensive process, and the time involved increases quickly as the number of features increases. Any steps to reduce the number of buffers created at one time will facilitate analysis, for example, by performing queries before buffering rather than afterwards.

### Append and merge

The **append** function is used to combine the features of two or more layers and place them into an existing target

Dissolved

Fig. 10.13. Use the dissolve option to prevent overlapping buffers
Source: ArcGIS

feature class (Fig. 10.14). The appended layers must have the same feature type as the target (i.e., both polygons, both lines, or both points). The two layers must also share the same coordinate system. Overlapping of layers is permitted.

Fig. 10.14. Append combines the features of two adjacent layers.
Source: ArcGIS

The treatment of attribute tables during an append requires some consideration. If you wish to combine the attribute information from both layers, the attribute fields must have the same definition and must occur in the same order in both tables. If the two tables differ, one can use the NOTEST option. In this case, if a field in the input layer has the same name and data type as the target layer, then the information will be copied into the target. Fields without matching names will not be carried into the target.

**Merge** is similar to append, except that it creates a new feature class and offers more flexible treatment of attribute tables. Instead of insisting that the two tables match, it allows the user to specify the fields to be included in the output feature class and which table they will come from.

## About ArcGIS

### About geoprocessing

GIS analysis involves many functions that operate on data objects, such as feature classes or tables. Geoprocessing applies one or more of these functions in sequence to solve a problem or investigate the properties of data sets. Most ArcGIS tools operate on any allowed data type. In performing a road buffer, for example, one may specify a coverage, a shapefile, or a feature class from a geodatabase. The tools make the necessary allowances to execute the function on the various data types. A few tools operate only on certain data types. An ArcGIS Advanced license, for example, provides the user with a suite of additional tools that work on coverages.

### Ways to run tools

Geoprocessing tools may be executed in several ways. *Toolbars and menus* provide interactive control of tools. Users may customize menus and toolbars to add frequently used tools that don't appear on the toolbars by default. Users can also create tools and place them on menus, on toolbars, or in ArcToolbox.

**ArcToolbox** organizes all of the installed tools into one central location. It provides a window with most of the functionality of ArcGIS, containing tools that can be run from ArcMap or from ArcCatalog. Each tool has a set of parameters, or inputs, that must be specified before the tool is run. Some parameters are required, but others are optional, which means that the software supplies a default value that can be changed, if necessary.

ArcMap has a command line that allows the user to type a command and its parameters rather than filling in boxes in a window. Typing requires greater familiarity with the tools, but for experienced users entering commands by typing may be faster and more efficient. Becoming familiar with the command line also paves the way to writing scripts to execute functions. The command line possesses a sophisticated interface to help beginners correctly enter the commands.

**Model Builder** provides a graphic canvas to string tools together and execute them in sequence. Models have several advantages. First, they record the steps and parameters used to execute an analysis in case questions on methodology arise later. Second, models can be used to explore the consequences of different parameters on the final outcome. Consider a model that calculates fire danger from precipitation, temperatures, vegetation, and structure locations through a series of tools. With a model, one could redo the analysis based on the current day's precipitation and temperature conditions more quickly than if all the steps were executed interactively, and with a lower chance of mistakes. Models can be shared with other users, and they permit a less experienced user to reliably repeat an analysis set up by an expert. Models can incorporate decision structures, conditional statements, and iteration functions if needed. Finally, models can be saved as scripts to provide a starting point for creating geoprocessing programs.

A **script** is a program that may contain conditional statements (if/then), iterative loops, and other control structures that permit sophisticated analysis. Scripts may be written in one of several programming languages, although in each case the geoprocessing commands are identical and only the control statements differ. ESRI has adopted the readily available language Python as the recommended scripting platform. A copy of Python is installed with ArcGIS, and all ESRI scripting examples and supplemental tools are written in Python. However, users may elect to use any COM-compliant language such as Jscript, VBScript, or Perl.

### Geoprocessing environments

The operation of tools and commands is impacted by **Environment settings** specified by the user. For example, the default coordinate system setting is *Same as Input,* meaning that the output file has the same coordinate system as the input file. The user could alternatively specify a particular coordinate system, such as UTM Zone 13 NAD 1983. Under this setting, every output feature class would have the UTM coordinate system regardless of its input system.

Figure 10.15 shows some of the Environment settings. The Workspace setting tells ArcGIS where to initially look for data and to save output data sets. It also sets a scratch workspace where temporary files are stored. The Output Coordinates setting specifies the coordinate system given to any outputs. If it is set to UTM Zone 13N, then every output data set will be stored in that coordinate system.

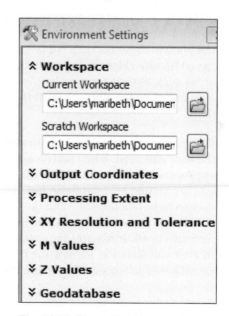

Fig. 10.15. Some Environment settings
Source: ArcGIS

Environment settings are hierarchical. Application settings are set using the main menu bar in ArcMap or in ArcCatalog, and they affect the operation and output of every tool that is run. (They will not cause changes to files that already exist.) The application settings are saved along with a map document, so it is possible for the settings to be operating without the user's awareness. It is not a bad practice to check environment settings if you are getting unexpected results from the tools. When a new map document is opened, all settings are returned to their defaults. However, settings made in ArcCatalog will become the default for all new map documents.

All tools in ArcToolbox have an Environments button that can be used to change the Environment settings temporarily for that one run of the tool. For example, the user might have set the application Environment processing extent to a small project area, but needs to buffer some streams and allow the areas to extend beyond the current study area. The Environments button in the Buffer tool can be used to temporarily override the extent setting. Models, scripts, and tools can all have their own Environment settings, which take precedence over the application settings.

The default Environment settings work fine for most applications. Users may change them when doing so provides more efficient or accurate work. For example, Amy might have to clip 10 different feature classes and project them from a geographic coordinate system to a projected coordinate system. Instead of running both the Clip and Project tools on each layer, she could set the Environment coordinate system to the desired projection and then run the clips, reducing the amount of work.

### Coordinate systems

Tools will accept any coordinate system or combination of coordinate systems for input, but rules of precedence dictate the coordinate system of the output. Outputs will be projected on the fly, if necessary, according to the following rules:

▶ If the output is placed in a feature dataset, the coordinate system will always match that of the feature dataset.

▶ If the coordinate system is set in the Environment settings and the output will not become part of a feature dataset, the Environment settings coordinate system is used.

▶ If the Environment setting is not set, the default rule applies—that the output will match the coordinate system of the first input to the tool.

When setting the output coordinate system in the Environment settings, you can choose Same as Input (the default), or Same as Display (the current data frame), or set it to match the coordinate system of a particular data set.

Geoprocessing tools are fundamentally spatial in nature and commonly manipulate areas and distances as part of the processing. When geoprocessing, it is best to always use a projected coordinate system, rather than a geographic coordinate system (GCS). As with spatial joins, it is not enough to set the data frame coordinate system to a projection. The tools operate on the saved feature class on disk, and the saved coordinate system is the one being used. When large areas such as continents or large countries are being analyzed, no projection can preserve both area and distance, and one must then choose the projection based on the type of analysis being done.

Often a GCS is employed by data providers because they cannot predict what coordinate systems the user will need. When downloading data, it will often be in a GCS, and the user must choose an appropriate coordinate system and project the data to it. Chapter 3 explains this process. When you are setting up a database for a mapping and analysis project, a suitable projection should be chosen based on the scale and the extent of the data being analyzed. It must be kept in mind that distortions present in the map projection, if significant, will affect distance and area measurements. It remains important to choose a suitable map projection with minimal distortion of area and distance. Review the coordinate system properties on the inside front cover and the coordinate system selection guidelines in Chapter 3, if necessary.

### Areas and lengths of features

Geodatabases and coverages automatically create and update fields containing the areas and perimeters of polygons or the lengths of lines. Shapefiles do not maintain this information. Users may create AREA or LENGTH fields for shapefiles and calculate the values manually. However, be careful when using information from these fields. If polygons in a shapefile are clipped, dissolved, or intersected or they undergo any other operation that changes their shape, the AREA fields will not be updated automatically. The user must manually update the fields again to ensure they are correct. Be cautious using an AREA or a PERIMETER or a LENGTH field in a shapefile unless you are sure they are correct. Chapter 6 introduced the Calculate Geometry tool for updating areas, lengths, and perimeters in a variety of units.

## Summary

▶ Map overlay resembles a spatial join, but it splits features when they partly overlap. This function enforces a one-to-one relationship between features when their attributes are joined in the output table.

▶ Map overlay comes in two basic types. A union keeps all the features from both layers. An intersect keeps all the features that are common to both input layers. Attributes from both layers are joined together in the output.

▶ A dissolve combines features within a data layer if they share the same attribute. This function can be used to convert many street segments into a single line feature or to remove boundaries between parcels with the same zoning.

▶ Buffers are polygon constructions that enclose the area within a certain distance of features. Buffers may be created for points, lines, or polygons, and they may be constructed as single or multiple rings.

▶ Append and merge allow feature classes with the same feature type to be combined as a single feature class, such as merging two adjacent quadrangles to make a single file.

▶ Map overlay and spatial analysis are generally best done using a projected coordinate system, especially if one anticipates determining areas and lengths as part of the analysis.

▶ Geoprocessing involves stringing together a sequence of commands during spatial analysis. Spatial functions may be executed from the menus, ArcToolbox, the command line, Model Builder, or scripts.

▶ Environment settings can be used to modify the behavior of functions and tools to achieve more efficient and reliable processing parameters.

▶ The Shape_Area, Shape_Length, and Shape_Perimeter fields in geodatabase feature classes are stored and updated automatically. Other area-based or length-based fields, such as Area, Acres, or Road_km, are not automatically updated.

▶ Lengths and areas of features stored in shapefiles must be updated manually.

### Important Terms

| | | | |
|---|---|---|---|
| append | Environment settings | merge | suitability analysis |
| ArcToolbox | erase | Model Builder | union |
| buffer | geoprocessing | parameter | XY tolerance |
| clip | intersect | script | |
| dissolve | map overlay | sliver | |

## Chapter Review Questions

1. Which different types of outputs are possible (points, lines, polygons) when performing intersect and union?

2. What is the most important difference between a spatial join and a map overlay?

3. What are slivers? Explain how they can be prevented.

4. What is a buffer? Why is a dissolve often performed when buffering features?

5. What function would you use to create a map of a study area such that all the features in the map stopped at the study area boundary?

6. What attribute fields will be present in a layer resulting from a dissolve?

7. Why is it usually advantageous to use a projected coordinate system when doing a map overlay?

8. How can you determine the areas of polygons in a geodatabase? In a shapefile?

9. What determines the coordinate system of the output when overlay is used?

10. What is geoprocessing? In what different ways can commands be executed?

# Mastering the Skills

## Teaching Tutorial

The following examples provide step-by-step instructions for doing basic tasks and solving basic problems in ArcGIS. The steps you need to do are highlighted with an arrow →; follow them carefully. Click on the video number in the VideoIndex.htm file to view a demonstration of the steps.

To demonstrate map overlays, we will do the problem described earlier in the chapter concerning delineating habitat for the rare Black Hills snail using overlays. This habitat will be defined by three criteria: on a limestone geology unit, in dense coniferous forest, and between the elevations of 1200 and 1600 meters.

> → Start ArcMap and open the map document ex_10.mxd.
> → Use Save As to rename the map document. Remember to save it often as you work.

**TIP:** You were directed to turn off background processing in Chapter 2, but choose Geoprocessing > Geoprocessing Options from the main menu now to be sure it is disabled.

As in Chapter 9, we don't want the new feature classes created during the analysis to become part of the original geodatabase, so we'll create another geodatabase to contain them. We'll also set the geoprocessing Current Workspace as the default location to place output data.

Source: ArcGIS

> 1→ Click the ArcToolbox tab, or the button on the main toolbar to open it. Dock it on the right side of the ArcMap window, if necessary.
>
> 2→ In ArcToolbox, expand the Data Management > Workspace entries and double-click on the Create File GDB tool.
>
> 2→ Click the Browse button for the Location and navigate to the mgisdata folder. Select the BlackHills folder by clicking it once (don't go inside it) and click Add.
>
> 2→ Enter chap10results for the output geodatabase name. Click OK.

> 3→ From the main menu bar, choose Geoprocessing > Environments.
>
> 3→ Expand the Workspace entry and use the Browse button to set the Current Workspace to chap10results (Fig. 10.16).
>
> 3→ Also set the Scratch Workspace to chap10results.
>
> 3→ Expand the Output Coordinates entry and check that the Output Coordinate System is set to Same as Input. Click OK.

Fig. 10.16. Setting geoprocessing environments
Source: ArcGIS

*Preparing to overlay*

Our first step is to create layers that contain polygons where each condition holds. We will begin with the Geology layer. The limestone areas include the Madison Formation and the Upper Paleozoic units. We will use an attribute query to place these units in a separate layer.

4→ Choose Selection > Select By Attributes from the main menu bar.

4→ Set the layer to Geology and enter the expression:

NAME = 'Upper Paleozoic' OR NAME = 'Madison Limestone'. Click OK.

4→ Right-click the Geology layer and choose Selection > Create Layer from Selected features.

4→ The new layer appears at the top of the Table of Contents. Rename it *Limestone*.

**TIP:** Creating a layer from a selection, although not necessary for processing, is often a good practice. It keeps the selection available if a question about the procedure arises later, or if a tool must be rerun due to an error.

Next, we will select the dense conifers and create a layer from them. We will do the query in two steps. First we will select the ponderosa pine (TPP) and white spruce (TWS). Then we will select the dense areas (class contains C or 5) from the already selected set.

5→ Turn on the Vegetation layer and turn off the Limestone layer.

5→ Clear all selected features.

5→ Open the Select By Attributes window again and set the layer to Vegetation.

5→ To select only the conifers, enter the expression:

COV_TYPE = 'TPP' OR COV_TYPE = 'TWS'. Click Apply.

6→ Change the selection method to *select from current selection*.

6→ Clear the expression and enter another one that says:

DENSITY96 = 'C'. Verify the expression and click OK.

6→ Right-click the Vegetation layer and choose Selection > Create Layer from Selected Features.

6→ Rename the new layer *Dense Conifer*. It should look like the polygons shown in Figure 10.17.

Fig. 10.17. Dense conifer
Source: ArcGIS

The elevation range has already been prepared. You are almost ready to intersect.

7→ Clear the selected features and turn off the Vegetation layer.

7→ Turn off all layers except the Elevation Range layer, the Limestone layer, and the Dense Conifer layer.

7→ Zoom in to the middle right of the map and examine the Dense Conifer polygons.

Notice that these polygons have many boundaries inside them because they are divided based on age and density as well as species. The amount of time needed to intersect a layer is proportional to the number of features in the layer. To streamline the intersect, we are going to remove the unnecessary boundaries between these polygons using the Dissolve tool.

8→ In ArcToolbox, expand the Data Management > Generalization entries and double-click the Dissolve tool to start it (Fig. 10.18).

When using a tool for the first time, it is wise to select Show Help to find out what the tool does and what the **parameters** mean.

8→ Click Show Help on the Dissolve tool and read the description.

8→ Click on the Input Features box. The Help message changes to describe the Input Features parameter.

Fig. 10.18. The Dissolve tool
Source: ArcGIS

8→ Click on and read the descriptions for the other input parameters in the window.

8→ To bring the tool description back, click on the gray tool area.

8→ Click on the Tool Help icon at the bottom of the tool for more detailed information. Read about the Dissolve tool.

8→ Close the ArcGIS Desktop Help window and click Hide Help on the Dissolve tool.

9→ Click the drop-down arrow in the Input Features box and set it to Dense Conifer (Fig. 10.18).

9→ Click the Browse button to place the Output Feature Class in the BlackHills/chap10results geodatabase and name it *DensConifDisslv*.

9→ Check the box to dissolve on the COV_TYPE field.

9→ Uncheck the box to *Create multipart features*. We want unconnected polygons to remain separate features.

9→ Notice the other options in the tool. We don't need to change any of them, however.

9→ Click OK.

**TIP:** When using a tool, be aware that some options may be out of sight, and you must scroll down to see them. Often the defaults are fine, but get in the habit of checking before you run it.

10→ Examine the output file and note that the intervening boundaries have disappeared.

10→ Right-click the Dense Conifer layer and choose Remove.

10→ Rename the DensConifDisslv layer *Dense Conifer*.

### Intersecting polygons

Now we are ready to overlay. We will use Intersect because we want to find the areas common to all three layers. If you have an ArcGIS Basic license, the tool can intersect only two layers at a time, so we will use it twice.

11→ Open the ArcToolbox > Analysis Tools > Overlay > Intersect tool (Fig. 10.19).

11→ Click Show Help and examine the entries.

1.  What are the options for the Join Attributes parameter? Which one do you think is best to use?

_____

_____

11→ Click on the drop-down button under Input Features and choose the Dense Conifer layer. It will be added to the list of Features.

11→ Click the drop-down button again to add the Limestone layer to the list.

11→ Enter the Output Feature Class, placing it in chap10results and naming it *conf_lime*.

11→ Set the Join Attributes option to NO_FID.

11→ The remaining options can be left with their defaults. Click OK.

11→ Examine the output. The conifers outside the limestone have disappeared.

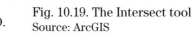

Fig. 10.19. The Intersect tool
Source: ArcGIS

**TIP:** The Feature ID, or FID, is a unique integer assigned to every feature for tracking. During an intersection, new FID values are assigned to each new feature. The NO_FID option eliminates the original FID values from the output table, since they have little use.

12→ Use the Intersect tool again to intersect conf_lime and the Elevation Range layers, naming the output *snailhab* and putting it in chap10results.

**TIP:** If you have an ArcGIS Advanced license, you can intersect more than two layers at a time.

13→ Zoom out to the full extent of the data.

13→ Rename the snailhab layer *Snail Habitat*.

13→ Remove the Limestone, Dense Conifer, and conf_lime layers.

13→ Zoom in to the extent of the Snail Habitat and turn on the Roads layer.

### Overlay of lines in polygons

The snails have a three-week breeding season in early June. During this period they seek the open areas offered by roads, and many get crushed. The Forest Service wants to consider closing primitive roads that traverse through snail habitat during the breeding season to lessen the number of crushed snails. They need to assess which roads must be closed. You will make a map with the proposed road closures. This process involves a line-in-polygon intersection.

14→ Use Select By Attributes to select the Primitive roads TYPE = 'PR'. (Make sure you set the method back to *Create a new selection.*)

15→ Intersect the Roads layer (with the primitive road selection) and the Snail Habitat layer. Name the output *propclose* and put it in chap10results.

15→ Rename the propclose layer *Proposed Closures.* Clear the selected features.

16→ Create a map similar to Figure 10.20, with the closure roads highlighted in red and the other roads in black. Turn off the Elevation Range layer. Set the transparency of the Snail Habitat layer to 50% to help bring out the road patterns.

Fig. 10.20. Map showing proposed road closures in snail habitat areas
Source: ArcGIS

16→ Save your map document.

## *Clipping layers*

The vegetation layer does not extend as far north as the rest of the data, but the map may give readers the false impression that the analysis is valid there. Clipping the Roads layer at the edge of the vegetation layer will prevent any misunderstanding. You need the Clip tool. Recall that you can use the Search window to find tools.

17→ On the main menu bar, choose Geoprocessing > Search For Tools.

17→ Type *clip* in the search box. Several suggestions appear, including Clip (analysis) and Clip (management). The words in parentheses refer to the area of the Toolbox.

17→ Ignore the suggestions for now and click on the magnifying glass button.

The list includes some exact matches (clip) and some similar functions, such as clipping rasters, along with a description of each.

17→ Click the Options button in the search window and choose Search Options.

17→ Click the General tab and check the box to *Show Pop-up window in search results.*

17→ Click OK and hover the cursor over the blue tool names to see the descriptions. The Clip (Analysis) appears to be the one to use.

17→ Click the blue Clip (Analysis) text to open the tool.

**TIP:** You can pin the Search window to the right tab bar for easy access later.

18→ Set the Input Features to Roads (or use the Browse button to find the usfsroads feature class in the Sturgis geodatabase).

18→ Set Vegetation as the Clip Features (or find the vegetation feature class in the Sturgis geodatabase).

18→ Store the output in the chap10results geodatabase as roadclip. Click OK.

18→ Turn off the Roads layer. If the roadclip layer was not added to the map automatically, add it from chap10results now.

> **TIP**: The Clip tool should have drop-down boxes to allow the section of layers from the map, and the output data set should be added automatically, but a bug in Versions 10.4 and 10.5 may prevent this. Saving the map document and restarting ArcMap should alleviate both issues.

You can quickly transfer the symbology of the old Roads layer to the clipped roads.

19→ Open the properties of the roadclip layer and click the Symbology tab.

19→ Click the Import button and choose to *Import symbology from a layer or layer file*.

19→ Click the drop-down box to set the layer to Roads. Click OK.

19→ The road classification is based on the TYPE field. The clipped roads use the same field, so leave the TYPE name in the box. Click OK and OK.

20→ Rename the roadclip layer Clipped Roads.

20→ Click and drag the Clipped Roads layer below the Snail Habitat layer.

20→ Remove the Roads layer.

### Working with buffers

To prevent snail death on the major roads, which cannot be closed, a Forest Service biologist has suggested thinning the tree stands within 200 meters of the roads to make these areas less attractive to snails and keep them away from roads. We can prepare a map showing the stands to be cleared and determine the total percentage of snail habitat that would be eliminated.

21→ Use Select By Attributes to select the primary and secondary roads from the Clipped Roads layer, using the expression TYPE = 'P' OR TYPE = 'S'.

21→ Create a layer from the selected roads and name it Major Roads. Clear the selected features.

> **TIP:** Buffers are time-consuming to create, so it is helpful to reduce the number of features before buffering. We will use the Elevation Range layer to clip the roads before buffering.

22→ Open the tool ArcToolbox > Analysis > Extract > Clip.

22→ Set Major Roads as the Input Features and Elevation Range as the Clip Features.

22→ Save the result in chap10results as majroadclip. Click OK to run the tool.

22→ Turn off the other road layers to examine the new file.

Now buffer the clipped roads.

23→ Open the tool ArcToolbox > Analysis > Proximity > Buffer.

23→ Set the Input Features to majroadclip and the Output Feature Class to roadbuf in chap10results.

23→ Set the Linear unit to meters and type 200 in the box.

23→ Scroll down, if necessary, and set the Dissolve Type to ALL. Click OK.

Now intersect the road buffers with the snail habitat to determine the potential areas to be cleared.

Fig. 10.21. Proposed thinning areas (in pink)
Source: ArcGIS

> 24→ Open ArcToolbox > Analysis > Overlay > Intersect.
>
> 24→ Choose Snail Habitat and roadbuf as the Input features.
>
> 24→ Specify the Output file as *proposedthin* in chap10results. Click OK.
>
> 25→ Clean up the map display to show the results. Turn off the roadbuf layer and make the proposedthin layer a symbol that contrasts well with the snail habitat (Fig. 10.21).

Now the final step is to determine the area of snail habitat that will be eliminated by the thinning. Geodatabases automatically keep track of the areas, lengths, and perimeters of features.

> 26→ Open the proposedthin table and examine the Shape_Area and the Shape_Length fields.

These fields were renamed Shape_1_Area and Shape_1_Length during one of the intersections, but they will contain the correct areas and perimeters for the feature class. The units of the Shape_Area field will be in the same coordinates as the map units of the feature class coordinate system, in this case, square meters.

> 26→ Use the Statistics function to determine the total area to be thinned.

2.  What is the total area of the proposed thinning in square kilometers? _____

3.  What is the total area of snail habitat in square kilometers? _____

4.  What percentage of the habitat would be eliminated by the proposed thinning? _____

## *Investigating relationships with union*

Intersect is valuable for finding areas where known criteria exist simultaneously. Union is a powerful tool for investigating relationships between two types of information, such as slope and geology combinations that affect geologic hazards such as landslides and sinkholes.

27→ Turn off all layers except Geology. Add the veg2006 feature class from the Sturgis geodatabase.

27→ Open ArcToolbox > Analysis > Overlay > Union.

27→ Choose Geology and veg2006 as the Input Features.

27→ Name the output feature class *geolveg* and put it in the chap10results geodatabase.

28→ Zoom to the full extent and turn off all layers except for geolveg.

First, notice that all areas of both input layers are included in the output, even though the Vegetation layer does not cover as much area as the Geology layer does.

28→ Click the Identify tool. Click on several of the large polygons at the top and examine the attributes.

28→ Click on several small polygons at the bottom.

Where both Vegetation and Geology were present, the attributes are stocked with information from both layers. However, the polygons at the top came only from Geology, so the vegetation fields contain blanks and zeroes. Let's eliminate these areas from further consideration, using the fact they they received a –1 FID value from the veg2006 layer.

29→ Close the Identify window and open the geolveg table.

29→ Open the Table Options > Select By Attributes window and enter the expression FID_veg2006 >= 0. Click Apply.

29→ Create a layer from the selected features and name it *Union Results*. Clear the selected features, close the Table window, and remove the geolveg layer.

The top areas of concern for sinks include limestone areas with low slopes. We will select the polygons that meet these criteria.

30→ Open the Select By Attributes window and select the polygons from the Union Results layer that meet the condition (NAME = 'Upper Paleozoic' OR NAME = 'Madison Limestone') AND SLOPE_PCT <= 20. (Be sure to include the parentheses to enforce evaluation in the correct order.)

30→ Create a layer from the selected features and name it *Sink Hazard Areas*. Clear the selected features.

Next we will find areas at risk for rock fall incidents, which include stiff rocks in steeply sloping areas.

31→ Open the Select By Attributes window and select the polygons from the Union Results layer that meet the condition (NAME = Precambrian' OR NAME = 'Madison Limestone') AND SLOPE_PCT >= 40.

31→ Create a layer from the selected features and name it *Rockfall Hazard Areas*. Clear the selected features.

Finally, the landslide hazard areas contain weaker rocks with moderate to high slopes.

31→ Open the Select By Attributes window and select the polygons from the Union Results layer that meet the condition (NAME = 'Lower Mesozoic' OR NAME = 'Cenozoic') AND SLOPE_PCT >= 25.

31→ Create a layer from the selected features and name it *Landslide Hazard Areas*. Clear the selected features.

Fig. 10.22. Geologic hazard areas
Source: ArcGIS

The map showing all three hazards should look similar to Figure 10.22.

### Working with slivers and tolerances

We will now work on a short exercise to demonstrate the problem of slivers and how to help prevent them using tolerances. We need a different data frame for this part of the tutorial.

32→ Choose Insert > Data Frame from the main menu bar. Collapse the Sturgis Area data frame to hide the layers you've been working with.

32→ Add the cd111 and counties feature classes from the mgisdata\Usa\usdata geodatabase.

32→ Zoom in to the San Francisco Bay Area in northern California (Fig. 10.23).

32→ Choose Bookmarks > Create Bookmark. Enter the name *San Francisco* and click OK.

Fig. 10.23. Zoom in to this area.
Source: ArcGIS

This area will be our experiment. We don't need to overlay the entire country, so to save processing time, we will use the Environment settings to constrain the output extent to this display window.

33→ From the main menu bar, choose Geoprocessing > Environments.

33→ Expand the Processing Extent entry.

33→ Change the Extent drop-down to read Same as Display. Click OK.

34→ Open the tool ArcToolbox > Analysis > Overlay > Union.

34→ Enter cd111 and counties as the input features.

34→ Name the output *union_notol* and put it in chap10results. Click OK.

**TIP:** Once set to an extent, the Processing Extent setting will continue to use the same range of *x*-*y* coordinates, so it will not be affected as you zoom in and out.

35→ Zoom to the full extent of the data.

35→ Notice that only a small group of polygons around San Francisco is included in the output. This is the result of the Processing Extent environment setting.

35→ Choose Bookmarks > San Francisco, and examine the output.

Fig. 10.24. Slivers from the union
Source: ArcGIS

Notice that many slivers occur (Fig. 10.24). The district and county boundaries coincide in real life, but these two data sets come from different sources, and the boundaries aren't identical. To specify a processing tolerance, it would be helpful to know the typical width of the slivers.

36→ Zoom in to a region with typical slivers.

36→ Click the Measure tool on the Tools toolbar. Set the units to meters.

36→ Measure across several slivers to estimate their widths. Close the Measure tool.

You will find that a typical width is about 1000 meters. A tolerance is a radius, so we should use about half the measured value. Let's repeat the union and use a tolerance of 500 meters.

37→ Open the ArcToolbox > Analysis > Overlay > Union tool.

37→ Set the input layers to cd111 and counties, as before.

37→ Name the output *union500* and place it in the chap10results geodatabase.

37→ Enter *500* meters as the XY tolerance. Click OK and examine the output.

37→ Turn the new layer on and off a few times to compare the boundaries with the union_notol layer. Many slivers are gone in the new layer.

Notice two things about the new layer. First, many slivers are still present. Our initial tolerance was not sufficient to remove them. Second, the district boundaries have lost some of their detail. The XY tolerance may not have fixed all the slivers, but it already is having an effect on the accuracy of the output. Nevertheless, we will try a larger tolerance.

38→ Repeat the union of cd111 and counties, but this time use an XY tolerance of 1000 meters and name the output *union1000*.

38→ Turn on just the union_notol and the union1000 layers and compare them by turning the top one on and off a few times.

Most of the slivers are now gone, but the loss of detail in the district boundaries is apparent. The county boundaries don't look much worse, but they were a lower resolution to begin with.

39→ Try one more union with a 1500-meter tolerance and compare.

Now the generalization of the boundaries is obvious, and a few stubborn slivers still remain. Further generalization and loss of accuracy are not worth getting rid of the few remaining slivers. The 1000-meter tolerance gave the best compromise performance. Before we go on, we must set the Environment settings back to their defaults.

40→ Choose Geoprocessing > Environments from the main menu bar.

40→ Expand the Processing Extent entry and set it back to Default. Click OK.

**TIP:** One challenge of using Environment settings is to remember to reset them after you are finished. Note that the Environment settings are saved with map documents.

### Geoprocessing with Model Builder (optional)

This section shows how to use Model Builder to record sequences of geoprocessing steps. Since the snail habitat problem is familiar, we'll use it as our example.

→ Open the original version of ex_10.mxd without any changes.

→ Use Save As to save it under a new name and save often as you work.

Model Builder often creates intermediate data sets while it works. You can specify a scratch workspace where these are placed, so that they don't interfere with the original data. You set the workspace using the Environment settings.

41→ Open Geoprocessing > Environments from the main menu bar.

41→ Expand the Workspace entry. Set both the Current Workspace and Scratch Workspace to the chap10results geodatabase. Click OK.

5. You already set these Environments earlier in the tutorial. Why do you have to set them a second time now?

### Creating and running a model

You store a model inside a geodatabase or in a toolbox. We will create a toolbox for it.

 42→ Click the Model Builder button on the main tool bar.

42→ Choose Model > Save in the Model Builder window.

42→ Navigate so that you are in the BlackHills folder.

 42→ Click the New Toolbox button in the Save window.

42→ Name the toolbox *BHtools* and click Enter.

42→ Double-click the new BHtools toolbox to enter it.

42→ Type *SnailHab* as the name for the model and click Save.

43→ Choose Model > Model Properties from the Model Builder menu bar.

Fig. 10.25. Naming and describing a model
Source: ArcGIS

311

43→ Label the model *Snail Habitat Analysis* and give it a description (Fig. 10.25).

43→ Examine the other settings as shown in Figure 10.25 and then click OK.

**TIP:** Storing relative pathnames for a model makes it work if the model and data together are moved to a different folder. Absolute pathnames only work if the data stays put.

In Model Builder, you paste the desired tool in the window, set arguments for the tool, and run it. Each step can be run separately as the model is built, or all steps can be run together. When building the model, it is helpful to execute the steps one at a time. In the beginning of this tutorial, we used Select By Attributes to select the desired polygons from the geology and vegetation layers. In Model Builder, we use the Make Feature Layer tool.

44→ Find the ArcToolbox > Data Management Tools > Layers and Table Views > Make Feature Layer tool. Click and drag it to the Model Builder window.

44→ Right-click the Make Feature Layer tool in the Model Builder window and choose Open.

44→ Set the Input Features to Geology and name the Output Layer *Limestone*, as shown in Figure 10.26.

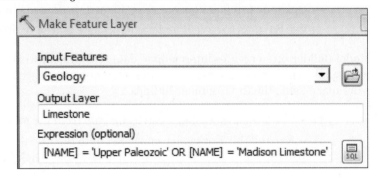

Fig. 10.26. Setting the arguments for the Make Feature Layer tool to create the Limestone layer
Source: ArcGIS

45→ Click the SQL button to enter the selection expression: [NAME] = 'Upper Paleozoic' OR [NAME] = 'Madison Limestone'. Verify the expression.

45→ Click OK to close the SQL window and OK to close the tool.

45→ The boxes in the model become colored, indicating the tool is ready to run. Choose Model > Run, or click the Run button to execute this part of the model.

45→ Close the dialog box after the tool completes successfully. A drop shadow behind the model shapes indicates that the tool has been run (Fig. 10.27).

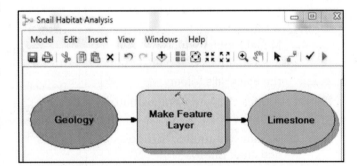

Fig. 10.27. The model after the first step is run
Source: ArcGIS

The blue ovals indicate inputs, and the green ovals indicate outputs. Both inputs and outputs have properties that can be set.

45→ Right-click the green Limestone oval and choose Add to Display. The new layer appears in the Table of Contents and is shown on the map.

Now we'll use the same tool to select the dense conifers. Because we're using a tool, it is easier to do the selection in one step.

46→ From ArcToolbox, click and drag another copy of the Make Feature Layer tool onto the model canvas.

46→ Right-click the tool to open it. Set the Input Features to Vegetation and the Output Layer to *Dense Conifer*.

46→ Click the SQL button and carefully enter the expression (with the parentheses):
([COV_TYPE] = 'TPP' OR [COV_TYPE] = 'TWS') AND [DENSITY96] = 'C'.

46→ Click the Verify button to ensure that you did not make a mistake in the query.

47→ Click OK to close the SQL window and OK to close the tool window.

47→ Click the Run button and close the dialog window after it finishes.

47→ Right-click the Dense Conifers output and choose Add to Display. Make sure that the selection looks correct (see Fig. 10.17).

**TIP:** Click on a model shape to select it. Use click and drag boxes or the Shift key to select multiple shapes. Click and drag selected shapes to move them around on the model canvas or to resize them. Use these techniques to arrange the model in neat patterns.

The Auto Layout button can be used to quickly arrange the model in a logical pattern.

48→ Click and drag the ArcToolbox > Analysis > Overlay > Intersect tool onto the model canvas.

48→ Right-click the Intersect tool on the model canvas and open it.

 48→ Set the Input Features to Limestone and Dense Conifer. When choosing the layers, be sure to choose the ones with the blue symbol, indicating that they are outputs from the model.

48→ Name the Output Feature Class *conf_lime* and store it in the chap10results geodatabase. Click Yes to overwrite the one that exists, if you are asked.

48→ The tool should look like Figure 10.28. Leave the other arguments set to their defaults.

Fig. 10.28. Setting the arguments for the first Intersect tool
Source: ArcGIS

Notice the red or yellow warning sign next to the Output Feature Class. Put the cursor on it to read the message that says that this output already exists. Of course it does, since we've done this analysis before. When running models, you generally want to be able to overwrite previous results. We can set this option.

49→ Click OK to close the tool.

49→ Choose Geoprocessing > Geoprocessing Options from the main menu bar.

49→ Check the box to *Overwrite the results of geoprocessing operations.* Click OK.

50→ Drag another copy of the Intersect tool from ArcToolbox to the model canvas.

50→ Open the second Intersect tool.

50→ Select the conf_lime model output from the Input Features drop-down box.

50→ Choose Elevation Range from the Input Features drop-down box.

50→ Name the output feature class *snailhab2* and place it in chap10results.

50→ Leave the rest of the arguments set to their defaults and click OK.

51→ Choose Model > Save to save the model in its current form. It should look similar to the model in Figure 10.29.

51→ Run the model. The two intersections you set up will both be executed.

51→ Right-click the snailhab2 output to add it to the display and examine it.

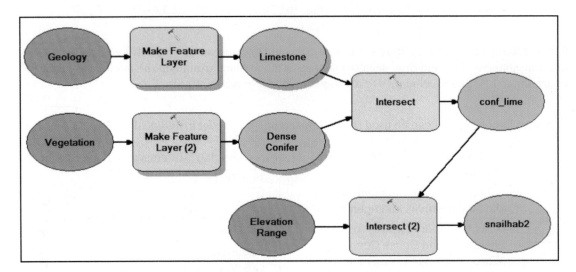

Fig. 10.29. The completed Snail Habitat model
Source: ArcGIS

### Changing and rerunning a model

The saved model provides a record of the inputs and parameters used to derive the output. It can also be edited and run again to try out different solutions. Imagine that a geologist colleague has commented that the Lower Paleozoic geological units also contain limestone and should be included with the other geology layers.

52→ Right-click the Make Feature Layer input box for the Geology query and open it.

52→ Click the SQL button and click at the end of the current expression to place the cursor there.

52→ Add the rest of the expression, OR [NAME] = 'Lower Paleozoic', so that all three units will be selected. Verify the query and then click OK to close the tool.

Notice that the model boxes lose their shadows (except the unchanged Vegetation query), indicating that they need to be run again with the new conditions.

53→ Right-click the Make Feature Layer tool from the Geology query again and choose Run to run only this step of the model to check your query.

53→ The previous Limestone output was removed from the display. Right-click its green oval in the model and choose Add to Display to uncheck it. Repeat to add it to the display a second time. Examine it.

54→ Click the Run button in the Model Builder toolbar to run the rest of the model steps.

54→ Add the snailhab2 output to the map and examine it to see how it changed.

54→ Save the new version of the model by choosing Model > Save.

**Creating a tool from a model**

Finally, models can be set up to run as tools by turning the inputs into **parameters**. This approach is helpful when the same series of steps is to be run with different inputs each time.

55→ Right-click the blue Vegetation oval and choose Model Parameter. A small P appears to the right of the blue oval, indicating that it is now a parameter.

55→ Right-click on the snailhab2 green oval and make it a parameter.

55→ Save the model and close it.

Before we can run the tool, we must add your toolbox to ArcToolbox.

56→ Open ArcToolbox. Right-click the ArcToolbox entry at the top and choose Add Toolbox.

56→ Navigate to the BlackHills folder and click on the BHtools toolbox. Click Open.

56→ Open ArcToolbox again and double-click the BHtools > Snail Habitat Analysis tool.

The model is opened as a tool, with input boxes for the two parameters. Examine the titles above the boxes, Vegetation and snailhab2. The titles are taken from the names assigned to the ovals in the model. Let's rename them so they make more sense to a user.

57→ Close the tool. Right-click on the model in the toolbox and choose Edit.

57→ Right-click the Vegetation oval and choose Rename. Call it *Input Vegetation Layer*. (You can slightly enlarge the box so that the text fits.)

57→ Rename the snailhab2 oval *Output Habitat Layer*.

57→ Save the model and close the model canvas.

57→ In ArcToolbox, double-click on the model to open it as a tool.

**TIP:** Double-clicking or choosing Open will open the model as a tool to be run. To open the model canvas, right-click the model and choose Edit.

There is a more recent vegetation feature class in the Sturgis geodatabase called veg2006. We'll run the model again using the new data to see if the habitat has changed in the last decade.

58→ Click the Browse button and find the veg2006 feature class in the Sturgis geodatabase. Add it.

58→ Set the output feature class to *snailhab06* and put it in the chap10results geodatabase. Click OK to run the tool.

58→ Turn off all layers except the Roads, snailhab2, and snailhab06.

58→ Compare the snailhab2 and snailhab06 layers, noting the change in habitat.

Recall that to create the tool we designated the vegetation input and the habitat output as parameters that could be set each time the tool is run. You can also set individual arguments as parameters. Imagine that we wanted to be able to select different types of vegetation each time we ran the tool.

59→ Right-click the Snail Habitat tool in ArcToolbox and choose Edit.

59→ Right-click on the Make Feature Layer box associated with the vegetation query and choose Make Variable > From Parameter > Expression. A new blue input oval appears for the expression.

59→ Rename the new oval *Vegetation Query*. Move or resize as needed.

59→ Right-click the Vegetation Query oval and choose Model Parameter. A P appears next to it (Fig. 10.30).

59→ Save the model and close it.

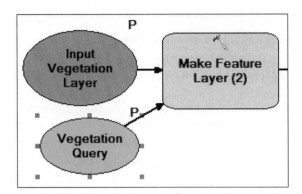

Fig. 10.30. Making the query a parameter
Source: ArcGIS

Now let's run the model again. Suppose we want to see the potential habitat if all densely treed areas are considered, not just conifers.

60→ Open the Snail Habitat model and set the input and output feature classes to Vegetation and snailhab_all.

60→ Click the SQL button next to the new Vegetation Query input box.

60→ Clear the existing query and replace it with DENSITY96 = 'C'.

60→ Click OK to run the tool again.

60→ Compare the old snailhab2 and new snailhab_all outputs to see the changes in habitat.

This brief introduction shows only a small part of what Model Builder can do. To learn more, read about Model Builder in the ArcGIS Help files.

**TIP:** Models can be saved as scripts in the Python programming language. If you learn Python, then you can create full-fledged programs with conditional statements (if/then), looping, and so on.

This is the end of the tutorial.

→ Close ArcMap. Save your changes.

## Exercises

Use the data in the BlackHills folder and Sturgis geodatabase to answer the following questions. Because solution methods can vary, your answers may be slightly different from the ones in the back of the book.

1. The Madison Limestone is an important aquifer. Citizens are concerned about gas or chemical trucks overturning on roads and causing contamination. What is the total length of roads crossing the Madison, in kilometers?

2. The infiltration rate of the geological substrate is a strong factor in evaluating the spill hazards for roads. Create a map of all the roads based on the infiltration capacity of the geology, using appropriate symbols to indicate hazard. Capture your map.

3. The use of a road is another factor in evaluating spill hazards. Create a new field in the roads feature class with a hazard rating of 3 for primary roads, 2 for secondary roads, and 1 for primitive roads. Create a map and capture it.

4. Now create new field for a hazard index based on both infiltration and use, adding them together. What is the range of values for the index? Create a map of the roads based on the hazard index. Capture the map.

5. Rangeland can cause significant impacts on surface water quality. Using the lulc and wsheds2b feature classes, create a table showing the area of rangeland in each watershed. (Include both herbaceous and mixed rangeland categories.)

6. Areas that are within 300 meters of both streams and primitive roads are the most popular places for fishing. Create a map showing these areas. Capture your map.

7. The fishing areas in #6 should be restricted to areas within the national forest (OWNER = 'NFS' in the vegetation feature class). Create a revised map showing the public access fishing areas in green and the private fishing areas in red. Capture your map.

8. You need a feature class that shows polygons based only on infiltration and only within the wshds2b area. Symbolize this layer based on infiltration. Capture your map.

9. The Forest Service is concerned about septic systems contaminating the Madison Limestone. Create a map showing private land (OWNER = PVT in the Vegetation layer) on the Madison. Capture the map.

10. The Forest Service is considering adopting a policy to not cut timber within 200 meters of a primary road (TYPE = P). Assuming that cutting timber includes all SSTAGE96 LIKE '4*', determine the total area in square kilometers lost to harvesting if this policy were adopted. Capture a map showing the off-limits areas.

## Challenge Problem: A Perfect Place

Imagine that you have always wanted a vacation home in an aspen stand with a little pond. Create a map showing all aspen stands (COV_TYPE = 'TAA') that lie on low infiltration geology units (INFIL = LOW) and are within 500 meters of a primary or secondary road (TYPE = P or S). Capture your map.

# Chapter 11. Raster Analysis

## Objectives

▶ Learning different raster analysis functions and techniques

▶ Understanding the raster data model and the storage of raster in ArcGIS

▶ Using Spatial Analyst for raster analysis and geoprocessing

▶ Controlling the analysis environment when using Spatial Analyst

**NOTE:** This chapter requires Spatial Analyst, an optional extension to ArcGIS. If you do not have it licensed on your computer, you will not be able to perform the functions described.

## Mastering the Concepts

### GIS Concepts

#### *Raster data*

The availability of two different data models, raster and vector, provides flexibility to options for data storage and analysis. (Readers may wish to review the sections in Chapter 1 on raster and vector data models.) Neither data model is intrinsically superior; they both have areas in which they excel and areas in which they are at a disadvantage. The GIS professional who has a grasp of both tools holds the keys to developing the most efficient and accurate analysis.

The analysis functions available for rasters number in the hundreds and include very sophisticated applications. This chapter introduces some basic concepts and analysis functions, but the topic spans a much greater breadth of theory and technique than we have time to explore here. The serious student of raster analysis will find plenty of additional material in the software documentation for Spatial Analyst.

Recall that spatial data come in two forms (Fig. 11.1). Discrete data, such as soil types or land use, represent spatial objects and form distinct features or regions on the map; they exist in some places and not in others. Continuous data, such as elevation, vary smoothly over a range of values and form a surface or field that extends everywhere.
The vector data model best stores discrete data. Rasters easily store and analyze continuous data.

Like vector data, rasters have a defined coordinate system with a projection and a datum. The raster is georeferenced by specifying the *x-y* coordinates of the center of the upper-left pixel and the cell size (the height/width of the pixels). Usually pixels are square. A 30-meter raster has pixels composed of $30 \times 30$-meter squares. The cell size is specified in the coordinate system's map units, and meters or feet are typically used.

Fig. 11.1. Discrete land use data and a continuous digital elevation model
Source: ArcGIS

Although one can store a raster in a GCS, this practice is not common because the distortion causes unsatisfactory data performance. Furthermore, many raster analysis functions incorporate distance and angle formulas that will give inaccurate results when applied to a raster in a GCS or in a projection with significant distortion. Therefore, raster data are normally stored in whatever projected coordinate system gives the best and most undistorted view of the data. The main exception occurs for data providers, such as the US Geological Survey, which provide topographic or other data in GCS format. Such data should be projected before they are used.

### *Raster analysis*

Raster processing is fundamentally different from vector processing. Vector analysis manipulates $x$-$y$ coordinates. Raster processing operates on stacks of aligned cells and can be very efficient. This chapter describes the most common functions, but many more are available.

### Map algebra

A raster is an array of numeric values arranged in rows and columns. Several stacked rasters representing different attributes can be analyzed using cell-by-cell formulas based on mathematical, logical, or Boolean operators. This technique is called **map algebra**. Consider the stack of two input rasters in Figure 11.2. These rasters can be added together cell by cell to produce a new raster with cells containing the sum of the corresponding cells in the inputs. If the cells of the input rasters are not already aligned, then they must be resampled to achieve this.

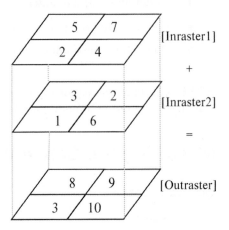

Fig. 11.2. Map algebra addition

The map algebra expression for the operation in Figure 11.2 would appear as [Outraster] = [Inraster1] + [Inraster2] with the raster names enclosed in brackets. Imagine a raster of precipitation in inches. The expression [Precip] * 2.54 produces an output raster with precipitation in centimeters. Expressions can include multiple rasters, functions, and values. A more complex expression to predict erosion potential as a function of rainfall, slope, soil erodability, and vegetative cover might read:

[Precip] * 2 + [Slope] * 4 / ( [Erosion] – [Vegcover]).

Map algebra also includes logical functions that can return **Boolean rasters**, which have values of true (1) or false (0) indicating where some condition holds. Figure 11.3 shows the result of entering an expression [Elevation] > 1400; the purple area shows the cells that meet this criterion.

### Boolean overlay

Overlay is commonly applied to **suitability analysis**, in which a landscape is evaluated to find which areas best serve a given purpose based on a set of factors. The snail habitat problem in Chapter 10 is an example, in which the purpose is to identify habitat for snails, and the factors included elevation, vegetation density, and geology. Raster overlay is especially well suited to suitability problems. First, it can incorporate conditions from continuous data sets, such as elevation or slope, which cannot be easily stored as vectors.

Fig. 11.3. The purple portion of this map shows the pixels where Elevation > 1400 meters.
Source: ArcGIS

Second, the process is faster because corresponding cells are simply added or multiplied together, instead of splicing the existing polygons and creating new ones. Third, raster overlay can process more than two layers at once.

**Boolean overlay** is similar to vector overlay, but it uses map algebra with Boolean rasters and operators. Chapter 8 introduced Boolean operators, which evaluate two input conditions, A and B, to determine a result that is true or false. In raster analysis, this function is performed on a cell-by-cell basis.

Figure 11.4 demonstrates the Boolean operators. The circles represent input rasters A and B, which contain true values inside the circles and false values outside. The squares represent two 4 × 4 input rasters, and the third

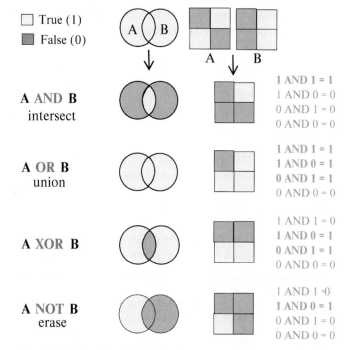

Fig. 11.4. The Boolean operators

column details the operator logic, where 1 indicates a true value and 0 indicates a false value. The AND operator finds the areas common to both inputs and is the equivalent of the vector intersect function. The OR operator result combines the areas of both and matches the union function. NOT is equivalent to the erase function and is the only Boolean operator that is not commutative. The XOR operator has no equivalent tool for vectors (although the same result could be achieved with intersect followed by erase).

Imagine using Boolean overlay to create a potential habitat map for lodgepole pine trees based on two criteria: areas above 1500 feet in elevation and with more than 60 cm of precipitation annually. In Figure 11.5a, the Boolean raster shows where the precipitation conditions for lodgepole are favorable. Areas where the precipitation is greater than 60 cm have the value true

Fig. 11.5. Boolean overlay: (a) a Boolean raster in which the value 1 (true) indicates where precipitation is > 60 cm; (b) a raster indicating elevations > 1500 meters; (c) the result of multiplying the first two rasters, showing where both conditions are true
Source: ArcGIS

321

(1), and areas labeled false (0) have precipitation <= 60 cm. Figure 11.5b shows the favorable conditions based on elevation, where the cells containing 1 have elevations > 1500 meters.

To find the lodgepole habitat, we apply the Boolean operator AND to the input rasters. If both inputs are 1, the output cell is 1. If either or both of the input cells have a 0, then the output will be 0. Figure 11.5c shows the output raster with the potential habitat areas identified as cells with the value 1 (true).

| | |
|---|---|
| 1 AND 1 = 1 | $1 \times 1 = 1$ |
| 1 AND 0 = 0 | $1 \times 0 = 0$ |
| 0 AND 1 = 0 | $0 \times 1 = 0$ |
| 0 AND 0 = 0 | $0 \times 0 = 0$ |

Fig. 11.6. Equivalence of AND and multiplication

AND operates the same way as multiplication of the two Boolean input rasters. Figure 11.6 compares the outputs of AND and multiplication of two Boolean inputs, and they are identical. Thus, either AND or multiplication can be used to find the lodgepole habitat. Since both of these operators are symmetrical, meaning that the order of inputs does not matter, they can be applied to three or more rasters as easily as two.

Map *addition* provides another way to model the habitat—to *rank* the probability of finding lodgepole. In this model, if both elevation and precipitation are ideal, then lodgepole is most likely to occur. If only one is ideal, then it might occur but is not as likely. If neither condition is ideal, then lodgepole is very unlikely. To implement this model, we add the two input maps instead of multiplying them. The final raster in Figure 11.7 has three possible values: 0 if neither condition is true, 1 if either condition is true, or 2 if both conditions are true. Thus, 2 indicates the highest probability of finding lodgepole, 1 a moderate probability, and 0 a low probability. Of course, we would wish to test our model of lodgepole habitat suitability by testing to see if we predicted where lodgepole is actually found.

Fig. 11.7. A Boolean addition overlay
Source: ArcGIS

### Weighted overlay

Boolean overlay has several drawbacks for suitability analysis. First and greatest, it demands that all variables be portrayed as true/false, when in reality it makes little sense that a slope of 10.0 is all right for a building site, but 10.1 is not. It is often more realistic to consider a gradation of values going from less ideal to ideal. Second, it assumes that all input layers are of equal importance. Yet in deciding on a building site, perhaps being close to an existing road is more important than slope, because the cost of developing a long driveway is greater than doing some extra bulldozing to prepare the site.

**Weighted overlay** analysis addresses these two drawbacks. Like Boolean overlay, it develops a set of K rasters ($R_1$, $R_2$, ..., $R_k$) specifying the suitability of a set of conditions. Instead of a simple true/false, however, each raster is ranked into ordinal values 1 to N from worst to best, where N is any positive integer. Each raster is also assigned a weight $W_k$ between 0 and 1, with all the weights summing to 1 ($W_1 + W_2 + ... + W_k = 1$). The output is then calculated as $W_1R_1 + W_2R_2 + ... + W_kR_k$, yielding a raster with values between 1 and N. High values in the output indicate a higher suitability.

Consider siting a landfill, for example (Fig. 11.8). Four factors (K = 4) might be considered: slope, soil infiltration, distance from streams, and distance from roads. A suitability index value of N = 3 is desired. First, one must **reclassify** each raster, converting its original values into a set

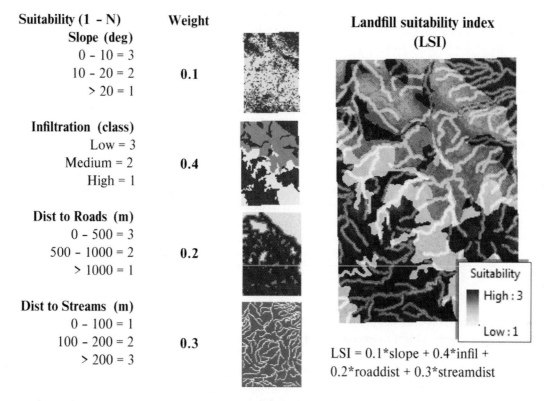

| Suitability (1 – N) | Weight |
|---|---|

**Slope (deg)**
0 – 10 = 3
10 – 20 = 2     **0.1**
> 20 = 1

**Infiltration (class)**
Low = 3
Medium = 2     **0.4**
High = 1

**Dist to Roads (m)**
0 – 500 = 3
500 – 1000 = 2     **0.2**
> 1000 = 1

**Dist to Streams (m)**
0 – 100 = 1
100 – 200 = 2     **0.3**
> 200 = 3

**Landfill suitability index (LSI)**

Suitability
High : 3
Low : 1

LSI = 0.1*slope + 0.4*infil +
0.2*roaddist + 0.3*streamdist

Fig. 11.8. Weighted overlay to determine suitability for a landfill site. The dark colors in each raster indicate the most suitable areas.
Source: ArcGIS

of new values from 1 to 3, creating four index rasters, one for each factor, as shown in Figure 11.8. The final raster is achieved by multiplying each index raster by its weight and adding them together. The final raster contains floating-point values between 1 and 3, with the highest values (and darkest colors) indicating the most suitable areas. Weighted overlay provides a more flexible and sophisticated approach to suitability modeling than Boolean overlay. Its main disadvantage is that assigning the suitability ranks and weights is usually a subjective process, and the final model is heavily dependent on the analyst's choices.

### Distance functions

Distance functions calculate a continuous range of distances from a specified set of objects. The **Euclidean distance** function produces a raster in which each cell represents the shortest distance from a set of specified objects, for example, streams (Fig. 11.9a). A related output is the Euclidean *direction* function, which records the direction of the closest object. At any given point, these two grids would show the distance from the closest stream and the direction you would have to travel to get to it.

The Euclidean distance function is also the means for creating buffers as raster data sets. Once a

Fig. 11.9. (a) Straight-line distance; (b) least-cost path between two points
Source: ArcGIS

323

distance surface is generated, it is a simple matter to identify the distances that are greater than or less than a specified buffer value, to produce a Boolean raster with a 1 value inside the buffer and a 0 value outside.

Cost-distance functions consider more complex and flexible approaches to travel. Consider a mountain with two points on opposite sides. The shortest path to the opposite point goes over the mountain. However, it is easier to go around the base, even though the path is longer, due to the effort (cost) of climbing over the mountain. Slope is one kind of cost. Heavy vegetation, marshes, and off-limits areas are other factors that increase the cost of travel.

In a cost-distance analysis (Fig. 11.9b), a raster is first prepared that specifies the cost of travel over a surface. A single cost factor, such as slope, could be used, or the raster could combine different types of costs. This cost raster is used to develop a cost direction raster, which tells a traveler which directions incur the least cost, and a cost accumulation raster to keep track of the total effort expended as the traveler moves. These rasters can be combined to calculate the least-cost path between two points, indicating the easiest journey rather than the shortest. In Figure 11.9, a slope raster was used to designate the cost, and a least-cost hiking trail was calculated.

## Density

A density function creates rasters showing the frequency of objects within a search radius specified by the user. The function visits each cell of the output raster, counts the points (cities) falling inside the search radius centered on that cell, and assigns the count to the cell. If an attribute field is specified, the counts are weighted by the attribute. Figure 11.10 shows a point density map of city populations in the United States. Without the attribute, the function would count the number of cities; with the attribute, it counts the number of people. The counts are given per unit area, as in people per square mile or cities per square kilometer. A large search radius produces a smoother map but will show less detail than a

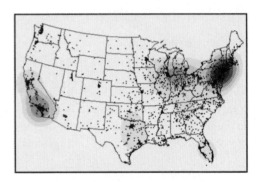

Fig. 11.10. Density of population from cities using a 500-mile search radius
Source: ArcGIS

smaller radius. Choosing the search radius can take some experimentation, but it should be large enough so that every cell will have at least one feature within the search radius; to meet this condition in Figure 11.10, a radius of several hundred miles would be required because of the low number of cities in the Mountain subregion.

A line density function counts line occurrences within the circle instead of points, and it weights them by length and optionally by attribute. This function could be used to calculate road density within a national forest. An attribute field might give stronger weights to primary roads and a lesser weight to primitive roads.

The point density function assumes that the entire weight value is present at each point, but this is not always true. A city cannot have 100,000 people living at a single point. A kernel density function spreads the population value smoothly over a circular area, with the maximum value in the center and decreasing values around it. The kernel density function provides a more realistic model of the population distribution, and the resulting raster will appear smoother than the result obtained with the point density function. The search radius value has less impact on the kernel density function, although larger radii will produce more generalized rasters.

## Interpolation

**Interpolation** takes measured values at points and distributes them across a raster, estimating the values in between the measurements. A common example involves taking precipitation measurements at weather stations and interpolating the precipitation values in between stations to create a continuous precipitation surface. Figure 11.11 shows the interpolation of the total precipitation measured at the stations over the Black Hills region, with the watersheds superimposed on the image.

Fig. 11.11. Rainfall raster interpolated from weather station measurements
Source: ArcGIS

Four different interpolation methods are common: inverse distance weighted (IDW), splines, trends, and kriging. Each method uses different assumptions and different means to obtain the result. More information about these four interpolators can be found in the Help documentation for Spatial Analyst. Interpolation is a complex topic and the subject of entire textbooks. GIS professionals should become familiar with the different methods in order to understand the benefits, drawbacks, and assumptions of each one. However, all methods rely on the assumption that values measured close together are more likely to have similar values than values at greater distances. This property is often called the First Law of Geography, or **Tobler's Law**. It is a common property of spatial data, but not a ubiquitous one.

Although interpolated and density maps appear similar, the functions work differently. Density functions count occurrences within a radius and divide by the area. Interpolation estimates a value for each raster cell based on nearby measurements.

## Surface analysis

**Surface analysis** includes a set of functions for calculating properties of a surface, such as slope and aspect. Although most commonly applied to elevation data, these functions can be used on any type of continuous raster. Using these functions on a DEM could help find an ideal location for a cabin, for example. A slope map, showing the steepness of the terrain, could help identify sites flat enough to build on. An aspect map, showing the compass direction of the slope, could locate south-facing hillsides to give the cabin a warm southern exposure. A hillshade map, which represents the surface as though an illumination source were shining on it (Fig. 11.12), can help the reader visualize the landforms. A viewshed analysis, which determines what parts of a surface can be seen from specified viewpoint(s) (Fig. 11.12), could help in choosing the site with the best view. Finally, the

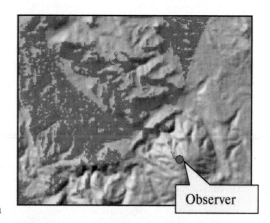

Fig. 11.12. A viewshed map shows the areas visible from a point on a hilltop in red. The background map is a hillshade.
Source: ArcGIS

Cut/Fill function calculates the differences between two input surfaces and is useful for developing estimates of the amount of material to be added or removed from a building site.

## Neighborhood statistics

**Neighborhood functions** operate on an area of specific size and shape around a cell, known as a neighborhood. A moving window passes over the raster and calculates a statistic from all the cells in the neighborhood, records the result, and moves on. Neighborhoods may have different shapes,

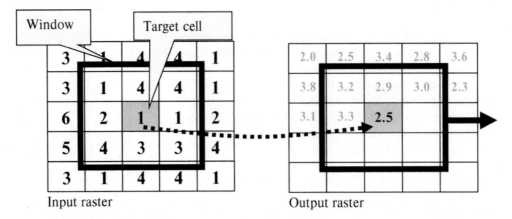

Fig. 11.13. A 3 × 3 averaging window averages the nine cells in the window and places the result in the target cell of the output raster.
Source: ArcGIS

including squares, circles, rectangles, and annuli. The statistical functions include minimum, maximum, range, sum, mean, standard deviation, variety, majority, minority, and median. Two types of neighborhood functions are used: block functions and focal functions.

A **block function** moves the window such that it never overlaps the previous window, but it produces a set of side-by-side adjacent windows that are independent of each other. Once the statistic is calculated for the window, every cell in the neighborhood receives the same value before the window moves on. Every cell is used only once. A block function can be used to resample a raster from a higher to a lower resolution, as might be needed to adapt a high-resolution land cover data set for a coarse climate modeling grid.

A **focal function** centers the window on a cell, called the target. It calculates the statistic for the neighborhood and assigns the value to the target cell, and then it repeats the process at the next adjacent cell. Consider a 3 × 3-cell averaging window (Fig. 11.13). For each target cell, the algorithm finds the average of the nine cells in the window and in the target cell of the output raster. The window then moves to the next cell. An edge cell is still evaluated but will have fewer cells in the neighborhood; the cell in the upper-left corner of Figure 11.13 has only four values to average: the target and its three neighbors. A **filter** is another name for a focal function.

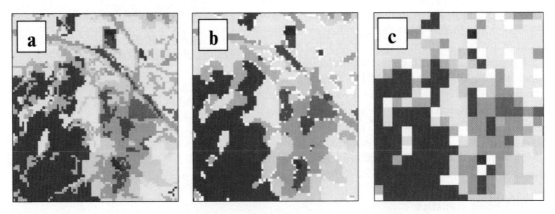

Fig. 11.14. A focal function versus a block function. (a) function: original land cover raster; (b) after a 4 × 4 focal majority function; (c) after a 4 × 4 block majority function. The white cells show where no majority was found.
Source: ArcGIS

Figure 11.14 shows the difference in output between a block and focal majority function applied to a land cover raster (Fig. 11.14a). In both cases, a 4 × 4-cell neighborhood containing 16 cells was used, and the statistic is the most frequent land use value found in the neighborhood. The white output cells occur where no majority was found. In Figure 11.14b, the focal function evaluated each cell with its 15 immediate neighbors and then moved to the next adjacent cell. In Figure 11.14c, the block majority function evaluates the 16 cells and then moves to the adjacent group of 16. Both generalize the original data set, but to a different degree.

## Zonal statistics

The **zonal statistics** function is sometimes considered another type of neighborhood function, but the neighborhoods are not a fixed size or shape. Instead, they are defined by another layer, called the zone layer. A **zone** is a region of a raster or a polygon feature class that shares the same attribute value. A zone might be a watershed, a parcel, or a soil type. The attribute might be a name, a category, or a unique identifier, such as a parcel-ID or even the table ObjectID.

The function gathers all the cells from an input raster that fall inside each zone and calculates cell statistics for the zone. For example, if a watershed layer defined the zones and the input raster contained slope, then zonal statistics would calculate the average slope for each watershed. Figure 11.15a shows a set of watersheds displayed on a hillshade map, and you can clearly see which watersheds have more variable topography and a higher average slope. When the zone function is executed, the slope values are averaged for each watershed.

Fig. 11.15. Using zonal statistics to find the average slope of watersheds: (a) the zone raster shown over a hillshade; (b) the watersheds colored by and labeled with the average slope
Source: ArcGIS

Zonal statistics functions may produce two types of output, a raster or a table. A raster output places the calculated statistical value into every cell in the zone. The table output contains the zone attribute field plus one or more statistics for each zone. This table could be joined to the original watershed polygons using the zone field and used to map the average slope of the watersheds (Fig. 11.15b).

Zones need not be contiguous like the watershed zones, however. The geology map in Figure 11.16 contains many unconnected areas with the same geological unit. The map has only six zones, one for each color.

The zonal statistics function can also assign values from a surface to line or point data, such as finding the average slope along a stream

Fig. 11.16. This geology map has six zones.
Source: ArcGIS

section or attaching an elevation value from a DEM to a layer of wells. Each line or point must have a unique name or ID value that identifies it as a distinct zone.

### Reclassify

The **reclassify** function changes the values of a raster according to a scheme designed by the user, such as classifying a slope map into three regions of low, medium, or high slope, or ranking suitability factors in a weighted overlay (Fig. 11.8). For a map with only a few values, such as land use codes, the user might specify a new value for each code individually. Perhaps a town has annexed a nearby planning unit that used different land use codes and must convert one set of codes to another; the code 143 becomes 347, and so on. With continuous data covering a range of values, such as slope, the values are typically classified into a discrete number of ranges. Classification uses a tool similar to the one described in Chapter 4 for classifying attribute data.

### Resampling

Often rasters must be converted to a raster with a different cell size. This process is called **resampling**. Users may resample purposely in order to align rasters with each other or to decrease the resolution of a data set. Resampling must occur any time two rasters with different cell sizes are analyzed together. Even if the rasters have the same cell size, the cell rasters might not be aligned with each other, requiring that resampling occur.

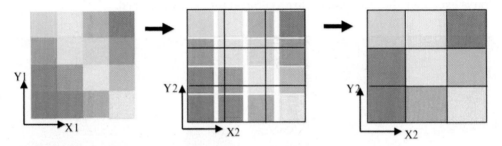

Fig. 11.17. Resampling raster cells using the nearest neighbor method
Source: ArcGIS

Figure 11.17 shows an example of resampling. The cell centers on the left are being transformed from coordinate system X1-Y1 to a slightly larger one, X2-Y2, introducing gaps between the cells. The cell grid for the new coordinate system is shown in the output by the black lines. The old cells may be resampled to the new cells using one of three methods. In **nearest neighbor** resampling, the new cell is given the value of the old cell that falls at or closest to the center of the new cell. In **bilinear** resampling, a distance-weighted average is taken from the four nearest cell centers. The **cubic convolution** resampling method determines a new value by a curve fit through the nearest 16 cell centers. Nearest neighbor is the fastest method and should always be used with categorical raster data. The bilinear method is best for continuous rasters like elevation.

The block neighborhood function provides an alternative resampling method that is more flexible than the three standard strategies, because different types of statistics can be used. In Figure 11.14c, a majority statistic was used to find the most frequent land cover value in the block, which provides a better estimate of the region's contents than using the value at the center of the block. Although it is not an option for automatic resampling, it can be helpful when a user needs to coarsen raster data and wants to do it using the best possible method.

Resampling can degrade the quality of a data set. Best practices dictate that the user retain full awareness of the coordinate systems and cell sizes of all input layers being used for analysis and endeavor to standardize them before processing occurs. Some GIS systems automatically resample without warning. Users must be especially careful with these systems if they are attempting to retain a high degree of spatial integrity in their rasters.

## Conversion

Users can easily convert from vectors to rasters, such as turning a geology shapefile into a raster. The user must choose *one* attribute field to populate the values of the raster, such as the geology unit name. Rasters can only store numbers. Numeric fields are converted as is, but text values are assigned arbitrary numeric codes for storage in the raster. Some raster formats place the original text values in a raster table for reference.

You may also convert discrete rasters to polygons as long as the number of polygons does not become excessive. Continuous data, such as DEMs, should never be converted to polygons, since each pixel is different from its neighbors and will become a separate polygon. If too many polygons will result from a conversion, the software may be unable to complete the task.

## About ArcGIS

### Storing rasters

ArcGIS can read many raster data formats and can read and write some of them. Prior to ArcGIS Version 10, Spatial Analyst could only operate on a format unique to ArcGIS, called a **grid**. Other rasters, such as TIFF files, SID files, and JPEG files, had to be converted to grid format before they could be analyzed, and all raster output generated by Spatial Analyst took the form of grids. At Version 10, Spatial Analyst reads any supported raster format and supports four different output formats: grids, geodatabase rasters, TIFF, and ERDAS Imagine. In this chapter, we will work almost exclusively with geodatabase rasters.

**TIP:** Rasters should NOT be stored in folders that have pathnames containing spaces, because some functions will not work properly. GIS users are advised to avoid spaces in ALL folder names, no matter what.

**TIP: NEVER copy, move, rename, or delete grids using Windows Explorer.** Grids have complex storage formats, and Windows cannot manage them properly. Always use ArcCatalog to manage grids.

Rasters come in two forms. Integer rasters store whole numbers in binary format. Floating-point rasters can store decimal values, but they require 2–4 times as much space as integer rasters. Thus, it is foolish to use floating-point rasters except when decimal values are required. When creating output, the raster type will usually be the same as the input. If you add or multiply two integer rasters, the result will also be an integer raster. If one of them is floating-point, the output will be floating-point. Some functions always create floating-point rasters, such as dividing or averaging a raster. (Floating-point rasters may be converted into integer rasters using the INT tool.)

**TIP:** Raster names must contain only letters, numbers, or an underscore. Grids have even stricter naming conventions and are limited to 13 characters. Consult ArcGIS Help for more information on supported formats and naming rules.

### Raster attribute tables

Rasters can have attribute tables, but they are slightly different from feature attribute tables. Each record lists one unique value present in the raster. The table has three standard fields: the ObjectID field, containing a unique ID for the table rows; the Value field, showing each unique cell value; and the Count field, indicating how many cells contain that value (Fig. 11.18). This table shows that the Precambrian rock covers the largest area because more cells (64,573) have that value than any other. The set of all cells sharing the same value in the raster is called a **zone**.

| geolgrid | | | |
|---|---|---|---|
| OID | Value | Count | NAME |
| 0 | 1 | 22607 | Cenozoic |
| 1 | 2 | 8624 | Upper Mesozoic |
| 2 | 3 | 38679 | Lower Mesozoic |
| 3 | 4 | 32925 | Upper Paleozoic |
| 4 | 5 | 28901 | Madison Limestone |
| 5 | 6 | 64573 | Precambrian |
| 6 | 7 | 24983 | Lower Paleozoic |

Fig. 11.18. The attribute table of a geology raster
Source: ArcGIS

The attribute table in Figure 11.18 comes from a raster that was converted from a polygon feature class using the rock unit Name field, so the Name is included in the table. The Value field contains numeric codes representing each rock unit, which were arbitrarily assigned at the time the feature class was converted. The Cenozoic unit is represented by the code value 1, the Upper Mesozoic by code 2, and so on. Attribute tables may also include fields added by the user.

Only integer rasters can have attribute tables. The tables of categorical rasters are the most useful, as they document a fairly small set of discrete values. However, an attribute table is automatically created for any raster with fewer than 500 unique values.

### NoData values

Rasters may contain a **NoData** value that indicates a null, or missing, value for a cell. Because rasters are always rectangles, the NoData value may be used to mask areas outside a nonrectangular area, as shown by the black cells in Figure 11.19. NoData values are typically displayed with black or transparent color. Some raster formats use 0 instead, which may cause skewing of image statistics. NoData values are usually ignored in processing, providing more consistent results.

Fig. 11.19. NoData values at the top of a DEM
Source: ArcGIS

Some tools allow the user to specify different methods for handling NoData values. If you choose to *Ignore NoData values*, then the tool tries to complete the computation using the values it has available; this is usually the preferred and default option. The other option is to assign NoData to the output cell if the input cells contain one or more NoData values. In other words, if any input values are NoData, the tool assumes that it does not have enough information to calculate a valid result for that cell.

### *Using Spatial Analyst*

Spatial Analyst is an extension to ArcGIS. It is installed by default, but the license must be purchased separately. It contains a set of tools for ArcToolbox, under the heading Spatial Analyst Tools (Fig. 11.20). The tools are organized by function type. Spatial Analyst tools are fully integrated into the geoprocessing environment. They can be run from the toolbox or the command line, used to build models in Model Builder, and incorporated into scripts.

The tools are well documented in the Help files, in many cases including complete discussions of the concepts behind each tool. Users interested in interpolation, for example, will find extensive information on, and references to, the various interpolation methods (IDW, kriging, trending, and splining). Reading about unfamiliar tools before implementing them is highly recommended.

### Environment settings

The Spatial Analyst tools obey the Geoprocessing Environment settings just like any other tool. Several settings are particularly useful during raster analysis and bear mention.

The **Workspace** settings specify the default input and output locations for data sets. Spatial Analyst produces many output rasters that may be intermediate steps toward a solution. Specifying a separate folder or geodatabase to hold the outputs helps keep them separated from permanent data collections and simplifies cleanup later. Setting the current workspace to this temporary place also streamlines the entry of output file names in tools.

The **Output Coordinates** setting specifies the output coordinate system. It is usually best left to the default, Same as Input. Projecting rasters consumes time and memory and should be done only when necessary, preferably as a separate and conscious step and not as a forgotten background process.

The **Processing Extent** controls the rectangular area used to preselect features or raster cells to be used in a tool. Each tool has its own default which is rarely documented, so in practice it is wise to explicitly control the type of processing you want with this setting. With rasters, this setting is particularly important because a raster must have a rectangular extent. If two data sets represented by the black squares in Figure 11.21 are input to a tool, the output raster might contain the entire blue area even though the cells outside the black lines will be NoData. The default for raster processing usually is the intersection of inputs option (the yellow area containing all cells with valid values from both inputs). You can also set it to match the display area, but the best practice is usually to specify a specific data set to which everything else should conform.

The **Cell Size** setting controls the output resolution of rasters (Fig. 11.22). It can be set to a specific value or to match an existing data set. The default is the maximum of the inputs. For example, if two inputs have cell sizes of 50 meters and 30 meters, the output will have 50-meter resolution. The cell size is always measured in coordinate system units and is typically in meters or feet. If a raster is stored in a GCS, however, the units will be degrees.

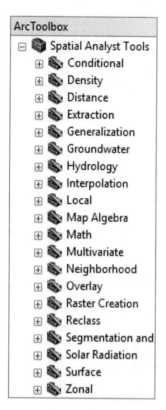

Fig. 11.20. The Spatial Analyst toolbox
Source: ArcGIS

Fig. 11.21. Extent settings
Source: ArcGIS

Fig. 11.22. Cell size settings
Source: ArcGIS

Some tools ask for a cell size even if the environment is set. In this case, the cell size box will be filled with a suggested value based on the map units and the analysis extent. The suggested size will usually create a raster with a sufficient, but not excessive, number of cells. If the suggested size is far different from what you expect or want, then double-check the extent settings and the map units before proceeding.

The **Mask** setting is used to specify a polygon feature class or raster that controls which cells will be processed. A **mask** raster contains NoData cells outside the region of interest, and all output cells there will be given NoData values. If a feature class is used as a mask, then any cells outside the feature class boundary will receive NoData values.

Using a mask is similar to using the clip function for vector data. In Figure 11.23, areas outside the watershed have NoData instead of elevation after the mask is applied.

Mask raster

Elevation after masking

Fig. 11.23. NoData cells in a mask raster are forced to be NoData cells in the output raster.
Source: ArcGIS

**TIP:** Environment settings persist through a work session unless changed, and they are saved with map documents. They are reset to the defaults when a new map document is opened. However, settings made in ArcCatalog determine the default values for new map documents.

### Coordinate system management

Raster functions treat layers as arrays of numbers, which can be processed as a stack. To perform raster analysis, all of the input rasters must have the same cell size and extent so that the cells in each stack align. If a command receives inputs with different cell sizes or extents, the rasters will automatically be resampled to match each other. By default, all rasters are resampled to the most coarse cell size present in the input rasters. When input rasters have different coordinate systems, Spatial Analyst will automatically project the rasters to match.

The resampling and projection process is invisible to the user and seems convenient; however, it does increase the processing time. Resampling or projecting very large rasters may overwhelm the computer and cause a program failure. Since resampling or projecting can also degrade the quality of a data set, it is unsettling to realize that it can happen without the user's knowledge. Storing all rasters to be analyzed together with the same cell size and extent can reduce unintentional resampling. Using the Environment settings to force every tool to produce a raster with the same extent and cell size is usually the best practice to follow during an analysis session.

If multiple coordinate systems are input for analysis, the output raster will be given a coordinate system based on this order of precedence: (1) the coordinate system specified by the Environment settings, and (2) the coordinate system of the first input to the tool.

## Summary

▶ Rasters consist of spatial data stored as individual cells in an array and can represent discrete features or continuous fields of information.

▶ Rasters have coordinate systems just like other spatial data. Many raster calculations are based on distance and area functions, so it is important to store rasters in a projected coordinate system with minimal distortion.

▶ Map algebra treats rasters as arrays of numbers that can be added, multiplied, and otherwise manipulated on a cell-by-cell basis. Arithmetic, logical, Boolean, and bitwise operators and functions offer a wide array of options for analyzing rasters and scalars.

▶ Raster overlay is generally faster and better incorporates continuous data than vector overlay for suitability analysis. Boolean overlay uses true/false rasters and Boolean operators. Weighted overlay uses reclassification and weights for greater flexibility in modeling suitability.

▶ Distance functions measure distances from a set of objects, and can calculate costs and paths associated with traveling over a surface.

▶ Density functions measure the number of objects occurring within a specified search radius and report it in features per unit area.

▶ Raster interpolation estimates a surface based on measurements at isolated locations, and should be employed only on data sets that follow Tobler's Law.

▶ Statistical measures can be calculated for rasters based on single cells, within a specified neighborhood, or within large, irregular zones.

▶ Rasters can be reclassified and given new values as needed.

▶ Resampling occurs automatically when rasters with different cell rasters are analyzed together but can degrade the quality of a data set. Users should attempt to minimize resampling when possible by aligning raster data before analysis.

▶ The analysis environment controls certain settings for all new rasters created during a session. The settings include the working directory, the extent and cell size of output rasters, and the analysis mask.

### Important Terms

| | | | |
|---|---|---|---|
| azimuth | filter | neighborhood functions | weighted overlay |
| bilinear | focal function | NoData | zenith angle |
| block function | grid | reclassify | zonal statistics |
| Boolean overlay | interpolation | resampling | zone |
| Boolean raster | map algebra | suitability analysis | |
| cubic convolution | mask | surface analysis | |
| Euclidean distance | nearest neighbor | Tobler's Law | |

## Chapter Review Questions

1. State Tobler's Law. Give three examples of spatial data measurements that would exhibit this behavior. Then give at least one example of a data set that would not.

2. Why is it best to store rasters in the same coordinate system in which you plan to use them?

3. You are adding two rasters together in the Raster Calculator using the formula raster1 + raster2. Raster1 is in UTM projection, raster2 is in State Plane, and the geoprocessing Environment setting for the output coordinates is USA Equidistant Conic. By default, what coordinate system will the output raster have?

4. Define a Boolean raster. What values can it have?

5. List and describe the five Environment settings that can be helpful in managing raster analysis. When should these be checked and set?

6. Imagine that a parcels map is used as the zone layer for a zonal statistics function. Explain how the results would differ if the zone attribute used were (1) the parcel-ID number or (2) the land use designation.

7. You are adding three rasters with cell resolutions of 30 meters, 50 meters, and 90 meters. List the four available options that would determine the cell size of the output raster and state what the output resolution would be in each case.

8. Describe why resampling should use the nearest neighbor method for geologic units and the bilinear method for elevation.

9. Imagine you have Boolean rasters for 10 different criteria regarding the siting of a landfill. If you add the rasters together, what is the potential range of values in the output raster?

10. What is the potential range of values in Question 9 if you multiply the rasters?

# Mastering the Skills

## Teaching Tutorial

The following examples provide step-by-step instructions for doing basic tasks and solving basic problems in ArcGIS. The steps you need to do are highlighted with an arrow →; follow them carefully. Click on the video number in the VideoIndex.htm file to view a demonstration of the steps.

> → Start ArcMap and open ex_11a.mxd in the MapDocuments folder.

> → Use Save As to rename the document and remember to save frequently as you work.

Fig. 11.24. Snail habitat determined by vector analysis
Source: ArcMap

To become acquainted with raster analysis, we will use rasters to solve the snail habitat suitability problem from Chapter 10 (Fig. 11.24). Spatial Analyst functions are all found in a Spatial Analyst toolbox. Before using the Spatial Analyst extension, however, it must be turned on.

> 1→ Choose Customize > Extensions from the main menu bar.

> 1→ Check the box to activate the Spatial Analyst extension, if it is not already checked.

> 1→ Close the Extensions window.

> 1→ Click the List By Drawing Order icon at the top of the Table of Contents.

As before, we will create a practice geodatabase to store our results and prevent them from mixing with the original data.

> 2→ Open the Catalog tab and expand the Folder Connections > mgisdata folder.

> 2→ Right-click on the BlackHills folder and choose New > File Geodatabase. Name it chap11results.

It is wise to set some Environment settings before starting. We will be producing many rasters, so it helps to have a default workspace to store them. We should also set a default cell size and extent, so that all outputs have the same area and resolution as our Elevation raster, 30 meters.

> 3→ Open Geoprocessing > Environments from the main menu bar.

> 3→ Expand the Workspace entry. Set both the Current Workspace and Scratch Workspace to the chap11results geodatabase.

> 3→ Expand the Processing Extent entry. Set the Extent drop-down to *Same as layer Elevation.*

> 3→ Expand the Raster Analysis entry. Change the Cell Size drop-down to *Same as layer Elevation.* Click OK.

**TIP:** It is good practice to form a habit of checking the Environment settings at the beginning of a work session. Forgotten settings from previous sessions may cause strange results.

In Chapter 10, we predicted snail habitat from vector layers of geology, vegetation, and elevation. We will now perform the same analysis using three Boolean rasters, each one representing one of the ideal snail conditions. We will then multiply them together to find the habitat. We'll begin by creating the elevation range from the DEM.

### Performing a Boolean overlay

The snails prefer elevations from 1200 to 1600 feet. We need to create a Boolean raster in which the cells meeting the elevation criteria have a value of 1 and the rest have a value of 0. The Reclassify tool is used to convert one set of raster values to another.

4→ Open ArcToolbox and dock it as a tab on the right side of ArcMap.

4→ Open the ArcToolbox > Spatial Analyst Tools > Reclass > Reclassify tool.

4→ Use the Input raster drop-down to select the Elevation raster.

4→ The Reclass field is already set to Value, the field containing the values stored in the raster. This is almost always the field you use when processing rasters.

You can edit the entries here directly (be sure to put spaces around the hyphens if you do). However, it is often easier to use Classify to set up the desired class breaks first, either manually or using one of the classification schemes described in Chapter 4.

4→ Click the Classify button. Click OK if you get a sampling error message.

4→ Change the Classification method to Equal Interval and the number of classes to 3.

4→ Change the Classification method to Manual.

5→ In the Break Values box, change the top number to 1200 and click Enter.

5→ Change the middle value to 1600.

5→ Look in the Classification Statistics window and note that the maximum value is 1958. Set the third class break value above it, to 2000.

5→ Click OK to close the Classification window.

5→ Enter the 0 and 1 in the New values boxes as shown in Figure 11.25, including 0 for NoData.

5→ Name the output *elevrange* and save it in chap11results. Click OK.

Fig. 11.25. Reclassify elevation
Source: ArcGIS

The new raster should appear as in Figure 11.26 (your colors may be different). This is a Boolean raster with two values: a 1 where elevation is in the desired range (green) and a 0 where it is not.

6→ Right-click the 0 value symbol for elevrange and set its color to Gray 10%. This helps you visualize the parts of the raster that meet the criteria.

6→ Turn off the Elevation raster and the Roads layer and turn on the Geology layer.

Fig. 11.26. Boolean elevation range
Source: ArcMap

Next, we will convert the Geology layer to a raster. Rasters store only one attribute value, so we must specify which field in the geology table will supply the data. We will use the NAME field.

| Input features | | |
|---|---|---|
| Geology | ▼ | 📁 |
| **Field** | | |
| UNIT | | ▼ |
| **Output raster** | | |
| \chap11results.gdb\geolraster | | 📁 |
| **Output cell size (optional)** | | |
| 30| | | 📁 |

7→ Choose ArcToolbox > Conversion Tools > To Raster > Feature to Raster.

7→ Select Geology as the input features and set the Field to NAME (Fig. 11.27).

7→ Name the output *geolraster* and place it in chap11results.

Fig. 11.27. The Feature to Raster tool
Source: ArcMap

7→ Check that the default value we set, *30*, is entered for the output cell size. Click OK.

7→ Turn off the Geology polygon layer.

We converted on a text field (NAME), but a raster can store only numbers. The legend for geolraster may show numbers or names, depending on your version of ArcMap, but in either case the cell values are numbers.

8→ Right-click the geolraster layer and choose Open Attribute Table.

In a raster table, the Value field shows the values stored in the cells (Fig. 11.28). A number was assigned to each different name, and the NAME field was added to the table to show what each value represents. The Count field indicates how many cells have that value.

| geolraster | | | |
|---|---|---|---|
| | **Value** | **Count** | **NAME** |
| | 1 | 62881 | Cenozoic |
| | 2 | 23984 | Upper Mesozoic |
| | 3 | 107355 | Lower Mesozoic |
| | 4 | 91459 | Upper Paleozoic |
| | 5 | 80237 | Madison Limestone |
| | 6 | 179391 | Precambrian |
| | 7 | 69367 | Lower Paleozoic |

1. What percentage of this raster consists of Madison Limestone? _____

Fig. 11.28. The geolraster table
Source: ArcGIS

**TIP:** To save space and reading effort, future tutorial references to "ArcToolbox > Spatial Analyst Tools > will be written simply as "Spatial Analyst >."

Because we are only interested in the limestone, we must reclassify geolraster to create a Boolean raster containing pixel values of 1 for the Madison and Upper Paleozoic rock units.

8→ Close the Table window.

9→ Open Spatial Analyst > Reclass > Reclassify.

9→ Set the Input raster to geolraster and the Reclass field to NAME.

9→ Change the New values to 1 for the Madison Limestone and the Upper Paleozoic units.

9→ Change the other values, including NoData, to 0.

9→ Name the output *limestone* and place it in chap11results. Click OK.

| Old values | New values |
|---|---|
| Cenozoic | 0 |
| Upper Mesozoic | 0 |
| Lower Mesozoic | 0 |
| Upper Paleozoic | 1 |
| Madison Limestone | 1 |
| Precambrian | 0 |
| Lower Paleozoic | 0 |
| NoData | 0| |

Source: ArcMap

**TIP:** If you wish to see the 0/1 values in the legend instead of the unit names, open the Symbology properties and set the Value field to Value.

Next, we will create a Boolean raster showing areas of dense conifers.

10→ Turn on the Vegetation layer and turn off the geolraster layer.

10→ Open the ArcToolbox > Conversion Tools > To Raster > Feature to Raster tool.

10→ Set the input features to Vegetation and the value field to DENSITY96.

10→ Name the output *density* and put it in chap11results.

10→ Leave the cell size set to *30* and click OK.

11→ Turn off the Vegetation layer.

11→ Open Spatial Analyst > Reclass > Reclassify.

11→ Set the Input raster to density and the Reclass field to DENSITY96.

11→ Change the New value to 1 for C, and the rest to 0.

11→ Name the output *trees* and put it in chap11results.

| Old values | New values |
|:---:|:---:|
| 0 | 0 |
| A | 0 |
| C | 1 |
| B | 0 |
| NoData | 0 |

Source: ArcMap

We are ready to overlay. We have three Boolean rasters, each containing 1 where the habitat condition holds and 0 where the condition is absent. When we multiply them, areas that meet all three conditions will be assigned the value 1, and all other cells will be 0.

12→ Choose Spatial Analyst > Map Algebra > Raster Calculator.

12→ Double-click the raster names and click the * operator button to enter the expression:

"elevrange" * "limestone" * "trees"

12→ Name the output *snailhab* and place it in chap11results. Click OK.

13→ Turn off all layers except for snailhab. (Use the Ctrl key to do them all at once.)

13→ Zoom to the extent of the snailhab layer and compare it carefully with Figure 11.29 (your colors may be different).

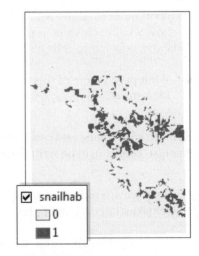

Fig. 11.29. Snail habitat raster
Source: ArcMap

In Figure 11.29, the snail habitat has the value 1 and appears purple. It is the raster equivalent of the snail habitat calculated in Chapter 10 by intersecting vector features. With rasters, though, we can choose to <u>add</u> them to get a <u>ranking</u> of suitability, rather than just a simple yes or no.

14→ Choose Spatial Analyst > Map Algebra > Raster Calculator.

14→ Double-click the raster names and click the + operator button to enter the expression:

"elevrange" + "limestone" + "trees"

14→ Name the output *snailhabrank* and place it in chap11results. Click OK.

14→ Open the Symbol properties for snailhabrank and change the color scheme to a light to dark green color ramp.

This time the best habitat receives the value of 3, and other areas are ranked by how many of the preferred conditions occur there (Fig. 11.30).

### *Distance functions and buffers*

Vector analysis uses buffers to evaluate proximity to features, such as areas within 200 meters of a road. Distance analysis using rasters is more powerful and flexible, partly because rasters can store continuous distance information, whereas vectors are limited to portraying discrete objects. Let's explore some of the raster distance functions.

Fig. 11.30. Ranked habitat raster
Source: ArcMap

15→ Collapse the legends and turn off all layers except for Roads.

15→ Open the Spatial Analyst > Distance > Euclidean Distance tool.

15→ Set the input feature source data to Roads (Fig. 11.31).

15→ Name the output distance raster *roaddist* and save it in chap11results.

15→ Leave the maximum distance blank.

15→ The output cell size should already be 30 from the Environment setting.

15→ Give the optional output direction raster the name *roaddir* and save it in chap11results. Click OK.

16→ Turn off roaddir for the moment to examine roaddist.

16→ Open the roaddist layer properties and click the Symbology tab.

16→ Click Classify and change the method to Jenks. Click OK and OK.

Fig. 11.31. Euclidean Distance tool
Source: ArcMap

Normally, equal interval is the proper classification for distance rasters, but in this case the roadless area in the upper left is skewing the distance measurements and the symbols.

17→ Click the Identify tool and click on cells of the raster to see the values they contain.

Each cell of the roaddist raster represents the distance of that cell to the nearest road (Fig. 11.32a). The distances are calculated in the coordinate system units, meters. When you click on a point, the pixel value is the distance you would have to travel to get to the closest road.

Fig. 11.28. The (a) roaddist and (b) roaddir rasters
Source: ArcMap

17→ Turn on the roaddir raster and examine the values in its legend.

17→ Set the Identify window to identify values from roaddir, and click on several cells. Close the Identify window when finished.

The roaddir raster cells record the direction to the closest road (Fig. 11.32b). The raster values are in degrees and range continuously from 0 to 360, with 0 (and 360) indicating north, 90 east, and so on. For any location in the map area, these two rasters tell you how far you are from the closest road and what direction you would have to travel to get there.

Chapter 10 introduced polygon buffers. A distance raster can be used to create raster buffers in the form of a Boolean raster with 1 inside the buffer and 0 outside it (Fig. 11.33). The Raster Calculator is a quick way to produce it. Imagine we want buffers showing areas within 300 meters of the roads.

Fig. 11.33. Boolean road buffers
Source: ArcMap

18→ Open Spatial Analyst > Map Algebra > Raster Calculator.

18→ Enter the expression "roaddist" <= 300.

18→ Name the output *roadbuf300* and save it in chap11results. Click OK.

**TIP:** The attribute table of a discrete raster tells you how many cells have a given value. Each cell has a given area defined by the cell size. You can use this information to calculate areas of rasters, where: Area = # cells × cell size × cell size.

2. Using the cell size (30 meters) and attribute table of roadbuf300, calculate the total area of the buffers in square kilometers. _____

We can also create multiple ring buffers with rasters. As an example, consider the snail habitat again. Because snails tend to get crushed on roads, the best habitat is further away from roads. Biologists have determined that the snails rarely travel more than 400 meters during their lifetimes. We can rank the habitat based on distance by using 200-meter ring buffers.

19→ Open Spatial Analyst > Reclass > Reclassify.

19→ Set the input raster to roaddist and the Reclass field to Value.

| Old values | New values |
|---|---|
| 0 - 200 | 1 |
| 200 - 400 | 2 |
| 400 - 600 | 3 |
| 600 - 9400 | 4 |
| NoData | NoData |

Source: ArcMap

19→ Click the Classify button. Set the number of classes to four.

19→ In the Break Values box, click the first entry and change it to *200*. Change the next two to *400* and *600*.

19→ Set the last break value above the maximum, *9400*.

19→ Click OK to close the Classify window.

19→ Make sure that the New values in the Reclassify tool go from 1 to 4, as shown here.

19→ Name the output *ringbuf200* and save it in chap11results. Click OK.

Now we have a raster with 1 signifying the least desirable distance and 4 the best. Let us multiply the ring buffers with the snail habitat raster. Recall that habitat has a value of 1 and non-habitat a value of 0, so multiplying by ringbuf200 will yield a new raster with values from 1 to 4, but only in the snail habitat.

20→ Use the Raster Calculator or the Times tool to evaluate the expression:

"snailhab" * "ringbuf200"

20→ Name the output *disthabitat* and save it in chap11results.

20→ Turn off all layers except disthabitat and roads.

20→ Change the color scheme of disthabitat so that 1 is red, 2 orange, 3 yellow, and 4 green (Fig. 11.34).

Fig. 11.34. Snail habitat ranked by distance to roads
Source: ArcMap

This map might raise new concerns, because very little of the habitat lies in the safer regions away from the roads.

It is important to note that the classified display of a distance raster (Fig. 11.32a) looks similar to buffers, but do not be fooled. A distance raster contains continuous distances, whereas a buffer raster contains discrete values of 0 or 1. You cannot create buffers by changing the symbology of a distance raster. Even though it may appear to give you buffers, the values are still distances.

21→ Compare the legends and rasters of the ringbuf200, roadbuf300, and roaddist layers.

Nor can you create buffers using the maximum distance feature in the Euclidean Distance tool. This is a common misconception.

22→ Open Spatial Analyst > Distance > Euclidean Distance.

22→ Set the input features to Roads, the output to *mindist*, and the maximum distance to *300*. Don't specify a direction raster; you don't need it. Click OK.

22→ Turn off all layers except mindist and roadbuf300. Zoom in and compare them carefully. Compare their legends in the Table of Contents as well.

Superficially, the mindist raster looks like buffers, because only the cells close to roads have values. However, the values are distances. The true Boolean roadbuf300 has 1 in the buffer cells. To turn mindist into buffers, you would still need to reclassify it. So setting the maximum distance does not save any effort, and the output will not work correctly for Boolean overlay.

### *Topographic functions*

Topographic functions form an important subset of raster tools that are often used with elevation data. In this section, we will use them to find low-slope, sunny areas as prospective sites to build a winter ski cabin.

23→ Save the map document and open a new, blank map.

23→ Save the map with a new name, and save frequently as you work.

23→ Add the raster mgisdata\BlackHills\Sturgis\topo30m to the map.

23→ Set the topo30m symbology to this stretched color elevation ramp.

You have opened a new document, so check the Environment settings.

Source: ArcMap

24→ Choose Geoprocessing > Environments. Expand the Workspace entry and set the Current and Scratch workspaces to chap11results.

24→ Expand the Processing Extent entry and set the Extent to *Same as layer topo30m*.

24→ Expand the Raster Analysis entry and set the cell size to *Same as layer topo30m* (which is 30 meters). Click OK.

The slope function calculates, in layperson's terms, how steep the surface is. In mathematical terms it calculates the rate of change of the surface.

25→ Open Spatial Analyst > Surface > Slope.

25→ Set the input surface to topo30m.

25→ Name the output *slope30m* and save it in chap11results.

25→ Keep the defaults for the other parameters (Fig. 11.35). The number and type of additional parameters varies depending on your version of ArcMap. Click OK.

25→ Turn off the topo30m and examine slope30m.

Fig. 11.35. The Slope tool
Source: ArcMap

3. What is the maximum slope of this raster? _____

**TIP:** From this point onward, only the output file name will be given. Make sure that all output is saved in chap11results unless otherwise indicated.

Let's find the areas that are flat enough to build a cabin, with slope less than 10 degrees.

26→ Open Spatial Analyst > Map Algebra > Raster Calculator.

26→ Enter the expression "slope30m" < 10 and name the output *lowslope*. Click OK.

26→ Set the 0 values of lowslope to light gray and the 1 value to a nice color.

To find sunny areas, we need an aspect raster. The aspect is the direction a slope faces—measured in degrees from north, with true north being 0 (and 360) degrees, east 90 degrees, and so on.

27→ Turn off the slope30m and lowslope rasters.

27→ Choose Spatial Analyst > Surface Analysis > Aspect.

27→ Set the input raster to topo30m, and name the output *aspect30m*. Click OK.

4.  The aspect map is automatically colored and labeled according to compass direction. Which color indicates a northern aspect? _____ A southern aspect? _____

Let's find the areas with a warm southerly exposure, where the aspect is between 130 and 230 degrees (50 degrees either side of south).

28→ Open Spatial Analyst > Map Algebra > Raster Calculator.

28→ Carefully enter this expression, including the parentheses:

(aspect30m > 130) & (aspect30m < 230)

28→ Name the output *southerly*. Click OK.

**TIP: The parentheses are required to force the calculator to** evaluate the two conditions (aspect > 130) and (aspect < 230) before it executes the Boolean AND (&).

Now we do a Boolean overlay to find potential cabin sites.

29→ Use the Raster Calculator or the Times tool to evaluate the expression:

"lowslope" * "southerly"

29→ Name the output *cabinsites* and click OK.

29→ Compare your results with Figure 11.36.

29→ Collapse all legends, and turn off all layers.

Fig. 11.36. Cabin sites
Source: ArcMap

Two other surface functions are worth investigating. The Hillshade function creates a raster that mimics the illumination of the surface from a light source at a specified direction (**azimuth**) and altitude (**zenith angle**), both measured in degrees. The result looks similar to what a person might see when flying over the surface.

30→ Choose Spatial Analyst > Surface > Hillshade.

30→ Set the input raster to topo30m, and name the output *hillshd30m*. Leave the other settings to their defaults. Click OK.

Compare the elevation and hillshade rasters, and notice how the Hillshade brings out fine topographic details. It makes an excellent base map.

The Viewshed function determines what areas are visible from a set of points. It is used for applications such as placing fire watchtowers or determining whether timber clear-cut areas are visible from a lookout. We will use it to determine the visibility offered by three fire towers.

31→ Add the towers feature class from the BlackHills\Sturgis geodatabase.

31→ Open Spatial Analyst > Surface > Viewshed.

31→ Set the input raster to topo30m, and the observer features to towers. Name the output towerview and click OK.

31→ Save the map document.

5. Examine the attribute table of towerview. What values does the raster actually store? Use the tool Help to determine what the values mean.

**TIP:** Notice that, as you accumulate rasters, you have to wait while each draws. For best performance, get in the habit of turning off or removing any rasters not currently in use.

### Neighborhood functions

A neighborhood function examines a target window in a raster and calculates statistics from cells in the window. Consider the idea of habitat variety.

32→ Add the landcover raster from the BlackHills\Sturgis geodatabase.

32→ Use the Identify tool to associate each color with its landcover class. Close the Identify window when done.

Some areas are homogeneous, with only one or two landcover units, but other areas are more variable. We can analyze the variety of landcover types in different areas using neighborhood functions. Let us calculate the variety of landcover types within a 10 × 10-cell moving rectangle.

33→ Open Spatial Analyst > Neighborhood > Focal Statistics.

33→ Set the input raster to landcover and name the output lcvarfoc (Fig. 11.37).

33→ Choose a rectangle neighborhood. Enter 10 cell units for both height and width.

33→ Select VARIETY for the statistic. Click OK.

33→ These are ordinal data, so change the display color scheme to a dichromatic color ramp—yellow to blue works well.

Fig. 11.37. Focal Statistics
Source: ArcMap

A value of 1 in the lcvarfoc raster indicates that only one landcover type is found within a 10 × 10-cell (300 × 300-meter) area; it is homogeneous. A value of 9 indicates that nine different types were found. This raster gives an objective estimate of landscape heterogeneity.

A block statistics function works much in the same way except that the target window, instead of moving cell by cell, moves over to a completely new area each time.

34→ Open Spatial Analyst > Neighborhood > Block Statistics.

34→ Use the same inputs as for the Focal Statistics tool, but name the output *lcvarblock*. Run the tool.

34→ Change the display color scheme to the same one as lcvarfoc. Compare the two rasters. Then turn them both off and collapse the legends.

The neighborhood functions can use different statistics. The focal majority function, for example, does a good job of simplifying a raster for conversion to features. This landcover data set was generated from raster satellite data, but perhaps you would like a vector version. In its original form it would make a complex set of polygons with many tiny spots and stringers, and needs simplifying before conversion. The Majority Statistic tool examines the target window and selects the landcover value that occurs most frequently. We will use the focal version to get a smoother result. The larger the window, the greater the generalization.

35→ Open Spatial Analyst > Neighborhood > Focal Statistics.

35→ Set the input raster to landcover and name the output *major5*.

35→ Select a 5 × 5-cell Rectangle window and the MAJORITY statistic. Click OK.

When simplifying, a smaller rectangle done two or three times usually works better than a larger rectangle, so run the majority filter once more.

36→ Open Spatial Analyst > Neighborhood > Focal Statistics.

36→ Set the input raster to major5 and name the output *major5x2*.

36→ Select a 5 × 5-cell Rectangle window and the MAJORITY statistic. Click OK.

To compare the final result, it would help if the colors matched. The landcover raster had a colormap stored with it, but it was not transferred to the new raster.

37→ Open ArcToolbox > Data Management Tools > Raster > Raster Properties > Add Colormap.

37→ Set the input raster to major5x2 and the input template to landcover. Click OK.

37→ Remove the major5x2 raster and add it again from chap11results to see the new colors. Turn off the major5 layer.

37→ Compare major5x2 (Fig. 11.38b) with landcover (Fig. 11.38a).

Notice the simpler appearance of the new map. It will convert to polygons more easily. However, keep in mind that you have modified the original resolution and accuracy of the data and made them more generalized. Keep this in mind when interpreting results based on the new data.

### *Converting rasters to features*

Since we've done the simplification, let's go ahead and convert the landcover to polygons.

38→ Open ArcToolbox > Conversion Tools > From Raster > Raster to Polygon.

38→ Set the input raster to major5x2 and the field to Value.

38→ Name the output *landcoverpoly* and check the box to *Simplify polygons*. (Otherwise, the polygon edges will follow the cell boundaries.) Click OK.

Fig. 11.38. (a) Original land cover and (b) land cover after two passes of a 5 × 5 majority filter
Source: ArcMap

The field you specified, Value, is transferred to the polygon table as an attribute named *grid code*. Unfortunately, the text field containing the names of the landcover classes, which was present in the original landcover raster, was omitted when the neighborhood majority tool was run. However, we can copy the field from the original landcover table using a join.

39→ Open the landcoverpoly table and examine the grid code field.

39→ Choose Table Options > Add Field and add a text field called Covertype with a length of 25 characters.

39→ Choose Table Options > Joins and Relates > Join.

39→ Make sure that the first drop-down reads *Join attributes from a table*.

39→ Set the first key field to grid code, the join table to landcover, and the second key field to Value. Click OK.

40→ Right-click the empty Covertype field and choose Field Calculator.

40→ The names are odd, but enter the second Covertype field name from the landcover image (appears as [r_1.img.vat.Covertype] or VAT_landcover.Covertype) in the expression box, depending on your software version. Click OK.

40→ Choose Table Options > Joins and Relates > Remove Join(s) > Remove All Joins.

40→ Examine the newly populated Covertype field and then close the Table window.

40→ Symbolize landcoverpoly using a unique values map based on the Covertype field.

The snailhab raster from the first analysis will serve to demonstrate another potential issue when converting rasters to polygons.

41→ Collapse and turn off all layers.

41→ Add the snailhab raster from chap11results.

Notice that the raster has 0 and 1 values. If we convert this raster, both 0 and 1 areas will be converted to polygons (Fig. 11.39), even though we only want the polygons with 1 values. We must first use Reclassify to convert the 0 values to NoData.

Fig. 11.39. Snail habitat polygons with 0 values
Source: ArcMap

346

42→ Choose Spatial Analyst > Reclass > Reclassify.

42→ Set the Input raster to snailhab and the Reclass field to Value.

42→ In the New values box, change the value for 0 to NoData. Change the 1 value to 1.

42→ Name the output *snailhab2* and save it in chap11results. Click OK.

42→ Turn off snailhab to better see the new raster.

| Old values | New values |
|:---:|:---:|
| 0 | NoData |
| 1 | 1 |
| NoData | NoData |

Source: ArcMap

43→ Open ArcToolbox > Conversion Tools > From Raster > Raster to Polygon.

43→ Set the input raster to snailhab2, and set the field to Value.

43→ Name the output *snailhab2poly* and check the Simplify box. Click OK.

43→ Examine the polygons. Then save the map document.

## *Interpolation and zonal statistics*

Interpolation calculates a raster surface from point measurements. For example, climate stations in the Black Hills record precipitation. In between the stations, we don't know what the precipitation is, but we can estimate it at each cell to create a precipitation raster.

44→ Open the ex_11b.mxd map document. Use Save As to save it under a new name.

44→ Open the Climate Stations attribute table and examine the fields. This data set contains monthly precipitation values in centimeters for 1997, with annual totals.

44→ Close the attribute table.

This data set covers a greater extent than the others, and keeping the cell size at 30 will generate rasters with large storage requirements. Nor do we need that much resolution at this scale. We will set the default resolution to 200 in the Environment settings.

45→ Choose Geoprocessing > Environments from the main menu.

45→ Expand the Workspace entry and set the Current and Scratch workspace to chap11results again.

45→ Expand the Raster Analysis entry, set the cell size to As Specified Below, and enter *200* underneath it. Click OK.

We will create a raster with an estimate of annual precipitation from the Sum field.

46→ Open Spatial Analyst > Interpolation > IDW.

46→ Set the input points to Climate Stations and the Z value field to SUM.

46→Name the output *annprecipcm*. Keep the defaults on everything else (Fig. 11.40).

46→ Click OK and examine the new raster.

6. Do you think the precipitation raster is an integer or a floating-point raster? _____ Examine its properties to check your answer.

Fig. 11.40. The IDW tool
Source: ArcMap

Although we needed all the stations to get the best possible interpolation, the estimate worsens farther from the stations. We also are only interested in areas inside the watersheds. We can extract part of a raster using a mask. A mask works like a clip, keeping the areas of interest and assigning NoData to the others. Either rasters or polygon features may be used as masks.

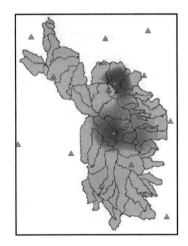

> 47→ Open Spatial Analyst > Extraction > Extract by Mask.
>
> 47→ Set the input raster to annprecipcm.
>
> 47→ Set the mask data to the feature class Watersheds.
>
> 47→ Name the output *precipmask* and click OK.

> 48→ Turn off annprecipcm.
>
> 48→ Symbolize the precipmask raster using the stretched method with a green-blue color ramp (Fig. 11.41).

Fig. 11.41. Precipitation raster after Extract by Mask
Source: ArcMap

The Raster Analysis Environment settings include a mask setting. You can specify a mask, and every new raster created will have NoData in the space outside the mask area.

> 49→ Choose Geoprocessing > Environments and expand the Raster Analysis entry.
>
> 49→ Set the mask to Watersheds and click OK.

We are doing a hydrology study and wish to calculate the total volume of precipitation in each watershed. We start by calculating the volume of precipitation per cell. The depth of received water in the cell in meters is precipmask* 0.01. The area of each cell is $200 \times 200$ meters. Therefore, $V(m^3)$ = precipmask (cm) * 0.01 m/cm $\times$ 200 m $\times$ 200 m (Fig. 11.42).

Fig. 11.42. Volume of water in a cell (height exaggerated)
Source: ArcMap

> 49→ Choose Spatial Analyst > Map Algebra > Raster Calculator and enter the expression:
>
> "precipmask" * 0.01 * 200 * 200
>
> 49→ Name the output *precipvol* and click OK.

The volume per cell ranges from about 16,000 to 35,000 $m^3$. Next, if we add all the cells in the watershed, we get the total volume of water in the watershed. Zonal statistics will sum the cell volumes in one raster (precipvol) based on zones defined by another raster or feature class (Watersheds). In the zone feature class, we must choose an attribute field that holds the unique values that define the zones. We will use the field SHEDNAME.

50→ Choose Spatial Analyst > Zonal > Zonal Statistics (Fig. 11.43).

50→ Set the input zone data to Watersheds and the Zone field to SHEDNAME.

50→ Set the input value raster to precipvol.

50→ Name the output *shedvolume* and set the statistic to SUM. Click OK.

50→ Open the symbol properties of shedvolume and give it a light to dark blue color ramp (Fig. 11.44). (The Stretched renderer in Version 10.4.1 may not display this raster properly. If the watersheds all show in one color, open the Symbol properties and use the Classified display option instead.)

Fig. 11.43. Zonal Statistics
Source: ArcMap

In the output raster, each cell in the watershed contains the summed volume over the watershed. The Zonal Statistics as Table tool provides an alternative way to calculate and view the totals.

51→ Choose Spatial Analyst > Zonal > Zonal Statistics as Table.

51→ Fill out the first three boxes as before (Fig. 11.43).

51→ Name the output table *shedvoltable*.

51→ Instead of the single SUM statistic, choose ALL. Click OK.

51→ Open the shedvoltable table and examine the statistics for the watersheds.

The output table lists more information, including the cell count, the area, and statistics about the water volumes (min, max, etc.). This table could be joined back to the Watersheds layer if needed.

7.  What is the name of the watershed that has the largest volume of water? What is the volume?

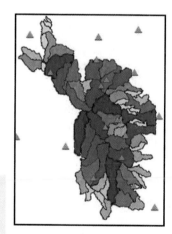

---

**TIP:** You can also use zonal statistics on zones defined by points or lines. You could calculate the elevation of a set of summits, for example, or the average slope of individual streams.

This is the end of the tutorial.

→ Close ArcMap. You can save your changes.

Fig. 11.44. Map showing volume of water in each watershed
Source: ArcMap

## Exercises

Use the data in the BlackHills\Sturgis geodatabase to answer the following questions. Set the Environment cell size to 30 meters for all problems and make sure any masks are off.

1. Create a Boolean raster showing areas where the density of vegetation is open (DENSITY 96 is 0 or A) and the land is owned by the Forest Service (OWNER is NFS). Capture the map.

2. Create a Boolean raster showing the areas that are within 1000 meters of a primary or secondary road (TYPE = P or S). Capture your map.

3. Use Block Statistics on the slope30m raster to calculate average slopes over 300 × 300-meter areas. Then create a Boolean raster showing where the average slopes are < 10 degrees. Capture your map.

4. Imagine that you are looking for a good landfill site. Use the rasters produced in Exercises 1 to 3 to create a Boolean raster showing the areas where all three conditions are met. What is the total area of these potential sites? Capture the map.

5. The site must be at least 1 sq km in area to be considered for the landfill. Create a helpful map of the sites meeting this criterion, labeled with the site area in square kilometers and including helpful information, such as streams and roads, for the final selection. How many potential sites have the minimum area? (**Hint:** Convert the sites to polygons.)

6. Which geological unit has the highest average slope? Which has the lowest? What is the average slope for each one?

7. The canopy raster contains the forest canopy percentage. Create a map of wshds2c showing the average canopy percentage over each watershed. Capture your map.

8. Prime harvestable timber in the Black Hills has SSTAGE96 = 4C or 5 and is more than 200 meters from a stream. How much good timber is left, in square kilometers? Create a map showing the timber and streams. Capture the map.

9. Which of the summits has the highest elevation? Which has the lowest?

10. Create a raster with an integer canopy index that ranges from 1 to 5. Create another raster with an integer slope index that ranges from 1 to 5. From these, produce a raster showing an erosion potential index based on canopy and slope. Capture the map. (**Hint:** Use Reclassify.)

## Challenge Problem: My Secret Hideaway

Imagine that you would like to build a cabin somewhere in the Sturgis area. Develop a combined Boolean/weighted index model to evaluate the possibilities. One condition is required: it cannot be on National Forest Service land (vegetation layer, OWNER = NFS) (**Hint:** Create a mask.)

The other factors can be ranked. The land cannot be too steep; having a southerly aspect is good; being close to an existing road is cheaper. It can't be too close to a stream, but it would be nice not to be too far away from one. A forested area is preferable to a meadow, bare rock, or already developed land. (Note: You may need to resymbolize rasters after reclassifying them in order to see your new assigned values correctly displayed.)

Create a model design similar to that shown in Figure 11.8, showing the five ranked factors, index classes/values, and weights. Capture the final result. Afterward, evaluate your model choices. Would you do something differently next time?

# Chapter 12. Editing and Topology

## Objectives

▶ Understanding topology and learning to preserve topological rules during editing

▶ Learning techniques for digitizing complex groups of polygons

▶ Using additional editing tools to help create and preserve topological relationships: merge, union, intersect, clip, modify, and reshape

▶ Editing shared lines and polygon boundaries with map topology

▶ Creating planar topology and fixing topological errors

# Mastering the Concepts

## GIS Concepts

Chapter 7 presented some basic techniques for creating new features in a feature class. In this chapter, we examine additional ways to form and modify features, with an emphasis on maintaining correct topology during editing.

### Topology errors

**Topology** concerns how the features are spatially related to each other. The four types of spatial relationships introduced in Chapter 1 included adjacency, connectivity, overlap, and intersection. One major goal during editing, other than simply getting features into a feature class, involves ensuring the **logical consistency** of features, in other words, making sure that features are free of geometric errors and that their topological relationships are adequate for the purpose intended. Some of the topological errors to avoid during editing are shown in Figure 12.1.

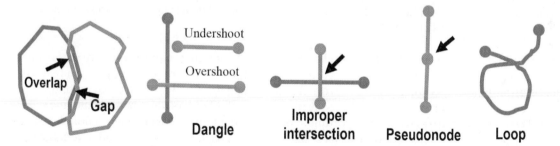

Fig. 12.1. Topological errors can occur when editing is not carefully performed.
Source: ArcGIS

### Basic topology rules

1. Adjacent polygons must share a coincident boundary that is exactly the same for both. There are no gaps or overlaps between adjacent polygons.

2. Lines should always end on other lines. Failure of two lines to meet is called a **dangle**. Dangles can occur when one line is not quite long enough to meet the other (**undershoot**) or when one line crosses too far over the other (**overshoot**).

3. Lines that intersect should each have a **node** (endpoint) at the intersection. Lines crossing without nodes are termed improper intersections.

4.  Nodes should exist only where three or more lines intersect. A node where only two lines meet is called a **pseudonode**.

5.  Lines or polygon boundaries should not cross over themselves and form loops.

6.  There should not be duplicate copies of any points, lines, or polygons.

Perfectly legitimate exceptions can occur to topological rules. For example, a dead-end street by definition does not meet another road on one end and must be a dangle. A highway overpass can cross over a street without intersecting it. Polygons showing areas sprayed with pesticide on different dates might have gaps or overlaps between them.

The effects of topological errors can vary. In some cases they are merely a visual nuisance. At times they can be a legal liability, such as an overlap between two parcels. In other cases they contribute to undesirable outcomes, like the slivers that can be formed during geoprocessing operations. In the worst case, they can cause a failure of the data set for its purpose—a street network containing improper intersections will not properly route traffic. Each user must evaluate the purpose of a data set and establish the topological relationships needed. In some cases these rules will include more than the six listed previously. However, these six should be considered a basic level of integrity to which every data set should conform (not including unavoidable exceptions).

Topological integrity can be established at the outset by careful editing of features to minimize the occurrence of errors. Most GIS software includes tools for ensuring basic topological integrity during editing, such as snapping and the Auto Complete Polygon tool discussed in Chapter 7.

### Topology rules and tools

Even careful editing cannot avoid all topological errors. Moreover, one does not always have control over the creation of a data set; one may be stuck with data that were poorly digitized without attention to topology. A GIS that is capable of storing topology will generally also possess tools that can be used to find errors and fix them.

Many topological errors are difficult to find by mere inspection; they require software algorithms to test and evaluate features at the $x$-$y$ coordinate level. These tools are generally run on feature classes after digitizing is complete. The errors that are found are fixed by additional editing, which can create new errors. So the process is repeated until all errors have been found and fixed. These routines are extremely important in developing a logical consistency report for the feature class metadata. This report describes the procedures used and tests applied to a data set to ensure that it meets the topological rules we have listed.

Some GIS systems contain advanced capabilities for defining topological rules and ensuring that they are met. The ArcGIS geodatabase model is one of these. It can create and store a network topology, which stores and manages connections between linear features and can be used for analyzing flow problems in networks such as roads or a water distribution system. Networks can have connectivity rules about how features connect. Feature datasets within geodatabases can also contain **planar topology**, which expands the use of rules to model a variety of topological relationships, not just within a single layer but between layers as well.

One class of rules applies to features within a single feature class. The types of rules available vary according to feature geometry. Each of the six topological rules cited above is represented by rules such as Must Not Have Gaps, Must Not Have Overlaps, Must Not Have Dangles, and

Must Not Have Pseudonodes. Figure 12.2 shows a planar topology in ArcGIS that uses these rules to test the topology of geology and faults.

Other rules specify relationships *between* feature classes, such as the Must Cover Each Other rule that describes the relationship between counties and states in which every bit of the state area must belong to a county and no part of a county should lie outside the state.

| Topology Properties | | | |
|---|---|---|---|
| General | Feature Classes | **Rules** | Errors |

| Feature Class | Rule | Feature Class |
|---|---|---|
| faultst | Must Not Have Pseudo Nodes | |
| faultst | Must Not Have Dangles | |
| geologyt | Must Not Overlap | |
| geologyt | Must Not Have Gaps | |

Fig. 12.2. Topology rules in a planar topology of geological units and faults
Source: ArcGIS

Figure 12.3 shows typical topological errors occurring between layers. The boundary of the Pine Ridge Indian Reservation (stippled area) should match the boundary between Shannon and Bennett counties, but it does not. Neither does it match the South Dakota–Nebraska boundary. A similar problem was encountered between census tracts and county boundaries at the end of Chapter 9 (Fig. 9.28).

Creating planar topology makes it easier to locate and eliminate boundary errors by using special Topology tools while editing. This approach is helpful because many errors are too small to see at normal viewing scales—in Figure 12.3 the county boundaries appear to match the state boundaries but might not actually do so. Such errors typically occur because data come from different sources or because some people do not understand or use snapping effectively.

Using planar topology requires an ArcGIS Standard or Advanced license. The tutorial for this chapter includes a demonstration of using topology to test for pseudonodes, dangles, gaps, and overlaps for users who have the appropriate license. However, it is nothing more than a brief introduction to an extensive and complex topic. Users interested in learning about topology should read the appropriate ArcGIS Help files.

Users with an ArcGIS Basic license have limited tools for working with topology and must take responsibility during editing to keep features as correct as possible.

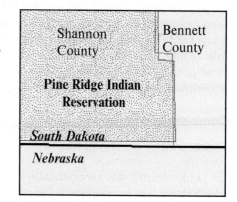

Fig. 12.3. Topological errors between layers
Source: ArcGIS

### Editing complex polygon shapes

Some map types, such as geology and soils, have a complex interweaving of different units. It takes some forethought and planning to efficiently create polygons with correct topology. Experience digitizing such maps is the most effective way to learn appropriate techniques and avoid common mistakes. When planning, the key constraints to remember are:

▶ A line created with the Auto Complete Polygon tool must completely enclose a region.

▶ A line created with the Cut Polygons tool must start and end outside of an existing polygon selection.

▶ An island polygon requires creating a separate feature, which will then overlap the larger polygon underneath. The overlap must be corrected afterward by clipping.

Consider the geology map shown in Figure 12.4. It would be easy to get "painted into a corner" and not be able to finish the sequence of polygons. For example, if one digitized polygons 2 and 5 first, polygon 3 would be a hole that would be a nuisance to fill. An island polygon, such as 13, cannot be created by the Auto Complete Polygon tool because it lies completely inside another polygon. It must be added after the polygon around it, which causes it to overlap the existing polygon underneath, violating a topology rule. The following are two ways to approach the editing. Both methods should help avoid gaps and overlaps and result in topologically correct polygons.

Fig. 12.4. Complex editing
Source: ArcGIS

### Divide and conquer

Divide and conquer starts by digitizing the outline of a large area, such as the square outline of this map (Fig. 12.4). The Cut Polygons tool is then used to carve out polygons from the larger region. One might start with polygons 1, 2, and 6 (creating a fake boundary between 6 and 7b that goes through 5). Then create 7 as a single unit, 8 as a single unit, and 9 as a single unit. Finally, use Create New Feature to create 5, 10, 11 to 13, and the remaining islands on top of the larger units. After creating each island, the Clip command (discussed in the next section) is used to eliminate the overlap with the larger polygon underneath. Occasionally with this method one is forced to create a superfluous boundary within a polygon, and then merge the two pieces afterward.

### Adding territory

Adding territory relies mainly on the Auto Complete Polygon tool. One might start with Create New Feature to create polygon 1. Switch to Auto Complete Polygon and add 2 and 3. Add 4, temporarily closing the narrow neck between it and 6. Add 5 and 6. Go back and merge 4 and 6. Add 7a and 7b, leaving a hole where 10 is. Add 11 and 12 using Auto Complete. Add 8a, 8b, and 9. Finally, close off 10. Go back to Create New Feature and create 13, making sure to clip out the underlying polygon area beneath it.

### *Combining features*

During editing, especially when editing complex shapes, creating a single feature at a time is sometimes not practical. We mentioned that creating island polygons generates an overlap with the polygon around it. Occasionally it is easier to create a polygon in two or more pieces, after which they must be combined. During editing, several useful techniques can aid in maintaining correct topology. The editing functions are similar to the geoprocessing tools, such as Clip and Intersect, but they operate on selected features, rather than on entire layers. Moreover, the attributes are treated differently. Instead of combining all the attribute fields from both inputs, the output feature simply has the same attribute fields as the layer being edited.

When using these combinations, the user should be sensitive to how attributes of the features are handled. In the case of a merge, a union, or an intersect, the resulting features will be given the attributes of one of the original features. In a geodatabase, the attributes of the feature selected FIRST will be copied to the output feature. For shapefiles and coverages, the feature with the lowest feature-ID will be copied. If the feature attributes matter, then the user must pay attention to which attributes are being copied and must correct any attributes that were not copied as desired.

## Merge

A merge takes two or more features and combines them into a single feature (Fig. 12.5). If the features are adjacent, then the boundaries between them are removed. If the features are separate, then a multipart feature will be created. This function might be used frequently to update a parcels map when an owner has purchased two adjacent parcels and combined them into one. It is also useful when one finds it easier to digitize a feature in two or more pieces and then combine them afterward.

## Union

A union performs the same operation as a merge, except that the original features remain unchanged and a new feature is created in addition to the originals.

## Intersect

An intersection creates a new feature from an area common to both original features (Fig. 12.6). A new feature is created, and the original ones are maintained. If two features are selected and an intersection is performed, the resulting new polygon will consist of only the areas shared by the original polygons. The new feature is created in addition to the original features, which are not changed or deleted. This function might be used to identify repeat infestations of pine bark beetle attacks. Two features representing infestations in two different years could be intersected to reveal the area attacked twice.

## Clip

A clip behaves in a cookie-cutter fashion. If one feature lies over another, the underlying feature will be cut along the boundaries of the overlying feature (Fig. 12.7a). Two options may be specified, preserving the area common to both features (Fig. 12.7b) or discarding the area common to both features and retaining what is outside the clip polygon (Fig. 12.7c). In either case, the feature used for clipping is retained unchanged, and the feature underneath is modified. As shown in Figure 12.7c, clipping is one way to create a "donut" polygon. Clipping is useful any time a polygon must have internal boundaries, such as a residential area with a park in the middle. It is also useful for creating an island polygon inside another. Clipping is crucial to enforcing the "must not overlap" rule by removing sections of a polygon that lie underneath another polygon.

Fig. 12.5. (a) Merging adjacent polygons to create a new feature; (b) merging separate polygons to form a multipart feature
Source: ArcGIS

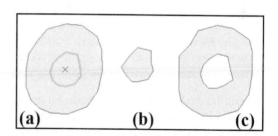

Fig. 12.6. Intersection of two polygons to create a third polygon
Source: ArcGIS

Fig. 12.7. Clipping: (a) ready to clip the outer polygon with the selected polygon; (b) option to preserve the common area; (c) option to discard the common area
Source: ArcGIS

## *Buffering features*

A buffer delineates the area within a specified distance of a feature and can be created from points, lines, or polygons (Fig. 12.8a–c). The output may be lines or polygons. The buffer distance must be specified in map units. In the case of buffering polygons, a negative distance may be used to reduce the size of the feature (Fig. 12.8d). Buffers are useful for such tasks as identifying setbacks from parcels, finding drug-free zones around schools, or creating road widths from a set of centerlines.

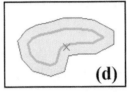

Fig. 12.8. Buffering:
(a) the original features;
(b) buffer output as lines;
(c) buffer output as polygons;
(d) a negative buffer
Source: ArcGIS

## *Stream digitizing*

Two methods can be used to place vertices for lines and polygons. So far you have used the point-by-point method in which each vertex is placed with a click (Fig. 12.9a). **Stream digitizing** offers another approach. The user chooses a tolerance that controls how far apart each vertex should fall. The first vertex is added with a click. Then the user smoothly moves the mouse pointer over the line to be digitized. Each time the tolerance distance is traversed, a new vertex is added automatically (Fig. 12.9b). Stream digitizing can be faster than point-by-point digitizing, but it does have some drawbacks.

Fig. 12.9. A polygon captured with (a) point-by-point and (b) stream digitizing
Source: ArcGIS

First, in point-by-point digitizing the user adds only as many vertices as needed. Straight edges require fewer vertices and curved edges need more closely spaced vertices. In stream digitizing, one tolerance is used for both. Thus, straight lines often have more vertices than needed and may not be as straight, and curved sections may not have enough vertices. As a result, the features may be less accurately captured. Furthermore, the file size is often larger due to the extra vertices. The choice of stream tolerance is critical for good results.

Second, when creating a line point by point, the user can place a vertex exactly on the inflection points of the feature to capture the shape with greatest accuracy (Fig. 12.9a). In stream digitizing, the vertex falls when it reaches the set tolerance and may be offset from the actual inflection.

The third problem is that a slight movement off the line will create unwanted vertices that must be cleaned up later. A steady hand is required for stream digitizing. In Figure 12.9b, the sketch shows three places where the curve deviates from the feature being digitized (arrows).

Most users find that they develop a preference for one kind over the other. The style of the features being digitized affects the choice as well. Feature classes with many long, straight lines and sharp corners will be much more efficiently digitized using the point-by-point method. Feature classes with many smooth curves can benefit from stream digitizing. It is also possible to switch back and forth between styles while digitizing a single feature, using stream for curved sections and point-by-point for straighter sections. A tablet-type computer with a touchscreen and pen are highly effective for stream digitizing.

## About ArcGIS

We now turn to a discussion of some additional tools and techniques for editing in ArcMap.

### *Changing existing features*

Once a polygon or line has been created, it can be changed by eliminating unwanted vertices, adding vertices, or moving existing vertices. Other modifications include flipping lines, trimming them, extending them, and performing a variety of other editing functions. These functions are useful for cleaning up errors from stream digitizing.

### Modifying features

The Editor toolbar contains an option to **modify** existing features by adding, deleting, or moving vertices. The Edit Vertices tool causes the sketch of the feature to appear. When the edits are complete, the feature is updated to the current form of the sketch. Figure 12.10 shows how an original, circular polygon is modified by moving vertices in the sketch.

Fig. 12.10. Modifying a feature
Source: ArcGIS

### Reshaping features

The Reshape Feature tool reenters a feature or a portion of a feature by using a new sketch to define the revised shape. The sketch must start and finish exactly on the original feature, so it is helpful to have vertex or edge snapping on. When the sketch is finished, the original feature is modified to follow the sketch. Figure 12.11 shows how to **reshape** parts of lines and polygons.

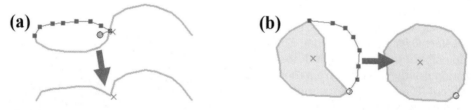

Fig. 12.11. Reshaping features: (a) reshaping part of a line; (b) reshaping a polygon
Source: ArcGIS

### Flipping lines

Although we are generally not aware of it, lines have a direction, beginning at a "from" node and ending at a "to" node, which is established according to the direction in which the line is initially digitized. Normally the direction of a line generates little concern, and either direction works just as well. In some situations, however, the direction matters. In geocoding, it matters because it defines which way the addresses increase along the street. A streams feature class used for network analysis uses the line direction to encode the flow of water. A similar situation would hold for constructing a network of water pipes or sewers in which the direction of flow must be constrained and recorded.

Fig. 12.12. A flipped line
Source: ArcGIS

The direction of lines may be viewed by drawing them with arrow-ended symbols. Once drawn, you can see the direction of flow and flip the direction, if necessary (Fig. 12.12).

### *Editing with map topology*

Building planar topology with rules for a feature dataset and using it during editing requires an ArcGIS Standard or Advanced license. However, users with ArcGIS Basic licenses or users who are editing shapefiles can use a function called **map topology** to edit features with shared edges or vertices. Map topology creates temporary relationships between features so that they can be edited together. Its purpose is to preserve existing coincident boundaries and nodes.

Topological editing uses a **cluster tolerance** to help enforce snapping and coincident boundaries. When two vertices of features being edited fall closer together than the cluster tolerance, the vertices are made coincident. Setting the cluster tolerance requires great care. If too large, it will collapse coordinates unnecessarily and degrade the precision and accuracy of the data. If too small, it does not help prevent topological errors. However, it is better to err on the side of too small than too large. The default tolerance is designed to preserve the accuracy of features rather than do extensive corrections and should be used unless you have reason to do otherwise.

The user creates map topology by selecting which feature classes will participate in it. The topology is created on the fly for the set of features currently in the data view. The endpoints of lines are called **nodes**, and lines or polygon boundaries are called **edges**. The user selects the node or edge to be edited using the Topology Edit tool, which is similar to the regular Edit tool, except that it selects shared edges or nodes for editing. When the Topology Edit tool is used to select a feature or part of a feature, all features sharing that boundary are also affected by the edits.

Figure 12.13 shows the use of the Topology Edit tool and the Reshape Edge tool to change the boundary between two polygons. First, the shared boundary is selected using the Topology Edit tool. The selection color of the Topology Edit tool is purple to distinguish the selection from those made with the Edit tool. The Reshape Edge tool is then used to draw the new boundary between the polygons. When the sketch is finished, the new boundary replaces the old one, and the change is applied to both polygons.

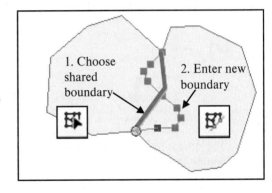

Fig. 12.13. Reshaping a polygon
Source: ArcGIS

The Topology Edit tool is also convenient when working with connected line features, such as roads. If one road node is moved, the roads attached to the node move also. In Figure 12.14, the road node was selected with the Topology Edit tool and then moved downward. The attached vertices of the other lines are also moved when the sketch is finished. In Figure 12.14, only the node is moved, causing kinks in the attached lines. For smoother adjustment, one can turn on an option to adjust the attached lines proportionally.

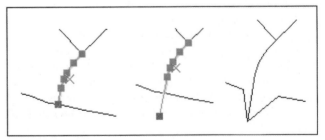

Fig. 12.14. When editing shared nodes, the node is moved, and the attached roads follow.
Source: ArcGIS

The Topology Edit tool and other topology editing tools should always be used for editing a map topology. They can be applied to moving features, reshaping them, and modifying them. However, the map topology must be created prior to using the topology tools.

### Editing with planar topology

With an ArcGIS Standard or Advanced license, feature classes within a feature dataset may be assigned to participate in a planar topology. The topology is created in ArcCatalog, and specific rules are invoked concerning the desired topological relationships. Once the rules are defined, **validation** is performed to assess adherence to the rules. Errors are found and marked during the validation process. Afterward, the user can view the errors and use specialized tools to fix them.

The available rules vary depending on what type of feature (points, lines, or polygons) they govern and whether they apply to a single feature class or to relationships between two feature classes. Figure 12.15 shows the rules that could be applied to lines (above) or polygons (below).

The topology is assigned a cluster tolerance when it is created. The cluster tolerance is the minimum distance allowed between vertices. During validation, if two or more vertices lie within the cluster tolerance, they will be snapped together.

By default, the cluster tolerance is the same as the feature class XY tolerance, or 0.001 meter. This value is designed to preserve coordinate accuracy. If the default is accepted, then virtually no vertices will be snapped together. Yet it is often desirable to set the cluster tolerance to a larger value, because then the topology can fix some errors automatically. The cluster tolerance should never be less than the XY tolerance.

Must Not Overlap
Must Not Intersect
Must Be Covered By Feature Class Of
Must Not Overlap With
Must Be Covered By Boundary Of
Must Not Have Dangles
Must Not Have Pseudo Nodes
Must Not Self-Overlap
Must Not Self-Intersect
Must Be Single Part
Must Not Intersect Or Touch Interior
Endpoint Must Be Covered By
Must Not Intersect With
Must Not Intersect Or Touch Interior With
Must Be Inside

Must Not Overlap
Must Not Have Gaps
Must Not Overlap With
Must Be Covered By Feature Class Of
Must Cover Each Other
Must Be Covered By
Boundary Must Be Covered By
Area Boundary Must Be Covered By Boundar
Contains Point
Contains One Point

Fig. 12.15. Rules for lines (above) and polygons (below)
Source: ArcGIS

Choosing the cluster tolerance can be tricky. Both the resolution of the data set and the size of the typical errors must be considered. If the tolerance is too small, then all edits must be manually done. If too large, it can degrade the data set accuracy. Imagine a road data set collected by GPS from a truck, with vertices collected every 25 meters. As lines are stopped and started at intersections, an offset of 3–5 meters typically occurs. These data are then placed in a topology. If the default cluster tolerance of 0.001 meter is employed, then none of the ends will be snapped, and each intersection must be edited manually. However, if the cluster tolerance is set to 6 meters, then most of the road intersections will be snapped automatically, leaving far fewer errors to edit. Since vertices are normally 25 meters apart, the cluster tolerance will not collapse actual measurements together.

During validation, several processes occur. First, cracking places an extra vertex on edges that fall within the cluster tolerance of another edge, vertex, or end, and then clustering collapses the vertices together. It is important to realize that the validation step can move the vertices of features from their original locations. Users can specify ranks for the different feature classes. Lower-ranked features will be moved to match higher-ranked features. If features have equal ranks, then both sets of vertices may be moved.

The validation produces a topology feature class that contains point, line, and polygon errors, which may be drawn and symbolized during an editing session. Figure 12.16 shows some topology errors for the Edwards aquifer geology and faults. The pink boxes indicate point errors; the pink line, a line error; and the red outline a polygon error.

The user examines and fixes the identified errors singly or in batches, as appropriate. Unavoidable errors, such as the dangles that occur where a fault ends, are marked as exceptions. Each type of error has different possible fixes that can be applied. Dangles may be snapped, extended, or trimmed. A gap must have a feature created to fill it. An overlap must have the overlap area removed.

Fig. 12.16. Topology errors
Source: ArcGIS

**TIP:** There is an excellent summary of topology rules and error fixes in ArcGIS Help. Search for "topology error fixes" and select "Geodatabase topology rules and topology error fixes."

After all errors are found and fixed, the user must validate again. Often fixing one set of errors produces new ones. The process of validating, editing, and validating again must be repeated until no new errors are found.

## Summary

▶ Careful editing helps create and maintain topological integrity between features. Topological rules establish requirements for feature adjacency, connectivity, overlap, and intersections.

▶ Planar topology establishes rules about the spatial relationships within and between layers, and it can be used to help locate and eliminate errors.

▶ Strategic planning helps when digitizing complexly arranged polygon feature classes.

▶ Features may be combined to create new ones using merge, union, intersect, and clip. Care must be taken to ensure that attributes are correctly copied during these transactions.

▶ New features can be created by buffering points, lines, or polygons.

▶ Stream digitizing can help create smooth curved boundaries, but it has some drawbacks.

▶ Modifying and reshaping are two ways to change existing features.

▶ Temporary topological relationships, called map topology, may be used when editing with an ArcGIS Basic license, allowing features that share vertices or boundaries to be edited simultaneously with the Topology Edit tool.

▶ Planar topology may be created for features in a feature dataset in a geodatabase if one has an ArcGIS Standard or Advanced license.

▶ Validation tests features against the assigned rules and automatically repairs some errors using a cluster tolerance. Unfixed errors may be visually inspected and repaired during editing.

## Important Terms

| | | | |
|---|---|---|---|
| cluster tolerance | map topology | planar topology | topology |
| dangle | modify | pseudonode | undershoot |
| edge | node | reshape | validation |
| logical consistency | overshoot | stream digitizing | |

## Chapter Review Questions

1. For each of the six topology rules listed, explain which of the four spatial relationships they deal with: adjacency, connectivity, overlap, intersection.

2. What is meant by the term *logical consistency*?

3. What two approaches are helpful in digitizing complexly related polygons?

4. What is the difference between modifying a feature and reshaping a feature?

5. What is the difference between a union and a merge?

6. What determines the direction of a newly created line?

7. Examine Figure 12.17. In (a), the polygons were created one after the other using Create New Feature. In (b), a clip was performed after digitizing the second polygon. They look identical, but they are not. What is the difference?

8. What is a cluster tolerance? What does it do?

9. What are the differences between using map topology and using planar topology during editing? How are they similar?

10. What is validation?

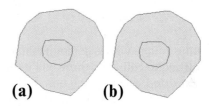

**(a)**     **(b)**

Fig. 12.17. What's the difference?
Source: ArcGIS

# Mastering the Skills

## Teaching Tutorial

The following examples provide step-by-step instructions for doing basic tasks and solving basic problems in ArcGIS. The steps you need to do are highlighted with an arrow →; follow them carefully. Click on the video number in the VideoIndex.htm file to view a demonstration of the steps.

1→ Start ArcMap and open ex_12a.mxd in the mgisdata\MapDocuments folder.

1→ Add the geology layer from the mgisdata\Austin\Edwards geodatabase created in the Chapter 7 tutorial.

1→ Use Save As to rename the document and remember to save frequently as you work.

### Editing complex polygon topology

Digitizing separated buildings as polygons offers no special challenges, but interlocking polygons requires experience and planning to avoid dead ends and ensure proper topology.

2→ Give the geology layer a unique values symbology based on the Unit field.

2→ Change the symbols so that Kkd is bright pink, Kkbn is purple, and Kgru is brown, as they are in the scanned map.

2→ Use the layer properties Display tab to make the geology layer 30% transparent.

2→ Give the geology layer 10 pt. bold labels based on the Unit field.

3→ Choose Start Editing from the Editor toolbar.

3→ Choose the Edwards geodatabase as the workspace to edit.

3→ Open the Attributes window and dock it beneath the Table of Contents.

Source: ArcMap

If the unique values geology map is made before starting to edit, then three edit templates should be created, one for each unit. If they were not created, then create them now by adapting the instructions in Steps 57–58 of Chapter 7.

3→ Zoom in to the pink and purple nested polygons near A in Figure 12.18, so that they fill the screen.

Nested polygons create a challenge because creating them can produce overlaps. These aren't fully nested because they share a boundary, so the Cut Polygons tool will work.

4→ Click the Kkbn template and digitize around the outer purple polygon.

 5→ On the Editor toolbar, click the Cut Polygons tool. The first polygon must still be selected.

Fig. 12.18. Polygon labels for editing
Source: ArcMap

5→ Starting outside of the shared upper boundary (#1 in Figure 12.19), digitize around the pink polygon, ending outside again at #2. Double-click to finish.

Now both polygons are selected.

5→ In the Attributes window, click one of the polygons to flash it. Make sure that the Unit code says Kkbn for the outer polygon, and type *Kkd* for the inner one.

The floating toolbar and solid polygon that may appear during digitizing are annoying. You can turn these off.

6→ Choose Editor > Options and click the General tab.

6→ Uncheck the box to *Use symbolized feature during editing*, and uncheck the box to *Show feature construction toolbar*. Click OK.

Fig. 12.19. Cutting to create an inner polygon
Source: ArcMap

Notice the tiny pink island in the purple polygon. It is almost too small to digitize, but it illustrates an important point. It is entirely enclosed by the purple polygon, so we cannot use the Cut Polygons tool as we did last time. We must create a new feature.

7→ Click the Kkd template and digitize around the pink island polygon. Finish the sketch. The new polygon should be highlighted.

7→ Turn off the EdwardsCrop map image for a moment.

7→ Click the Edit tool, click inside the new polygon, and drag it away from the other polygons. Notice that the purple unit underlies it.

7→ Choose Edit > Undo Move from the main menu bar.

The new feature rests upon the old one like the top layer of a cake, creating an overlap. We must use Clip to delete the overlapping area.

8→ <u>While the new polygon is still selected</u>, choose Editor > Clip from the Editor toolbar.

8→ Make sure that the button is filled to *Discard the area that intersects*, and click OK.

8→ Drag the new polygon away from the others again, noting that there is now a hole.

8→ Choose Edit > Undo Move from the main menu bar.

8→ Turn the image back on again.

8→ Choose Bookmarks > Main Area from the main menu bar.

Although you might be tempted to digitize the next group of nested polygons north of the fault, an experienced editor would realize that it would produce a problem connecting the two groups. You usually want to continue using Auto Complete rather than starting new polygons.

9→ Zoom in to polygons B and C below the nested ones, shown in Figure 12.18.

9→ Click the Kkd template and select the Auto Complete Polygon construction tool from the lower panel of the Create Features window.

9→ Digitize polygon B, taking care to snap to the vertices of the adjacent polygon.

9→ Choose Editor > Save Edits. Also save the map document.

### Stream digitizing and section editing

Point-to-point digitizing gives the best control, but some people prefer stream digitizing. Let's try it on the next polygon.

10→ Choose Editor > Options and click the General tab.

10→ Set the Stream tolerance to 100 map units to start with and click OK.

10→ Click the Kkd template and choose the Auto Complete Polygon tool.

10→ Enter the first vertex of polygon C, snapping to the lower-left corner of polygon B.

10→ Right-click off the sketch to open the Sketch menu. Choose Streaming.

10→ Enter the next vertex to start streaming. Then move the cursor, without clicking, along the boundary of the polygon. Notice how closely spaced the vertices are.

10→ Right-click and choose Delete Sketch.

The initial stream tolerance was too small, producing more vertices than needed and increasing the file size. Let's try a larger tolerance.

1. If you planned to create topology for the geology, what size should the cluster tolerance be relative to the stream tolerance? _____

11→ Use Editor > Options to set the Stream tolerance to 300 map units.

11→ Click on the vertex of polygon B to start. Streaming is still on, so you will begin streaming after that first click. Go all the way around, even if you mess up.

11→ End by snapping to the lower-right vertex of polygon B. Double-click to finish.

Chances are that your polygon has wobbles in it. Stream digitizing takes some practice. Notice how many vertices were entered. They serve an important purpose on the curve edges but are more than needed on straight edges along the faults. You can switch back and forth between stream and point modes during a single sketch using the F8 function key.

12→ Click the Delete key to delete the streamed polygon.

12→ Click on the starting vertex again, and then click the F8 key to turn streaming off.

12→ Digitize the straight edge along the fault, using only the vertices you need.

12→ At the corner where the contact leads away from the fault, pause and press F8.

12→ Click once to start streaming, and then move the cursor along the boundary, pausing when you get to the lower fault.

12→ Press F8 to stop streaming. Enter the straight boundary along the fault.

12→ Press F8 to start streaming again and digitize the contact between the faults.

12→ When you get to the upper fault, press F8 to stop streaming. Digitize the remaining vertices and finish the sketch.

**TIP**: If your computer has no function keys, you must turn streaming on and off by right-clicking to open the Sketch menu. Streaming works wonderfully with a tablet pen instead of a mouse.

Chances are that, even with this new method, you will see spots where your streaming wandered from the actual boundary. This is a good place to learn about the Edit Vertices tool.

13→ Make sure that the new polygon is still selected.

13→ Click the Edit Vertices tool from the Editor menu.

13→ The sketch returns, showing the individual vertices.

13→ Hover over a vertex. The cursor changes to a little box, allowing you to click and drag the vertex to a new location.

13→ To add a vertex, right-click a segment between vertices and choose Insert Vertex from the context menu.

13→ To delete a vertex, right-click it and choose Delete Vertex.

13→ Edit the vertices until you are satisfied. Then right-click on the sketch and choose Finish Sketch.

Streaming is faster but not as easy to control. Let's try another one.

**TIP:** If you forget that streaming is on and mess up the start of a polygon, just right-click and choose Delete Sketch so you can start over.

14→ Pan, if necessary, so that you can see all of purple polygon D.

14→ Click the Kkbn template. Choose the Auto Complete Polygon tool.

14→ Start on the lower corner of the Kkd polygon just finished, on the fault.

14→ Snap the first vertex and enter more vertices along the fault (with streaming off).

14→ Turn streaming on at the end of the fault portion, and stream your way around.

Source: ArcMap

14→ At the end, click the mouse to snap to the corner vertex of polygon B.

**TIP:** When using Auto Complete with streaming, always click once on the snap point at the end to ensure that the polygon actually closes.

Sometimes you stream well around most of the polygon but have a poor section that you would like to fix without having to redo the entire polygon or edit each vertex. The Reshape Feature tool is perfect for this task. Consider the southern end of polygon C. Imagine that you would like to reenter just the portion between the faults.

15→ Click the Edit tool on the Editor toolbar and select polygon C.

15→ Pan/zoom in to the section to be edited so that you can see it clearly.

15→ Choose the Reshape Feature tool from the Editor toolbar.

15→ Click on the upper end of the section, being sure to snap to the vertex of the polygon.

15→ Click F8 to turn streaming off so that you can be careful where you place the section vertices.

It is important to have snapping on to start the new section, because Reshape will not work if you don't start and end exactly on the existing section. However, snapping will probably interfere with where we place the new vertices, as some will probably be close to the old ones.

16→ If the Snapping toolbar is not already open, choose Editor > Snapping > Snapping Toolbar.

16→ Click the toggle switches for vertex and edge snapping to turn them off.

16→ Enter the new section of the polygon, pausing before you place the last vertex (Fig. 12.20).

16→ Click the toggle switch to turn vertex snapping back on.

16→ Enter the last vertex of the new section, snapping it to an existing vertex. Finish the sketch.

17→ Pan to the other end of polygon C that was digitized using streaming.

17→ Reshape this boundary also, for practice.

17→ Save your edits.

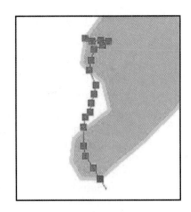

Fig. 12.20. Reshaping a section of polygon C
Source: ArcMap

Both the Edit Vertices and the Reshape Feature tools are handy for editing part of an existing feature. They are useful for fixing problems with streaming but can be used at other times also.

From this point on, you can choose whether you continue to use streaming. You can use it all the time, only for some sections, or not at all, depending on your preference.

### Digitizing by adding territory

We have been using an "adding territory" approach to digitizing using the Auto Complete Polygon tool. We will continue this method for a little longer, even though the polygons are getting more complicated. Since we will be using the Auto Complete Polygon tool nearly all the time, let's change the default construction tool for the templates.

18→ Right-click the Kkd template and choose Properties. Change the Default Tool to Auto Complete <u>Polygon</u> (not Auto Complete Freehand). Click OK.

18→ Change the default tool for the other two templates as well.

18→ Save the map document, so the changes to the templates are saved.

19→ Pan or zoom, as necessary, to see all of polygon E (Fig. 12.21).

Look at polygon E. It is large, so take a moment to follow the boundary around so that you know where to start and stop. We will use the dashed fault on the upper end as a boundary.

19→ Use template Kkd and the Auto Complete Polygon tool to digitize polygon E.

19→ Use Edit Vertices or Reshape Feature, if necessary, to fix any glaring errors.

Fig. 12.21. Complex polygon topology
Source: ArcMap

20→ Pan or zoom to see polygons F and G, as necessary.

20→ Click the Kkd template and digitize the pink polygon F.

2. Why did we choose to do Polygon F next, rather than G? _____

21→ Click the Kkbn template and digitize the purple polygon G. On the top right edge, ignore the fault this time and include the two little triangles with the rest.

22→ Pan/zoom to see polygons H–J.

22→ Click the Kgru template and digitize polygon H. Note that it stops against polygon I at the upper dashed fault, so you don't need to go all the way around on top.

23→ Click the Kkbn template and digitize polygon I. Go all the way around, leaving the pink island of Kkd (polygon J) enclosed in the middle.

24→ Choose the Cut Polygons tool from the Editor toolbar. Start outside, cross the shared boundary, and digitize all the way around polygon J to the outside again.

24→ In the Attributes window, find the inside polygon and change the Unit field to Kkd.

The last ribbon of brown Kgru is so narrow that it would be easier to digitize it with snapping off.

25→ Click the Kgru template and enter the first vertex, snapping to the Kkbn polygon.

25→ Click the toggle switches to turn off vertex and edge snapping.

25→ Digitize the boundary of the Kgru polygon, stopping before the last vertex.

25→ Click the toggle switch on the Snapping toolbar to turn vertex snapping on again.

25→ Enter the last vertex, snapping it to the Kkbn polygon.

We have been treating the faults as geological contacts between polygons. However, some geologists do not want faults to separate polygons with like units. We can use the Merge command to eliminate false boundaries created by faults.

26→ Choose Bookmarks > Main Area from the main menu bar.

26→ Click the Edit tool on the Editor toolbar.

26→ Select the two pink Kkd polygons at the southern end (B and C).

26→ Choose Editor > Merge from the Editor toolbar.

A window appears with both polygons listed. You need to choose one polygon as the host. The host will absorb the other one, and all attributes in the host will become the attributes of the absorbed polygon. In this case they are equivalent, so it does not matter which you choose.

26→ Click one polygon in the list, and then the other. The flash tells which is which.

26→ Click one of them to highlight it, and click OK.

27→ Select the two adjacent purple Kkbn polygons (D and G) and merge them.

28→ Select the two adjacent pink Kkd polygons (E and F) and merge them.

28→ Save your edits.

**TIP:** Depending on your software version, you may see the AutoComplete Freehand tool, a new addition to the Construction Tools set. Experiment with it if you wish.

### Digitizing by divide and conquer

Instead of the "adding territory" approach, some people prefer to "divide and conquer." In this method, we digitize the boundary of an area containing many polygons and then cut it up.

29→ Zoom to the area shown in Figure 12.22.

29→ Click the Kkd template.

29→ Use the Auto Complete Polygon tool to create the polygon highlighted in Figure 12.22, starting at the lower arrow, following the boundary along the faults, and closing it off against the brown Kgru polygon at the upper arrow.

30→ Click the Cut Polygons tool on the Editor toolbar.

30→ Cut out the lowest purple polygon. Be sure to start and end outside of the "conquer" area.

30→ After it is done, use the Attributes window to figure out which of the two selected polygons is the purple one and enter *Kkbn* in the Unit field.

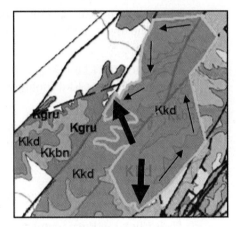

Fig. 12.22. Divide and conquer area
Source: ArcMap

31→ Choose the large pink Kkd polygon next to it as the next piece to carve out. Remember to snap to or start outside of the polygon that is being cut.

31→ Make sure the attributes are correct.

**TIP:** The polygon to be cut must always be selected before you start cutting. You can turn the geology layer off and on if you need to see the colors underneath.

32→ Continue carving up the "conquer" area. There are various ways to go about it. If you need to create extra polygons to get a shape cut out, do it and merge them afterward. Try to find the most efficient way that requires the fewest merges.

**TIP:** If you end up with an island that does not share a boundary, use the Polygon construction tool to create it as a top "layer cake" and then use Clip to remove the overlap.

33→ When you are done with carving, merge any polygons that require it.

34→ Zoom to Bookmarks > Main Area.

34→ Turn off the EdwardsCrop image so that you can see the polygons clearly. They should look similar to the polygons in Figure 12.23.

34→ Save your edits, and save the map document.

### *Using map topology to edit polygons*

So far we have demonstrated some techniques for creating polygons with logically consistent relationships with a minimum of effort—certainly less effort than it takes to fix topology errors later. However, creating polygons with correct topology is only the first part of the battle. One must also be careful that future editing does not introduce new topology errors. For this goal, editing with a map topology provides the answer.

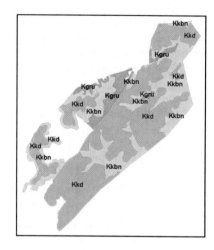

Fig. 12.23. The final polygons
Source: ArcMap

35→ Choose Editor > More Editing Tools > Topology.

The Topology toolbar provides functions for editing topology. First we must create a map topology, a temporary topology created on the fly.

35→ Click the Select Topology tool on the Topology toolbar.

35→ Check the box to have geology participate in the topology.

35→ Expand the Options button so you can examine the Cluster Tolerance. Accept the default value and click OK.

The regular Edit tool selects an entire feature for editing. The Topology Edit tool selects a shared edge or node for editing.

36→ Turn the EdwardsCrop image on.

36→ Zoom in to the nested polygons (H, I, and J) in the central part of the map.

36→ Find an edge between the nested polygons that could have been more accurately digitized.

36→ Zoom in closer to the section to be reshaped.

First we will see what happens when we edit adjacent polygons *incorrectly* using the Edit tool.

37→ Click the Edit tool on the Editor toolbar and select one of the polygons next to the edge to be reshaped.

37→ Click the Reshape Feature tool on the Editor toolbar.

37→ Click to enter the first vertex of the section to be reshaped, snapping to a vertex.

37→ Turn off vertex snapping and digitize the rest of the section. Turn snapping on again to enter the last vertex. Double-click it to finish the sketch.

Chances are that you introduced a visible gap between the two polygons. You may have also created overlaps, although they can't be seen yet.

37→ Click the Edit tool again, hold down the Shift key, and select the other polygon adjacent to the reshaped edge.

369

With both polygons selected, you can clearly see two boundaries: the original boundary of the polygons and the reshaped edge (Fig. 12.24). The Edit tool selected a single polygon feature, and as a result you edited its boundary alone, introducing gaps and overlaps.

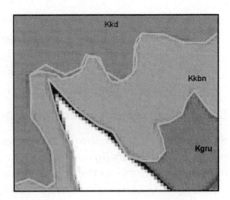

Fig. 12.24. Overlaps and gaps introduced by improper editing
Source: ArcMap

37→ Choose Edit > Undo Reshape from the main menu bar.

Now let's do it properly.

38→ Click the button on the Standard toolbar to clear the selected features.

 38→ Click the Topology Edit tool on the Topology toolbar.

38→ Click on the boundary between the two polygons. It will turn purple instead of blue to indicate that it is a selected shared boundary.

To reshape this time, you will use a different tool. It is similar to the Reshape Feature tool, but it works on edges.

 38→ Click the Reshape Edge tool on the Topology toolbar.

38→ Click the first vertex of the section to be reshaped, as before.

38→ Turn off snapping, enter the rest of the section, and turn snapping on to add and double-click the last vertex.

39→ Click the Edit tool and select both polygons—no gaps and overlaps this time. Clear the selected features.

39→ Find another section that needs reshaping.

39→ Select it with the Topology Edit tool and reshape it with the Reshape Edge tool. Double-click to finish the sketch.

The Modify Edge tool lets you edit individual vertices along a shared boundary.

 40→ With the boundary selected, click the Modify Edge tool on the Topology toolbar.

40→ Edit a few vertices; then right-click on the sketch and choose Finish Sketch.

 41→ Click the Shared Features button on the Topology toolbar. It shows you which polygons share the selected edge.

41→ Close the Shared Features window and click off the polygons to clear the edge selection.

41→ Choose Bookmarks > Main Area.

41→ Save your edits and stop editing. Also save the map document.

### Using map topology to edit lines

Editing roads presents a different set of challenges.

→ Open the ex_12b.mxd map document from the Map Documents folder.

→ Save it with a new name and remember to save often as you work.

42→ Choose Editor > Start Editing.

42→ Zoom to the area outlined in Figure 12.25.

42→ Use the layer properties for Streets to give them 10-pt. green labels using the STREET_NAM field.

42→ Zoom in to BRENDA St.

42→ Turn off all snapping except for end snapping.

Notice that the intersections and centerlines are offset from the center of the street. We will try the <u>wrong</u> way first to demonstrate what happens.

43→ Right-click the Streets layer in the Table of Contents and choose Selection > Make This The Only Selectable Layer.

43→ Click the Edit tool and select BRENDA between MIRANDA and ESTHER.

43→ Click the Edit Vertices tool on the Editor toolbar.

43→ Move the vertex at the intersection of BRENDA and ESTHER to the center of the intersection. Right-click and finish the sketch.

Fig. 12.25. Zoom to these roads.
Source: ArcMap

Now the BRENDA segment is in the centerline, but the three other segments remain where they were. We've lost the connection between BRENDA and the other roads. We would have to edit the other three segments to match them up again. Map topology makes this editing easier.

43→ Choose Edit > Undo Modify Feature from the main menu bar.

43→ Clear the selected features.

44→ Open the Topology toolbar, if necessary, and position it in a convenient spot.

44→ Click the Select Topology button on the Topology toolbar.

44→ Choose Streets and Parcels as the layers to participate, and accept the default Cluster Tolerance under Options. Click OK.

45→ Open Editor > Editor Options. Click the General tab and make sure that the box is checked to *Stretch geometry proportionately when moving a vertex.* Click OK.

45→ Click the Topology Edit tool and click in the intersection of the roads to select the junction. If you get an edge, try again.

45→ Move the Edit tool cursor on top of the junction. The cursor will change to a move symbol. Move the junction to the center of the intersection.

With map topology, all four lines stay attached to the node, which is far easier than editing them separately. We will turn end snapping off so it doesn't snap to the existing location.

45→ Click the toggle to turn off end snapping on the Snapping toolbar.

45→ Select the junction at the intersection of BRENDA and DEBORAH and center it.

45→ Select and center the intersection of BRENDA and HUNTLAND.

45→ Pan west to ISABELLE St and center its junctions as well.

Even with the junctions centered, ISABELLE does not follow the street centerline. Its vertices need editing.

46→ Select the lower edge of ISABELLE with the Topology Edit tool.

46→ Click the Modify Edge tool from the Topology toolbar.

46→ Edit the vertices to center ISABELLE. You will need to insert a few vertices as well as moving existing ones. Right-click and choose Finish Sketch when done.

**TIP:** If nearby vertices are moving too much, use Editor > Options to switch the stretch geometry option off again. Use it as needed for particular tasks.

That worked, but it was a bit tedious. Reshaping can provide a quicker alternative.

47→ Pan/zoom to MIRANDA between ISABELLE and BRENDA. Turn on end snapping.

47→ Select the edge of MIRANDA with the Topology Edit tool.

47→ Click the Reshape Edge tool on the Topology toolbar.

47→ Click on the intersection vertex to start, enter the vertices of the new MIRANDA, and be sure to snap and double-click to the vertex at the other end.

This is a good opportunity to show off the capabilities of another Sketching tool.

48→ Still using the Reshape Edge tool, snap a vertex to one end of MIRANDA.

48→ On the Editor toolbar, click the Sketching tools drop-down button and select the Arc Segment tool.

48→ Click several vertices along MIRANDA and end by snapping to the other vertex and finishing the sketch. Smooth curves will be created between each segment.

Source: ArcMap

49→ Zoom out slightly, if needed, to view the entire block of ESTHER.

49→ Select ESTHER with the Topology Edit tool and reshape it using the Reshape Edge and Arc Segment tools.

49→ Save your edits and your map document, and stop editing.

The next part of the tutorial requires an ArcGIS Standard or Advanced license. If you do not have one, you can stop the tutorial here and go on to the exercises.

### Editing with planar topology (optional)

In this tutorial, we will see how planar topology can help locate and fix topology errors. Imagine that someone has partially digitized the faults and geology of the Edwards aquifer. To create the topology, you must put them in a feature dataset.

50→ Click the New Map File button to start a new, blank map.

50→ Open the Catalog tab and navigate to your mgisdata\Austin folder.

50→ Right-click the Austin geodatabase and choose New > Feature Dataset.

50→ Name it Edwards and click Next.

50→ Import the coordinate system from one of the feature datasets in the Austin geodatabase. Click Next.

50→ Do not specify a vertical coordinate system. Click Next.

50→ Accept the default tolerance and resolution values. Click Finish.

51→ In the Catalog tab, right-click the new Edwards feature dataset and choose Import > Feature Class (Multiple).

51→ Navigate to the mgisdata\Austin\Topology folder and select both shapefiles. Click Add and OK to start importing.

Now we need to build the topology.

52→ In the Catalog tab, right-click the Edwards feature dataset and choose New > Topology. Click Next.

52→ Accept the name, Edwards_Topology, and the default cluster tolerance. Click Next.

52→ Check both boxes so that faultst and geologyt will participate. Click Next.

52→ The geologyt lines are narrower on the image and therefore more accurate, so rank them 1 and the faultst 2. Click Next.

Next we do the rules. We want to ensure that the geology polygons have no gaps or overlaps. The faults should not have pseudonodes unless they join different line types. Faults very close together should meet, so we don't want dangles.

53→ Choose Add Rule. Select the faultst layer and set the first rule as *Must Not Have Pseudo Nodes*. Click OK.

53→ Choose Add Rule again. Select the faultst layer and choose the second rule, *Must Not Have Dangles*. Click OK.

53→ Add a rule that geologyt *Must Not Overlap*.

53→ Add a rule that geologyt *Must Not Have Gaps*.

When finished, the rules window should look like Figure 12.26.

| Feature Class | Rule |
|---|---|
| faultst | Must Not Have Pseudo Nodes |
| faultst | Must Not Have Dangles |
| geologyt | Must Not Overlap |
| geologyt | Must Not Have Gaps |

53→ Click Next and Finish. Click Yes when asked to validate the new topology.

Fig. 12.26. The topology rules
Source: ArcMap

54→ When it is finished, add the Edwards _topology layer to the map. Choose NOT to add the participating feature classes (they are already there).

54→ Choose Editor > Start Editing.

54→ Choose Editor > More Editing Tools > Topology to open the Topology toolbar, if necessary.

55→ Right-click the Edwards_topology layer and choose Properties.

55→ Click the Errors tab and choose Generate Summary. You will see a total of 63 errors listed. Examine them and click OK.

We will work on the fault errors first. We can look only at the errors we are working with.

56→ Turn off the geologyt layer for now.

56→ Right-click the Edwards_Topology layer and choose Properties. Click the Symbology tab.

56→ Uncheck all boxes except Point Errors. Highlight the Point Errors entry, and choose to *Symbolize by error type*.

56→ Right-click the Dangle symbol and select the green color. Choose blue for the Pseudo Node color. Click OK.

The pseudonodes fall into two groups. Some separate a fault with a solid line from one with a dashed line and need to be retained and marked as exceptions. The others connect two faults of the same type and can be eliminated.

57→ Create a unique values map of faultst based on the Linetype field, using dashed and solid symbols for the respective types.

57→ Zoom in to the blue pseudonode in location A as shown in Figure 12.27. This pseudonode connects a dashed and a solid fault and needs to be retained.

58→ Click the Fix Topology Error tool on the Topology toolbar.

58→ Click on the pseudonode. It will turn black, indicating that it is selected.

58→ Right-click on the selected error and choose Mark as Exception.

Fig. 12.27. Editing locations
Source: ArcMap

59→ Zoom to the full extent of the map and then zoom in to the next group of pseudonodes (B).

59→ Click the Fix Topology Error tool. Hold down the Shift key and select each of the four pseudonodes that connect different fault types.

59→ Right-click one of the selected pseudonodes and choose Mark as Exception.

59→ Find the pseudonode that connects two dashed faults at an angle. They are the same type but are different faults, so mark it as an exception also.

59→ Leave the pseudonodes joining similar fault types alone for now.

60→ Continue zooming in to each group of pseudonodes, marking the ones that connect two fault types as exceptions, until all are marked.

60→ Zoom to the full extent when done.

Now that the exceptions are marked, we can automatically fix the others by merging the two like faults on either side. The Error Inspector lets you search for and fix errors of a specific type.

61→ Click the Error Inspector button on the Topology toolbar. It is convenient to dock the window to the lower edge of the ArcMap window.

61→ Set the drop-down to Show: Faultst – Must Not Have Pseudo Nodes.

61→ Uncheck the box for *Visible Extent only*. Click Search Now.

61→ Click the top row. Then scroll down to the bottom, hold the Shift key, and click the bottom row, so that all are selected. Right-click one of the rows and choose Merge to Largest. All pseudonodes should now be gone.

For the dangles, some simply indicate the end of the fault, and these will be exceptions. Others are errors that need to be fixed.

62→ Zoom to the five dangles at the southwestern end of the fault map (C in Figure 12.27).

62→ Click the Fix Topology Error tool.

62→ Click on the three dangles that are just the ends of faults and select them all.

62→ Right-click on one error and examine the menu choices (Fig. 12.28). Then choose Mark as Exception.

62→ Zoom to the remaining two dangles in this view.

63→ Use the Fix Topology Error tool to select the lower of the dangles that appears to need snapping to the adjacent fault.

63→ Right-click it and choose Snap. Type 500 into the box and click Enter.

Fig. 12.28. Suggested fixes for dangles
Source: ArcMap

The other one might be the end of a fault, or it might need to be connected. We should look at the original map to be sure.

64→ Add the EdwardsCrop.png image from the Austin folder.

64→ We should connect the fault. Turn off the image until the next time we need it.

64→ Select the dangle with the Fix Topology Error tool and right-click it. Choose Extend and enter 1000 in the box that appears. Click OK.

65→ Zoom in to the next group of five dangles (D in Figure 12.27). These appear to be exceptions.

65→ Click and drag with the Fix Topology Error tool to make a rectangle to select all five. Unfortunately, more than dangles are selected.

We can set the selectable errors to ensure that we get only what we want.

66→ Click away from the selected errors to clear the selection.

66→ Right-click the Edwards_Topology layer and choose Properties.

66→ Click the Selection tab. Uncheck all the boxes except for *Select Errors* and *Must Not Have Dangles*. Click OK.

66→ Select the dangles with the Fix Topology Error tool rectangle again. This time only the five dangles should be selected.

66→ Right-click one and choose Mark as Exception.

67→ Continue the process with the rest of the dangles, using one of the error fixes or marking them as exceptions. Consult the image if you are not sure what to do.

**TIP:** If a Snap or Extend does not work, try entering a larger tolerance in the box. If two dangles need to be snapped together, select both and snap instead of trying to do them one at a time.

68→ Save your edits and zoom to the extent of geologyt.

68→ Turn geologyt back on and make it 50% transparent.

68→ Open the Properties for Edwards_Topology again and click the Symbology tab.

68→ Check the boxes to display the area and line errors as well as the point errors.

68→ Highlight the Area Errors entry and choose to *Symbolize by error type.*

68→ Click the Selection tab. Turn on the checkboxes for *Must Not Overlap* and *Must Not Have Gaps.* Click OK.

69→ Use the Fix Topology Error tool to select the pink line that goes all the way around one set of polygons. It will turn black to indicate that it is selected.

69→ A glance at the Error Inspector panel reveals the type of error.

The rule is Must Not Have Gaps. The gaps occur because the outer polygons are not touching other polygons. This arrangement can't be avoided—the map has to end somewhere. So this outer boundary, and the one around the other polygon group, must be marked as exceptions.

69→ Right-click the selected black boundary and choose Mark as Exception.

69→ Select the other outer boundary and mark it as an exception also.

Now we can concentrate on the smaller errors.

70→ In the Error Inspector, set the Show box to geologyt – Must Not Overlap and click Search Now. Five errors appear.

70→ Right-click the first entry and click Zoom To. The selected feature is highlighted with a black boundary.

This one appears to be an island error, where the digitizer forgot to clip. The Create Feature fix will create a new polygon and remove the overlap.

Source: ArcMap

70→ Right-click the error and choose Create Feature. It is fixed and disappears from the Error Inspector.

70→ In the Error Inspector window, right-click the top entry and choose Zoom To.

70→ Right-click the error and examine the menu choices for fixing it.

Source: ArcMap

Subtract will eliminate the overlapping area entirely. Create New Feature will add another one on top. The proper solution is to merge the overlap with one polygon or the other.

70→ Turn on the image to see which edge best represents the boundary. Zoom out a little if you need to see more to decide.

70→ Right-click the error and choose Merge. Flash the two polygons in the Merge window and choose to merge the sliver with the one that looks like the best choice.

71→ Continue working in the Error Inspector until all of the Overlap errors are merged.

72→ Change the Show box in the Error Inspector to geologyt – Must Not Have Gaps.

72→ Uncheck the box to search in the *Visible Extent only*.

72→ Click Search Now. You will find four errors.

Gaps have no polygon covering them. The usual fix is to create a new polygon to fill in the space and then merge it with one of the adjacent polygons.

73→ Right-click the first error and zoom to it.

73→ Right-click the error again and examine the choices. Then choose Create Feature to fill in the space.

73→ Use the Edit tool to select the filled space and one of the adjacent polygons. Choose Editor > Merge.

74→ Fix the other three gap errors using the Error Inspector. You may need to click Search Now again in the Error Inspector window after using the Edit tool.

75→ Zoom to the full extent of the map and verify that all errors are gone.

75→ Save your edits.

75→ Click the Validate Topology in Current Extent button from the Topology toolbar.

A few new point errors appear, which often happens after validating. You are not really done with topology until the validation step produces no new errors. You can fix the new errors now if you wish, or do them later. It may take several iterations.

This is the end of the tutorial.

→ Save your edits and stop editing. Save the map document.

## Exercises

1. The Travis–Hays county line cuts the Edwards geology map approximately in half. Digitize all of the geology south of the county line. Capture a unique values map of the polygons.

2. Finish adjusting the streets in the streets feature class to match the aerial photo.

3. Create a layout showing your adjusted roads, with the aerial photography in the background, and create a PDF map from it.

Exercises 4–5 require an ArcGIS Standard or ArcGIS Advanced license.

4. If you have a Standard or Advanced license, construct planar topology for the geology and faults feature classes that you digitized, using the same rules from the tutorial. Examine the error summary (as in Step 55) and capture it as a screenshot. Write a critique of your editing ability. How many errors of each kind did you have? Why are they present? Do you need to modify your editing strategies?

5. Validate the topology and fix any errors that you find. Capture the error summary again after you fix the errors to verify that they have all been fixed.

## Challenge Problem: Isn't That Interesting!

Find a map image on the Internet showing an interesting topic that would make a good addition to your home state geodatabase (or another project you are working on). Georeference the image if necessary, create one to three feature classes, and digitize features from the map. Remember to create an Item Description with citations for the georeferenced image and each feature class. Capture a map showing the image and your digitized features.

# Chapter 13. Extending Geodatabases

## Objectives

▶ Understanding the geodatabase model

▶ Developing a design for a geodatabase

▶ Creating geodatabases and feature datasets; importing data

▶ Setting up and using attribute domains and split/merge policies

▶ Creating subtypes and using them during editing

▶ Creating and editing annotation stored in geodatabases

# Mastering the Concepts

## About ArcGIS

As we have learned, the ArcGIS software has a long history of development, and coverages were the first data model used. Later, shapefiles were developed for use in the ArcView package. The geodatabase model arrived with the release of ArcGIS 8. The new model offers several advantages over both coverages and shapefiles. Like coverages, it can store topological relationships between features for added functionality in modeling real-world phenomena, but the geodatabase is simpler in construction and more robust in general usage. Like shapefiles, the features of a geodatabase are simply constructed and difficult to corrupt, but they provide many more benefits than shapefiles. Another advantage of geodatabases over shapefiles is the automatic tracking of lengths and areas for features. Geodatabases also provide behavior and validation rules that control how features behave, how they can connect, and what attributes are allowed. A user can also set up planar topology for features and construct topological rules, such as specifying that a state boundary must always match county boundaries or that counties must never cross a state boundary. We've used feature classes extensively in this text, but this chapter describes the feature dataset and the geodatabase capabilities available to users with an ArcGIS Basic license. More advanced and powerful features are available to those with an ArcGIS Standard or Advanced license; some of these are mentioned as well.

### About geodatabases

The geodatabase model is constructed on the architecture of standard database systems and can be implemented with several commercially available database software packages. Three types of geodatabases may be created.

*Personal geodatabases* are designed for use by individuals or small workgroups and are stored in a single Microsoft Access file. This file is limited to 2 GB in size and works only in the Windows operating system.

*File geodatabases* are also designed for individuals and small groups, but each data set is stored as a separate file within a system folder, and each file can be up to 1 terabyte (TB) in size. File geodatabases can be accessed by different operating systems and are best for cross-platform operations, such as a company that supports Windows, Linux, and Unix computers.

*Enterprise geodatabases*, once known as *SDE (Spatial Data Engine) geodatabases*, store GIS data within one of several commercial relational database management systems (RDBMS), such as Oracle®, SQLServer®, and IBM Informix®. Enterprise geodatabases are designed to meet security and management needs for large data sets accessed by many simultaneous users. For example, at least three groups of users might need to edit parcel records in a large city database: the tax department, the city surveyors, and the deeds office. Instead of maintaining three sets of data, an enterprise geodatabase allows users to "check out" certain portions of data for editing and merges the changes back into the central database. Potential conflicts caused by simultaneous editing by different users are avoided. An additional software package called ArcGIS Portal (previously named ArcGIS for Server) is required to translate geographic data into a consistent and customized format in the large database and share it with large groups.

Enterprise databases have distinct advantages when the spatial data sets are large and/or are accessed by many people. For example, such a system might be developed for the planning department of a large city, which may have dozens of employees in several departments needing access to the information, both for reading and for editing. An enterprise setup allows many people simultaneous access to data sets, provides security and permission handling, and manages cases of multiple simultaneous editing of the data. It also makes it possible to set up version tracking for added data security and protection.

An additional advantage of ArcGIS Portal is that it allows the administrator to share geographic data and geoprocessing tools over the Internet. It can publish data services like those found in ArcGIS Online and make data or geoprocessing tools available through web pages or mobile devices. It is not easy or cheap to set up an enterprise geodatabase, but it does offer cost benefits. For example, it may be that 100% of the employees in an organization need to use the GIS data, but 80% of them could perform their tasks using a web interface with a few simple tools. Offering this capability on a web page means that the organization can purchase fewer ArcGIS Desktop licenses. Many people are discovering productivity improves by moving some of their work to tablets or smartphones; now this option is available for GIS data and tools as well.

However, the additional work in setting up an enterprise geodatabase only makes sense when these issues are present. In this chapter, we will work exclusively with file geodatabases, but most of the commands and functions are the same for the enterprise version.

As we learned in Chapter 1, a geodatabase may contain a variety of objects (Fig. 13.1). Feature classes are used to store the spatial information, and we've used them extensively in this book already. Geodatabases may also contain rasters, tables, layers, relationships, geometric networks, and feature datasets. A feature dataset is a collection of related feature classes with the same coordinate system. The oregondata database in Figure 13.1 contains two feature datasets: Transportation and Water. The Water feature dataset has been expanded to show the feature classes, including rivers and waterbodies (lakes). The geodatabase also contains standalone feature classes such as cities and counties, as well as rasters like gtopo1km. In this case, the standalone feature classes and rasters share the same coordinate system, but they do not have to.

Fig. 13.1. A geodatabase
Source: ArcGIS

Geodatabases must be created as empty shells to which feature classes and other database objects are later added. Feature classes may be loaded from a variety of sources. Coverages and shapefiles may be imported into the geodatabase, either as standalone feature classes or as part of a feature dataset. INFO and dBase tables may be loaded. Users may also create empty feature classes and use editing to add features, as we did in Chapters 7 and 12. Rasters, tables, annotation, and relationships can also be stored inside a geodatabase. Sometimes new feature classes are added to a geodatabase as a result of the operation of a tool or command.

Many databases are developed by importing shapefiles or coverages of existing data. This process is usually straightforward. The entire data set, including features, attribute tables, and coordinate systems, is transferred into the database. In the case of coverages, the attribute fields must be converted from the coverage field types to the geodatabase field types. This conversion is handled automatically, although the user can customize the interpretation of certain fields if desired. The user can also request a coordinate system reprojection as part of the transfer process.

## Designing and creating geodatabases

Good databases do not happen by accident but are the result of careful planning. A project database is often accumulated on the fly with no terrible repercussions, but a long-term repository of data for an organization benefits from forethought. The designer needs to list the feature classes to be included and the attributes that they will have. She must decide whether feature datasets are needed and how they will be organized, and whether any of them will have topology or networks. The designer must decide on coordinate systems, data formats, and even what individual feature classes and files will be named.

It is even important to think through the contents of various attribute fields and decide on content. Will street names be all uppercase or a mix of upper- and lowercase? How many different categories of buildings are needed? At every step, decisions must be enforced to ensure that fields and names are consistent and easily understood.

Developing a written plan is a must for a good geodatabase. Writing the plan facilitates the organization itself, and the written plan serves as a guide to the creator, who may have forgotten a particular design decision several months later and needs to look it up, as well as to other workers responsible for implementing the design.

Planning may be done as the task of a single individual or by an entire committee. The most important task is to first develop a clear vision of the purpose of the database: what it should be able to do and who is going to use it. There is no one way to design a database, but the general sequence might follow a process similar to the one listed below. As you perform each step, you must continually keep the purpose of the database in mind and ask how each piece will support that purpose. As you work through the plan, type it into a document that will serve both as your planning guide and as a source of documentation once the geodatabase is assembled.

▶ Define the overall purpose of the geodatabase and its intended users. Determine what type of database it will be (personal, file, enterprise), where it will be stored, and whether all the equipment and software needed to implement it are available.

▶ Talk to the intended users, if they are available. Ask them what they expect to be able to do, what they must be able to do, and what they would like to be able to do. Then make a list of the key tasks that the database should support.

▶ Define the geographic characteristics. Determine the geographic extent of the area of interest and the scale at which most users will spend their time. Decide on a suitable coordinate system and projection that will support the intended uses.

▶ Make a list of feature classes and rasters that are required, and notate them as critical or supporting. Start a preliminary organizational diagram, showing the feature datasets the geodatabase will have, if any, and how the feature classes will be organized within them. Remember that feature classes must be in the same feature dataset to participate in a network or planar topology.

▶ For *each* feature class and raster, define the specific uses it will support and the required scale or resolution it needs. Make a list of the attribute fields that are required, including the field types and properties. If **subtypes** are to be used, list them and create a table showing the default values for each subtype. Define whether a planar or network topology is required and which feature classes will participate.

▶ For *each* feature class, specify where it is to come from. If you have specific sources in mind, list them with citations. If they will be developed from scratch, indicate how (digitizing from a scanned map, collecting using GPS, etc.).

▶ Create a list of **domains** to be created for the geodatabase, including the field types and lists of all items in coded domains and the upper and lower bounds of range domains.

▶ Decide what type and level of metadata will be produced and to which standard it will adhere. Decide on a process to ensure that the necessary data for producing the metadata will be collected and preserved during the data assembly phase.

▶ If it is a long-term project, develop priorities for which parts of the geodatabase must be developed first, and which can wait. Develop a time-line and project goals with completion dates.

▶ Once the plan is fleshed out, then the assembly work can begin. However, no database plan is ever a static, finished thing. As the database evolves, the plan can evolve as well, and it should be updated accordingly.

**TIP:** Designing and creating good geodatabases involve many issues far beyond the scope of this book. Users planning to work with geodatabases should read more about them.

The design and structure of a geodatabase, including the datasets and feature classes, the fields in the tables, the standalone tables, the relationship, and other objects, are together termed the database **schema**. The description of all the objects in a geodatabase can be stored and saved without any actual data and then reused to generate multiple geodatabases with the same structure. This approach offers many advantages for developing large, complex databases with a well-designed and tested schema. A city planning department that is relatively new to GIS might choose to use a published schema and populate the database with the local information rather than trying to design the geodatabase from scratch. Schemas for various industries and applications may be downloaded free from the ESRI web site.

However, using a published schema does not excuse you from planning. A downloaded schema may have many more feature classes and capabilities than you can afford to implement. Your major tasks will include becoming familiar with the contents of the schema, determining how the data will be assembled, and then deciding which portions of the schema will be developed, and whether unused portions will be deleted or saved for potential future use.

### Organizing feature classes

An inexperienced database developer might have difficulty at first in deciding how to organize feature classes and what each should contain. Faced with a set of objects to be stored, such as interstates, highways, roads, trails, and sidewalks, it might not immediately be obvious whether these objects should be stored in five feature classes, one feature class, or something in between. Moreover, the optimum design depends strongly on the specific uses to which the data are put. An organizational scheme might work very well for one task, but be clunky and difficult for another. The designer must take into account the anticipated use of the data and develop a schema that will support the most important tasks.

In general, it is better to keep similar features together in a single feature class, provided that all the attributes can be assembled into a single table of fields. For some reason, beginners seem to proliferate feature classes, for example, creating a geology map with separate feature classes for each geological unit or each quadrangle. This strategy of separation is usually unnecessary. Queries make it easy to separate a subset of features from a feature class when it is needed. If one wants a map showing only the volcanic rocks, a definition query or a layer selection can accommodate that wish. The Select By Location or Clip functions solve the issue of focusing on a smaller region.

When features are stored separately, it is a much harder task to integrate them. It is a nuisance, for example, to have to load and symbolize 20 feature classes of different geological units to see a single geological map. The Append and Merge tools are valuable but tend to have unsatisfactory results unless all the inputs have identical attribute tables; otherwise, attributes may be lost when bringing the classes together. It can be tedious work to manually reorganize the data tables so that the correct information is ported to the merged table. In addition, gaps and overlaps may be generated between adjacent units and will take time and effort to eliminate.

In short, it is easy to extract subsets of features when they are all stored together, but it is problematic and time consuming to combine separated features. When in doubt, keep them together and separate them later if needed.

### *Creating feature datasets*

A feature dataset is a collection of feature classes that are related to each other in some way (Fig. 13.2). For example, a transportation feature dataset might contain feature classes for roads, railways, airways, and interstates. These features could be used to construct a geometric network so that the movement of people or goods along the network could be modeled. Similar models could be constructed for utilities by using feature classes representing water mains, water sources, sewer lines, and treatment plants. Feature datasets may also contain planar topological relationships between the feature classes, allowing for easier identification and correction of errors, such as gaps or overlaps between polygons. (See Chapter 12 for a description of planar topology.)

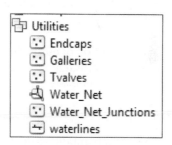

Fig. 13.2. A feature dataset contains related layers with a common coordinate system.
Source: ArcGIS

#### About the spatial reference

The feature classes in a feature dataset must all share the same spatial reference, which includes the coordinate system, the X/Y domain, and the resolution. The coordinate system was discussed in Chapter 3.

The **X/Y domain** is the range of allowable $x$-$y$ values that can be stored in a feature class. Integers are used because the computer can process them several orders of magnitude faster than floating-point values; in the case of enterprise databases, they can also be compressed for more efficient use of storage space. The software converts the integers to real numbers on the fly when it needs them (e.g., when reporting the current $x$-$y$ cursor location in ArcMap).

The **resolution** refers to an underlying coordinate grid to which values are snapped. The units are the same as those for the defined coordinate system. A resolution of 0.001 meter in a projected coordinate system stores values to the nearest thousandth of a meter. A resolution of 0.000001 in a GCS stores values to the nearest millionth of a degree.

The user chooses the coordinate system for a feature dataset, and the software suggests appropriate values for the domain and resolution. The default resolution and domain are determined based on the coordinate system chosen by the user and will be fine for most applications. Users requiring very high resolution, for example, in laying out a surveying grid, can increase the resolution at the expense of a smaller domain.

## Default values

One advantage of using geodatabases lies in setting up default values for attributes. For example, if a user is digitizing roads, local residential streets (LOC) are usually the most common type and typically have two lanes and a speed limit of 25 mph. The user can set default values for each of these fields (Fig. 13.3). Then when each road is added, the road type, lanes, and speed limit are automatically set to LOC, 2, and 25, respectively. The information needs editing only if the new road differs from the default. A user could even set the defaults one way, digitize all the local roads, and then close the layer, change the defaults, open the layer again, and digitize the next group. This approach saves editing time and helps reduce attribute errors.

| Field Properties | |
|---|---|
| Alias | ROADTYPE |
| Allow NULL values | Yes |
| Default Value | LOC |
| Domain | RoadType |
| Length | 3 |

Fig. 13.3. Default values for attributes save time during data entry.
Source: ArcGIS

The introduction of editing templates in ArcGIS 10 provides another way to work with default values. Different templates can be created for each type of feature, with different default values stored in the template. This technique was introduced in Chapter 7.

## Domains

One great advantage of geodatabases is their tools for helping maintain correct attributes. An attribute **domain** constrains the values that may be entered for a particular attribute. Such constraints help ensure consistent data entry and keep out incorrect values. For example, imagine an attribute field named PipeDiam that contains the diameters of water pipes. Water pipes don't come in an infinite variety of sizes; in fact, a town might use only 1-inch, 3-inch, 6-inch, and 12-inch pipes. Setting up a domain for the PipeDiam field provides a "pick list" of choices for the field and prevents any user from mistakenly entering a 4-inch or 5-inch pipe. These constraints can enforce data entry rules in situations when more than one person in situations when many people are entering data and not everyone knows or can remember all the rules. The pick lists also speed data entry, whether it is on a desktop computer, in the field on a GPS unit, or in a data collection application on a smartphone.

Domains come in two types. The first type is a **range domain**, which specifies the lowest and highest possible values but lets the values take any number in between. Range domains are applicable only to numeric data. For example, a range domain could be applied to a student Grade

Point Average (GPA) field. A GPA ranges between 0 and 4.0. A range domain would allow any value between 0 and 4.0 but would exclude negative values or values greater than 4.0. The second type is a **coded domain**, which allows only certain values taken from a list. The PipeDiam domain mentioned previously is a coded domain because it allows only four specific values from a list: 1, 3, 6, or 12.

Domains are created and maintained in ArcCatalog as a property of the geodatabase rather than as a property of a single feature class or attribute field. Therefore, a domain can be used more than once, and it can be used in multiple feature classes. For example, a range domain called Percent, which allows values from 0 to 100, could be used in many feature classes whenever percentage values were being stored in an attribute.

Figure 13.4 shows how the properties of a percentage domain would be set. The field type is specified as short integer. The domain type is range, and the minimum value of 0 and the maximum value of 100 are entered. A domain can also have split and merge policies assigned to it, which we will cover in the next section.

| Domain Properties: | |
|---|---|
| Field Type | Short Integer |
| Domain Type | Range |
| Minimum value | 0 |
| Maximum value | 100 |
| Split policy | Default Value |
| Merge policy | Default Value |

Fig. 13.4. Properties of a range domain
Source: ArcGIS

The field type of the domain must always match the field type of the attribute to which it will be assigned. (See Table 6.1 if you need a review of field types in ArcGIS.) Thus, a short integer domain can be used only with a short integer attribute field and not with a long integer or float field. Users must pay careful attention to the fields for which the domain is intended in order to select the appropriate field type for the domain. If a database had some fields that stored decimal percentages and others that stored integer percentages, then two different Percent domains would need to be created, one with a Float field type and one with a Short field type.

An example of the PipeDiam domain is shown in Figure 13.5. The field type is short integer, and the domain type is set to coded values. The codes themselves are typed in below. The code is the value actually stored in the attribute field. The description appears in legends and tables during an ArcMap session or in a data collection application, providing easily understood information to the people working with the attributes. However, the descriptions do need to fit well inside the menus and legends, so it is best to keep them short.

| Domain Properties | |
|---|---|
| Field Type | Short Integer |
| Domain Type | Coded Values |
| Split policy | Duplicate |
| Merge policy | Default Value |

Coded Values:

| Code | Description |
|---|---|
| 1 | 1-in |
| 3 | 3-in |
| 6 | 6-in |
| 12 | 12-in |

Fig. 13.5. Properties of a coded domain
Source: ArcGIS

Coded domains provide a valuable way to combine the ease of numeric codes with the luxury of understandable text information for people. Land use planning, for example, often uses numeric codes to indicate various zoning types, such as 32 = Residential and 14 = Commercial. Numeric codes are advantageous because they are less susceptible to typing errors and take less space to store. However, it is difficult for people to remember the meanings of many codes. With a coded domain, the zoning codes may be stored as integers, but the information presented to the user takes the form of the textual description (Fig. 13.6).

### Split and merge policies

Domains offer a way to control the updating of attributes during a split or a merge operation. The default policy of making copies or entering blanks may be overridden by alternate and more useful actions. Split and merge policies are associated with each attribute domain. Setting up these policies correctly can save time during editing.

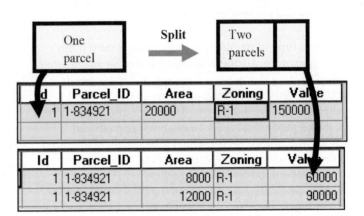

Fig. 13.6. Using domains to link numeric codes to understandable descriptions
Source: ArcGIS

#### The split policy

The **split policy** is applied when a feature is split into two. The policy choices include Default, Duplicate, and Geometry Ratio. If Default is chosen, both split feature attributes will contain the default value for that field. If Duplicate is used, then both new features retain the value of the original feature. If Geometry Ratio is used, then the new value is assigned based on the relative size of the original and split features.

Examine an example of splitting a parcel (Fig. 13.7). The Parcel_ID field is nominal data, so it is not practical to assign a domain. Instead, the normal rules for splitting and merging features without domains would apply. At least one of the new parcels will need a new number.

The Area and Value fields use the Geometry Ratio policy. The proportion of areas in the two new polygons is 60-40; thus the smaller parcel is assigned 40% of the original area and value, and the larger parcel is assigned 60% of the original area and value. Finally, the Default policy is applied to the Zoning field, again assigning the most common value of R-1.

Fig. 13.7. The attributes of two parcels created by splitting one parcel according to the split policy
Source: ArcGIS

In deciding whether to use split and merge policies, one must consider the number of attributes involved and the amount of editing that a data layer receives. Setting up and applying domains to attributes takes some time and thought. Unless the layers are being frequently edited, the effort may not be worth it. When appropriately used, however, domains can save a substantial amount of time and prevent many incorrect data entries.

**TIP:** A user can cause a field to receive a blank during a split or merge by setting the policy to Default and leaving the default value blank.

#### The merge policy

A **merge policy** in ArcGIS is not evaluated during editing. It exists so that programmers writing advanced applications can take advantage of them, but they will not help the average user. Thus, only a brief sketch of merging policies is given.

By default in coverages or in shapefiles, the merged parcel will contain the attributes of the first feature found in the database. In geodatabases, it will contain the attributes of the first feature selected before merging. The merge policy overrides this haphazard method by specifying a different action for each field.

Fig. 13.8. Attributes from two parcels are updated according to the merge policy when they are combined into one.
Source: ArcGIS

Three different policies may be assigned: Default, Weighted Average, or Sum. Default works the same way as in the split policy. Weighted Average assigns the new value based on the relative areas of the two features, and Sum assigns the sum of the two input features. In Figure 13.8, the Zoning field was assigned the Default policy. The Area and Value fields would use the Sum policy.

### Subtypes

**Subtypes** are another way to facilitate the entry and validation of attribute data. Normally, all features within a feature class have the same properties and behave the same way. Subtypes allow features within the same feature dataset to have different properties (such as separate default values) or behave differently (such as having varied ranks when validating a planar topology). A subtype must be based on categorical data, and each feature in the class must belong to one of the categories. Roads, for example, may belong to one of several types, such as interstates, highways, connectors, or local roads (Fig. 13.9). On the surface, subtypes are not that different from a categorical data field. However, subtypes have additional requirements and capabilities.

Fig. 13.9. Subtypes are formal classes of features within a layer.
Source: ArcGIS

A subtype must be based on a short integer field that contains a numeric code representing each category (1 for interstates, 2 for highways, etc.). Like a domain, the subtype also has a text description that lets the user see what the code stands for in plain words. The subtype essentially defines its own coded domain for the feature class.

Subtypes provide more functionality for a feature class than a simple categorical field with a domain, however. You can think of them as subfeature classes within a feature class. First, when a data layer is loaded into a map, the subtypes are automatically displayed with different symbols. When edit templates are created, each subtype has its own template created by default (Fig. 13.10).

Fig. 13.10. Subtype editing templates
Source: ArcGIS

One of the most advantageous properties of subtypes is that each subtype can have its own default values for attribute fields. In the case of the roads, each road type has characteristic values that are different from those of the other types. A ramp typically has one lane, a local street has two lanes, a connector has four lanes, and an interstate has four lanes. Speed limits can also be assigned for each type, such as 25 for local streets, 35 for connectors, 65 for highways, and 75 for interstates. When a subtype is established, different default values can be specified for each of the types.

The defaults are automatically entered in the attribute table when new features are created. To add local streets, the user would select the Local template and begin adding streets. As they were digitized, each street would immediately receive the Local default values: two lanes, 25 mph speed limit, and so on. To enter a connector, the user would select the Connector template, and subsequently added streets would receive the default values for connectors. Subtypes streamline both data entry and the assignment of attributes to features. Occasionally, features may deviate from the default values, such as a connector having six lanes, but such exceptions may be corrected during editing.

Subtypes are an intermediate alternative between placing different groups of features into different feature classes and keeping them in one feature class with a field that distinguishes them. Because subtypes require effort to create and manage, one must consider carefully the design question of whether to use them.

Consider the roads. If one plans to do any network analysis, it would clearly be a mistake to save the different road types in separate feature classes. However, the different types have distinct differences in other attributes (lanes, speed limits) that make it attractive to be able to assign different sets of default values to each road type. Roads are a good candidate for subtypes.

Next consider census boundary designations: blocks, block-groups, tracts, and so on. These certainly seem like different categories of census polygons. However, they cover each other in overlapping sets, like a layer of jigsaw puzzles with ever larger pieces. Trying to map or display them together in the same feature class might be a mess. Census areas definitely need to be in separate feature classes.

Finally, consider vegetation polygons. Vegetation type is certainly categorical data, but there are so many different types of vegetation, generally, that one would need dozens of different categories, making it very tedious to construct the subtypes. Moreover, it is unlikely that other fields in the feature class would have values dependent on the vegetation type alone, so that there is little opportunity to make use of defaults. Vegetation is not a good candidate for subtypes.

Before commiting to using subtypes, make sure that they have a purpose and will actually aid in the construction or use of the geodatabase. Novices tend to overuse them, so always stop and consider whether a single feature class with some categories and a domain would not accomplish as much or more for the intended use of the geodatabase.

### Geodatabase annotation

In Chapter 5, we learned about dynamic labels and creating annotation that exists as simple text within a map document. **Annotation** can also be stored in a geodatabase, in which case it can be used in multiple map documents. Annotation feature classes must be modified using the Editor.

Annotation is another type of feature class, like a line or a point feature class, except that it stores labels. Like any other feature class, it has an attribute table, and the table stores properties of the annotation, such as the text string, the font, the size, and the formatting (Fig. 13.11).

| roadsAnno | | | | | | | |
|---|---|---|---|---|---|---|---|
| | Status * | TextString | FontName | FontSize | Bold | Italic | Underline | VerticalAlign |
| | Placed | IVY AV | Arial | 8 | No | N | | |
| | Placed | BIRCH AV | Arial | 8 | No | N | | |
| | Placed | MICHIGAN AV | Arial | 8 | No | N | | |
| | Placed | HAWTHORNE AV | Arial | 8 | No | N | | |
| | Placed | E IOWA ST | Arial | 8 | No | N | | |
| | Placed | HOEFER AV | Arial | 8 | No | N | | |
| | Placed | IDAHO ST | Arial | 8 | No | N | | |

Fig. 13.11. Annotation feature class in a geodatabase
Source: ArcGIS

Annotation may be created as standard annotation, where each text string is a standalone feature not connected to anything else. Alternatively, it may be created as feature-linked annotation, in which the text string is linked to the feature that it belongs to. With feature-linked annotation, deleting the feature deletes the annotation, and vice versa. Feature-linked annotation requires an ArcGIS Standard or Advanced license, but standard annotion can be edited with ArcGIS Basic.

Creating annotation often begins by creating dynamic labels with the desired font size and other properties. When the labels are converted to annotation, they are stored in a geodatabase. Labels that overlap are stored in the attribute table as unplaced annotation, and they can be accessed and placed at another time. Annotation can also be manually created one label at a time.

After the labels are created, editing can modify the location, style, and font attributes for each label. Editing annotation is not that different from editing other types of features. Annotation has editing templates and a variety of construction tools (Fig. 13.12). Annotation can be placed horizontally or on a straight line at an angle. It can be placed along a curve or set to follow along a feature.

Fig. 13.12. Annotation construction tools
Source: ArcGIS

Annotation is always given a **reference scale**, which is the scale at which the labels appear at their assigned size. The user specifies the reference scale at the time that the annotation is created. If the reference scale is 1:24,000, for example, and the annotation size is 10 pt., then it only appears as 10-pt. text when the map is shown at 1:24,000. If the user zooms in, the labels will increase in size; if the user zooms out, the labels will decrease in size. Before creating the annotation, users should view the map at the desired reference scale to ensure that the font and size selected for the labels are acceptable.

## Summary

▶ A geodatabase is the native data model developed for ArcGIS. It uses commercially available standard database formats and technology.

▶ Geodatabases offer significant advantages over shapefiles and coverages, including simpler and more robust implementation, the storing of topology, geometric networks, and behavior and validation rules.

▶ Geodatabases are created as empty containers and then filled with feature classes, feature datasets, layers, tables, relationships, and other objects.

▶ Designing a geodatabase must start with the purpose and intended users, and should evolve into a written plan for future reference.

▶ Feature datasets contain related feature classes with the same spatial reference. The X/Y domain specifies the range of $x$-$y$ coordinates allowed in the dataset, and the resolution specifies the underlying grid to which features are snapped.

▶ Attribute domains define allowed values for attributes in geodatabases. Attribute domains may cover ranges of numeric values, or they may have specific coded text or numeric values.

▶ Domains also have split and merge policies, which define the assignment of attributes when features are merged or split.

▶ Subtypes allow features in a layer to be classified into categorical groups. Each group may have its own set of default attribute values.

▶ Annotation stored in a geodatabase is available for use in many map documents. Standard annotation is simply a type of feature, such as a point or a polygon. Feature-linked annotation is connected to its feature and is deleted when its feature is deleted, or vice versa.

### Important Terms

| | | | |
|---|---|---|---|
| annotation | merge policy | resolution | subtypes |
| coded domain | range domain | schema | X/Y domain |
| domain | reference scale | split policy | |

## Chapter Review Questions

You may need to consult the Skills Reference section to answer some of these questions.

1. List three advantages of the geodatabase model.

2. What is the function of a geodatabase schema?

3. How does a personal geodatabase differ from a file geodatabase and an enterprise geodatabase?

4. What is the difference between a feature class and a feature dataset?

5. Why is a resolution of 0.001 appropriate for a parcels feature class stored in UTM but not for one stored in a GCS?

6. Can you have a geodatabase that contains one feature dataset in UTM Zone 13 and another feature dataset in South Dakota State Plane? Is it a good practice to do so?

7. You have a forest feature class with polygons showing individual stands of trees. It contains the three attribute fields: TreeSpecies, CanopyCover, and Acres. TreeSpecies contains the dominant type of tree (e.g., ponderosa pine, aspen). CanopyCover gives the percentage of the stand covered by tree crowns. The Acres field contains the area of the stand. For each of these attributes, list the most appropriate split policy and merge policy.

8. For each of the following zoning attributes: Zoning, StreetAddress, Value, and PercentImperviousArea, state whether it would be suitable for a domain. If yes, then say whether a coded or range domain would be more appropriate. In each case, explain your reasoning.

9. Domains are set up as properties of a geodatabase rather than as properties of feature classes. Why is this arrangement an advantage?

10. How does one establish a default value for an attribute in a feature class? In a shapefile?

# Mastering the Skills

## Teaching Tutorial

The following examples provide step-by-step instructions for doing basic tasks and solving basic problems in ArcGIS. The steps you need to do are highlighted with an arrow →; follow them carefully. Click on the video number in the VideoIndex.htm file to view a demonstration of the steps.

We will build a geodatabase of Rapid City by importing existing data layers from the mgisdata\Rapidcity\archive folder. We will start with ArcCatalog, since we have a lot of data management to do before we need to use ArcMap.

> 1→ Start <u>ArcCatalog</u> and navigate to the mgisdata\Rapidcity folder.
>
> 1→ Right-click the Rapidcity folder and choose New > File Geodatabase.
>
> 1→ Type in *rcdata* as the name of the geodatabase and press Enter.

1.  What coordinate system do the feature classes in the archive folder use?

The new geodatabase will be in UTM NAD 1983. As we work, we will be careful that each coordinate system is converted during the import process.

### *Creating feature datasets*

Most of the feature classes will go in feature datasets, which are required to have a predefined coordinate system, and which we will carefully create using NAD 1983 UTM Zone 13N. Feature classes imported to these feature datasets will automatically be converted to match.

> 2→ Right-click the rcdata geodatabase and choose New > Feature Dataset.
>
> 2→ Name it *Admin* and click Next.
>
> 2→ For the coordinate system, expand Projected Coordinate Systems > UTM > NAD 1983 and select UTM Zone 13N.
>
>  2→ While the UTM Zone 13 coordinate system is still highlighted, click the Add to Favorites button, so that it will be easily available next time. Click Next.
>
> 2→ Do not set a vertical coordinate system; click Next.
>
> 2→ Accept the default tolerances, resolution, and domain, and click Finish.
>
> 3→ Use the same procedure to create another feature dataset in rcdata named *Environmental*. (This time choose NAD 1983 UTM Zone 13 from the Favorites.)
>
> 4→ Create a third feature dataset named *Transportation*.
>
> 5→ Create a fourth feature dataset named *Watersystem*.

### Adding coverages to a feature dataset

We will begin by importing two coverages, the city boundary and the land use data, to the Admin feature dataset. Recall that coverages contain multiple feature classes, so we must specify polygons as the class to import.

6→ Right-click the new Admin feature dataset and choose Import > Feature Class (single) tool (Fig. 13.13).

6→ Click the Browse button for the Input Features, navigate to the mgisdata\Rapidcity\archive folder, and double-click the landuse coverage to see its feature classes.

6→ Select the polygon feature class and click Add.

6→ Enter landuse for the Output Feature Class Name.

Fig. 13.13. Importing a feature class
Source: ArcCatalog

The Field Map section lists the attributes in the feature class and controls how they are imported. Fields may be removed or added, or the order can be changed. With coverages, some fields are part of the coverage storage structure and contain no useful information after the conversion. It is best to delete them.

7→ Click on the AREA field to highlight it, and then click the X button to delete it. (It will not be deleted from the original coverage.)

7→ Also delete the PERIMETER and LANDUSE# fields from the map.

7→ Leave the rest of the options set to their defaults.

7→ The window should appear as in Figure 13.13. Click OK.

7→ Use the Preview tab to examine the newly imported polygons.

Everything looks in order. Next we import the citybnd coverage.

8→ Right-click the Admin feature dataset and choose Import > Feature Class (single).

8→ Set the input feature class to the polygons of citybnd in the Archive folder.

8→ Set the output name to citybnd.

8→ Highlight and delete the AREA, PERIMETER, and CITYBND# fields.

8→ Click OK to start importing.

Often, users have multiple feature classes to import into the same geodatabase. Another tool allows multiple imports to be set up and run in batch mode.

9→ Right-click the Environmental feature dataset and choose Import > Feature Class (multiple).

9→ Expand the archive folder in the Catalog Tree. Click on the watersheds shapefile and drag it over to the Input Features box on the tool. The feature class is placed in the list to process.

9→ Click and drag the stategeol shapefile to the Input Features box. Click OK.

9→ Expand the Environmental dataset and examine the new feature classes.

Notice that, when using the tool for importing multiple feature classes, you cannot set options individually for each one, as you could with the single input tool. We continue importing.

10→ Right-click the Admin feature dataset and choose Import > Feature Class (multiple).

10→ Drag these three shapefiles from the archive folder to the Input Features box: connects, parcels, and buildings. Click OK.

10→ Expand the Admin dataset and check the new feature classes.

11→ Right-click the Transportation dataset and choose Import > Feature Class (single).

11→ Import the rc_roads shapefile from the archive folder. Name the output feature class *roads*. Don't change the Field Map settings. Click OK.

We also would like to add schools, but the name of the feature class gives us pause.

12→ Click the sdschools shapefile in the archive folder and click the Preview tab above the display panel, if it is not already clicked.

As suspected, this shapefile contains schools for the whole state, although we want only the ones in the Rapid City area. We need to use the Clip tool this time. The geology map has a good boundary for clipping what we need.

12→ Open ArcToolbox > Analysis Tools > Extract > Clip.

12→ Click and drag the sdschools shapefile into the Input Features box on the tool.

12→ Click and drag the stategeol shapefile into the Clip Features box.

12→ Name the output *schools* and place it in the Admin feature dataset.

12→ Accept the default XY Tolerance and click OK.

Finally, we plan to have a feature class to contain water and sewer lines. This feature class does not yet exist—we will digitize it. We must create the empty feature class.

13→ Right-click the Watersystem feature dataset and choose New > Feature Class.

13→ Name the feature class *waterlines* and set the type to Line Features. Click Next.

13→ Accept the default configuration key word options. Click Next.

13→ We are not going to create the fields for this feature class yet. Click Finish.

### Creating attribute domains

During the planning phase for this geodatabase, it was determined that it should have domains. A table was developed, listing the domains and their properties (Table 13.1). With the planning done, we are ready to start adding the domains to the geodatabase.

TIP: ALWAYS, ALWAYS click Apply after EACH domain is added, before adding the next one. If you make an error in one domain but add more before clicking Apply, then it may not be clear which one has the error. You'll often have to delete them all and start again.

14→ In ArcCatalog, right-click the rcdata geodatabase and choose Properties. Click the Domains tab.

14→ Type the domain name, *RoadType*, into the first empty box, and enter the description *Type of road* (Fig. 13.14). (Don't press the Tab key here; you need to continue entering the properties of this first domain, not to add another one.)

14→ Use the mouse to click in the Field Type box next.

14→ Set the Field Type to Text.

14→ Set the Domain Type to Coded Values.

14→ Set the split policy to Duplicate.

14→ Set the merge policy to Default Value.

15→ Type the three-letter codes into the Code column and enter each description in the Description column: INT = Interstate; HWY = Highway; RMP = Ramp; CON = Connector; and LOC = Local. You can use the Tab key to move from cell to cell within this part of the window.

15→ Click Apply.

| Domain Name | Description |
|---|---|
| RoadType | Type of road |
| | |
| | |

Domain Properties:

| Field Type | Text |
|---|---|
| Domain Type | Coded Values |
| Split policy | Duplicate |
| Merge policy | Default Value |

Coded Values:

| Code | Description |
|---|---|
| INT | Interstate |
| HWY | Highway |
| RMP | Ramp |
| CON | Connector |
| LOC | Local |

Fig. 13.14. Setting up the RoadType domain
Source: ArcCatalog

Table 13.1 summarizes the characteristics of all the domains to be added. Use it for reference, if needed, as you continue adding the domains in Steps 16–23.

Table 13.1. Domains for the geodatabase

| Domain | Description | Field | Type | Values |
|---|---|---|---|---|
| RoadType | Type of road | Text | Coded | INT, HWY, RMP, CON, LOC |
| Directions | Cardinal directions | Text | Coded | N, S, E, W, NE, NW, SE, SW |
| StreetAbbrev | Common street abbreviations | Text | Coded | AVE, ST, RD, DR, BLVD, CT, LN, WAY, CIR, PL |
| Diameter | Pipe diameter in inches | Short | Coded | 1, 3, 6, 12, 24 |
| Materials | Pipe construction material | Text | Coded | COP, PVC, LEA, TIL, CON, OTH |
| Flow | Flow range in gpm | Short | Range | 1–2000 |
| LengthFloat | Generic length for split policy | Float | Range | 0–9999999 |
| Use Duplicate and Default Value for the split and merge policies for all domains except LengthFloat. | | | | |

2. Why do you suppose the Materials domain uses three-letter codes? _____

16→ Click in the next open Domain Name box and enter the name *Directions*.

16→ Type *Cardinal directions* for the Description.

16→ Click in the Field Type box and set it to Text; then set the split and merge policies to Duplicate and Default Value.

16→ Enter the eight codes from Table 13.1. Be sure to also enter the Description for each one using the same values, such as N.

16→ Click Apply.

| Code | Description |
|---|---|
| N | N |
| S | S |
| E | E |

Source: ArcCatalog

**TIP:** You must *always* enter a description, even if it will be the same as the code. If the description is left blank, then the values will appear blank when you try to view or edit them. ArcMap always displays the descriptions, not the codes.

17→ Create the StreetAbbrev coded domain from the information in Table 13.1. Enter the street codes in the Code boxes, also entering the full-word equivalent for each one.

17→ Click Apply.

| Code | Description |
|------|-------------|
| AVE | Avenue |
| ST | Street |
| RD | Road |

18→ Create the Diameter coded domain from the information in Table 13.1. Enter the number for the Code (it is a Short Integer field type, but in the Description enter 1-in, 3-in, and so on).

18→ Click Apply.

| Code | Description |
|------|-------------|
| 1 | 1-in |
| 3 | 3-in |
| 6 | 6-in |

19→ Create the Materials domain from the information in Table 13.1. Enter the three-letter code values for Code, but for Description enter the full-length words: copper, PVC, and so on.

19→ Click Apply.

The last two domains are range domains, so we won't need to type any codes—just the minimum and maximum range values.

| Code | Description |
|------|-------------|
| COP | copper |
| PVC | pvc |
| LEA | lead |
| TIL | tile |
| CON | concrete |
| OTH | other |

Source: ArcCatalog

20→ Create the Flow domain. Make sure it uses the Short Integer type.

20→ Set the minimum value to 1 and the maximum value to 2000, and set the policies as Duplicate and Default Value, as for the others.

20→ Click Apply.

This last domain is intended to provide suitable split and merge policies for feature lengths, such that if you split a line each piece will receive a proportional value of the original length. Thus, it will use different policies than the other domains. The Range is irrelevant, but it needs to be large enough to hold any reasonable length measure.

21→ Create the LengthFloat domain. Make sure it uses the Float type.

21→ Set the minimum value to 0 and the maximum value to 999999999.

21→ Give it Geometry Ratio for the split policy, and Sum Values for the merge policy.

21→ Click Apply.

21→ Click OK to close the Database Properties window.

3. What would you do if you wanted to apply a similar length domain to fields with a Double field type? _____

**TIP:** Domain policies and codes may be edited after the domain is created. However, the field type and domain type, once set, cannot be altered. To change them, you must delete the domain and create a new one. To delete a domain, select it and press the Delete key.

## *Assigning domains and default values*

The roads feature class already exists, and we can assign domains to the appropriate fields.

22→ Open the Properties for the roads feature class in the Transportation feature dataset.

22→ Click the Fields tab.

22→ Click in the little gray box to the left of the PREFIX field to highlight it.

22→ Click in the Domain drop-down box below and select the Directions domain.

The Directions domain actually applies to two fields in this table, the PREFIX and SUFFIX2 fields. Domains can be reused in multiple feature classes.

22→ Highlight the SUFFIX2 field and set the domain to Directions.

22→ Highlight the SUFFIX field and set the domain to StreetAbbrev. Click Apply.

Next, the road type field does not yet exist in this feature class, so we must create it. We will also assign a default value, since local residential streets outnumber the other types.

23→ In the next available field name box, type ROADTYPE. Make it a Text field. Give it a length of 3 and set the domain to Roadtype.

23→ In the Default Value box, enter LOC, the code for local roads. Click OK.

Later in the lesson, we will see how these domains function during editing.

## *Creating subtypes*

The waterlines feature class will have four subtypes: water mains, water laterals, sewer mains, and sewer laterals. Subtypes must be based on a short integer field, so we must create this field in the feature class in addition to the other fields. We will also assign a domain to each field, if one is available for it.

24→ Open the Properties for waterlines in the Watersystem feature dataset.

24→ Click the Fields tab.

24→ Type Linetype in the first available field name box and set the Data Type to Short Integer. Click Apply. This will be the subtype field.

25→ Type PipeDiam in the next box and set the type to Short Integer. Click in the Domain box and choose the Diameter domain. Click Apply.

25→ Type PipeMaterial in the next box and set the type to Text. Set the length to 3 and the domain to Materials. Click Apply.

26→ Type InstallDate in the next box and set the type to Date. Click Apply.

26→ Type MaxFlow in the next box and set the type to Short Integer. Set the domain to Flow. Click Apply.

26→ Type Feet in the next box and set the type to Float. Set the domain to LengthFloat. Click OK.

**TIP:** Note that the list of domains varied for the fields. The domain field type must match the attribute field type. A short integer domain cannot be applied to a long integer attribute.

Before we create the subtypes, we must know the default values to be assigned. If domains are used, then the defaults must list the actual code value, not the description (6 rather than 6-in, for example). For some fields, such as the length in feet or the install date, we cannot predict a typical value ahead of time, so there is no point in setting a default for it.

27→ Open the properties for the waterlines feature class.

27→ Click the Subtypes tab in the Feature Class Properties window (Fig. 13.15).

27→ Set the Subtype Field to Linetype.

27→ In the first Subtype box, enter 1 for the code and *Water Main* for the description.

27→ Highlight the top Water Main row to edit the boxes below.

28→ Under Default Values and Domains, enter the value 6 in the Default Value field for PipeDiam.

28→ Enter *COP* in the Default Value field for PipeMaterial.

28→ Enter *700* in the Default Value field for MaxFlow.

28→ Make sure that the window appears as in Figure 13.15 and click Apply.

Fig. 13.15. Entering the first subtype
Source: ArcCatalog

Table 13.2. Default values for each subtype

| Attribute | Water main | Water lateral | Sewer main | Sewer lateral |
|---|---|---|---|---|
| PipeDiam | 6 | 3 | 12 | 6 |
| PipeMaterial | COP | PVC | CON | PVC |
| MaxFlow | 700 | 125 | 1500 | 230 |

29→ Enter the Code 2 and the description *Water Lateral* in the second row.

29→ Click on the second row to highlight it.

29→ Enter the default values for water laterals as listed in Table 13.2. You don't need to assign domains; they were assigned for the Water Main subtype. Click Apply.

30→ Enter subtype 3, *Sewer Main*, and then enter the default values. Click Apply.

30→ Enter subtype 4, *Sewer Lateral*, and the default values. Click Apply.

30→ Make sure that the Default Subtype drop-down box at the top is set to Water Main. The final list of subtypes should appear as shown to the right.

30→ Click OK to close the Feature Class Properties window.

| Code | Description |
|------|-------------|
| 1 | Water Main |
| 2 | Water Lateral |
| 3 | Sewer Main |
| 4 | Sewer Lateral |

Source: ArcCatalog

### *Using domains while editing*

Now we are ready to see how domains and subtypes work during an editing session, which is when they are most noticeable.

31→ Close ArcCatalog.

31→ Start ArcMap with a new, blank map. Set the default geodatabase to the mgisdata\Rapidcity\rcdata geodatabase.

31→ Add the roads layer from the Transportation feature dataset.

31→ Open the Editor toolbar, if necessary, and choose to Start Editing.

32→ Use Selection > Select By Attributes to select all the roads where:

STREET = 'MOUNT RUSHMORE'. Close the window when done selecting.

32→ Click the Attributes button on the Editor toolbar. Dock the window in a convenient location and resize it, if needed, to see the field values.

32→ Click one of the entries in the list and examine its attributes.

First, notice that the SUFFIX field entries are listed as "Road" rather than RD, which is the code actually in the field. This resulted from giving this field the StreetAbbrev domain.

33→ Click in the space next to the SUFFIX field.

A coded value domain has a drop-down list showing the codes for that attribute (Fig. 13.16). It prevents the user from entering an inappropriate value such as AV instead of AVE, and it allows the user to see a real word rather than a cryptic code.

The ROADTYPE field was newly added to the feature class and thus contains no values at all. Hence, the <Null> value appears in the box. We can set it to the appropriate value now.

33→ Click the roads heading at the top, so that an edit will apply to all selected records.

33→ Click in the RoadType box and select the Connector road type, setting it for all the selected records.

It would be reasonable to assume that roads designated as "Circle" or "Court" are local roads. We can select them and assign the road type.

34→ Open Select By Attributes from the main menu bar and clear the expression.

Fig. 13.16. Coded domains appear as drop-down lists during editing.
Source: ArcMap

34→ Start entering the expression SUFFIX = and then click Get Unique Values. Notice that both code and description are shown.

34→ Continue adding the full expression SUFFIX = 'CIR' OR SUFFIX = 'CT' by double-clicking the appropriate values in the Get box. Notice that only the code is placed in the expression. Click OK.

34→ Click the top entry in the Attributes window so that edits apply to all the selected features. Set the ROADTYPE to Local.

Unfortunately, this task must be completed to assign a road type to all of the roads. This is why designing your database ahead of time is so important. Adding domains and fields after the features exist creates extra work. We'll leave this task for another time.

Recall that we assigned a default value, Local, to the ROADTYPE field. Our reasoning was that most of the roads in a city are local ones, so whenever we add a local road we do not have to enter the type.

35→ Clear the selected features.

35→ Click the roads template in the Create Features window, and add a line somewhere.

35→ Scroll down and examine the ROADTYPE field in the Attributes window.

Most of the fields are <Null> for this new road, but the ROADTYPE is automatically set to Local. If this were indeed a local road, we would not have to enter the value. If it were not, we would. Thus, defaults are useful only when a particular value occurs often enough that it saves time overall.

35→ Click the Edit tool on the Editor toolbar. Press the Delete key on the keyboard to delete the road that you just added.

35→ Save your edits.

### Using subtypes while editing

Now let's examine how subtypes make our work easier.

36→ Add the buildings, connects, and parcels feature classes from the Admin feature dataset. Turn off the roads.

36→ Zoom to the extent of the parcels layer, and then zoom out a little bit.

36→ Add the waterlines feature class from the Watersystem feature dataset.

This feature class is empty, so no features appear on the map, but each subtype appears with its own symbol in the Table of Contents.

| Linetype |
| --- |
| Sewer Lateral |
| Sewer Main |
| Water Lateral |
| Water Main |

37→ Modify the waterlines symbols so that the mains are thick lines and the laterals are thinner lines. Make the water lines blue and the sewer lines orange.

Source: ArcMap

No editing templates have been created for waterlines yet. Let's do it now.

38→ In the Create Features window, choose the Organize Templates button.

38→ Click New Template, and make sure waterlines and connects are checked.

400

38→ Click Finish and close the Organize Feature Templates window.

Now let's digitize some water lines.

39→ Open the Snapping toolbar, if necessary, using Editor > Snapping > Snapping Toolbar. Turn off all types of snapping.

39→ Click the Water Main template and add a water main that runs north of the block. The precise location is not important. See Figure 13.17.

39→ Examine the attributes in the Attributes window.

The defaults that we set for water mains (6-inch copper pipe with a flow rate of 700 gpm) are already entered in the attribute fields, saving us the trouble.

39→ Click the Sewer Main template and add a sewer main above the water main.

39→ Examine the Attributes window. Different defaults were entered for the sewer, just as we set up.

Fig. 13.17. The water and sewer lines
Source: ArcMap

40→ Click the Water Lateral template and turn on edge and point snapping.

40→ Add a lateral from the connection point on each house to the water main, being sure to snap from the point to the water main edge. Note the default attributes.

41→ Click the connects template and add another connection point to each house, next to the first one, snapping to the edge of the building.

41→ Click the Sewer Lateral template and add a lateral from each second connection point to the main sewer line. Note the default attributes.

42→ Open the attribute table for waterlines.

Look at all the information entered in the table, even though we never typed a word. You may have thought that the subtypes were a hassle to create, but now perhaps you begin to see how they are worth the effort! Now we just need to assign the attributes that did not have defaults.

42→ Clear the selected features, if any.

42→ Right-click the Feet field and choose Field Calculator. The map units are meters, so enter the expression SHAPE_Length * 3.281 and click OK.

42→ Right-click the InstallDate field and choose Field Calculator. Enter the expression "3/17/1997" including the double-quotes. Click OK.

42→ Close the Table window and save your edits.

We still have not seen the split/merge policies at work.

| | |
|---|---|
| OBJECTID | 1 |
| SHAPE_Length | 221.0599 |
| Linetype | Water Main |
| PipeDiam | 6-in |
| PipeMaterial | copper |
| InstallDate | 3/17/1997 |
| MaxFlow | 700 |
| Feet | 725.2975 |

43→ Select the water main with the Edit tool and examine its attributes, particularly the Feet field (Fig. 13.18). Your lengths will differ.

43→ Click the Split tool on the Editor toolbar, and click on the water main about one-third of the way from its left end.

43→ Select the short end with the Edit tool and view the attributes.

43→ Select the long end with the Edit tool and view the attributes.

Fig. 13.18. Water main attributes before splitting
Source: ArcMap

Most of the fields were copied from the original, using the assigned Duplicate split rule, which makes sense. If the pipe was 6-inch copper before it was split, it should not have changed. However, the Feet field was assigned the Geometry Ratio split rule from the LengthFloat domain, and the length was automatically updated using the proportion to the shape length.

Imagine that this section of water main ruptured and was repaired using a bigger PVC pipe. You are updating the record.

44→ In the Attributes window, change the PipeMaterial to pvc.

44→ Click the InstallDate box and click the Today link at the bottom of the calendar.

44→ You intend to enter 850 for the new flow rate, but pretend to make a mistake and type *8500* instead.

44→ Clear the selected features and save your edits.

Now let's see how to take advantage of the range domain error checking. It is done by means of a separate step called validation.

45→ Use the Edit tool to click and drag a rectangle that selects all of the water lines but none of the other features.

45→ Choose Editor > Validate Features from the Editor toolbar.

If errors are found, it reports the number in a dialog window and places the incorrect records in the selected set, ready for editing. Only attributes with domains will be tested. In this case it found your incorrect entry of 8500, and the offending water main remained selected.

45→ Change the flow rate to *850* in the Attributes window.

45→ Save your edits and stop editing.

### Editing annotation (optional)

Annotation can be stored in a feature class within a geodatabase, which makes it available for use in many map documents. We will create road name annotation for part of the roads layer.

46→ Turn on the roads layer. (You may see an offset between the roads and parcels depending on your version of ArcGIS. Ignore it.)

46→ Zoom out from the parcels to the area shown in Figure 13.19.

46→ Turn off all layers except the roads.

46→ Choose Bookmarks > Create Bookmark from the main menu bar. Enter the name *Anno* and click OK. Now you can return to this extent whenever you wish.

Creating annotation begins by creating dynamic labels with the desired properties.

47→ Open the Labels tab of the roads layer properties.

47→ Create labels using the ROADNAME field with 8-pt. Arial font. Click OK.

47→ Set the map scale to exactly 1:10,000.

Fig. 13.19. Area to zoom to for editing annotation
Source: ArcMap

This will be your **reference scale**, the scale at which the annotation appears at the 8-pt. size. You want to make sure that the labels appear as you wish at this scale before you create the annotation. It looks fine, so let's continue.

47→ Right-click the Layers data frame name and choose Convert Labels to Annotation.

47→ Choose to store the annotation in a database. Notice the reference scale (Fig. 13.20).

47→ Fill the button to create annotation for features in the current extent.

47→ Uncheck the box to create Feature Linked annotation.

Fig. 13.20. Creating an annotation feature class
Source: ArcMap

47→ Notice the Destination box; the annotation will be saved in the Transportation feature dataset of the rcdata geodatabase with the name roadsAnno.

47→ Click Convert. The roadsAnno feature class is added to the data frame.

The annotation is stored as a separate feature class in the Transportation feature dataset, as shown in Figure 13.21.

Fig. 13.21. Annotation
Source: ArcMap

**TIP:** If you made a mistake and want to start over, delete the annotation feature class from the geodatabase using the Catalog tab.

First, let's examine the effect of the reference scale.

48→ Zoom in to a smaller area and watch the annotation get larger. Zoom out and watch it get smaller. Then return to 1:10,000. You can zoom in a little more, if you wish, to see and work with the annotation better.

49→ Start Editing.

49→ An editing template for roadsAnno should appear in the Create Features window, but if not, create one using the Organize Templates button.

Selecting annotation for editing requires a different tool than feature editing.

    49→ Click the Edit Annotation tool on the Editor toolbar.

49→ Select E OAKLAND ST.

When annotation is selected, there are four handles with it (Fig. 13.22). The shape of the cursor changes when it is placed on top of a handle, indicating the action that will occur on a click and drag. The two blue handles on the ends rotate the annotation. The red triangle enlarges or shrinks the text. The black cross allows the annotation to be moved.

Fig. 13.22. Action points on selected annotation
Source: ArcMap

50→ Place the cursor near the top middle until it changes to a two-ended arrow. Click and drag to enlarge or shrink the text. Leave it large.

50→ Place the cursor near the bottom middle until it changes to a four-ended arrow. Click and drag to move the text to a clear spot.

50→ Place the cursor near the end until it changes to a rotate tool. Click and drag to rotate the text.

50→ Shrink the text back to its approximate original size and move it back to its starting location.

Unplaced labels are put in the table, with the designation "Unplaced" in a Status field. The Overflow window does not open automatically, but you can open it to place the annotation.

51→ Choose Editor > Editing Windows > Unplaced Annotation.

51→ Set the Show drop-down box to roadsAnno and click Search Now.

51→ Right-click the row WALNUT DR, and click Place Annotation. It appears on the map, selected. (If it is not in the list, choose another name.)

51→ Adjust the location or angle, if necessary, using the Edit Annotation tool.

51→ Right-click the next row and place the annotation. Adjust the location.

51→ Close the Unplaced Annotation window.

We will not continue placing annotation right now, but any time you want to continue placing the annotation you can open this window and search for unplaced labels.

The Attributes window has additional tabs and options when working with Annotation.

52→ Open the Attributes window and dock it in a convenient place.

52→ Select the E OAKLAND ST annotation with the Edit Annotation tool.

52→ Click the Annotation tab in the lower panel of the Attributes window and examine the settings. You can set the font properties and retype the text, if necessary.

52→ Examine the Attributes tab. Each piece of annotation can have all its properties set individually.

52→ Return to the Annotation tab. Click the drop-down box, choose Default for the symbol, and click Apply. The E OAKLAND ST text will go back to original size.

53→ The Edit Annotation tool should still be active. Hold down the Shift key and click on another nearby piece of annotation to select it also. Now both annotation features appear in the Attributes window.

53→ Click on the roadsAnno entry at the top of the Attributes window.

**TIP:** You may need to enlarge the lower panel of the Attributes window to see the font size and formatting buttons, or undock the entire window.

53→ Click the B button for boldface type and click Apply. Both annotation items are updated to bold font.

53→ Click B again to return to normal font and click Apply. Leave the Attributes window open so you can use it as needed.

Now we will create new annotation for some streets in the southeast corner. These probably have unplaced labels waiting, but we will create new ones, just for the practice.

54→ Click the roadsAnno template, noticing that the default construction tool is Straight. The Annotation Construction window appears.

54→ Type SIDNEY DR in the open area of the Annotation Construction window.

54→ Click on the center of SIDNEY DR, shown in Figure 13.23.

54→ Move the cursor to rotate the text to the desired angle and click again to place it.

54→ If you don't like where it was put, press the Delete key to get rid of it and then place it again. (You don't need to type the text again.)

Fig. 13.23. Street names in the SE corner of the map
Source: ArcMap

Observe an unlabeled street running north into PARK HILL DR. You can use Identify to determine its name.

55→ Click the Identify tool and click the unlabeled street. Leave the Identify window open but tuck it out of the way.

55→ Click the roadsAnno template again and type CHURCHILL DR in the Annotation Construction window.

55→ Place the annotation on the unlabeled street.

55→ Save your edits so far, and save the map document.

56→ Click the Edit Annotation tool on the Editor toolbar.

56→ Click on E MEADE ST to select it. Move it to the center of the block.

56→ Right-click E MEADE ST and choose Copy.

56→ Right-click anywhere and choose Paste. Choose to paste it into roadsAnno. A copy of E MEADE ST will be placed on top of the original.

56→ Click and drag the copy to the eastern part of the street below E ST FRANCIS ST. Rotate and adjust its position until you are satisfied.

Flipping annotation, like flipping a line, makes it go in the opposite direction.

57→ Select BALSAM AV with the Edit Annotation tool.

57→ Right-click it and choose Flip Annotation. Now it matches the other nearby annotation for easier reading.

57→ Select E TALLENT ST and move it west of HAWTHORNE AV.

Source: ArcMap

Placing curved annotation is slightly more complicated, but not much.

58→ In the Create Features window, click the Curved construction tool

58→ Type *E TALLENT ST* in the Construction box.

58→ Find the curved eastern end of E TALLENT ST and click underneath on the left, the middle, and the end to enter a curve that follows the street. Double-click to end the sketch.

58→ If you don't like the result, delete it and try again.

Source: ArcMap

Annotation can be made to follow a particular feature, either curved or straight. First, you must set the following options.

59→ Select ROBBINS DR with the Edit Annotation tool.

59→ Right-click the annotation and choose Follow > Follow Feature Options.

59→ Choose Curved and set the offset to 15 map units. Click OK.

You only need to set the options the first time. Now place the annotation.

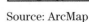

59→ Right-click on the curved section of road that you want ROBBINS DR to follow and choose Follow This Feature. The text is moved to the point you clicked and rotated to follow the street.

59→ Place the cursor on the text until you see the four-ended arrow. Click and drag the text above and below the street, and then leave it below.

Source: ArcMap

59→ Click and drag the text along the street, noticing how it hugs the feature. When you are satisfied with the placement, release the mouse button.

You can also create new annotation that follows a feature.

60→ Click the roadsAnno template and select the Follow Feature construction tool.

60→ Type *City Bus Route* in the Text box.

60→ Click along the inclined segment of E Meade St to select the feature on which to place the annotation.

60→ Move the mouse until you are satisfied with the position and then click again to place the annotation.

60→ Click the Annotation tab of the Attributes window and change the font to italic. Click Apply.

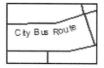

Source: ArcMap

61→ Click the Edit Annotation tool and select MAYWOOD DR.

61→ Press the O key on the keyboard to open the Follow Feature Options.

61→ Fill the button for Curved and click OK.

61→ Right-click on the center of MAYWOOD DR and choose Follow This Feature.

61→ While the annotation is still selected, place the cursor on it and click and drag it above, below, or along the road to get the best placement and curvature. Release the mouse button when you are satisfied with the placement.

62→ Select the Horizontal construction tool.

62→ Type *Robbinsdale School* in the Text box.

62→ Click in the center of the large, open area west of IVY AV to place the text.

62→ In the Attributes window, change the font to 10-pt. Bold Arial and make it red. Click Apply.

You can stack annotation for better placement.

63→ Click the Edit Annotation tool and click on the Robbinsdale School annotation to select it.

63→ Right-click the selection and choose Stack.

63→ Click and drag the annotation to a good position.

This tutorial demonstrated only a few common annotation editing tasks. You can learn more using ArcGIS Help.

Source: ArcMap

63→ Save your edits and stop editing. Also save the map document.

→ Exit ArcMap.

This is the end of the tutorial.

## Exercises

1. Consider a buildings feature class such as the one you started in the Crestview geodatabase for Austin. Make a list of likely attributes that such a feature class might have.

2. Take the list of buildings attributes and decide which ones could have domains and what types of domains and values they should have. Construct a table similar to Table 13.1 showing these domains.

3. Divide the buildings into categories that would make suitable subtypes. Consider the different default values that each subtype would have, and construct a table similar to Table 13.2.

4. Using the information assembled in Exercises 1 through 3, create a buildings feature class in the Crestview geodatabase. Set up the geodatabase domains, create the fields for the feature class, and assign default values.

5. If you have an ArcGIS Standard license, create subtypes for the buildings with the default values you determined in Exercise 3. If you do not, create editing templates instead of subtypes, with the default values stored in the templates. Consult the Help and learn how to save and load editing templates.

6. Digitize approximately 50 buildings from the Crestview area, distributed among your subtypes. Capture a map showing the buildings and a screenshot of the attribute table showing the various fields of information.

7. Finish editing the roadsAnno annotation in the rcdata geodatabase to make an aesthetic and clearly legible set of labels. Capture a map showing the streets and annotation.

8. Create geodatabase annotation for the streets in the Crestview geodatabase and edit them for optimum placement and legibility. Capture a map showing the streets and annotation.

## Challenge Problem: My Alma Mater

The Challenge Problem in Chapter 7 asked you to create a campus database and digitize building footprints. We will add to this project now in several ways.

▶ Digitize the roads and sidewalks as separate polygon feature classes. (For sidewalks, you may find it easier, if less accurate, to digitize the centerlines and then create buffers.)

▶ Create and place feature annotation for the roads and buildings.

▶ Create a point feature class of objects to be collected using a GPS or smartphone (trees or trash cans, for example). Include several attributes to collect and create coded domains as appropriate, so that they will appear as pick-lists on the devices.

▶ Remember to create an Item Description for all new feature classes.

▶ (Optional) If you have a smartphone or tablet, and publication privileges in ArcGIS Online, publish the point feature class as an editable feature service. Collect at least 20 data points using your mobile device and the ArcGIS app. (You can sign up for a 30-day trial subscription to ArcGIS Online if you don't have an account with your university.)

**Write a short report**, with screen captures, documenting your work.

# Chapter 14. Metadata

## Objectives

▶ Knowing the dominant metadata standards and their relationship to each other

▶ Gaining a basic understanding of data quality issues

▶ Understanding the layout and main sections of CSDGM and ArcGIS/NAP metadata

▶ Learning to use the metadata editor and templates to create metadata

# Mastering the Concepts

## GIS Concepts

Metadata is "data about data"—it is critical to GIS, providing the ability to discover, access, and use geospatial data sets and services. Providing metadata is a part of creating GIS data sets. It is a professional obligation when data will be distributed to clients or the public. Best management practices dictate that metadata should also be prepared for in-house data, even if it is nothing more than the simple Item Description, so that critical information about the source and processing of a data set will not be lost. Figure 14.1 shows metadata developed for the radar image georeferenced in Chapter 3, which you will create in this chapter.

Preparing complete metadata that complies with a standard is a daunting task. However, even partial metadata is better than none. This chapter provides a basic introduction to the most important elements of geospatial metadata and how they can be developed.

Fig. 14.1. Metadata example
Source: ArcGIS

### The metadata standards

To facilitate sharing of metadata across the many GIS formats and applications, the Federal Geographic Data Committee (**FGDC**) and the International Organization for Standardization (**ISO**) have worked to develop metadata standards that specify the structure and content of geospatial metadata records. The FGDC is "an interagency committee that promotes the coordinated development, use, sharing, and dissemination of geospatial data on a national basis" (www.fgdc.gov), and it serves as a liaison to the international body of standards. The FGDC web site contains a wealth of information on metadata and geospatial data in general.

A **metadata standard** lays out the specific information fields that should be included and defines which ones are optional and which are mandatory. It also specifies the exact format and organization of the information so that programs and metadata interpreters can always find the information in the same place and so that the metadata are transportable from system to system.

Currently, metadata standards in the United States are in a state of flux. For many years, users in the United States have used a standard prepared by the FGDC in 1994 and updated in 1998: the Content Standard for Digital Geographic Metadata (**CSDGM**). Companies that produce data for federal agencies are required to provide metadata that complies with the CSDGM standard, and in turn, data produced by under contracts to many government agencies must be compliant as well. Much effort has been expended in development, software, and training for CSDGM, and much of the data that you download and use today will employ this content standard.

However, the international community has been using a more generic standard, called the ISO 19139 standard, for nearly as long. It is not particularly suited for geographic data, so another standard, the ISO 19115, has been established for geographic data. ISO standards allow a community of users to develop and publish a **profile**, a customized version of a content standard that is designed to meet the needs of that community, as long as they follow rules that ensure that the profile will maintain compatibility with the original standards.

The United States and Canada geographic data committees agreed to jointly develop a profile now known as the North American Profile of ISO 19115, or NAP. It has been adopted by Canada. The FGDC has formally endorsed the ISO 19115 standard, and federal agencies are being encouraged to move to the ISO standard as they are able, although this process may take years.

The practical upshot of this situation is that users are likely to encounter both CSDGM and NAP metadata for some time and must be familiar with both standards (as if metadata weren't difficult and confusing enough before). Much of the information contained in the two standards is similar, but the organization is different, and the cross-walk is not perfect or simple. Some, but not all, of the content of CSDGM content can be recorded using ISO 19115. For example, documentation of attribute tables and attribute fields cannot be recorded using ISO 19915 compliant metadata. Many of the ISO standards are being updated as time goes on, while CSDGM has been unchanged since 1998. Because of these complexities, many organizations are still using CSDGM but are gradually adopting ISO standards as they (and their implementations) are improved.

The following publications are recommended for users who need to learn about metadata standards and implementation. All but the last refer to the CSDGM standard. Users are encouraged to explore the FGDC web site http://www.fgdc.gov/metadata for updated information and new publications (http://www.fgdc.gov/metadata/metadata-publications-list).

***National Geospatial Data Assets (NGDA) Metadata Guidelines* (2016).** Provides practical and specific advice on key metadata fields, particularly for data sets utilizing web services for distribution.

***Geospatial Metadata Fact Sheet* (2011).** Brief explanation of what metadata is and how it is organized. Includes a discussion of the new NAP standard and why it is being implemented.

***Geospatial Metadata Quick Guide* (2005).** The famous "Don't Duck Metadata" publication, which describes some of the basic types of information that should be included.

***Top Ten Metadata Errors* (2006).** A short, fun read.

***CSDGM Essential Metadata Elements* (2008).** The guide for those who are looking for guidance on producing CSDGM content that only includes the most critical information.

***Content Standard for Digital Geospatial Metadata Workbook* (2000).** Too exhaustive for a beginner but an excellent reference for the serious metadata enthusiast. Contains some good examples that beginners might find helpful.

***Preparing for International Metadata* (2011).** A discussion of the development and layout of the North American Profile of the ISO 19115, and recommendations on how CSDGM metadata records can be created in way that simplifies the conversion to ISO 19115.

***Institutionalize Metadata Before It Institutionalizes You* (2005).** Useful discussion of templates and incorporating metadata creation into the organizational workflow.

Some of the information in metadata is easy to understand and document, such as who produced the data. Many of the terms used in the standards and literature presuppose an understanding of geospatial data topics, such as logical consistency or completeness. Users will develop expertise in these areas as they travel the path to becoming geospatial professionals. Merely trying to understand these terms as you try to produce metadata will make you a better creator of data as well. Some sections of metadata are not applicable to a particular data set and can be ignored. As you gain experience reading and creating metadata, you will understand more and more about geospatial data, where it comes from, what it is good for, and what happens to it during its lifetime. You will also gain experience in evaluating the quality of data and deciding whether it is good enough for a particular project or application.

## Data quality issues

One fundamental purpose of metadata is providing sufficient information for a user to decide whether a data set is of sufficient quality for a particular application. Assessing the quality of a data set is not a simple judgment of good or bad but a complex analysis of whether a particular data set is fit for an intended purpose. A data set that is inadequate for one application might be completely suitable for another. The data quality information in metadata is provided so that a potential user can evaluate the data set in light of what he or she wants to do with it. During this process, six key issues must be considered: (1) lineage, (2) positional accuracy, (3) attribute accuracy, (4) logical consistency, (5) completeness, and (6) temporal accuracy. The following sections discuss these issues and provide examples of how a user might document the information in the metadata.

Source: City of Austin

### Lineage

**Lineage** is concerned with documenting the original **source** of the data set (such as the City of Austin web site that provided most of the Austin data for this text) and recording the operations and transformations that have occurred between the source and the final product. Questions to be answered about the data include who collected the data, how and why they were collected, what was the original scale or accuracy, how they were converted into digital form, and what operations have been performed on it since.

The metadata include sections on the **process steps**, which list the sequence of operations that have been performed in producing the data. In particular, the impacts of any of the processing steps on the attribute or positional accuracy need to be recorded. For example, if the lines are generalized to the nearest 500 meters to prepare a detailed states map for a faster-drawing national map, then that information should be recorded. If the feature class is the result of intersecting two layers with an XY tolerance of 25 meters, then that information, with the potential impact on the feature accuracy, must be noted.

> ▶ You digitize published maps of aquifer outcrop areas from two adjacent quadrangles generated by the US Geological Survey. You merge the two quadrangles and match the edges between them. In the metadata, you include the publication information and a complete citation for both maps. You record a process step for digitizing the maps, including the RMS errors reported for registering the maps on the digitizer. You report a second process for edge-matching and merging the two maps, including the RMS error associated with that step.

Source: ArcGIS

**Positional accuracy**

**Positional accuracy** is an assessment of how closely features in the data set correspond to their actual locations in the real world. Impacts on positional accuracy occur at several stages in the development of a data set: (1) errors present in the source data or in the original data collection, (2) errors associated with transformations or georeferencing, and (3) errors associated with subsequent processing.

1. Source data errors are generated in many ways, but the most common are by digitizing a paper map or scan, by surveying, by photogrammetry or remote sensing, or by GPS measurement. Each of these methods has different sources of error that impact the positional accuracy.

At minimum, one assumes that paper maps are accurate to within one line width, or approximately 0.5 mm. This value can be converted to ground units using the scale, as demonstrated in Chapter 1. A 1:24,000 scale map has an effective resolution of 12 meters, a 1:100,000 scale map has an effective resolution of 50 meters, and so on. If a map is digitized, its accuracy can be estimated from the scale. This estimate serves as a minimum, however, for it neglects potential inaccuracies in assembling the map data. Some phenomena or objects can be less precisely located than others. A road is easy to locate, but a soil map has an inherent uncertainty in placing the boundary between two soil types that grade into each other.

Surveying and photogrammetric methods follow rigorous standards during development and are usually subjected to quality assurance and control steps along the way. In general, base data sources, such as topographic maps and derived digital products, can be assumed to conform to the National Map Accuracy Standards. These standards require testing by comparing at least 20 checkpoints composed of well-defined and locatable points to corresponding locations on a reference map of higher accuracy. No more than 10% of the points can be off by more than 1 in 10,000 units. In practice, few maps are actually tested due to the time and cost constraints associated with mapping projects, although the procedures developed to create the maps have been, and therefore they provide some assurance that most of the products meet the standard.

The accuracy of remotely sensed information varies greatly due to the differences in resolution between satellites and the differences in geometric correction that have been applied. GPS units also vary widely in their accuracy.

2. A paper map or a scanned digital image must be georeferenced, transforming the digitizer, scanner, or screen coordinates into a real-world coordinate system (see Chapter 3). The RMS error associated with this transformation should be recorded and included with the metadata.

3. Various geoprocessing operations, such as generalizing lines or intersecting feature classes, can degrade positional accuracy. Processing tolerances, such as the XY tolerance used when intersecting to remove slivers or a cluster tolerance applied when correcting topology errors, provide an estimate of the impact. These tolerances should be recorded when listing the process steps in the metadata.

   ▶ You digitize the power lines on a 1:24,000 scale topographic map. You infer the source map accuracy to be 12 meters. The RMS error reported when you registered the paper map on the digitizer was 4 meters. Report both of these in the metadata.

   ▶ You go out in an ATV and collect GPS data for the boundaries of prairie dog towns. Previous testing indicated that your GPS unit has an accuracy of 10 meters, and you estimate that you can drive within 5 meters of the town boundary without damaging the burrows. Report the accuracy as 15 meters.

Source: ArcGIS

## Attribute accuracy

**Attribute accuracy** assesses how well the values in the attribute fields correspond to the true values in the real world. Does the label on Rapid Creek represent its real name? Does the 40% crown cover value of a tree stand really reflect the actual density of the crown foliage? Because there are so many different kinds of attributes, assessing accuracy is complex, variable, and unguided by any sort of standard methodology. However, some of the issues that should be considered are as follows.

For categorical data, attention should be given to the classification scheme. How detailed is it? Are the crown-cover categories divided into 10 groups of 10% each or into three groups? Do the land use categories represent all possible land use types? Do the geological units provide sufficient detail to locate potential gold horizons or are they simply grouped by age or gross rock type?

Issues related to all data types include the following: What is the expected rate of error in assigning values to the field? How often is something misclassified or wrongly recorded? How much heterogeneity exists within the measurement unit? The sand percentage in a soil polygon, for example, cannot be expected to be uniform over the entire area. Have any attempts been made to assess the variability within a unit?

If any information on attribute accuracy is known or if any tests have been conducted to assess the accuracy, these should be reported in the metadata. If nothing is known, then it is acceptable to put Unknown. Here are some examples.

▶ A land use map has been generated from a Landsat satellite image. You have followed industry practice and tested the classification against a set of ground data points and calculated the overall accuracy, producer's accuracy, and user's accuracy for the classification. Include these three values in the attribute accuracy report.

▶ Your survey agency has surveyed all the counties in the state on their approval of the governor's performance and is reporting an approval rating for each county. The survey method has an accuracy of ± 3 percentage points. Include this accuracy in the metadata.

▶ You have digitized a geological map from a plate in a master's thesis. The author did not include any accuracy assessment, and you have no funds to evaluate the accuracy of the map. Put Unknown in the attribute accuracy report in the metadata but include the scale of the original map and the RMS error of the map registration to points with known real-world coordinates.

Source: ArcGIS

## Logical consistency

**Logical consistency** is a measure of how well the features in the data model correspond to their counterparts in the real world and of how well the data model is able to model the relationships in the real world. This is another complex and difficult area that is rarely tested in full. However, it is usually possible and desirable to test the internal logical consistency of a data set and how well it fulfills the topological rules that have been defined for it. These topological rules were described in Chapter 12. Polygons should not have gaps or overlaps, lines should intersect at nodes and not cross each other, and so on. The logical consistency report can include the topological rules that were tested and the results of the test.

▶ You develop a planar topology for a Transportation feature dataset and check for dangles, self-intersection, and improper intersections. All identified errors were resolved. Include a statement of the topological rules that were tested, and report that the feature dataset does not contain violations of those rules in the logical consistency report.

▶ You digitized a geological map as a shapefile using snapping and Auto Complete Polygon, and you are reasonably sure that most of the polygons have no gaps or overlaps. However, you don't have an ArcGIS Standard license and cannot test the topology. Report that snapping and coincident boundary protocols were followed during digitizing and that the data layer is free or nearly free of gaps and overlaps.

Source: ArcGIS

## Completeness

The **completeness** of a data set refers to how well it has captured every possible instance of the objects in the data set. Completeness for polygons is usually easy to assess; do the polygons completely cover the area of interest or not? Lines and points can be less certain. Did the Department of Transportation digitize every known road in its GPS survey, or could it have missed some? Did it digitize only public roads, or are private roads also included? Did it include gravel and dirt tracks or only paved roads? Does the table of oil and gas wells in the state include every known well, or have some historical records been lost or misplaced prior to the assembly of the data?

▶ You assembled a database of septic systems by identifying houses on digital orthophotos taken five years ago. You realize that you may have missed systems for houses that were built after the photos were taken or for houses under heavy tree cover that were not visible in the photo. State these limitations in the completeness report.

Source: ArcGIS

## Temporal accuracy

**Temporal accuracy** is the time period for which a data set is considered valid. Some geospatial data, such as a map of greenness condition or a NEXRAD radar rain map, represent ephemeral conditions valid for a few minutes to days. Other data sets represent phenomena that change more slowly, and data sets might be considered valid for several years. Population, for example, is usually referenced to a census year or census update. Land use data change slowly as cities grow and new developments spring up. Some data can be considered fairly stable. Geology, major road networks, rivers, and international boundaries change infrequently on the scale of human affairs.

For data that will change significantly over the lifetime of the data set, it is important to designate how frequently the data provider intends to update the information. Some data are undergoing continual revision and are considered to be in progress, with updates planned at irregular or regular intervals. Census data, for example, are gathered every 10 years, and estimated updates are available every 2 to 5 years for some portions of the data set. However, a lot of GIS data is simply produced once, and no updates are ever planned.

Two sections in the metadata refer to the time period, that of the source and that of the data product being documented. These two are often the same but can be different. For example, the city planning office might have digitized the original platting records that were updated manually until 1995 to produce its parcels feature class (the source), but continued updates bring the feature class time period to within the past year (the product).

▶ Your office performed a GPS survey during 2015 and 2016 to locate and classify the existing off-road trails in the national forest, both official and unofficial. No funds are provided to locate and map new trails created in the future. The temporal period of both the source and the product is 2015–2016, and no updates are planned.

▶ Your office maintains the county parcel records that were digitized from plat maps by a consultant in 2001. Every sale or modification is reported to the county as required by law through the office of deeds and land titles. You process the updates and make changes to the internal database as they come in. You push the updates to the public version of the database on your web site every three months. In this case, the time period of the source is 2001, the time period of the data set is the date it was last pushed to the public site, and the update frequency is quarterly.

## Metadata format

CSDGM metadata are usually stored either as a simple text file or as a text file in eXtensible Markup Language (**XML**). XML is similar to the more familiar HTML used to develop web pages. XML documents can be associated with an "XML stylesheet" which specifies page layout and text formats to be used by web browsers and other applications (such as ArcGIS) to display the content in a human-readable form.

The CSDGM metadata content standard includes seven main sections (Fig. 14.2). The North American Profile of ISO 19115 has its own section organization. In some cases they correspond fairly closely with the CSDGM sections, but not in others. The following descriptions refer to the CSDGM sections, with the NAP sections that contain similar information noted in parentheses.

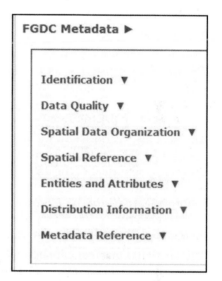

Fig. 14.2. CSDGM sections
Source: ArcGIS

### 1. Identification (Identification, Maintenance)

This section contains basic information about the original source of the data: who created them, what they contain, and any restrictions that have been placed on their use. The ***Abstract, Purpose***, and ***Supplemental Information*** communicate what the data set contains and notifies the user of any important information. The ***Currentness Reference*** indicates the time period that the data represent, and the ***Status*** says how frequently they are updated. The ***Data Set Credit*** and ***Citation Details*** provide information on who created the data product and how it should be cited in the literature. The ***Spatial Domain*** and ***Keywords*** help anyone searching for data to determine the location and the content of the data. The ***Access Constraints*** and ***Use Constraints*** specify who can use the data and whether they can be given to others, and usually contain standard liability and release statements to protect the data provider from lawsuits.

### 2. Data Quality (Quality)

This section provides information for the user to judge data fitness for a given purpose. Sections are included on the ***Logical Consistency*** and the ***Completeness*** of the data set as a whole. There is a section to report on the ***Attribute Accuracy*** and the ***Positional Accuracy***. The original scale and accuracy of the ***Source*** data must be given, including the temporal accuracy; the accuracy of the source may be different (better or worse) than the final data set. A complete ***Source Citation*** should also be included. Finally, there is a ***Process Step*** section that records the processing steps between the source and final product and the impact of these steps on the accuracy of the data.

### 3. Data Organization (Spatial Representation)

This section contains information on the format and organization of the data, such as whether they are vector or raster data. ArcGIS fills out this section automatically.

### 4. Spatial Reference (Reference System)

This section contains the complete **spatial reference** information on the coordinate system, including the GCS used, the map projection and its parameters, and the map units. ArcGIS also fills out this section automatically.

### 5. Entity Attribute (Content)

This section describes the attribute fields present in a vector data set or the Value field in a raster data set. This section is critical for giving data users information, such as what the land use codes represent or whether the AREA field is in square kilometers or square miles. The user can choose one of two methods to describe the attributes. The ***Overview Description*** can be used when all of the fields are similar or self-explanatory and can be described together in a paragraph or two. For example, a city feature class containing decadal population values since 1800 is easy to explain. A land use map with only a couple of fields might also fall into this category. The ***Detailed Description*** is used when each field requires a different explanation. The metadata creator enters separate information for each field.

### 6. Distribution (Distribution)

This section describes how the data are made available, by whom, and under what conditions. It states the ***Distributor*** and provides information for contacting that person or organization. It describes the format of the data (DVD, CD, FTP download) and the ***Standard Process*** by which a user should request the data, such as the ordering information or download instructions. It also reports the ***Available Time Period*** when the distributor intends to make the data available.

### 7. Metadata Reference (Metadata, Maintenance)

This section is metadata about the metadata, including who developed them, in what format they are stored, and if any restrictions are placed on the distribution of the metadata (usually, the metadata are freely distributable even if the data set is not). Some of this section is filled out by ArcGIS.

Standards identify certain mandatory elements for compliant metadata. There are several categories: mandatory, mandatory if applicable, and optional. Table 14.1 provides a brief overview of mandatory elements in NAP and FGDC; users should strive to cover these sections at a minimum. Some metadata editors provide guidance on required versus optional fields.

### *Acquiring metadata*

Many projects utilize data sets that come from another organization, which, if it follows industry practice, has already supplied metadata for the data set. In this case, implementing the metadata becomes easier. The user needs only ensure that the metadata are properly stored with the data set in ArcGIS and that any subsequent processing or modification after acquisition has been recorded in the Process Step elements.

For example, if the user has downloaded wetlands from the National Wetland Inventory and extracted the features in a particular county for a project, he or she only needs to record the extraction. This information would go into the Completeness Report and the Process Step sections of the metadata. The rest of the metadata need not change. In fact, it should not change;

it is important to retain the originators of the data set so that they receive proper credit and so that the prior information about the data is not lost.

However, this process may be complicated by the fact that not all organizations store the metadata as part of the data set, as ArcGIS does. The metadata may exist in a completely different document close to, but not with, the data they represent. They might be somewhere on the web site where the data are downloaded, in another folder on their source DVD, or occasionally even further afield. In many cases, the separate file will be in a standard FGDC format (text or XML) and can be easily imported to ArcGIS. Sometimes, however, the format is different or not quite correct, and the user will have a more difficult time implementing the metadata. After working with metadata for a while, you will learn to recognize

Table 14.1. Comparison of mandatory elements

| North American Profile | FGDC |
|---|---|
| Metadata Information<br>  Date Stamp<br>  Contact<br>  Metadata Standard<br>  File Identifier<br>  Language<br>  Character Set<br>  Parent Identifier<br>  Hierarchy Level<br>  Locale | Metadata Information<br>  Date<br>  Contact<br>  Metadata Standard<br>  Metadata Version |
| Identification Information<br>  Citation<br>    Title<br>    Date<br>    Responsible Party<br>  Status<br>  Language<br>  Topic Category<br>  Extent | Identification<br>  Citation<br>    Title<br>    Publication Date<br>    Originator<br>  Description<br>  Status<br>  Time Period of Content<br>  Spatial Domain<br>  Keywords<br>  Access Constraints<br>  Use Constraints |

the formats that are suitable and that can be imported into ArcGIS with a minimum of fuss. A number of programs and utilities have been developed to facilitate this process; for example, a program called mp (metadata parser) was created by the U.S. Geological Survey to read text-based CSDGM metadata and convert it to XML so that it can be ingested by ArcGIS; it is included in ArcGIS Desktop and can be run in the Python window (see ArcGIS Help for more information).

### Metadata development

For data sets developed by you or your organization, the metadata process must start at the beginning. Although entering the metadata may be one of the final steps in producing the data set, the process begins earlier, when the data are being assembled—critical information must be recorded throughout the process so that it can be added to the metadata when the time comes.

Imagine that you are digitizing a historical map of the Lewis and Clark journey. You initially need to record information about the map: author, publisher, year of publication, citation, scale, medium on which it was drafted, and so on. You would look for information about the positional accuracy or time period. Additional information is recorded as you process the data, including the RMS error of the georeferencing step, techniques or tests used to ensure topological integrity, processing tolerances, or information about content in the attribute fields.

If you wait until the data set is complete, then critical information may be lost. It is a good practice to develop a form on which notes can be recorded during the creation of the data, organized so that the information can be easily translated to the metadata at the end. You can

Table 14.2. Sample notes template for recording information about data layers

| Project: | | | Feature Layer: | |
|---|---|---|---|---|
| Compiler: | | Start Date: | | End Date: |
| ___Download | ___From Client | ___ CD/DVD | ___Other: | |
| Original Scale: | | | Original Format: | |
| Original Coordinate System: | | | | |
| Originator Information (name/agency, contact info, and/or URL): | | | | |
| Publication Reference (title, author, date, publication, version, edition, series, etc.): | | | | |
| Horizontal Accuracy: | | | Vertical Accuracy: | |
| Field Collection Notes or Other Notes: | | | | |
| Processing Steps and RMS When Applicable: | | | Notes on Attributes/Units/Definitions: | |

record these notes on paper or electronically. Even if all of the information recorded does not make it into the metadata, the notes will provide a valuable resource for the future. Table 14.2 shows a sample of a form that might be used. A copy of this form is stored in the mgisdata\Metadata folder for your use.

## About ArcGIS

### *Metadata management in ArcGIS*

Metadata functionality has been included in ArcGIS for many years. Although metadata software that works apart from ArcGIS is available, metadata managed within ArcGIS is convenient because it is stored as part of geospatial data sets. In addition, ArcGIS can automatically populate and update metadata elements from the data set properties (coordinate system, number of features, etc.). ArcGIS 9 had two metadata editors, one for FGDC metadata and one for ISO 19115, with a translator to convert it (somewhat imperfectly) to ISO 19139. No capability existed to produce NAP metadata at ArcGIS 9. On the other hand, ArcGIS 9 supported import and export of text-based format of FGDC metadata and applying it to data sets.

In ArcGIS 9, metadata storage and editing capabilities were closely bound to the standard being used, making it difficult to meet the needs of all users in North America and internationally. In ArcGIS 10, a major redesign of metadata management was done to support multiple standards and to adapt to changes over time. This goal was accomplished by storing metadata information in an internal ArcGIS format, then translating and exporting it to a standards-compliant XML file when needed. Users who are not required to produce data accompanied by a specific metadata format can simply stay within the ArcGIS metadata framework. Two different levels of metadata support are provided for these users.

A streamlined set of descriptive information, called the **Item Description**, can be filled out in a few minutes. This approach helps solve a long-standing problem—metadata were so complex and difficult that many people never bothered documenting a data set. The Item Description is easy to understand and to edit, and although it falls far short of a complete metadata record, it is much better than nothing. An Item Description can be created for any item type in ArcGIS, not just geographic data sets but also models, tables, scripts, and more. This minimum Item Description is required for every data set shared on ArcGIS Online.

Users desiring greater levels of documentation can access and edit additional metadata information by specifying a Metadata Style in the application options (Fig. 14.3). The style is based on one of the metadata standards, and it controls which fields you will see when editing metadata. It also defines how the metadata are validated and exported to a standard.

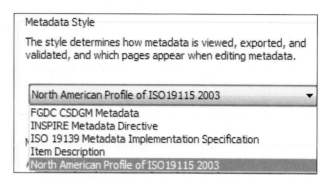

Fig. 14.3. Metadata styles
Source: ArcGIS

Users who are tasked with creating metadata conforming to a standard must edit the ArcGIS metadata content and then export it to a separate XML file, which can be posted on the Internet or submitted to a data clearinghouse. Metadata created in ArcGIS 9 using the FGDC or the ISO editor can be viewed in ArcGIS 10 but must be upgraded to be edited. This upgrade will be available (using an Upgrade button) if ArcGIS detects the presence of these metadata formats.

In this chapter, we will cover the basic editing of ArcGIS metadata items using the North American Profile of ISO 19115, more as an introduction to understanding metadata and some of their key components rather than as a serious effort to train a metadata producer.

**TIP:** The relationship among CSDGM metadata, ArcGIS metadata, ISO metadata, and other styles is complex. You can ignore this complexity if you are simply creating new metadata, but if you must work with data sets containing FGDC metadata developed prior to ArcGIS 10, you will need more guidance. See the ArcGIS Help under Manage Data > Data Types > Metadata.

### *The metadata editor*

In Chapter 2, you learned how to view the item description in the Catalog tab. To move beyond the simple description, you must specify a metadata style. Then the Metadata toolbar (Fig. 14.4) allows you to view, edit, import, export, and validate metadata (compare them to the requirements of the standard) using that style.

Fig. 14.4. The Metadata toolbar in ArcCatalog
Source: ArcCatalog

Clicking the Edit button takes you to the metadata editor (Fig. 14.5). It is easy to use, but its organization may take some time to learn. There are three main sections: an Overview of the data set, a Metadata section containing information about the metadata themselves, and a Resource section containing information about the data set. There is no set order; the sections can be visited in any order and at any time. Each section has multiple items to be filled out. In some cases they already exist, and the user simply fills them in. In other sections, users must create containers first, which are then filled with content.

Certain metadata fields are required by the standard in order to produce minimally compliant metadata. These boxes are filled in red to assist you in completing the information. In addition, error messages at the top of the editor let you know when an item is missing. Clicking the hyperlink may open the relevant item if it is not currently visible (although this feature may not "drill down" all the way to the item flagged due to known ArcGIS software limitations).

Some sections allow the user to add multiple items in an information category. For example, a data set might have used three data sources which all need to be documented (Fig. 14.6). Documenting at least one source is required information, and the entry boxes appear automatically. The Down arrows expand a section so that you can see its contents, and the Up arrows hide it again. You can type a general description above. You can also add items, such as a full-fledged citation, by clicking the + New Citation text. To add a second source, you would click the + New Source text to provide another container. The small red X symbol lets you delete a container if you decide you don't need it after all.

The ArcGIS 9 editors had places to enter all types of data quality reports, such as logical consistency, horizontal accuracy, vertical accuracy, and completeness, which in some cases were not applicable to a data set or the user had no useful information to impart. In the ArcGIS 10 editor, a report entry is created only if you need it. You add the report, specify the type, and fill in the relevant details. Thus, the final metadata that you see contain only useful information, not rows of blank entries.

### Metadata templates

Metadata standards require an impressive amount of information to be entered. Fortunately, much of this information is repeated verbatim in different data sets produced by the same organization. This fact makes the use of templates attractive. A **metadata template** contains the information that will be common to every data set produced by an organization or for

Fig. 14.5. The Metadata editor
Source: ArcGIS

Fig. 14.6. Adding source information
Source: ArcGIS

a particular project. The contact information and access and use constraints, for example, are metadata fields that are likely to be consistent within an organization.

Metadata templates can take a hierarchical approach (Fig. 14.7). Information common to every data set for the organization serves as the starting point. The template is then edited to include common information for a specific project (such as the contact information for the project manager). The usual practice for a company might be to make all data available only to clients, but if a certain project is funded as an educational effort, then the usual access constraints might be edited to make the data freely redistributable. Templates might have three levels of specificity:

***Organizational level:*** Contains fields that are the same for every data set produced by the organization. Examples are the organizational contact, the access and use constraints, and the metadata distribution constraints.

***Project level:*** Contains fields that are the same for every data set produced for a particular project. Some of the organizational-level fields might be edited for this level. Examples are the Purpose, the Data Set Credit, the organizational contact, and some of the key words.

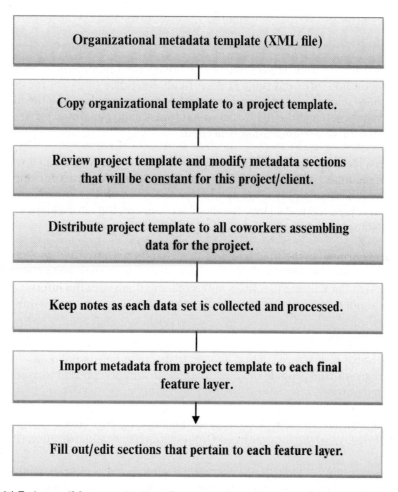

Fig. 14.7. A possible organizational process for using metadata templates

***Data set level:*** Contains fields that must be entered for each data set, such as the Summary and Description, the citation information, and accuracy and completeness reports.

The overall process might look something like Figure 14.7. The blue box refers to the organizational-level template, the orange boxes to the project-level template, and the green boxes to the data set-level metadata. The user starts by creating an empty XML file to serve as the metadata template. He then fills out all the sections common to every data set produced by the organization. This data set becomes the basic source for all subsequent metadata.

When a new project is initiated, the project manager copies the organizational template to a project template. She reviews the metadata and edits any portions that will be constant for the project. This might include changing some of the organizational fields or adding new information to the template. Once the project template is complete and has been reviewed, it is distributed to all of the workers who are producing data for the project. As each data set is assembled, the workers take notes on the source and processing. When the data set is ready to have its metadata created, the worker imports the project template into the metadata for the new feature class. The data set-specific information is added to the metadata to make them complete.

For guidance in filling out NAP metadata, the metadata editor itself provides tags that are visible when the cursor is placed on a particular item. In addition, many of the NAP and CSDGM items are similar, and suggestions for one are equally valid for the other.

### Metadata maps

The end of this chapter includes ArcGIS/NAP and CSDGM metadata maps that show the structure of the metadata in outline form and suggest what fields should be filled out for the organizational template, which ones might be edited or added for the project template, the fields that must be filled out for each data set, and the fields that are automatically generated by ArcGIS. They provide an overview of what information is needed.

When developing a metadata template, it is easier to compose the material in a word processor rather than in the metadata editor itself; it is much easier to see the structure, and it is more convenient to e-mail it to an individual or committee to review and get comments back using standard editing/tracking methods. Once the draft is complete, the information can be transferred to the metadata editor by copying and pasting into the relevant sections.

The metadata maps printed in the final pages of this chapter are provided as Word and .RTF documents in the mgisdata\Metadata folder, if you wish to use them to compose metadata templates by deleting the italicized sections and replacing them with the information for the metadata. (You cannot see the files in the Catalog, but you will see them in Windows Explorer.)

## Summary

▶ Metadata records provide information needed to discover geospatial data, document the history of a data set, and provide the data quality information needed to ascertain its fitness for an intended purpose.

▶ Metadata standards and formats have been established by the Federal Geographic Data Committee (FGDC) and the International Organization for Standardization (ISO).

▶ United States metadata standards are currently in flux as users migrate from the current FGDC CSDGM standard to a new international standard, the North American Profile of ISO 19115. International users are more likely to use the ISO 19139 or 19115 standard.

▶ Six key data quality issues are lineage, positional accuracy, attribute accuracy, logical consistency, completeness, and temporal accuracy.

▶ CSDGM metadata are organized into seven main sections, including Identification, Data Quality, Data Organization, Spatial Reference, Entity Attributes, Distribution, and Metadata Reference. NAP metadata have a different organization, but many of the sections are similar.

▶ ArcGIS automatically fills out some sections of metadata.

▶ Establishing templates can reduce the work needed to prepare metadata.

**Important Terms**

| | | | |
|---|---|---|---|
| attribute accuracy | Item Description | positional accuracy | temporal accuracy |
| completeness | lineage | process step | XML |
| CSDGM | logical consistency | profile | |
| FGDC | metadata standard | source | |
| ISO | metadata template | spatial reference | |

## Chapter Review Questions

1. What is the mission of the FGDC?

2. What is the difference between the source and the product when preparing metadata?

3. How much metadata is required to share data on ArcGIS Online?

4. Why have formal guidelines for testing attribute accuracy not been established?

5. What aspects of logical consistency are most frequently tested and reported?

6. Contrast the temporal accuracy of a rivers feature class with that of a congressional districts feature class.

7. Why will US users need to be familiar with two metadata standards for the near future?

8. Which section of metadata would you consult to determine how to order the data and how much they cost?

9. Which section of metadata would you consult to determine if the ROADLENGTH field of a roads feature class is in kilometers or miles?

10. You download a data set from the web and discover that it has no metadata when viewed in ArcGIS. What is your next step?

# Mastering the Skills

## Teaching Tutorial

The following examples provide step-by-step instructions for doing basic tasks and solving basic problems in ArcGIS. The steps you need to do are highlighted with an arrow →; follow them carefully. Click on the video number in the VideoIndex.htm file to view a demonstration of the steps.

In this tutorial, we will first examine some metadata examples to see what good metadata look like. We will then create metadata for the weather map that we georeferenced in Chapter 3. For an extended metadata editing session, some find ArcCatalog easier to use than the Catalog tab in ArcMap, but either may be used as preferred.

→ Start ArcCatalog, or open the Catalog tab in ArcMap and open the Item Description window.

### Examining metadata

Most of the feature classes in the usdata geodatabase came from the ESRI Data and Maps download that is distributed with the ArcGIS software. The ESRI metadata provide excellent examples to learn from.

1→ Navigate to the mgisdata\Usa folder and expand the usdata geodatabase.

1→ Click the quakehis layer to highlight it, and click the Description tab.

1→ Skim the information here, making sure you scroll to the end.

By default, ArcCatalog shows only the basic descriptive information here, the **Item Description**. To view and work with more extensive metadata, the setting must be changed.

2→ Choose Customize > ArcCatalog (or ArcMap) Options from the main toolbar.

2→ Click the Metadata tab.

2→ Change the metadata style to *North American Profile of ISO 19115 2003*. Click OK.

2→ Click the Preview tab and then click the Description tab to update the view.

2→ Scroll down again and notice how much more information there is, including two sections for ArcGIS Metadata and FGDC Metadata.

Styles may be based on a standard such as ISO 19115. ArcGIS stores the metadata information internally in its own format. The metadata style controls which information items are displayed and edited. The metadata can be exported to the standard format if needed.

**TIP:** The large-font blue headings are section headings and may be collapsed (side arrow) or expanded (down arrow). Use these headings to hide or view the information as directed in the next steps.

**ArcGIS Metadata ▶**

Topics and Keywords ▼

Citation ▼

Source: ArcGIS

3→ Start by collapsing all the headings under ArcGIS Metadata, if necessary, so that you can see the organization of the 17 sections.

3→ Expand the Topics and Keywords section and skim through the information.

3→ Close it when you are finished, by clicking the Topics and Keywords heading again or by using the Hide hyperlink at the bottom of the section.

3→ Examine the Citation section.

1. What was the publication date for this data set? _____

3→ Examine the Citation Contacts section.

2. Who originated this data set?_____ Who is the publisher? _____

4→ Examine the Resource Details section.

4→ Examine the Resource Points of Contact to find out whom you might contact with questions about this data set.

4→ Examine the Resource Maintenance section to find out how often this data set is updated.

Let's look more closely at where this data set came from.

5→ Expand the Lineage section and look at the process steps

3. How many process steps are listed? _____ When did they occur? _____

4. What source was used to compile this data set? _____

5. Can you find the original scale (scale denominator) of this data set? _____

6→ Examine the Spatial Reference section to see the coordinate system.

6→ Examine the Spatial Data Properties.

7→ Expand the Data Quality section. There are five data quality reports for this data set.

7→ Examine the names and content of the five reports, comparing them to the data quality issues discussed in this chapter.

8→ Examine the Distribution section.

6. How does one obtain a copy of these data? _____

8→ Examine the Metadata Details and Metadata Contacts section.

7. Who produced the metadata? _____ What standard was used?_____

8→ Examine the Fields section.

8. What is the earthquake depth unit of measurement? _____

Clearly, it takes a lot of effort to document data sets to this level of detail. Not all data sets are this well documented, but all data sets need at least minimal information. As you continue to learn about metadata, you can go back to the data sets in this usdata geodatabase for good examples of how metadata are organized and what types of information go into each section.

### Creating a project template

You have been provided with an organizational template from a fictional company called Re-GIS, Inc., which distributes weather and climate data sets by FTP download, usually at a cost. Imagine that you are the project manager for a new initiative to provide a freely distributable monthly summary of precipitation based on National Weather Service radar data. You will create a project template for use with the data sets to be developed for this project.

9→ In ArcCatalog, navigate to the mgisdata\Metadata directory and click on the REGIS_OrgTemplate XML file to highlight it.

9→ Click the Description tab, if necessary, and examine the various sections of ArcGIS metadata.

9→ Scroll through the metadata, noting the type of information that is already filled out.

Since this is only a template, many of the sections are still blank. The spatial information, for example, is only applicable to spatial data and is updated automatically by ArcCatalog. Therefore, it is not present in this XML template.

The organizational template contains information that will be the same for all data produced by the company. Your project template will include this information, as well as information that will be the same for all data sets produced for the project. The sections unique to each data set will be left blank, to be filled in as each data set is created.

10→ Right-click the REGIS_OrgTemplate XML file and choose Copy.

10→ Right-click the Metadata folder entry and choose Paste.

10→ Click on the new REGIS_OrgTemplate(1) XML file twice slowly, and rename it Radar_Project_Template.

**TIP:** To create a completely empty XML file to start your own organizational template, right-click a folder and choose New > XML Document, and give it a name.

 11→ Click on the Radar_Project_Template file to highlight it and click the Edit button on the Description tab.

11→ Examine the outline of sections on the left side of the editing window.

The Overview section contains a general description of the data set. The Metadata section contains information about the metadata, such as who created them and whom to contact with questions about them. The Resource section contains detailed information about the data set.

**TIP:** When you place the cursor in a box, a pop-up at the bottom of the ArcCatalog window tells you the type of information that it should contain.

11→ The Title, Tags, and Description will be different for each data set, so leave them blank.

11→ For the Summary, enter *This data set was developed for educational use for studying the spatial distribution of rainfall events in the United States.*

11→ For the Credits, type *Based on data provided by the National Weather Service. This data set is freely distributable for noncommercial purposes with inclusion of citation and full metadata.*

11→ Set the sliders for the Appropriate Scale Range from Continent to State.

11→ Save your edits and start editing again.

**TIP:** A red boundary around an item indicates that the content is required by the standard currently specified in the Metadata Style.

12→ Click on Overview: Topics & Keywords in the outline.

12→ Check the Atmospheric Sciences box.

12→ Set the Content Type to Downloadable Data.

12→ Click on the + New Theme Keywords text.

12→ Enter *precipitation, NEXRAD, radar* as key words.

13→ Click on + New Place Keywords and enter *United States.*

13→ Click on + New Temporal Keywords and enter *2013.*

**TIP:** The dates to be entered reflect when this tutorial was written. You may change them to dates that make sense for when you are doing the tutorial, if you wish.

Next you will enter the Citation information, showing how it should be cited by others.

14→ Click on Overview: Citation. Notice the messages at the top, telling you that this information is required by the metadata standard.

14→ Enter the Citation Title as *Monthly Summary of Radar Precipitation for 2013.*

14→ Click on + New Presentation Form and set the Presentation Form drop-down to Digital Image. Set the FGDC Geospatial Data Presentation Form to Raster Digital Data.

14→ Expand the Dates entry. The Created date will be unique for each data set, so leave it blank.

14→ The data are scheduled for publishing on 3/31/2014. Click on the Published calendar icon and click the *Date, Year* heading on top of the calendar to move quickly to March 2014. Enter 3/31/2014.

14→ Notice that the error messages at the top have disappeared, as has the red X on the Citation section.

14→ Save your edits and start editing again.

**TIP:** Save your edits frequently!

15→ Click on the Citation Contacts section.

15→ Expand the Contact and the Contact Information sections.

This information was present in the organizational template, saving you some typing. It could be customized with your name, but generally it is wiser to keep it generic, to a position rather than a person, in case of future personnel changes. The next section provides more information about the metadata themselves.

16→ Click on the Metadata: Details section.

16→ Click Create to create the File Identifier.

16→ Set the Function to Information.

16→ Set the Date Stamp to today's date.

16→ Set the Language to English and the country to UNITED STATES.

16→ Set the Hierarchy Level to Dataset.

17→ Click the Metadata: Contacts section. Again, this information was provided by the organizational template.

The previous contact information you saw referenced the data set. This one references the metadata. In this case they are the same.

18→ Click the Metadata: Constraints entry. Expand the Down arrows. The information from the organizational template is in here and does not need updating.

18→ In the Security Constraints section, set the Classification to Unclassified.

18→ Click Save and start editing again.

The Resource section contains information about the data set itself. Some of it will seem familiar because you already set it in the Metadata section.

19→ Click on the Resource: Details entry.

19→ Set the Language to English, and the Country to the United States.

19→ Leave the other information blank, to be filled in for each data set.

20→ Examine the Resource: Extents entry. This part will be filled in by ArcGIS for each data set.

20→ Examine the Resource: Points of Contact. Again, the template took care of it.

**TIP:** Notice that the metadata require multiple contacts to be filled in, often with the same information. The Overview: Contacts Manager and the Load a Contact button let you reuse previously entered contacts to save typing. The drop-down list seems to have a bug and does not show the actual names, which makes it confusing. Use it for your own metadata with caution.

21→ Click on the Resource: Maintenance entry.

21→ Change the Update Frequency to Not Planned. The rest of the information would be filled out if you were planning to update the data after publication, but that is not the case for this project.

22→ Click on Resource: Constraints and expand the General Constraints entry.

22→ Delete the first sentence of the Use Limitation, since the free distribution policy for this project is different from the usual company policy.

22→ Save and start editing again.

23→ Click the Resource: Spatial Reference entry. This section will be filled in by ArcGIS.

23→ Click the Resource: Spatial Data Representation entry. This section will be filled in by ArcGIS.

The Content, Quality, and Lineage sections will be different for each data set produced by the project, so these are left blank for now. Going on, the Distribution section explains how someone can obtain the data.

23→ Click on the Resource: Distribution section.

23→ Click the + New Distribution Format entry.

23→ For the Format Name, enter *ArcGIS File Geodatabase*.

23→ For the Format Version, enter *Version 10.5* (or whatever you are using).

The actual distributor is the manager of the company web sites, so the next section is for the company web master.

24→ Expand the Distributor entry.

24→ Expand the Contact entry and the Contact Information entry.

24→ Click the + New Online Resource text. You may need to expand the Distributor and Contact Information entries again.

24→ Enter *http://www.ReGIS.MonthlyRadar.com* for the Linkage.

24→ Enter *ftp* for the Protocol.

24→ For Name, enter *Monthly Summary of Radar Precipitation*.

24→ For Description, enter *Download site for educational data sets of monthly radar precipitation*.

24→ Set the Function to Download.

25→ Scroll down, if necessary, and click the + New Ordering Process entry. Change the Fees to *Free*.

25→ You intend to release the data at the end of March 2014. Change the Available Date to *3/31/2014*.

25→ Enter *Download* for Ordering Instructions.

25→ Enter *Immediate* for Turnaround.

25→ Click Save.

The project template is now complete.

### *Importing the template*

Now you will create metadata for your first data set for the project. The nwsradar.img image in migsdata\Metadata was downloaded and georeferenced from the National Weather Service (as you did in Chapter 3). It will serve as an example of the data used to generate the final product and is to be included in the publication data set. We will create metadata for it.

**IMPORTANT TIP:** It is recommended to turn off background processing before running the Import/Export tools.

26→ Choose Geoprocessing > Geoprocessing Options from the main menu bar.

26→ <u>Uncheck</u> the Enable box for background processing, if necessary. Click OK.

27→ Click on the nwsradar.img raster in the mgisdata\Metadata folder to highlight it. Note the virtually empty metadata.

27→ Click the Import button on the Metadata toolbar.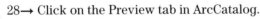

27→ For the Source Metadata, click the Browse button and navigate to the mgisdata\Metadata folder.

27→ Choose the Radar_Project_Template XML file and click Add.

27→ The metadata you've been entering are stored in the internal ArcGIS format, so change the Import Type to FROM_ARCGIS.

27→ Click OK to start importing.

28→ Click on nwsradar.img in the Table of Contents.

28→ Scroll down and examine the metadata now.

The metadata for this raster, which were essentially blank a moment ago, have been updated with the information from your template. ArcCatalog also filled in some information, for example, about the spatial storage format and the coordinate system.

28→ Click on the Preview tab in ArcCatalog.

28→ Click the Create Thumbnail button.

28→ Click the Description tab again.

The metadata have been updated to include the thumbnail you just created.

### *Entering metadata for an item*

Now we will finish entering the metadata for the nwsradar image. However, you will need to know some information about the source data. You had one of your coworkers investigate the web site where the information was obtained and talk to the people at the National Weather Service who provide the radar data. She has assembled the information in Table 14.3 for you.

Table 14.3. Information about the radar image for inclusion in the metadata

| Source information for NWS radar example | |
|---|---|
| Citation title | NWS Radar Mosaic<br>Time and date: 1718 UTC 06/20/2013 |
| Originator | National Weather Service<br>1325 East-West Highway<br>Silver Spring, MD 20910 |
| Resolution | 0.02 degrees |
| Scale of source | Approximately 1:4 million |
| Info | National mosaic map of radar sectors for the conterminous United States. Units in DBZ from –25 to 75 |
| URL | http://www.weather.gov/radar_tab.php |
| Estimated accuracy | Source resolution 0.02 degrees<br>RMS error of transformation = 0.002 degree |

29→ Click the Edit button and click the Overview: Item Description section.

29→ For Tags, enter radar, NEXRAD, precipitation, storms.

29→ For the Description, enter This data set provides an example of the radar measurements used to develop the monthly precipitation and storm summaries. Each color increment represents an increase of approximately 5 DBZ (decibels). Largest scale when displaying the data is approximately 1:4 million.

29→ Click Save and start editing again.

This is where the value of the template is obvious. Most of this section was already filled in.

30→ Click the Overview: Citation section.

30→ Expand Titles if necessary and replace the file name with Monthly Summary of United States Precipitation 2013 for the Title.

30→ Expand Dates and change the Created Date to the current date. The Published date is already entered from the template.

30→ Click Save and start editing again.

The rest of the Overview section should also remain the same as set up in the project template. Next we turn to documenting the Content, Quality, and Lineage sections. The data set is a raster representation of a map rather than a directly measured image like a Landsat scene, so we will use the Coverage type to describe the content.

31→ Click on the Resource: Content section.

31→ Click the red X to delete the Image Description entry.

31→ Click the + New Coverage Description entry.

31→ For the Attribute Description, enter NEXRAD Radar in DBZ (decibels) and set the Content Type to Image.

We now turn to the Quality section. We only need to enter a report for quality information that we actually have, which is an estimate of the positional accuracy.

32→ Click the Resource: Quality section.

32→ Set the Level Scope to Dataset.

32→ Click + New Report and set the Report Type to Relative Internal Positional Accuracy.

32→ Set the Dimension to horizontal.

33→ Click + New Quantitative Result.

33→ For Value Type, enter *Georeferencing RMS Error*.

33→ Enter *0.02* for the Value and select *plane angle: degree deg* for the Value Unit.

33→ Save and return to editing.

Next we need to document the Lineage of the data, which includes the sources and processing. When entering sources, one can be satisfied with a general statement and/or source description, or one can construct a detailed citation. We will do both, although some of the fields will be blank. This section is somewhat more extended than the others.

34→ Click the Resource: Lineage section.

34→ In the Statement box, type *Radar image was downloaded from the National Weather Service, georeferenced, and projected.*

34→ Click the + New Data Source entry.

34→ In the Source Description box, type *NWS Radar Mosaic Series, downloads available every 10 minutes. This item downloaded 17:18 UTC 07/28/2013.*

34→ For Medium Name, choose On-line.

34→ For Scale Denominator, type *4000000*.

35→ Click + New Source Citation. Enter the Title *NWS Radar Mosaic*.

35→ Click + New Presentation Form and choose Digital Map.

35→ For FGDC Geospatial Data Presentation Form, choose Map.

35→ Click + New Identifier and enter *National Weather Service/NOAA* for Code.

35→ Expand Dates and put *7/28/2013* for both Created and Published Date.

36→ Click + New Contact. Leave the Name blank.

36→ Enter *National Weather Service* for the Organization. Set the Role to Originator.

36→ Click + New Contact Information and choose Both for the Address Type.

36→ Enter the NWS address from Table 14.3.

36→ Click + New Online Resource (under the Email box) and enter the URL from Table 14.3 for the Linkage.

36→ Enter *HTTP* for the Protocol, and *Doppler Radar National Mosaic* for the Name, and select Download for the Function.

Next we enter the two processing steps: georeferencing and projecting. Although one could enter all kinds of information about who did it and when, we will just use a general statement.

37→ Scroll down and click the + New Process Step entry. For the Process Description enter *Downloaded NWS Mosaic radar image and georeferenced it to the GCS NAD 1983 coordinate system with RMSE of 0.02 degrees using ArcGIS 10.1.*

37→ Fill the Date with the processing date, *7/28/2013.*

38→ Click + New Process Step and enter the Process Description as *Projected the georeferenced raster to North America Equidistant Conic coordinate system with nearest neighbor resampling and output resolution of 2000 meters.*

38→ Fill in the process date with *7/28/2013.*

38→ Click Save and start editing again.

You are almost finished, but notice that three sections still have a red X, indicating that there are a few missing items to cover.

39→ Click the Metadata: Details section and click Create for the File Identifier.

39→ Click the Resource: Spatial Reference section and set the Dimension of the Reference System to horizontal.

There is one more error listed. The image has a value attribute table (VAT) containing the colormap of the image values. It is used internally by the program and does not need to have a description in the metadata.

40→ Click the Resource: Fields section.

40→ Click the red X next to the Details entry to delete it.

40→ Save your edits.

40→ Scroll down and admire your metadata.

→ Exit ArcCatalog.

## Exercises

Imagine that your class includes a GIS project (as it may). Search the Internet for two data sets that you might use for your project. Print MetadataNotes (.docx or .rtf) in the mgisdata\Metadata folder and fill it out for each data set (or fill it out electronically and print afterward).

## Challenge Problem

Prepare a metadata template for your class project (or use your home state geodatabase). Save a copy of one of the metadata maps (see below) in the mgisdata\Metadata folder. Determine which sections will be filled for the template, delete the italicized text, and replace it with actual information. Leave the italicized sections alone if the template will be blank for that section. Print and turn in for your instructor to review. Afterward, enter the information into an XML document to create the template.

## A NAP Reference Map

Use this map as a guide to creating metadata templates and filling in the Metadata sections using the ArcGIS 10 Editor using the North American Profile of ISO 19115 style. Word and RTF documents of this map are provided in the mgisdata\Metadata folder.

## Key

**To be filled in by developer for each layer**
**Provided by Organizational Template but might differ for Project Template**
Provided by Organizational Template
Automatically generated by ArcGIS

# 1. Overview

## Item Description

**Title**
*The name of the data set*

**Tags**
*Comma-delimited list of key words to aid in searching for the data*

**Summary**
*What is the intended use for this data set? Under what conditions is it valid?*

**Description**
*Provide a brief description of the information contained in the data set. This is also a good place to put the largest scale recommendation. Largest scale when displaying the data: 1:xxxxx.*

**Credits**
*List those who created or contributed the data. This is a good place to list modifications made to imported data, such as subsetting it to a study area from an ESRI data set. Other things may be added as appropriate.*

## Topics & Key Words

**Topic categories**
*Choose from a predefined list of general topics, such as environment or transportation.*

**Theme key words**
*Search key words relating to the content or theme, such as geology or census*

**Place key words**
*Search key words relating to the location, such as United States, Austin, or Oregon*

**Temporal key words**
*Key words relating to the relevant time period of the data, such as the 2010 Census, 1990, etc.*

## Citation

**Titles**
*The name used to identify the data set*

**Identifiers**
*The authoritative reference, such as the USGS or the National Weather Service*

**Dates**
*Dates when the data set was created, published, and if applicable, revised*

**Edition, Series**
*Information regarding the edition of the data, if multiple versions are available, and/or a collection or series of which the data are a part*

**Responsible Party**
*The person, position title, or organization responsible for or associated with the resource. At least one of the three must be provided, with appropriate contact information, including address, phone, e-mail, hours of operation, etc.*

## Citation Contacts

### Citation Contact(s)

*The person, position title, or organization responsible for or associated with the resource. At least one of the three must be provided, with appropriate contact information, including address, phone, e-mail, hours of operation, etc.*

## Locales

### Locale

*The language in which the data set is published and the country to which it pertains*

# 2. Metadata

## Details

### File Identifier

*Name of the metadata file and a date stamp*

### Language and Character Set

*Language and computer character set used to create the metadata*

### Hierarchy Level

*The scope to which the metadata apply—are they for a data set, software, model, etc.*

### Metadata Standard Name

*The name of the metadata standard and profile, if applicable, used to generate the data. These items can be left blank if the metadata are intended to remain within ArcGIS.*

## Contacts

### Metadata Contacts

*The person, position title, or organization responsible for the metadata. At least one of the three must be listed. Contact information should also be entered.*

## Maintenance

### Update Frequency and Scope

*Are there future updates planned for the metadata? At what intervals? When is the next scheduled update? Additional contacts may also be listed here.*

## Constraints

### Use Limitation

*Describe any potential limitations as to the use of the metadata.*

### Legal Constraints

*Specify who may legally use the metadata. Are they freely available or must they be purchased? This is the place for an organization to enter any statement limiting liability or restricting the use of the metadata to certain entities. Includes information on copyrights, trademarks, etc.*

### Security Constraints

*Use to indicate data for which security clearance or other restrictions are needed.*

# 3. Resource

## Status

### Status

*Is this in progress or a final version?*

### Credit

*List those who created or contributed the data. This is a good place to list modifications made to imported data, such as subsetting them to a study area from an ESRI data set. Other things may be added as appropriate.*

### Language and Character Set

*Language and computer character set used to create the metadata*

### Spatial Representation Type

*Vector data? Grid? Table? Video? What is the intended scale resolution for vector data, or distance resolution (cell size) for raster data?*

### Processing Environment

*What operating system was used to generate the data?*

# Extents

This entire section is updated automatically by ArcGIS.

# Points of Contact

### Contacts

*The person, position title, or organization responsible for or associated with the data set. At least one of the three must be provided, with appropriate contact information, including address, phone, e-mail, hours of operation, etc. If there is an online point of contact, then the URL linkage must be listed.*

# Maintenance

### Update Frequency and Scope

*Are there future updates planned for the data set? At what intervals? When is the next scheduled update? Additional contacts may also be listed here.*

# Constraints

### Use Limitation

*Describe any potential limitations on the use of the data set.*

### Legal Constraints

*Specify who may legally use the data set. Is it freely available or must it be purchased? This is the place for an organization to enter any statement limiting liability or restricting the use of the metadata to certain entities. Includes information on copyrights, trademarks, etc.*

### Security Constraints

*Use to indicate data for which security clearance or other restrictions are needed.*

# Spatial Reference

This entire section is updated automatically by ArcGIS.

# Spatial Data Representation

This entire section is updated automatically by ArcGIS.

# Content

### Coverage, Image, or Feature Description

*Add a section based on whether the data set is a coverage (gridded thematic data such as a geology raster), an image, or a feature catalog (vector data set). Then details about the content will be added, dependent on the type. For example, the gridded NEXRAD radar data are a coverage with a physical measurement content type. An image description would contain information such as the number of bands, the illumination angle, quality codes, processing levels, etc.*

# Quality

### Level Scope

*Level to which the data quality reports refer, such as the data set, a table, an attribute, a model, etc.*

### Reports

*One or more reports may be added as applicable, including common reports on:*

***Logical Consistency Report:*** *Logical consistency is largely concerned with topology. Did you test to see if the data contain dangles, gaps, or overlaps? What tests were applied, and what were the results?*

***Completeness Report:*** *Provide information about omissions, selection criteria, generalization, and other processes that might impact how complete a data set is. Are all spatial entities included? For example, did you get all the wells or might some be missing? Did you subset the data from their original source? Were any criteria used in deciding which features to include (public versus private roads, for example)?*

***Attribute Accuracy Report:*** *Summarize processes used to establish the accuracy of the attribute(s), for example, known detection limits of analyses. Evaluate detail or completeness of categorical classes. Provide known information about problems with any attributes. If no accuracy data are available, enter "Unknown."*

*Horizontal Accuracy Report:* If data are from a standard US federal data product, enter the national map accuracy standard of 1 in 10,000. If data are surveyed or obtained via GPS, enter the known or estimated positional accuracy of the survey or GPS unit. If data have been georeferenced, transformed, or spatially adjusted, also include the RMS error.

# Lineage

### Data Source(s)

A general statement or detailed citation of the source(s) of data used to create the resource. Multiple sources and citations may be added. You may include original publication details of the data set: title, originator, publication date, edition, and so on.

### Process Step(s)

Describe one or more processing actions taken on the data before reaching its final form, for example, importing STDS quads and joining elevation attributes, merging into a single data set, projecting to current coordinate system, and clipping to study area boundary. Enter software version, process date, and the name of the person who did the processing. Include impacts on accuracy, such as an RMS error associated with georeferencing.

# Distribution

### Distribution Format

Format in which the data are distributed, for example, zip file, DVD, download, specifications, etc.

### Distributor

Enter the agency or person responsible for distributing the data and their contact information. It also includes information on how to order the data, download options, URL linkages, costs, etc.

# Fields

### Entity and Attribute Details

A listing of data fields with specific field definitions. Most of the information is supplied by ArcGIS, but the user can add important descriptions. Describe the main characteristics, such as animal counts in number of animals, stocking densities in cow/calf units per acre, or production in bushels.

# Geoprocessing History

This entire section is updated automatically by ArcGIS.

## A CSDGM Reference Map

Use this map as a guide to creating or understanding FGDC CSDGM metadata. Word and RTF documents of this map are provided in the mgisdata\Metadata folder.

### Key

**To be filled in by developer for each layer**
**Provided by Organizational Template but might differ for Project Template**
Provided by Organizational Template
Automatically generated by ArcGIS

# 1. Identification

## General

### Abstract
*Provide a brief description of the information contained in the data set.*

### Purpose
*What is the intended use for this data set? Under what conditions is it valid?*

### Supplemental Information
*This is a good place to list modifications made to imported data, such as subsetting it to a study area from an ESRI data set. This is also a good place to put the largest scale recommendation. Largest scale when displaying the data: 1:xxxxx. Other things may be added as appropriate.*

### Access Constraints
*Who is allowed to access the data? The public? Only those who have licensed it?*

### Use Constraints
*Generally, this contains a legal statement concerning those who may access the data, whether and under what conditions they may be redistributed, and a liability clause.*

### Data Set Credit
*Credit the originator of the data set (USGS? Contractor? Map author?).*
Native Data Set Environment
Native Data Set Format

## Contact

### Point of Contact Details
*Provide the person or organization to be contacted with questions about the data set.*

## Citation

### Citation Details
*Provide information on the originator of the data set or publication as completely as possible, including publication references, URLs, agency address/phone, and so on.*

## Time Period

### Currentness Reference
*For what time period is this layer valid? Is it unlikely to change (topography), good only for a short time period (evapotranspiration), or somewhere in between (population)?*

## Status

### Status
*Is this in progress or a final version? How often are updates, if any, intended?*

## Spatial Domain
Bounding coordinates

## Key Words

### Key Word and Thesaurus
*Provide at least two key words for searching, including place name key words.*

## 2. Data Quality

### General

#### Logical Consistency Report

*Logical consistency is largely concerned with topology. Did you test to see if the data contain dangles, gaps, or overlaps? What tests were applied, and what were the results?*

#### Completeness Report

*Provide information about omissions, selection criteria, generalization, and other processes that might impact how complete a data set is. Are all spatial entities included? For example, did you get all the wells or might some be missing? Did you subset the data from their original source? Were any criteria used in deciding which features to include (public versus private roads, for example)?*

### Attribute Accuracy

#### Accuracy Report

*Summarize processes used to establish the accuracy of the attribute(s), for example, known detection limits of analyses. Evaluate detail or completeness of categorical classes. Provide known information about problems with any attributes. If no accuracy data are available, enter "Unknown."*

#### Value/Explanation

*For each attribute with a known accuracy, enter the data value and any explanation. For example, for a TMDL reading, enter "TMDL" as the value and "Detection limit xx mg/l, accuracy +/1 yy mg/l" for the Explanation.*

### Positional Accuracy

#### Horizontal Accuracy Report/Value/Explanation

*If data are from a standard US federal data product, enter the national map accuracy standard of 1 in 10,000. If data are surveyed or obtained via GPS, enter the known or estimated positional accuracy of the survey or GPS unit. If data have been georeferenced, transformed, or spatially adjusted, also include the RMS error.*

### Source Information

#### General

*The Source Scale Denominator is the original scale of the data set, such as 24,000 for quad data. Type of source media would be paper for a digitized map, data download/CD for a DLG, GPS unit, and so on. The Source Citation Abbreviation would be DLG, DRG, and so on.*

#### Source Citation

*Enter the original publication details of the data set: title, originator, publication date, edition, and so on.*

#### Source Time Period of Content

*Enter Currentness reference as Ground Condition (at time of measurement) or Publication Date. Also enter whether data were collected all at one time, at multiple times, or in a range of times and enter dates. For example, if you sampled wells from January 08 to March 09, enter Range and the dates.*

### Process Step

#### Process Description

*Describe one or more processing actions taken on the data before reaching their final form, for example, importing STDS quads and joining elevation attributes, merging into a single data set, projecting to current coordinate system, and clipping to study area boundary. Enter software version, process date, and the name of the person who did the processing. Include impacts on accuracy, such as an RMS error associated with georeferencing.*

## 3. Data Organization

### General

This entire section is updated automatically by ArcGIS.

## 4. Spatial Reference

### General

This entire section is updated automatically by ArcGIS.

## 5. Entity Attribute

*Fill out either the Detailed Description tab or the Overview Description tab, whichever makes the most sense for documenting the attribute values of that particular data set.*

### Detailed Description

Entity Type

**Attribute**

General

*This section is mostly filled in for you, but, for each attribute important to the data set, you should enter a **Definition**, including units. For a field titled "TMDL," you should enter "Total Maximum Daily Load in mg/l." For an Area field, you might enter "Feature area in sq km." If you wish to enter accuracies for fields, you can, but this is optional. The **Definition Source** describes who defined a particular attribute. If you are using Anderson landcover categories Level I or Level II, for example, you would enter it. If it is very generic or obvious, such as acres or population, just leave this part blank. You don't have to fill in a Definition for every field, but you should do it for anything that is not obvious and/or needs interpretation, units, and so on. Think about the information YOU would want to know if you had to use this data set.*

### Overview Description

*This is the place to save time if you have many fields in a data set with the same type of data. For example, an agricultural table might have 20 fields of different animal counts and 30 fields of agricultural production for different crops in acres. In some cases, you might have a large table of values that are not obvious. If you are using zoning codes and the table does not include a text description, you will need to provide that information as a separate table in the database. Here you could reference the name of that table so that people know where to find out what the codes mean.*

**Data Set Overview**

*Describe the general purpose of the table, for example, "This data set lists animal counts and production of various crops by county."*

**Entity and Attribute Overview**

*Describe the main characteristics, such as animal counts in number of animals, stocking densities in cow/calf units per acre, or production in bushels. You can put them all in one entry or add multiple entries using the + button at the bottom, such as one for animal counts, one for production, and so on.*

## 6. Distribution

*This section describes whether and how the data could be ordered from a vendor/distributor, if applicable. It also is where you put your standard liability clause.*

**General**

*Provide general instructions on how to access the data, costs, if any, and what form the distribution takes.*

**Distributor**

*Enter the agency or person responsible for distributing the data and the contact information.*

**Standard Order Process**

*Enter the procedures to be followed to order or access the data.*

**Available Time Period**

*This is the time when the data will be available to the clients or public. Enter starting date of data availability and end date of availability, if known.*

## 7. Metadata Reference

General

*Provide metadata date, contact info of person or organization that created the metadata, and access and redistribution limits on metadata, if any.*

# Skills Reference

# BASICS

## General

### *Starting ArcMap or ArcCatalog*

Source: ArcMap

1.  Look on the computer desktop or taskbar for an icon for ArcCatalog (left) or ArcMap (right). Double-click it to start the program.

2.  If no icon is present on the desktop, click the Start button on the computer's menu bar. Navigate to All Programs > ArcGIS and choose the name of the program desired.

3.  From ArcCatalog, launch ArcMap by clicking the ArcMap button in the menu bar.

4.  From ArcMap, launch the ArcCatalog window, which sits inside ArcMap and provides access to some of the ArcCatalog functions but does not open the standalone program.

### *Managing windows*

Both ArcMap and ArcGIS have numerous windows that can be opened when needed and closed or tucked away at other times. Windows may float above the main program window, or they may be docked in one of several docking locations. When docked, they may also be pinned, which causes them to open when the mouse is placed on them and to automatically hide when not in use. A window will generally return to the place where it was previously docked or pinned after it is closed and reopened.

### Docking a window

Docking is lengthy to explain because there are so many options, but it is really relatively simple. Watch the video, or experiment until you are comfortable using the docking functions.

Source: ArcGIS

1.  Click on the bar at the top of the window and drag it to one of the blue docking icons that will automatically appear. One icon will turn darker blue to show in which location the window will dock when you release the mouse.

2.  If a window is already docked at the location, another set of four docking icons will appear next to the original. The second set shows where the new window will be docked relative to the existing window (left, right, below, or above) on that side of the main window. If the window is dragged to the center icon, then both old and new windows will be placed in the same spot with tabs underneath for switching between them.

Source: ArcGIS

3.  To undock a window, click on its top bar and drag it away from the docked location. Dock it somewhere else.

4.  To make a window float again, click its top bar, drag it off any of the docking icons, and release the mouse.

**Auto Hide a window**

A docked window will have an Auto Hide button on its bar.

Fig. 1. A pinned window
Source: ArcCatalog

 1. Click the Auto Hide button to pin the window to the docking location (Fig. 1). It will become a tab on that side of the main window.

2. To access the window, hover the cursor on top of the tab. The window should open automatically, but you can click it if it is stubborn.

3. When you move the cursor off the window, it will automatically hide in the tab. Sometimes you must actually click elsewhere before it hides.

4. Click the X on the window bar to close the window and remove the tab.

 5. Click the pin again to unpin the window. It will remain open in the docked location.

**Adjusting window sizes**

1. The size of a floating window can be adjusted by clicking and dragging the corners or edges to the desired size.

2. The width of a docked toolbar can be adjusted by placing the cursor on the edge until a double-headed arrow appears (Fig. 2). Click and drag the edge to the desired width.

3. The height of a docked toolbar matches the window edge available. However, if two or more windows are docked above/below each other, then the relative width may be adjusted by placing the cursor on the edge between the windows. When the cursor changes to the double-headed arrow, click and drag the edge to the desired position.

4. Some windows have multiple panels, each of which is individually adjustable. Look for the gray line in between the panels. Click and drag it to adjust the relative size of the panels.

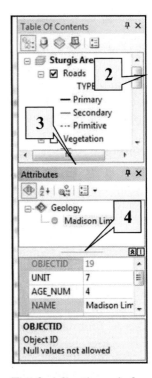

Fig. 2. Adjusting windows
Source: ArcCatalog

*Managing toolbars*

Dozens of toolbars are available and may be opened and closed as needed. Toolbars may be docked on any of the four edges of the program window, or they may float above the window (Fig. 3).

1. To move a floating toolbar, click and drag on the bar at the top to move it to a new location. To dock it, move it to one of the edges of the program and it will dock when it finds an open spot.

Floating

Docked

Fig. 3. Toolbars
Source: ArcCatalog

446

2. To move a docked toolbar, click on its handle (the dashed gray line on its left edge) and drag it to the desired position elsewhere on the GUI.

3. To move a docked toolbar off the GUI, use the handle to drag it off the GUI. It then becomes a floating toolbar. It can be docked again by dragging it to a location on the GUI.

4. To open a toolbar that is not present, choose Customize > Toolbars from the main menu and select the desired toolbar.

5. To close a floating toolbar, click on the X box in its upper-right corner.

6. To close a docked toolbar, right-click on a toolbar or empty space and choose the toolbar to close from the list.

7. To find out the name of a tool or button, hold the cursor on top of it for a moment without clicking, and the name will pop up.

### Creating/deleting folder connections

In order to access data files, you must set up a connection to the folder in which they reside. Connections may be established from ArcCatalog or ArcMap and will be available to both programs once created. Connections can be removed when they are no longer in use.

The Connect to Folder button may be found on the Standard toolbar in ArcCatalog, on the Catalog window in ArcMap, or on the Add Data window in ArcMap.

 1. To connect to a folder, click on the Connect to Folder button.

2. Navigate down the directory tree and click on the folder (or drive letter) to connect to, and click OK (Fig. 4).

3. The connections are stored in ArcCatalog in an entry called Folder Connections.

Fig. 4. Connecting to a folder
Source: ArcGIS

4. To delete a folder connection, right-click the folder connection in ArcCatalog or in the Catalog window and choose Disconnect Folder. This step only removes the connection; it does not delete the data from the disk.

### Setting options

You can control the way the programs perform various actions, display information, and other options. Each program has different options to set. Specific settings for various tasks are discussed under the appropriate headings.

### ArcCatalog Options

1. Click Customize >ArcCatalog Options from the main menu bar to access the following option tabs: General, File Types, Contents, Connections, Metadata, Tables, Raster, CAD.

2. Click Geoprocessing>Geoprocessing Options from the main menu bar to access options such as background processing, overwriting of existing files, and how long results are kept.

### ArcMap Options

1. Click Customize >ArcMap Options from the main menu bar to access the following option tabs: General, Data View, Layout View, Metadata, Tables, Raster, CAD, Display Cache.

2. Click Geoprocessing>Geoprocessing Options from the main menu bar to access options such as background processing, overwriting of existing files, and how long results are kept.

### Connecting to an Internet service

Using ArcCatalog, you can connect to and access data from external databases or from spatial database services. A Database Server or Database Connection provides access to a tabular database such as Oracle. An ArcGIS or ArcIMS Server provides access to spatial data. You must know the type of service and the URL, such as http://www.geographynetwork.com. Secure services will also require a login and password.

1. In ArcCatalog, scroll to the bottom of the Catalog Tree to find the server connections (Fig. 5). In ArcMap, open the Catalog window or click the Add Data button and find the server connections in the file drop-down.

2. Expand one of the entries in Figure 5, depending on the type of service to be added

3. Double-click one of the Add icons.

4. Type in the URL of the service in the top box.

5. If the server requires a user login and password, enter these at the bottom of the window and click OK.

6. Predefined connections are provided to ArcGIS Online and Geography Network.

Fig. 5. Connecting to a service
Source: ArcCatalog

### Opening/saving data sets

#### Opening a data set

1. Click the Browse button to access the Input/Output data window (Fig. 6). A few buttons are different between ArcMap (shown) and ArcCatalog.

2. Click the drop-down to pick from a list of other folders or connections.

3. Click the Up One Level button to move up a folder.

4. Click the Go To Home Folder button to go where the map document is stored.

5. Click the Default Geodatabase button to navigate to the default geodatabase.

6. Click the Contents View Type button to change how the listed files appear: as large icons, small icons, detailed list, and so on.

7. Click the Folder Connections button to add a new folder connection.

8. Click the New Folder button to create a new folder in the current folder.

9. Click the New File Geodatabase button to create a new geodatabase in the current folder.

10. Click the New Toolbox button to create a new toolbox in the current folder.

11. Change the Show of type: box if the desired type of data is not being displayed.

12. Click the desired data set to highlight it, or type its name in the Name box, and click Add.

Fig. 6. The Input Data Set or Feature Class Window
Source: ArcMap

#### Saving a data set

13. Type or edit the pathname of the data set to be saved, or click the Browse button.

14. Use buttons 2–10, if necessary, to navigate to or create the space for the file to be saved.

15. Type the name for the data set in the Name box. Extensions usually aren't needed.

16. Change the Save as type: box to change the type of data set saved, if necessary.

## Using the Zoom/Pan tools

The Zoom/Pan tools are available in Preview mode in ArcCatalog on the Geography toolbar, and all are available in ArcMap on the Tools toolbar.

 1. To zoom in to an area, click the Zoom In tool. Position the cursor to the upper left of the target area and then click and drag a box to encompass the area desired. Release the mouse button to finish.

 2. To zoom out, click the Zoom Out tool and click once at the center zoom point. Draw a large box to zoom out slightly or a small box to zoom out a great deal.

 3. To zoom in by a fixed amount while remaining centered within the current extent, click the Fixed Zoom In tool (ArcMap only).

 4. To zoom out by a fixed amount while remaining centered within the current extent, click the Fixed Zoom Out tool (ArcMap only).

 5. To move the map around in the window, click the Pan tool, click on the map, and drag it to the desired area.

 6. Click the Previous Extent button to return to the extent just before the current one. This one is useful if you make a mistake when panning or zooming.

 7. Click the Next Extent button to return to the extent after the current one (this button is dimmed unless the Previous Extent button has been clicked).

 8. Click the Full Extent button to view the full area of all the layers, or to an extent specified by the user.

**TIP**: Use the Data Frame tab in the Data Frame Properties to specify an alternate extent that will be used by the Full Extent button.

## Layers and layer properties

A layer includes a pathname reference to a spatial data set plus various properties that control how that data set is displayed and used. In ArcMap, layers are held in memory and saved as part of the map document. In ArcCatalog, layers are stored in layer files.

### Setting layer properties

Properties set for a layer do not modify the referenced data set. Detailed instructions for setting various properties are found under the appropriate headings.

1. Right-click a layer file (ArcCatalog) or a layer (ArcMap) and choose Properties.

2. Click the tab of the properties to be managed (Fig. 7).

3. Set the properties and click Apply to set the properties and leave the window open, or OK to apply the settings and close the window.

Fig. 7. Tabs for layer properties
Source: ArcCatalog

### Creating a layer file in ArcCatalog

1. Right-click the reference feature class and choose Create Layer.

2. Specify the name and location of the file to be saved. The .lyr extension will be added if necessary. Click Save.

3. Set the properties of the layer file.

### Creating a layer file in ArcMap

1. Set the symbology, labeling, and other layer properties of the data or group layer.

2. Right-click the data layer name in the Table of Contents and choose Save As Layer File from the context menu.

3. Specify the name and location of the file to be stored.

Source: ArcCatalog

## *Group layers and layer properties*

### Creating a group layer file in ArcCatalog

1. In the Catalog Tree, click the folder or geodatabase containing at least two of the feature classes to be part of the group layer.

2. Click the Contents tab to show the data in the folder. Hold down the Ctrl key and click two or more feature classes that will be included in the group layer to highlight them all.

3. Right-click one of the highlighted feature classes and choose Create Layer.

4. Specify the name and location of the file to be saved. The .lyr extension will be added.

### Creating a group layer in ArcMap

1. Set the symbols, labels, and other display properties of all the layers to be included in the group. Set the Table of Contents to Display by Draw Order.

2. Click the first group member in the Table of Contents to highlight it. Use Ctrl-click to also highlight the other layers in the group (Fig. 8).

3. Right-click one of the group members and choose Group from the context menu.

4. Click on the group name to highlight it and then click it again to type in a name for it.

5. Right-click the group name and choose Save As Layer File from the context menu.

6. Enter the location and file name of the group layer and click Save.

Fig. 8. Creating a group layer
Source: ArcMap

**Managing group layer properties**

1. Right-click the new group layer (ArcMap) or layer file (ArcCatalog) to open its properties, and click the Group tab (Fig. 9).

2. Click to highlight one of the feature classes and choose Properties to open and set its layer properties.

3. Click Add to add a new layer to the group.

4. Highlight the layer and choose Remove to remove it from the group layer (the file will remain on the disk).

5. Use the Up/Down arrows to change the draw order of the layers. Drawing always starts with the bottom layer.

6. Use the General and Display tabs to set properties for the entire group, such as a scale range or transparency.

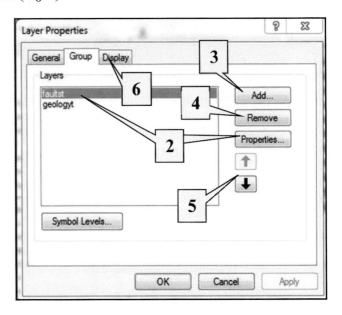

Fig. 9. Managing group layer properties
Source: ArcMap

*Identifying features*

Identify is used to quickly view attribute information for one or several features. The windows are slightly different in ArcMap and ArcCatalog but they function much the same way. You must be using the Preview tab to use Identify in ArcCatalog.

1. Click on the Identify tool on the Geography toolbar (ArcCatalog) or the Tools toolbar (ArcMap). The Identify window will open.

2. Click on the map feature or raster cell to identify.

3. If more than one layer are present in the map, you can set which layer or layers will be identified (Fig. 10).

4. Clicking an identified feature in the top panel will cause it to flash briefly.

5. If more than one layer were identified, click on a layer to view the attributes for that feature. Click on a different layer to see its attributes.

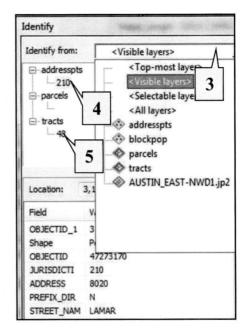

Fig. 10. Identify window (ArcMap)
Source: ArcMap

**TIP:** The field shown in the upper panel is the *display expression field* for the layer, which can be changed in the layer properties on the Display tab.

### *Finding features*

1. In ArcMap, click the Find tool on the Tools toolbar or in the Table Options menu. In ArcCatalog, choose Find from the Table Options menu when previewing a table.

2. Type the search text in the first box (Fig. 11a).

3. Choose the layer or set of layers to be searched.

4. Check the box for an exact match or for features that are similar to or contain the string.

5. Choose to search all fields, or set the field to search.

6. Click Find to begin the search. Records that match will be listed below.

7. Right-click one of the found records to display a context menu with options for flashing, zooming to, identifying, bookmarking, or selecting the found feature (Fig. 11b).

Fig. 11a. Finding features
Source: ArcMap

Fig. 11b. The Find list
Source: ArcMap

Fig. 12. Finding locations by name
Source: ArcMap

8. Click the Locations tab to find a location based on a place name (Fig. 12).

9. Select one of the locator services provided by ArcGIS Online. You must have an Internet connection.

10. Type the name of the location and click Find.

### *Using the Search window*

The same search window can be used to look for map documents, data sets, and tools. It can also be set to search the local drives only, or assigned to search a server or other Internet service.

Fig. 13. Searching for data sets (left) or tools (right)
Source: ArcMap

1. Click the Search button in ArcMap or ArcCatalog from the Standard toolbar. Once open, the window can be docked or pinned like any other window (Fig. 13).

2. Click the type of information to search for: All, Maps, Data, or Tools.

3. Type the search string indicating the name or key words for which to search.

4. Hover over blue text to find out more about the item or to view its item description.

5. Click on the green underlined text to follow a link to the item in ArcCatalog or the Catalog window.

### *Environment settings*

Environment settings control the operation of tools and functions in both ArcMap and ArcCatalog. Environment settings may be set at the program level, where they affect all tools, or they may be set for a single run of a particular tool. Environment settings are organized into functional groups, some of which are shown in Figure 14.

Use ArcCatalog to specify environments that will be used as defaults by all tools and map documents. Environments changed within a map document in ArcMap will be saved with that map document.

Fig. 14. Environment settings
Source: ArcMap

Environments set for a tool will temporarily override the Catalog and map document settings but will cease as soon as the tool is finished.

1. In ArcCatalog or ArcMap, choose Geoprocessing> Environments from the main menu bar. In a tool, click the Environments button.

2. Use the double-arrow switch to open/close the functional group you want to set.

3. Click Show/Hide Help to get information about settings.

4. Click in a setting box to find information about that setting, and Tool Help for more detailed information about that setting.

Set the desired settings; leave the others blank or set to their defaults. Click OK.

### Starting and using ArcToolbox

ArcToolbox is a dockable window that sits inside ArcCatalog or ArcMap.

1. Click on the ArcToolbox icon in either ArcCatalog or ArcMap to open the window.

2. After being opened, it may be docked or pinned like any other window.

**Using a tool**

3. Double-click the tool in ArcToolbox to open it (Fig. 15).

4. If using a tool for the first time, click the Show Help button to read about it.

5. With the Help showing, click on a parameter box to get more information about it.

6. Click the Tool Help button to get full help information about the tool.

7. Green dots indicate a required parameter.

8. Input/output features may be set by clicking the Browse button and locating the spatial data set, or by making a choice from the drop-down list (ArcMap only). If the drop-down list is used, only the selected features will be used in the tool.

9. Click the Environments button to change the Environment settings.

Fig. 15. Running a tool
Source: ArcMap

10. If a yellow warning or a red error icon appears after entering a parameter, place the cursor over it to see the message about the error or potential error.

11. Click OK to start the tool.

12. If running in background, a blue scrolling status bar will appear at the bottom of the program (Fig. 16), and a notification will appear to indicate that the tool has finished.

13. In foreground processing, a status window will appear with messages indicating progress. Close the window when it is finished, or check the box to close the window automatically on successful completion. If errors occur, the box will remain open so you can read the messages.

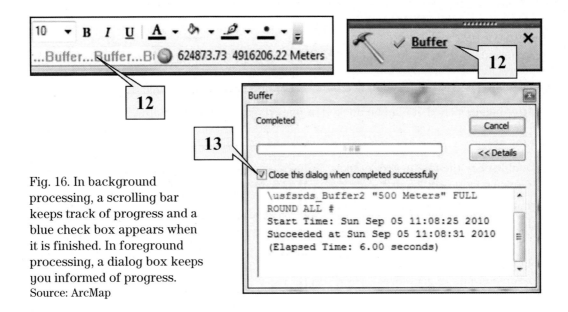

Fig. 16. In background processing, a scrolling bar keeps track of progress and a blue check box appears when it is finished. In foreground processing, a dialog box keeps you informed of progress.
Source: ArcMap

# ArcMap Basics

### *Opening an ArcMap document on start-up*

Start ArcMap using the methods described in ***Starting ArcMap or ArcCatalog***.

Fig. 17. ArcMap start-up dialog
Source: ArcMap

1. As ArcMap starts, it displays a dialog screen (Fig. 17). Choose to start with a blank map, open a map template, or open an existing map.

2. Click to change the appearance of the templates as a list, small icons, or large icons.

**To open a new map**

3. Click the New Maps text, or on one of the template headings below it.

4. Click on the Blank Map icon under My Templates, or scroll down and choose one of the predefined template sizes or layouts.

5. Click the Browse button to assign a default geodatabase for the map document if you want it to have a home location to look for and save files. The drop-down shows a list of recently used geodatabases.

6. Click OK.

### *Switching to a different map document*

**To open an existing map document**

1. Click the Open button on the Standard toolbar.

2. In the Open window, navigate to the right folder and double-click the desired map document, or click it and choose Open.

**To open a new map document**

1. Click the New Map File button on the Standard toolbar.

2. The same window appears as at ArcMap start-up. See ***Opening an ArcMap document on start-up***.

### *Adding data*

1. Click the Add Data button on the main toolbar, or choose File > Add Data from the main menu bar.

2. Use the Add Data dialog box to navigate to the folder containing the data to open (see ***Opening/saving data sets***).

3. Click on the data item to select it. Select multiple items by holding down the Ctrl or Shift key while clicking additional data sets.

4. Click Add. The data will be added to the active data frame.

**TIP:** If the disk or directory containing the data does not appear in the list, you probably need to add a folder connection to it (***Creating/deleting folder connections***).

### *Removing data*

Removing data removes them from the map document but does not delete them from the disk.

1. Right-click the data item in the Table of Contents and choose Remove.

### *Saving a map document*

1. Click on the Save button, or choose File > Save from the menu bar.

2. If the document has not previously been saved, a dialog box appears for the user to select a location and a name for the map document. Navigate to a directory in which to save it and then enter a name for the document.

3. Choose File > Save As from the menu bar to save the map document under a new name, leaving the original version unchanged. Specify the location and name as for Step 2.

**TIP:** The names of map documents, spatial data sets, and folders should not contain spaces or special characters such as #, @, &, *. Use the underscore character _ to create spaces in names, if necessary. Although ArcCatalog and ArcMap allow spaces in names, spaces can create problems for certain advanced functions. It is wise to avoid them entirely.

### *The Table of Contents window*

The Table of Contents controls many of the display and other functions available in the program. It contains the layers and data frames that make up the map document. It offers four different views of the data, which are controlled by a set of icons on the top of the window. Each view has different settings and different options in the context menus. A few of the more common tasks are described below for each view.

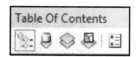

Source: ArcMap

### List By Drawing Order

List By Drawing Order lists the layers according to the order in which they are drawn, from bottom to top (Fig. 18). It is the default view.

Fig. 18. Table of Contents views
Source: ArcMap

1. To turn a layer on/off, click the check box.

2. Right-click an item name to open a context menu. The menu choices depend on the type of item clicked.

3. Double-click the item name to open its properties, or right-click it and choose Properties from the context menu.

4. Click twice slowly on a layer or legend item to rename it. Click Enter when done.

5. Hide the legend for a data layer by clicking the minus sign in the box next to the layer. Click the plus sign to expand it again.

6. Click on a layer symbol to change the symbol using the Symbol Selector.

7. Right-click on a layer symbol to change the color of the symbol.

8. Right-click a layer and choose Copy to make a copy of it. Then right-click one of the data frames and choose Paste Layer(s) to paste the copy into a data frame.

9. Right-click a layer and choose Remove to remove it from the map.

10. To change the draw order, click and hold on the layer name and drag it higher or lower in the list. Release the mouse button to finish.

**TIP:** If the check box is dimmed, then the current map scale is outside the display range set for the layer. Try zooming to another scale or reset the scale range in the layer properties.

**TIP:** If the layers don't seem to go anywhere when you drag them, check to make sure that the List By Draw Order view is selected at the top of the Table of Contents.

### List By Source

List By Source organizes layers and tables according to the folder or geodatabase that contains them (Fig. 18). The full pathname of the folder is followed by the data items from that folder. This is the only view in which standalone tables can be accessed. Tasks 1–9 listed for the List By Drawing Order view also work in List By Source view.

### List By Visibility

List By Visibility groups layers according to whether they are visible or not (Fig. 19). It also allows you to view groups of features that have the same symbol.

Fig. 19. List By Visibility
Source: ArcMap

 1. Click the icon to make a layer visible or not visible.

 2. Click the Selection toggle to make a layer selectable or not.

 3. Click the double-arrow to expand or collapse the categories/classes of a layer.

4. Click a category or class and the corresponding features will be flashed on the screen.

5. Right-click a layer to open a context menu.

### List By Selection

List By Selection provides tools for working with queries (Fig. 20).

 1. Click the Visibility icon to make a layer visible or not.

 2. Click the toggle switch to make a layer selectable or not.

 3. Click the clear button to clear the selected features.

4. Right-click a layer to open a context menu.

Fig. 20. List By Selection
Source: ArcMap

### *Setting a scale range*

The visible scale range controls at which scales a layer will be drawn.

1. Right-click the layer name in the Table of Contents and choose Properties.

2. Click the General tab.

3. Fill the button to *Don't show layer when zoomed* (Fig. 21).

4. Select one of the scales using a drop-down, or type the scale denominator in the box. Set a minimum scale, maximum scale, or both.

5. To clear a scale range, fill the button to *Show layer at all scales*.

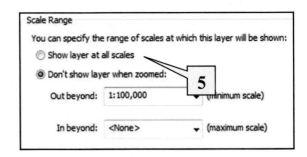
Fig. 21. Setting the scale range
Source: ArcMap

**TIP:** You can also set or clear the scale ranges by right-clicking the layer in the Table of Contents and choosing Visible Scale Range.

### *Using bookmarks*

A bookmark lets you return to a predefined map extent many times.

Fig. 22. Zooming to a bookmark
Source: ArcMap

1. To set a **bookmark**, zoom to the desired area and select Bookmarks > Create from the main menu. Type in a name for the bookmark.

2. To zoom to an existing bookmark, click Bookmarks and choose the desired bookmark (Fig. 22).

3. To delete an existing bookmark, choose Bookmarks > Manage. Select the bookmark name by clicking on it, and then click the Remove button.

4. To zoom to the reference scale, right-click the frame name and choose Reference Scale > Zoom to Reference Scale from the menu.

### *Measuring features*

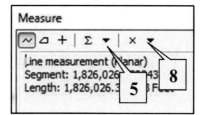

Fig. 23. The Measure tool
Source: ArcMap

1. Click the Measure tool (Fig. 23). By default it measures distance.

2. Click on the map to start the line. Click along the path to be measured. Double-click to end the line. Both total length and the length of the last segment are shown.

3. Click the Measure Area button to measure a polygon area. Click to create vertices for the polygon and double-click when finished.

4. Click the Measure Feature button, and click on a feature to get its measurement.

5. Use the Choose Units button to set the units for length and area.

6. Use the Show Total button to keep a running count of all features measured.

7. Use the Clear/Reset Results button to clear running totals and reset the measurements.

8. Use the Measurement Type button to change the type of distance measure from Planar (the default).

**TIP:** Snapping will normally be on and will snap to features when measuring. Use Snapping > Snapping Options on the Editor toolbar to turn it off.

## *Setting MapTips*

1.  Right-click the layer name in the Table of Contents and choose Properties from the context menu.

2.  Click the Display tab and check the box to *Show MapTips using the display expression.*

3.  Change the Field box to the desired attribute field to show in the MapTips.

4.  Uncheck the box to turn MapTips off.

Display Expression

Field:     NAME

☑ Show MapTips using the display expression

Source: ArcMap

## *Setting data frame properties*

1.  Right-click the data frame name and choose Properties from the context menu.

2.  Click the tab containing the properties to set (Fig. 24).

3.  Enter the appropriate information and settings.

| Feature Cache | Annotation Groups | Extent Indicators | Frame | Size and Position |
|---|---|---|---|---|
| General | Data Frame | Coordinate System | Illumination | Grids |

Fig. 24. Data frame properties tabs
Source: ArcMap

# ArcCatalog Basics

### The Catalog Tree

The Catalog Tree is a dockable window in ArcCatalog that lists folders and other items in the catalog, such as database connections and toolboxes (Fig. 25). Many types of items are visible in the Catalog Tree, and the options and actions of the Catalog depend on the type of item.

1. Click the + box to expand the contents of an item, such as a folder or geodatabase.

2. Click the – box to collapse the contents of an item.

3. Right-click an item to open a context menu with various actions. The list of actions will vary depending on the type of item.

4. Right-click an item and choose Properties to view or set its properties. Double-clicking an item is a shortcut to opening its properties.

5. To rename an item, click twice slowly on the item name. It will be highlighted and allow you to type or edit the name. Click Enter when done.

Fig. 25. The Catalog Tree
Source: ArcCatalog

**TIP:** You cannot select multiple items in the Catalog Tree, such as for copying or deleting. Use the Contents tab for this purpose.

**TIP:** You must establish folder connections before you can see hard drives with GIS data.

### Managing ArcCatalog items

Most of these options can be performed in ArcCatalog or on the Catalog window in ArcMap. ArcToolbox also contains tools for accomplishing these tasks.

**Creating new items**

1. Right-click on the folder or geodatabase to contain the new item.

2. Choose New > and the type of item to create (Fig. 26). The options will depend on what item was right-clicked.

3. A window or wizard will appear. Follow the instructions for creating the item. Different items require different parameters.

Fig. 26. Options for the New command
Source: ArcCatalog

463

## Copy and paste items

1. Right-click the object to be copied and choose Copy.

2. Right-click the folder or geodatabase to place the object in and choose Paste.

## Renaming files

1. Click twice slowly on the item name to highlight it. Type in the new name and click Enter.

## Deleting an item

1. Navigate to the folder or geodatabase containing the item to be deleted.

2. Right-click the item and choose Delete.

3. Say Yes, when prompted, to confirm deleting the item.

### *Viewing modes*

ArcCatalog has three different viewing modes for exploring data sets: Contents, Preview, and Description. Click the tab to use the desired viewing mode. Each mode is described in detail below.

Source: ArcMap

## Viewing the contents of a folder

1. Click on a folder or geodatabase in the Catalog Tree.

2. Click on the Contents tab in the content window.

3. Choose one of the display options from the toolbar: Large icons, List, Details, or Thumbnails (Fig. 27).

Fig. 27. Display options
Source: ArcCatalog

## Previewing data sets

Preview mode allows you to examine data to see what they look like. You can zoom/pan around them and even perform some actions on tables such as sorting. Only feature classes, rasters, layer files, and tables can be previewed. Items with multiple items inside, such as folders or geodatabases, cannot be previewed.

1. Click on the data set to highlight it in the Catalog Tree (Fig. 28).

2. Click the Preview tab.

3. Spatial data sets default to Geography mode, which shows the map.

4. Use the Zoom, Pan, Full Extent, or Identify button to explore the geography preview.

5. Click the Preview drop-down button to change from Geography mode to Table mode.

Fig. 28. Previewing a layer
Source: ArcCatalog

6. In Table mode, additional options are provided by a Table Options menu (Fig. 29). See the Tables section for more information about these options.

7. In Table mode, navigation buttons can be used to go to the beginning or end of a table, or to step through the rows.

8. In Table mode, additional options are available by right-clicking a field name. The options are a subset of those available when working with tables in ArcMap. See the Tables section for more information about these options.

Fig. 29. Options available when previewing tables
Source: ArcCatalog

### Viewing the item description

The item description is a brief overview to give the user basic information about an item (Fig. 30). A description may be created for many kinds of items, including data sets, tools, scripts, and more. Users can view and edit descriptions in ArcCatalog.

By default, only a brief description is shown and available for editing. However, the Description tab is also the portal for developing full-fledged metadata. See the Metadata section for more information.

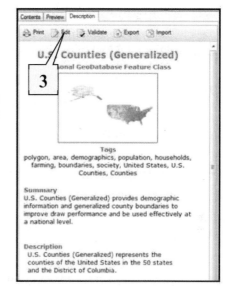

1. Click on the item in the Catalog Tree for which you wish to view the description.

2. Click the Description tab.

3. Click the Edit button to edit an item's description, or Print to print it.

4. The Description toolbar has other buttons that allow you to validate, export, and import descriptions. See the Metadata section and/or Chapter 14 for more information on how to use these tools.

Fig. 30. Item description
Source: ArcCatalog

### *Viewing and setting data set properties*

Data sets, including geodatabases, feature classes, rasters, and tables, have properties that can be modified in ArcCatalog. The types of properties will vary with the type of data set.

1. To access the data set properties, right-click on the layer/table name in the Catalog Tree and choose Properties. Double-clicking the layer/table also opens its properties.

2. Click the tab containing the properties to change.

**Geodatabase** tabs include General and Domains.

**Shapefile** tabs include General, XY Coordinate System, Fields, and Indexes.

**Geodatabase feature class** tabs include General, Fields, XY Coordinate System, Domain/Resolution/Tolerance, Indexes, Subtypes, Relationships, Representations, and Editor Tracking.

**Coverage** tabs include General, Projection, and Extent.

Other sections describe some of these properties and how to set them.

### Creating and viewing thumbnails

1. Click the layer in the Catalog Tree to highlight it and then click the Preview tab.

2. If desired, use the Zoom and Pan buttons to modify the appearance of the layer.

3. Click the Thumbnail button in the Geography toolbar.

4. To view all the thumbnails in a folder, click on the folder, make sure the Contents tab is clicked, and choose the Thumbnail button from the Standard toolbar.

5. To view the thumbnail for a single layer, click on the layer in the Catalog Tree to highlight it and make sure the Contents tab is clicked.

**TIP:** Thumbnails will be placed into an item's description or metadata.

### Dragging and dropping files to ArcMap

After searching through data in ArcCatalog, or on the Catalog tab, you can drag data into ArcMap for use.

1. Position the ArcCatalog and ArcMap windows so that both are visible.

2. Click the file in ArcCatalog that you want to open in ArcMap. Hold down the mouse button and drag it into the ArcMap window (you must be in the map window or the Table of Contents—not on a menu bar). Release the mouse button to drop the file into ArcMap.

3. You can also drag files from the Catalog tab to the map window in ArcMap.

**TIP:** If ArcMap is not visible on the screen, drag the file onto the ArcMap icon on the computer's Start Menu bar. Hold it there until ArcMap opens and then drag it into ArcMap.

### Preventing script errors when opening geoprocessing tools

ArcCatalog has a recurring problem that a script error occurs when opening a tool in ArcToolbox. It happens if the Description tab, or Item Description window in the Catalog tab, has been visited. The next time one tries to open a tool, a script error is generated and the program hangs and must be stopped and restarted. If you would like a detailed description of why this error occurs and what can be done to prevent it, see https://gisnuts.com/terra/blog/2013/01/22 /arcgis-geoprocessing-tools-scripting-error.

For a more expedient solution, try the following, with ArcMap and ArcCatalog both closed.

1. Open Internet Explorer and choose Tools > Internet options.

2. Click the Security tab.

3. Click the Custom Level button.

4. Scroll down and find the *Allow previously unused ActiveX controls to run without prompting* option, and choose Enable (Fig. 31).

5. Click OK in all the windows and close Internet Explorer.

6. Start ArcMap or ArcCatalog. The script error should no longer occur.

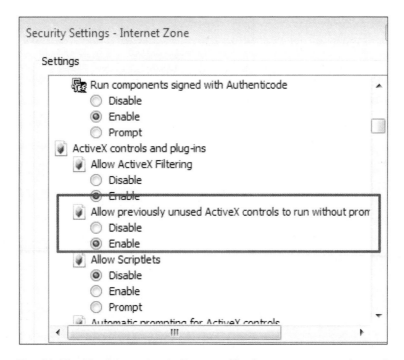

Fig. 31. Enable this option in Internet Explorer to prevent the script error problem.
Source: ArcCatalog

# DATA MANAGEMENT

## Files and Geodatabases

### Creating a geodatabase

1. In ArcCatalog, or in the Catalog window, navigate to the folder in which the geodatabase will be created.

2. Right-click the folder and choose New > Personal Geodatabase or New > File Geodatabase from the menu.

3. The geodatabase appears in the display window. Type in the name of the new geodatabase and press Enter.

### Creating a shapefile

1. In ArcCatalog, or in the Catalog window, right-click the folder in which you wish to create the shapefile and choose New > Shapefile.

2. Type the name of the new shapefile.

3. Select the feature type from the drop-down button (Fig. 32).

4. Click the Edit button to specify the spatial reference.

5. Select a spatial reference from one of the folders.

**TIP:** For more information and details about the spatial reference, see *Specifying a spatial reference*.

6. Click OK to create the shapefile.

Fig. 32. Creating a new shapefile
Source: ArcCatalog

## *Creating a feature dataset*

1. Determine the desired coordinate system and resolution for the feature dataset since it can't be changed once defined. ArcGIS will assign default values for the resolution and domain based on the coordinate system. If you have another feature class or feature dataset with the same properties, you can import it from that one.

2. In ArcCatalog, navigate to the geodatabase to contain the feature dataset.

3. Right-click the geodatabase and choose New > Feature Dataset.

4. Enter the feature dataset name (Fig. 33). Click Next.

5. Choose the coordinate system to use for the feature dataset. If you have another feature class to import the spatial reference from, choose Import. Locate the feature class, select it, and choose Add. Click Next.

6. If your feature dataset needs a vertical coordinate system (3D data only), select one. Click Next.

7. Examine the default tolerance. It is recommended to accept it.

8. It is recommended to accept the default resolution and domain extent.

9. Click Finish to complete creating the feature dataset.

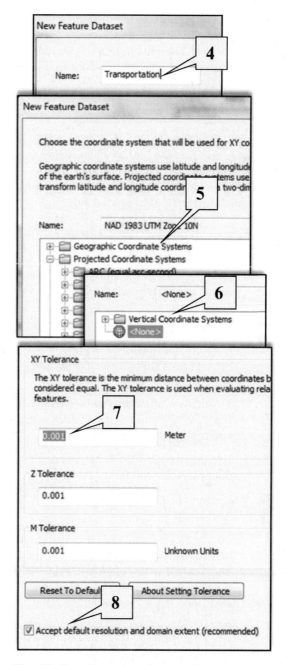

Fig. 33. Creating a new feature dataset
Source: ArcCatalog

### Creating a feature class

1. In ArcCatalog, or the Catalog window, right-click the geodatabase or feature dataset that will contain the feature class and choose New > Feature Class.

2. Enter the name of the feature class (Fig. 34).

3. Choose the Type of feature. Click Next.

If you are creating the feature class inside a feature dataset, the spatial reference will already be defined. If it is a standalone feature class, you must define the spatial reference.

4. Specify the coordinate system for the feature class. Click Next.

5. Examine the XY tolerance. The default is fine for most applications unless very high precision is needed.

6. Accepting the default resolution is usually recommended. If you uncheck the box, you'll get another panel to enter the resolution. Click Next.

7. To add new attribute fields, click in the first available box in the Field Name list and type in the field name.

8. Use the drop-down to choose one of the data types for the field.

9. Change the field properties as necessary.

10. To add fields from another feature class or shapefile, click the Import button. Locate the feature class and select it. All fields in the layer will be added to the new one.

11. Click Finish to create the feature class.

Fig. 34. Creating a new feature class
Source: ArcCatalog

### Importing feature classes

The following directions apply to importing feature classes as standalone classes or as parts of a feature dataset.

1. In ArcCatalog, navigate to the folder containing the geodatabase.

2. Right-click the geodatabase or feature dataset, choose Import, and select an import option based on the type of data to bring in (Fig. 35).

Fig. 35. Options for importing items
Source: ArcCatalog

The following directions assume that you have chosen the Feature Class (single) tool. The other choices will be similar.

3. Click the Browse button and navigate to the directory containing the data to import (Fig. 36). Select it and click Add.

4. Enter the name of the new feature class in the geodatabase.

5. Enter an SQL expression to bring only selected records into the geodatabase (optional).

6. Examine the fields list. Add or delete fields from the list, if necessary.

7. Click the Environments button to set the geoprocessing environment defaults, if desired. For example, use the coordinate system setting to project the data to a new coordinate system as they are imported.

8. Click OK to start importing the feature class.

Fig. 36. Importing a shapefile into a geodatabase
Source: ArcCatalog

**TIP:** To import several feature classes at the same time, choose Import > Feature Class (multiple). The options to rename the feature class, perform selections, and edit field names are not accessible with the multiple feature class tool, however.

## Exporting features

Exporting a layer takes the selected records or features and creates a new data set on the disk with them.

1. In ArcMap, perform a query, if desired, to select the desired records.

2. Switch to View by Drawing Order or View by Source mode in the Table of Contents, if necessary.

3. Right-click the layer and choose Data > Export Data.

Fig. 37. Exporting features
Source: ArcCatalog

4. Choose what to export (Fig. 37). You may export all features, the selected features, or the features in the current view extent.

5. Choose which coordinate system to use.

6. Click the Browse button to specify a name and location for the output feature class. Click OK.

7. Change the *Save as type* to a shapefile or geodatabase feature class, as desired. Click Save and OK.

### Creating attribute domains

Attribute domains are properties of the geodatabase and are set at that level.

1. In ArcCatalog, navigate to the folder containing the geodatabase.

2. Right-click the geodatabase and choose Properties.

3. Click the Domains tab.

4. Click in the first available Domain Name box and enter the name of the domain (Fig. 38).

5. Enter a longer description of the domain for documentation purposes.

6. Set the attribute type (Short Integer, Long Integer, Text, etc.). The domain can only be applied to attributes of the same type.

7. Click in the Domain Type box and choose a Range or a Coded Value domain.

8. For a Range domain, enter the minimum and maximum range values.

9. For a Coded Value domain, enter each value code and description in the lowest set of boxes. The descriptions should not be too long (<15 characters is probably best).

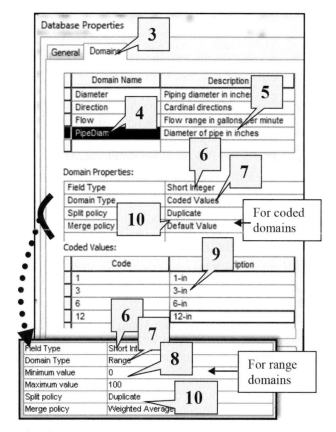

Fig. 38. Creating a domain
Source: ArcCatalog

10. Click in the Split and Merge policy boxes and choose the type of policy desired.

11. Click Apply to create the domain. It will be checked for errors and added if OK.

12. To enter additional domains, repeat Steps 3–10 or click OK if finished.

13. To delete a domain, click on the gray area to the left of its name to select it and press the Delete key.

**TIP:** Domain policies and codes may be edited after the domain is created. However, the field type and domain type, once set, cannot be altered. To change them, you must delete the domain and create a new one with the desired properties.

**TIP:** Always click Apply between one domain and the next. That way if you make a mistake, you know which domain to delete and reenter.

### Assigning default values or domains

Default values will automatically be entered into attribute fields when a new feature is created during editing. Domains must already exist in the geodatabase before they can be applied to a feature class.

1. In ArcCatalog, right-click the feature class containing the attribute and choose Properties.

2. Click the Fields tab (Fig. 39).

3. Select the field by clicking on the gray area to the left of its name. Its properties appear underneath.

4. Click the box for Null values and choose Yes to allow null values or No to disallow them.

5. Click in the Default Value box and type in the default value.

6. To assign a domain to an attribute, click in the Domain box and select a domain from the list.

7. For best results, click Apply after completing the entries for one field, before doing another.

8. If desired, enter default values for additional items by repeating Steps 3–7.

9. Click OK to effect the changes and close the window.

Fig. 39. Setting domains and default values for attributes
Source: ArcCatalog

**TIP:** The domain type must match the field type. A long integer domain cannot be applied to a short integer field, for example. Only valid domains will be listed for a field.

### Creating subtypes

Subtypes allow you to specify different defaults and behaviors for groups of features in a single feature class. Before starting, make sure that you have planned the subtype categories, created the other fields contained in the feature class table, and entered any domains to be used by the fields. Also plan the default values to be defined for each subtype.

1. In ArcCatalog, open the properties of the feature class to be given the subtypes.

2. Click the Fields tab. Create a short integer field that will be used to define the subtype categories. See ***Adding or deleting fields***. Do not set any defaults here. Click Apply to ensure that the field is created before going on to the next step.

3. Click the Subtypes tab (Fig. 40).

4. Set the Subtype Field to the short integer field created in Step 2.

5. Click in the first empty Code box and enter the value 1 (or any other short integer).

6. Enter a description for the first subtype.

7. Click in the gray area to the left to highlight the subtype row.

8. Enter a default value for each field in the table that will use one. If a field uses coded domains, be sure to enter the code value actually stored in the field, not the description value that the user sees (e.g., COP instead of copper).

9. Click Apply to finish the first subtype.

10. Repeat Steps 5–9 to add more subtypes.

11. Choose the default subtype.

12. Click OK when finished.

Fig. 40. Entering the first subtype
Source: ArcCatalog

## Creating a network

Building a network requires an ArcGIS Standard or Advanced license. If you want to try it, follow this example using the road network from the usdata geodatabase.

1. The features forming the network must be feature classes in a feature dataset in a geodatabase.

2. Right-click the Transportation feature dataset in the mgisdata\Usa\usdata.mdb geodatabase and choose New Geometric Network. A wizard will appear. Read the information and click Next.

3. Enter a name for the network (Fig. 41).

4. Enter No to indicate that the features should not be snapped. This data set already has the ends of roads snapped to each other. See the tip about snapping below. Click Next.

5. Check the box to include majroads in the network. Do NOT include interstates. Click Next.

6. Choose No so that all network features are enabled. Click Next.

7. Leave the role for majroads set to Simple Edge. Leave Sources & Sinks set to <None>. Click Next.

Fig. 41. Building a network
Source: ArcCatalog

**TIP:** To set weights, you must know the data type of the field used for the weight. Examine the fields in ArcCatalog before creating the network to obtain this information.

We will assign two distance weights based on both miles and kilometers (Fig. 41). They are single-precision Float fields named MILES and KM.

8. Click New to add the first weight (Fig. 42).

9. Enter *Distance-MI* for the weight name and choose the Single Type. Leave the Bitgate Size alone. Click OK.

10. The Distance-MI weight is entered in the weights list. Click on it to highlight it.

11. In the panel underneath, click in the box under Field Name and choose MILES as the attribute field to associate with the weight.

12. Click New to add the second weight. Name it *Distance-KM* and set the type to Single. Click OK.

13. Highlight the Distance-KM weight in the upper box. Select KM in the lower field name box for the attribute field. Click Next.

14. Review the summary and choose Finish.

Fig. 42. Assigning weights
Source: ArcCatalog

**TIP:** For the network to operate correctly, all edges must meet each other at junctions with no gaps or dangles. This adjustment may have already been done using planar topology, in which case you don't need to snap when creating the network. If not, you may be able to automatically fix the dangles by indicating a snap tolerance in map units. It must be large enough to fix the errors but not so large that features are incorrectly snapped together. Setting a good tolerance can be tricky, which is why snapping is often done interactively using topology. Choosing Yes may make changes to your data, so be careful and have a backup copy just in case.

**TIP:** If you make a mistake or need to change the network, the easiest way is to delete it and start over. Right-click the Transportation_Net entry in the feature dataset and choose Delete. It will remove the logical network but leave the majroads feature class as is.

### *Creating a planar topology*

A planar topology uses rules to define the desired spatial relationships within and between feature classes. It can test features for adherence to the rules, and it provides tools for finding and fixing errors. A planar topology can only be created for a feature dataset, and only feature classes in the feature dataset can participate.

1. Right-click the feature dataset to contain the topology and choose New > Topology. Click Next to begin.

2. Enter a name for the topology (Fig. 43).

3. Change the default cluster tolerance if desired. Read about the cluster tolerance or the Help before changing the default. Click Next.

4. Check the boxes for the feature classes that will participate in the topology. Click Next.

5. Change the default ranks if you have features that are more accurate than others that should not be moved during clustering. More accurate features can be ranked higher and will not move. Click Next.

6. Click Add Rule. For a single feature class rule, choose the feature class and the rule. For a two-layer rule, choose the first feature class, the rule, and the second feature class. Click OK to add the rule. Continue until all rules are added.

Fig. 43. Creating a topology
Source: ArcCatalog

7. Rule sets can be saved and used again for another topology. Click Next.

8. Read the summary and click Finish to create the topology.

9. Choose Yes if you want to validate the topology immediately.

**TIP:** Validation includes clustering, which can move features and degrade feature accuracy depending on the cluster tolerance. Always have a backup before you validate.

# Tables

## *Opening a table*

### In ArcCatalog

1. In ArcCatalog, tables may be viewed using the Preview tab. Some, but not all, of the functions described below are available in Preview mode.

### In ArcMap

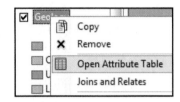

Source: ArcMap

1. Right-click on the spatial data set name in the Table of Contents and choose Open Attribute Table from the context menu.

2. To open a standalone table, first make sure that the table is visible by clicking on the List By Source tab at the top of the Table of Contents.

3. Right-click on the table name and choose Open from the context menu.

## *The Table window*

The Table window shows the contents of tables and provides buttons and commands for working with them.

1. The Table Options menu provides access to many commands and functions that affect tables (Fig. 44).

2. Buttons at the top of the table access the most commonly used commands, including, from left to right, Related Tables, Select By Attributes, Switch Selection, Clear Selection, Zoom To Selection, and Delete Selection.

3. Right-click a field heading to open a context menu with actions that use or affect the field (see Fig. 45).

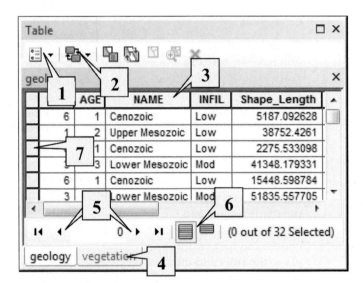

Fig. 44. The Table window
Source: ArcMap

4. Normally the window shows one table. If you open more, they will appear as tabs in the lower-left corner. Click on a tab to view that table.

5. The navigation buttons can take you to the beginning, the previous record, the next record, and the end of the table.

6. These buttons toggle between showing all records and showing only selected records.

7. Click the gray area to select a record from the table.

## *Changing a table's appearance*

You can change certain ways in which the table displays data. Note that none of these operations modifies the actual data on the disk, only the way in which the data are presented.

Fig. 45. The Table window
Source: ArcMap

**Adjusting field width**

1. Use the cursor to hover over the right edge of the field to resize until it changes to a bar with a double arrow (Fig. 45).

2. Click and drag the edge to the desired width and then release the mouse button.

**Sorting on a field**

3. Right-click on the field heading to sort by and choose either Sort Ascending or Sort Descending.

4. Use Advanced Sorting to sort on more than one field at a time.

**Displaying selected records**

A table can show all records with the selected ones highlighted, or it can show only the selected records.

5. Click on the *Show selected records* button at the bottom of the table. Notice that it also reports how many records are currently selected.

6. Click on the *Show all records* button to show all records again.

### Turning one field off

7. Right-click the field heading and choose Turn Field Off. The field will no longer be shown in the table.

**TIP:** To turn a field back on, you must use the Fields tab in the layer properties.

### To freeze/unfreeze a field

Freezing a field anchors it to the left side of the Table window so that it remains there when the table scrolls to the right.

8. Right-click the field heading and choose Freeze/Unfreeze Column.

9. To unfreeze a frozen column, right-click it and choose Freeze/Unfreeze Column.

**TIP:** An unfrozen column will not go back to its original place in the table automatically. Use the Fields tab of the layer properties to reestablish the default field order.

### Changing field properties

10. Right-click the field heading and choose Properties (Fig. 46).

11. Change the alias of the field, if desired.

12. Change the display characteristics.

13. Change the formatting properties of numeric fields.

### Changing table colors and fonts

The user can change how a table looks by changing its font and the colors that are used to highlight selected records.

1. With the table open, click the Table Options menu in the upper-left corner of the table and choose Appearance.

2. Click the color squares to choose a new color from the drop-down palette for the selection and highlight colors.

3. Choose a font, font size, and color.

Fig. 46. Field properties
Source: ArcMap

**Formatting field values**

You can change the alignment, decimal places, and other formatting characteristics of numeric fields.

1. Right-click a field heading and choose Properties.

2. Click on the ellipses next to the Numeric text in the Field Properties window.

3. Choose the numeric category, such as currency or percentage (Fig. 47).

4. Specify the number of decimals or significant digits, the alignment, and other options.

5. Click OK.

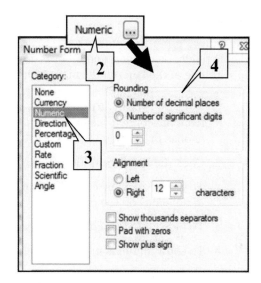

Fig. 47. Formatting field values
Source: ArcMap

*Managing multiple tables*

The Table window shows one table at a time by default. If more tables are opened, they appear as tabs in the bottom of the window. You can also create tab groups to view multiple tables in the window. A tab group can itself contain multiple tables.

1. Click a tab to view that table.

2. To close one table, right-click the tab and choose Close.

3. To view two tables side-by-side, click the Table Options button and choose Arrange Tables > New Vertical Tab Group (Fig. 48).

4. To view tables one on top of another, click the Table Options button and choose Arrange Tables > New Horizontal Tab Group.

5. To adjust the relative size of the tables within the window, click and drag on the bar in between them.

Fig. 48. Using a vertical tab group
Source: ArcMap

When more than one tab group are present, one must be the active one, which is designated by a darker blue title bar. Actions taken using the Table Options menu or buttons will affect the table visible in the active tab group. Newly added tables will be placed in the active tab group.

6. Click the blue title bar of the table in the tab group to make it the active tab group.

7. To move a table from one tab group to another, click on its tab to make it visible and to make its tab group active. Choose Table Options > Move to Previous Tab Group (or Move to Next Tab Group).

## *The Fields tab*

The Layer Properties window has a Fields tab that allows the user to modify the appearance, visible fields, field order, and other properties. Like all layer properties, they affect the appearance of the layer in the map document and do not affect the original file.

1.  If the table is an attribute table, right-click on the data set name and choose Properties from the context menu. If it is a standalone table, right-click on the table name and choose Properties.

2.  Click the Fields tab (Fig. 49).

3.  To create an alias, click the field name in the left panel to highlight it, and type the alias in the panel on the right.

4.  To hide a field, uncheck the box next to its name. Check the box to show it again.

5.  To show all fields in the table, click the *Turn all fields on* button.

6.  To turn off all fields, click the *Turn all fields off* button.

7.  To change the order of a single field, click to highlight it and use the Move Up or Move Down button to change its position. Use the drop-down on the Move Up or Move Down button to move a field to the top or bottom of the table.

Source: ArcMap

8.  Choose Options > Sort Ascending or Sort Descending to sort the fields alphabetically by field name.

9.  Choose Options > Reset Field Order to put the fields back in the original order as stored in the file (a good way to put an unfrozen field back in position).

10. Select whether field names or field aliases will be shown in the table and in menu items that list field names (such as Select By Attributes).

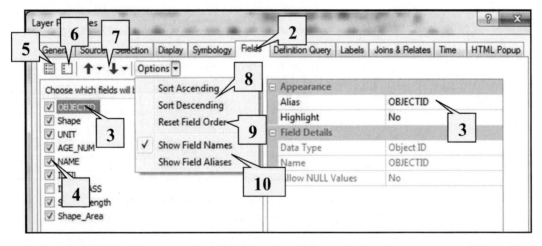

Fig. 49. The Fields tab of the Layer Properties window
Source: ArcMap

### Exporting a table

Exporting a table means saving it under a new name. There are two ways to export a table.

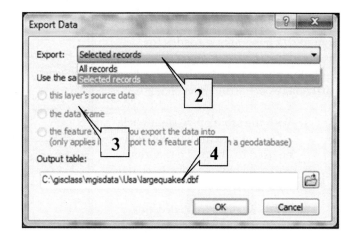

Fig. 50. Exporting records from a table
Source: ArcMap

1. If the table is open, click the Options button and choose Export. If it is a standalone table, you can right-click the table name in the Table of Contents and choose Data > Export.

2. To export the entire table, choose *All records* from the drop-down box. To export a previously selected subset, change the drop-down box to say Selected records (Fig. 50).

3. The choices regarding the coordinate system are dimmed because you are exporting a table instead of a spatial data set.

4. Choose the directory and name of the output file by typing it in or clicking the Browse button. Click OK.

### Getting statistics for a field

Statistics are useful for learning more about the data in fields. Statistics may be calculated for a selected set of records or for the entire table (Fig. 51).

Fig. 51. The Statistics window
Source: ArcMap

1. If desired, use a query to select a subset of records requiring statistical analysis.

2. With the table open, right-click on a numeric field on which to get statistics and choose Statistics from the context menu.

3. Examine the statistics and frequency diagram in the Statistics dialog box (Fig. 51). If the table has records selected, then statistics will be calculated only for the selected records.

4. Look at statistics for a different field by choosing it from the Field drop-down box.

### Summarizing on a field

Summarize generates statistics for groups of features in a given field, such as finding the average magnitude of earthquakes occurring in each state.

1. With the table open, right-click on a categorical field that will be used to group the records and select Summarize from the context menu.

2. In the Summarize dialog box, verify that the chosen field appears in the first box (Fig. 52).

3. In the second box, choose one or more statistics to calculate. Click on the plus sign to expand the list of statistics for a field, and then check the box next to the desired statistic(s).

4. Specify the name of the output table. Click on the Browse button to change the directory where the file will be saved. Click OK.

5. Click Yes to add the table to the map.

Fig. 52. The Summarize window
Source: ArcMap

### Editing fields in a table

Fields can be edited only during an editing session. For more information, consult Chapter 7.

1. If not already in an editing session, click the Editor Toolbar button on the Standard toolbar to open the Editor toolbar.

2. Choose Editor > Start Editing from the Editor toolbar.

3. Open the table, if it is not already open, by right-clicking it and choosing Open.

4. Click in the field space of the record to edit and start typing in the changes.

5. Choose Editor > Save Edits from the Editor toolbar to save during an edit session. When finished typing the changes, choose Editor > Stop Editing and answer Yes when asked whether to save your changes.

### Adding or deleting fields

Fields may be added to tables in two ways. The first way uses ArcCatalog or the Catalog window to enter and define new fields in the data set properties. This way is most efficient when you are entering many fields at once. You can also add a field in the Table window in ArcMap.

**IMPORTANT TIP:** Note that field names must have 13 or fewer characters; may include letters, numbers, and the underscore character (_); and should not contain spaces or special characters such as @, #, !, $, or %. Field names must also start with a letter, not a number.

**Using the data set properties**

1. If the table is currently open in ArcMap, use the Catalog window. Or use ArcCatalog after making sure that the table is not open in ArcMap.

2. Navigate to the table in the Catalog Tree in the ArcCatalog window or the Catalog window.

3. Right-click the table in the Catalog Tree and choose Properties from the context menu.

4. Click the Fields tab (Fig. 53).

5. Scroll down the list of fields, if necessary, to the first empty space at the bottom. Click on the empty space and type in the name of the new field.

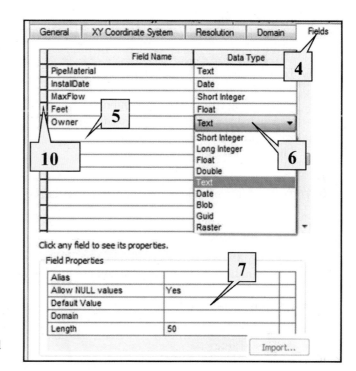

Fig. 53. Adding fields in ArcCatalog
Source: ArcCatalog

6. Click on the Data Type box to the right of the new field, and a drop-down list of field types will appear. Choose the appropriate type.

7. View the field properties in the lower part of the dialog box and change any that need it, such as the length of a text field, a default value, or a domain.

8. Repeat Steps 5–8 until all the fields have been added.

9. Click OK or Apply to enter the changes in the Properties dialog box.

10. To delete a field, follow Steps 1–4 to bring up the table's Fields tab. Select the field by clicking on the gray tab next to its name, and press the Delete key. Warning: You cannot undo this step.

**Managing fields in ArcMap**

1. With the table open, click the Table Options button and choose Add Field from the menu.

2. Type in the name of the field (Fig. 54).

3. Choose the type of field from the drop-down box.

4. View the field properties in the lower part of the dialog box and change any that need it, such as the length of a text field, a default value, or a domain.

5. Click OK. The field will be added to the end of the table.

6. To delete a field, right-click on the field name in the Table window and choose Delete Field from the context menu. Because this action cannot be undone, you will be prompted to confirm the action. Click Yes to delete the field.

Fig. 54. Adding a field in ArcMap
Source: ArcMap

**TIP:** You cannot add or delete a field during an edit session; you must stop editing first.

### Creating a new table

New standalone tables are best created in ArcCatalog or the Catalog window. Attribute tables are generated automatically when the program creates the associated spatial data set.

**Creating a dBase table**

1. In the Catalog Tree, right-click on the folder that is to contain the new file and select New > dBase Table.

2. Type in the name of the new table. Make sure the table name does not contain spaces or special characters such as @, #, %, ^, or &. Underscore characters are permitted.

3. To add fields to the table, see the instructions under *Adding or deleting fields*.

**Creating a table in a geodatabase**

1. Right-click the geodatabase in the Catalog Tree and choose New > Table. A wizard will appear.

2. Enter the name of the table in the box and click Next.

3. Add fields to the table following the instructions under *Adding or deleting fields*. Click Finish when done.

## Calculating fields

Use the Field Calculator in ArcMap to enter information into many records at once, either the entire table or a selected set. The information entered can be as a single number, a text value, or a mathematical expression. Calculations may be done outside of an edit session, but then they cannot be undone.

**TIP:** If the field is not empty before you perform a calculation, be VERY sure that you intend to erase the existing values in the field and replace them with the calculated values.

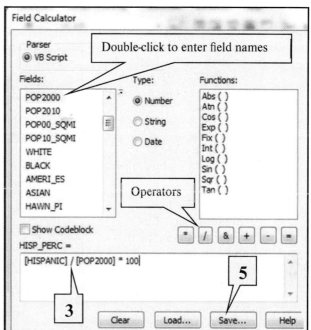

1. With the table open, right-click on the (usually empty) field to be calculated and choose Field Calculator. The Field Calculator will appear (Fig. 55).

2. Enter the expression in the box. Double-click a field name to enter it in the box. Click on operators to include those. Double-click functions to enter them. Numbers may be entered from the keyboard.

3. For example, the expression in Figure 55 was created by double-clicking [HISPANIC] in

Fig. 55. Calculating a field
Source: ArcMap

the Fields box, clicking the / operator, double-clicking [POP2000], clicking the * operator, and typing *100*.

4. Click OK.

5. Complex expressions that are used frequently may be saved as *.cal files and loaded later. Press Save, navigate to the desired directory, and enter a file name. To reload the expression, click Load, navigate to the file location, and select the desired expression file.

### Creating joins and relates

Joins and relates are used to link information from two different tables, based on a field that is common to both. The fields may be numeric or string fields, and they may have different names.

In a join, information from the source table is appended to the destination table. The destination table will then have its original fields plus the fields from the source table.

1. In the Table of Contents, right-click on the destination layer or table, OR click the Table Options button in the Table window.

2. Choose Joins and Relates > Joins from the menu. The Join Data dialog box appears (Fig. 56).

3. Make sure that Join Attributes from a Table is the selection action in the first drop-down box.

4. Enter the field in the destination table that will be used as the key to join the records.

5. Choose the source table from the list of tables in the map document, or use the Browse button to locate a file on disk.

6. Choose the field in the source table that will be used as the key.

7. Choose to keep unmatched records (in which case the appended fields will contain <Null> values) or drop the unmatched records from the table.

Fig. 56. The Join Data window
Source: ArcMap

8. Click OK.

9. To remove a join, right-click the destination table or layer in the Table of Contents or the Table Options menu and choose Joins and Relates > Remove Join(s). Then select a specific join to remove, or remove all joins on that table.

**TIP:** Exporting a joined table will put both the destination and the source table information in the new table.

488

## *Opening Excel Data*

### Preparing the data

Using Excel data is not difficult, but it requires careful preparation. First, make sure that the spreadsheet meets the following requirements.

- ▶ The first row of the worksheet contains the field headings.
- ▶ Field headings must start with a letter, must have no more than 13 characters, and must not contain spaces or shift characters such as %, $, #, and so on.
- ▶ No blank lines occur between the headings and the data or within the data rows.
- ▶ The spreadsheet contains no formulas.
- ▶ The bottom of the spreadsheet has no extraneous data, such as column totals.
- ▶ There are no merged or split cells, and every column contains consistent data types.
- ▶ Numeric columns have only numeric values in them, without text characters like "x" or "n/a" to indicate missing values. If present, missing values should have a numeric NoData marker value such as –99.
- ▶ Text fields contain no commas, unless all the text is enclosed in quotes.
- ▶ It is helpful if each column has been specifically formatted as text data or as numeric data with a specified number of decimal places.

An Excel workbook may have multiple worksheets inside. You must open each one separately. If they have not been named, they will appear as Sheet1$, Sheet2$, and Sheet3$, with Sheet1$ containing the data. If they have been named, the names will appear.

1. Click the Add Data button and navigate to the folder containing the Excel document.

2. Double-click the desired spreadsheet document to look inside it. Select the named sheet you want, or select Sheet1$ if the worksheets have not been named (Fig. 57). Click Add.

3. Use the spreadsheet as if it were any other table (except it is read-only).

4. To convert it to dBase format or save it in a geodatabase, right-click the spreadsheet table and choose Export. See the Skill Reference entry in this chapter on Exporting.

Fig. 57. An Excel workbook with multiple worksheets
Source: ArcMap

**TIP:** You must CLOSE the spreadsheet in Excel before you can open it in ArcMap.

**TIP:** If you are unable to open a file in ArcMap, try saving it in an earlier Excel format.

You can also save a spreadsheet in .csv or .dbf format and open it from ArcMap that way.

### *Displaying x-y data from a table*

Displaying $x$-$y$ data from a table creates an event theme, which is a temporary display rather than a permanent file. To save the event layer permanently as a shapefile or feature class in a geodatabase, you need to export it. See ***Exporting features***.

1. Make sure the table has fields for $x$ and $y$ coordinates and that you know the coordinate system and datum of the coordinates.

2. Right-click the table in the Table of Contents and choose Display XY Data.

3. Specify which fields contain the $x$-$y$ coordinates (Fig. 58). Recall that longitude is the $x$ coordinate and latitude is the $y$ coordinate.

4. Click the Edit button to specify the coordinate system of the points in the table. See ***Specifying a spatial reference***.

5. Click OK to create the points.

**TIP:** Be sure to set the coordinate system to match the $x$-$y$ values stored in the <u>table</u>, not the map currently in the data frame. For example, if the table contains latitudes and longitudes, the coordinate system must be set to a GCS even though the map you are viewing may be in UTM or State Plane.

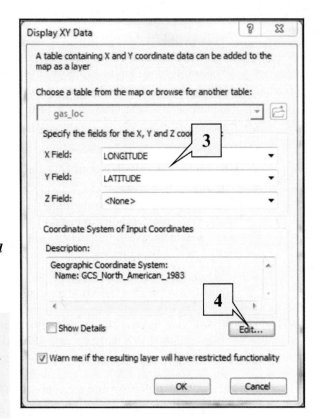

Fig. 58. Adding $x$-$y$ data to a map
Source: ArcMap

# Coordinate Systems

### *Examining the coordinate system*

Every data layer has $x$-$y$ coordinates that must be in some type of coordinate system. The coordinate system is stored as part of the information about a data set, the coordinate system label. It is critical that the label match the coordinates actually stored in the file.

### In ArcCatalog

1. Right-click the feature class in the Catalog Tree and choose Properties.

2. Click the XY Coordinate System tab (for feature classes) or the Projection tab (for coverages) (Fig. 59).

3. WARNING: Using the folders or buttons in this window to change the coordinate system will modify the label attached to the data set and should only be used on feature classes with an <u>unlabeled or incorrectly labeled</u> coordinate system. See ***Specifying a spatial reference***.

### In ArcMap

1. Right-click the layer name and choose Properties.

2. Click the Source tab.

3. Read the coordinate system information (Fig. 60).

The top panel shows the extent, or range of $x$-$y$ coordinates stored in the file. The lower panel shows the source data set for the layer and the coordinate system label.

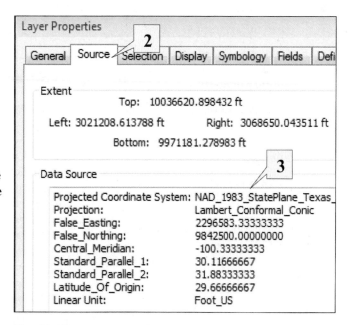

Fig. 59. The XY Coordinate System tab
Source: ArcCatalog

Fig. 60. Viewing a coordinate system in ArcMap
Source: ArcMap

### *Specifying a spatial reference*

The spatial reference is the complete description of a data set's coordinate system (CS). The same sets of windows are used in several places when a coordinate system needs to be specified. Coordinate system definitions are stored as text files with a .prj extension.

1. Examine the current spatial reference name and details.

The Details panel shows the parameters defined for the selected CS (Fig. 61).

2. Browse through the folders to choose the desired CS. The files are organized into geographic and projected coordinate systems (Fig. 62).

3. In ArcMap, the Layers folder lists the CS's of the current layers.

4. Once a CS is seleced, you may click the Favorites button to add it to the Favorites folder. Right-click one in the Favorites folder to remove it.

5. Type a search term and click the Search button to search for a CS.

6. Click Clear Search to clear the search term.

7. Click this drop-down to set or clear a spatial filter, which looks for a CS appropriate to the current geographic extent.

8. Click Add Coordinate System > Import to copy a CS from another data set.

9. Click Add Coordinate System > New to define a custom CS. See ***Defining a custom coordinate system.***

10. Click Add Coordinate System > Clear to set the CS to Unknown.

**TIP:** Always choose the plain version of a datum unless you have reason to do otherwise, such as NAD 1983 instead of NAD 1983 (HARN).

Fig. 61. Specifying a spatial reference
Source: ArcMap

Fig. 62. Organization of coordinate system folders
Source: ArcMap

### Defining a custom coordinate system

ArcGIS comes with many predefined coordinate system files, but sometimes it is necessary to define a custom projected coordinate system. (Defining a custom geographic coordinate system is rare and not recommended for beginners.)

1. Choose Add Coordinate System > New from the Spatial Reference window (Fig. 61) and select Projected Coordinate System. See *Specifying a spatial reference*.

If you are starting with a standard coordinate system as a model, shift the custom window aside so that you can see the original coordinate system parameters in the previous window, and you'll be able to refer to it when entering the parameters.

2. Type a name for the new coordinate system. Include both the datum and projection in the name (Fig. 63).

3. Select the type of projection to be used.

4. Set the parameters such as the central meridan or false easting. See Chapter 3 for a discussion of these parameters.

5. Choose the linear unit to be used to store the $x$-$y$ coordinates.

6. Click Change to choose a different geographic coordinate system than the default.

7. Click OK.

8. The CS is added to the Custom folder in the Spatial Reference window. To save the definition as a .prj file, right-click it and choose Save As.

Fig. 63. Defining a custom coordinate system
Source: ArcGIS

### Setting the coordinate system of a data frame

Use this feature to display data sets with different coordinate systems in the coordinate system of your choosing. The data sets are projected on the fly to whatever coordinate system you specify. The original data are not affected.

1. In ArcMap, right-click the data frame name and choose Properties from the context menu.

2. Click the Coordinate System tab (Fig. 64).

3. Use the folder tree provided to select a coordinate system.

4. You may use the Layers folder to select a coordinate system from the layers already in the data frame.

5. If you use the chosen coordinate system frequently, you can choose Add to Favorites to add it to the Favorites folder for quicker access next time. Highlight it in the Favorites folder and click Remove From Favorites to get it out of the folder.

6. Type a search term and click the Search button to search for a coordinate system. Click the Clear Search button to cancel the search.

7. Click the Spatial filter button to look for a CS based on the current extent; click again to turn the filter off.

8. If more than one GCS are being used in the data frame, you may click the Transformations button to select the method used to transform one datum to another.

9. For more information on using this window, see *Specifying a spatial reference*.

Fig. 64. Setting the data frame CS
Source: ArcMap

494

## *Defining coordinate systems of data sets*

Defining a coordinate system for a data set allows you to display it with other data. It is important to set the coordinate system to match the actual $x$-$y$ coordinates contained in the file. This tool should be used only if the data set coordinate system is undefined or incorrectly defined.

### Using the Define Projection tool

1. Open ArcToolbox, if necessary.

2. Choose the Define Projection tool from the section Data Management Tools > Projections and Transformations.

3. Click the drop-down button (ArcMap only) or the Browse button and select the data set to define (Fig. 65).

4. A warning symbol will appear if the layer already has a spatial reference. You should only change the spatial reference of a data set if the coordinate system label is currently Unknown or incorrect.

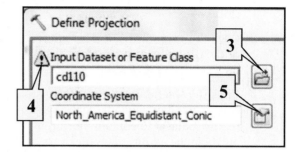

Fig. 65. The Define Projection tool
Source: ArcMap

5. Click the icon to open the Spatial Reference Properties window. Select, import, or create a coordinate system definition. See ***Specifying a spatial reference***.

6. Click OK.

### Using ArcCatalog

1. Right-click the data set in the Catalog Tree and choose Properties from the menu.

2. Click the XY Coordinate System tab.

3. Select, import, or create a coordinate system definition. See ***Specifying a spatial reference***.

---

**TIP:** If a layer is open in ArcMap, then use ArcToolbox or the Catalog window to define the coordinate system. A file lock will prevent ArcCatalog from changing a data set that is open in ArcMap.

### *Projecting feature classes*

Projecting data changes both the coordinate system definition AND the $x$-$y$ coordinates of the features, creating a new file in the process and leaving the original unchanged.

1. In ArcToolbox, choose the Project tool from the Data Management Tools > Projections and Transformations > Feature section (Fig. 66).

2. Click the drop-down (ArcMap only) or the Browse button to select the data set to be projected.

3. Click the second Browse button to specify a folder, a geodatabase, or a feature dataset to contain the output and give the output shapefile or feature class a name. Click Save to enter the new name. Click Next.

4. Click the button to open the Spatial Reference Properties window to select a coordinate system to which to project the data. See *Specifying the spatial reference*.

5. A green dot will appear by the Geographic Transformation box if one is required to convert from one datum to another. Select one or more transformations from the drop-down box.

6. Click OK to project the data.

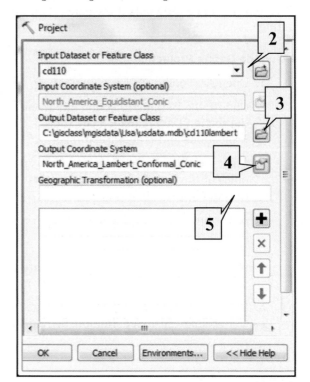

Fig. 66. The Project tool
Source: ArcMap

**TIP:** Projecting coverages requires an ArcGIS Advanced license. If you only have an ArcGIS Basic license, a red X button will appear by the Input Dataset box.

### Projecting rasters

1. If converting from a GCS to a projected coordinate system in meters or feet, calculate the target output cell size in meters and convert to feet, if necessary. Round the result to an even number. You can also usually take the cell size suggested by the tool and round it.

Cell size in degrees × 111.3 km/degree × 1000 = output cell size in meters

2. Open the ArcToolbox > Data Management Tools > Projections and Transformations > Rasters > Project Raster tool (Fig. 67).

3. Click the drop-down (ArcMap only) or use the Browse button to select the input raster to be projected.

4. Enter the name and location for the output raster.

5. Select the output coordinate system. See *Specifying a spatial reference*.

6. If a green button appears by the Geographic Transformation box, a datum transformation is required. Choose one from the drop-down list.

7. Set the resampling technique to NEAREST if the raster contains discrete or categorical data. Set it to BILINEAR if the raster contains continuous data. See the discussion in Chapter 3 for more information.

8. Set the output cell size as a value, or use the Browse button to match it to an existing data set.

9. Select a registration point if you wish for the cell grid to be snapped to a specific location.

10. Click OK.

Fig. 67. The Project Raster tool
Source: ArcMap

### Georeferencing rasters

Georeferencing takes a raster with an undefined and unknown coordinate system and assigns a coordinate system to it by means of matching points with a reference data set.

1. Determine the likely projection of the map in the raster. It need not be exact, but the closer it is to the correct shape the more accurate the georeferencing will be.

2. Find a reference data set that has a known coordinate system, with features that can be identified and matched to locations on the raster.

3. Load the reference data set into ArcMap and set the data frame to the coordinate system chosen in Step 1.

4. Zoom and pan until the reference data set occupies approximately the same position and region in the ArcMap window as the raster map does in its image.

5. Add the image to the map document. It will usually not be visible yet.

Fig. 68a. Georeferencing
toolbar
Source: ArcMap

6. Open the Georeferencing toolbar (Fig. 68a). Set the image to be georeferenced, if more than one image are part of the map document.

7. Choose Georeferencing > Fit To Display from the Georeferencing toolbar. If the image requires, use Flip or Rotate to position the image with north up.

8. Click the Control Points button.

9. Enter pairs of control points that link a point on the raster to a point on the reference layer.

Always enter the pairs in the same order (raster: reference or vice versa). The display will update as points are entered. Zoom and pan between pairs to get the points as accurate as possible. Place points near the four corners of the image at minimum, and more if needed.

10. When no further improvement is seen as points are entered, open the Link Table to examine the control points and residual errors (Fig. 68b).

11. A point with a high residual error relative to the others can be highlighted and deleted.

12. Change the Transformation method if a higher-order transformation is needed.

13. Save the control points as a text file, if desired, for future reference.

14. Record the total RMS error for inclusion in the data set metadata.

15. Click Update Georeferencing. The image transformation will be saved as a world file with the image. Use Rectify to save a new copy of the raster. See **Rectifying rasters**.

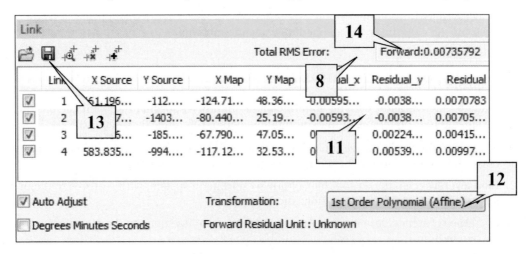

Fig. 68b. Georeferencing a raster
Source: ArcMap

### Rectifying rasters

Rectifying saves a georeferenced raster as a new raster file with the georeferenced coordinate system.

1. Georeference the raster. See ***Georeferencing rasters***.

2. Choose Georeferencing > Rectify from the Georeferencing toolbar.

3. Change the cell size, if desired (Fig. 69).

The suggested size is given in the raster coordinate system units (degrees in this case) and is an exact transformation of the cell size of the original raster to the new coordinate system. You can use it as is, or round it slightly if you wish.

Fig. 69. Rectifying a raster
Source: ArcMap

4. Set the resampling technique to NEAREST if the raster contains discrete or categorical data, or if it is an image. Set it to BILINEAR if the raster contains continuous data. See the discussion in Chapter 3 for more information.

5. Specify the folder or geodatabase in which to place the output raster.

6. Give the output raster a name. Leave the extension blank.

7. Set the output format if the raster is being saved in a folder.

8. Adjust the compression type and quality if desired.

9. Click Save.

**TIP:** Rectifying often changes the colors shown in an image raster. To reset it, use ArcToolbox > Data Management Tools > Raster > Raster Properties > Add Colormap to copy the colormap from the original image to the new image.

# Metadata

### *Setting the metadata style*

The default metadata style is the simple Item Description. To view or edit metadata based on a metadata standard, the style must be changed.

1. In ArcCatalog, choose Customize >ArcCatalog Options from the main menu bar.

2. Click the Metadata tab (Fig. 70).

3. Choose the desired metadata style from the drop-down list.

4. Usually, you will leave the box checked to *Automatically update when metadata is viewed.*

5. Click OK.

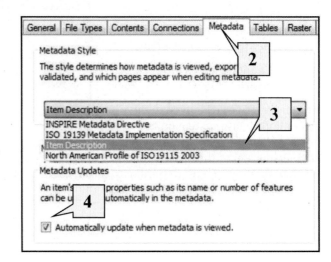

Fig. 70. Setting the metadata style
Source: ArcCatalog

### *Viewing metadata*

1. In ArcCatalog, navigate to the data item in the Catalog Tree and click on it to highlight it.

2. Click the Description tab.

3. If a standard-based metadata style is being used, click the text to expand the additional sections of metadata, including the ArcGIS Metadata or FGDC Metadata.

### *Creating a metadata template*

1. Determine the fields that will be common to all the data sets for which the template will be used and decide what will be put into them.

2. In ArcCatalog, navigate to the desired folder in the Catalog Tree.

3. Right-click it and choose New > XML Document.

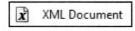

Source: ArcCatalog

4. Click to highlight the XML file in the Catalog Tree and click the Description tab. Click Edit to begin editing the template. See *Editing metadata*.

### *Importing or exporting metadata*

**TIP:** Because of a problem in the initial release, you must turn off background processing to import/export metadata. Use Geoprocessing > Geoprocessing Options from the main menu bar.

1. In ArcCatalog, navigate to the data set to receive the metadata. Highlight it in the Catalog Tree.

2. Click the Description tab, if necessary.

3. Click Import from the menu bar in the Description panel.

Import

Source: ArcCatalog

4. Click the Browse button and select the data set or XML file from which the metadata are to be imported (Fig. 71).

5. Set the Import Type. If you are importing from an XML file or a data set produced using ArcCatalog 10, choose FROM_ARCGIS.

6. The target metadata should already be set.

7. Click OK.

Fig. 71. Importing metadata
Source: ArcCatalog

### Editing metadata

1. In ArcCatalog, navigate to the data set to be documented and click to highlight it in the Catalog Tree.

2. Click the Description tab.

3. Click the Edit button to open the Metadata Editor (Fig. 72).

4. Click one of the sections in the outline on the left panel. A green check means the section is complete; a red x means it is missing one or more items.

5. Fill in the desired metadata for each field. Fields with red outlines are required for the metadata according to the standard specified in the metadata style.

6. Use the arrow buttons to expand or collapse items within a section.

7. Use the + button to add an item container such as a report.

8. Use the X button to delete an item container.

9. Click Exit to exit without saving changes.

10. Click Save to exit and save changes.

**TIP:** Click Save and start editing again before switching to a new section.

Source: ArcCatalog

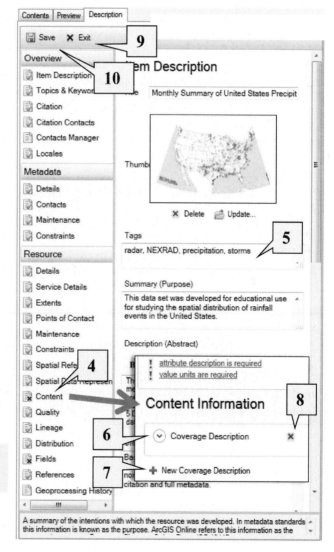

Fig. 72. The Metadata Editor
Source: ArcCatalog

# Editing

### Begin an editing session

1. Open ArcMap and add the data layers you wish to edit. For best results, add only as many layers as needed.

2. If the Editor toolbar is not already displayed, click the Editor Toolbar button on the Standard toolbar.

3. Choose Editor > Start Editing from the Editor toolbar.

4. If data from more than one folder or geodatabase appear in the map document, you will be prompted to choose which one to open for editing (Fig. 73). Choose a layer from the upper panel or a workspace from the lower panel.

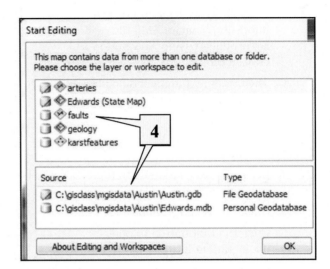

Fig. 73. Choosing an edit workspace
Source: ArcMap

5. If more than one coordinate system are represented in the Edit folder or geodatabase, a warning will be given. Generally, it is best to stick to one coordinate system; however, there are some situations where it is appropriate to have more than one. Just be careful.

### Saving edits

1. Choose Editor > Save Edits from the Editor toolbar.

2. Continue editing.

### Stopping an edit session

1. Choose Editor > Stop Editing from the Editor toolbar.

2. When prompted, indicate whether to save any current changes or discard them.

### Switching to another folder or geodatabase

1. Choose Editor > Stop Editing from the Editor toolbar.

2. Choose whether to save the changes from the current session.

3. Make sure that the layers from the new folder or geodatabase to be edited appear in the Table of Contents.

4. Choose Editor > Start Editing from the Editor toolbar.

5. Select the new folder or geodatabase to edit. Click OK.

### Controlling snapping

Snapping is turned on by default. When the cursor gets close to an existing feature, a Snap Tip pops up to indicate the layer and the type of snapping that will occur on the click. The Snapping toolbar provides access to manage snapping.

1. Choose Editor > Snapping > Snapping Toolbar from the Editor toolbar (Fig. 74).

2. Click the toggle switches to turn the different types of snapping on or off: (from left to right) point, end, vertex, and edge.

3. Choose Snapping > Use Snapping to uncheck it and turn snapping off entirely.

4. Choose one of the less frequently used snapping methods by selecting it from the Snapping menu: intersection, midpoint, or tangent snapping.

5. Choose Snapping > Options to control more aspects of snapping. You can also access this menu using Editor > Snapping > Options from the Editor toolbar.

6. Increase or decrease the snapping Tolerance from the default of 10 pixels, if desired (Fig. 75).

7. Check/uncheck the Show Tips box to control whether the Snap Tips are displayed.

8. Check/uncheck the boxes to control which parts of the Snap Tips are shown.

9. Click the Text Symbol box to specify a different font for the Snap Tips.

Fig. 74. The Snapping toolbar
Source: ArcMap

**TIP:** Keep the Snapping Environment window open to facilitate making changes to the snapping settings as you edit.

### Selecting things to edit

Several tools are used for selecting the features or other items to be changed during editing. Features may also be selected using the Select By Attributes and Select By Location.

Fig. 75. Snapping options
Source: ArcMap

 The Edit tool, on the Editor toolbar, is used to select entire features for editing, and it is the most commonly used tool.

 The Topology Edit tool, on the Topology toolbar, is used to select a shared node or edge for editing. It is only used when editing a map topology or a planar topology.

 The Fix Topology Error tool, on the Topology toolbar, is used to select topology errors and choose from a set of actions for fixing them. It can only be used when editing a planar topology.

 The Edit Annotation tool, on the Annotation toolbar, is used to select annotation.

### Using an editing tool

The editing tools are sensitive to many of the same techniques used for interactive selection, including setting the selectable layers, changing the selection method, and changing the selection options, as described in Chapter 8.

1. Click the editing tool on the appropriate toolbar.

2. Click the feature to be selected or click and hold down the mouse button to drag a box around the features to be selected (not advised for use with the Topology Edit tool).

3. To select multiple features, hold down the Shift key and click on each feature in turn.

4. Click off of any feature to clear the selection, or click the Clear Selection button on the Tools toolbar.

### *Moving features*

1. Select the feature(s) to be moved by using the Edit tool on the Editor toolbar.

2. Place the cursor over the objects until crosshairs appear.

3. Click and drag the feature to a new location.

### *Rotating features*

1. Select the feature(s) to be rotated by using the Edit tool on the Editor toolbar or by using another selection method (Fig. 76).

 2. Click the Rotate tool on the Editor toolbar.

3. Click and drag anywhere on the screen to rotate the features about the center of rotation (marked with an X).

4. To change the center of rotation, place the cursor over the X. It will change to a crosshairs symbol. Click and drag the center to a new location.

5. Click and drag anywhere outside the features to rotate them about the new center.

### *Deleting features*

1. Select the feature(s) to be deleted by using the Edit tool on the Editor toolbar or by using another selection method.

2. Make sure the Edit tool on the Editor toolbar is clicked.

3. Press the Delete key to delete the features.

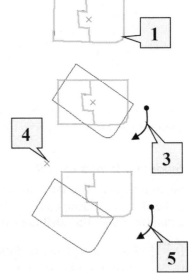

Fig. 76. Rotating features
Source: ArcMap

### *Creating features with templates*

Choose Editor > Editing Windows > Create Features to open the window containing the templates (Fig. 77). Templates usually appear for each layer in the Table of Contents. If a layer is displayed with a unique values map, a template will be created for each category.

### Using a template

1. Click the desired template for the new feature to be created.

2. Click a different construction tool than the default one, if desired.

3. Begin adding points or vertices of the feature to be added.

4. If creating lines or polygons, a provisional sketch will be created. Keep adding vertices until the sketch is complete. Double-click it, or right-click and choose Finish Sketch.

5. To create features in a different layer, change to a different template and select the construction tool.

### Creating a template

6. Click the Organize Templates button at the top of the Create Features window (Fig. 77).

7. Click the New Template button (Fig. 78).

8. Check the box for the feature class for which to create templates.

9. Click Finish and Close.

### Editing template properties

1. Right-click a template in the Create Features window and choose Properties.

2. Change the name, description, or tags.

3. Select the default construction tool for the template.

4. Use the buttons to change the order of the fields, either in layer order or sorted alphabetically. The View button controls whether field names or aliases are shown.

5. Enter values for any attribute fields to be placed in the table when the feature is created.

Fig. 77. The Create Features window
Source: ArcMap

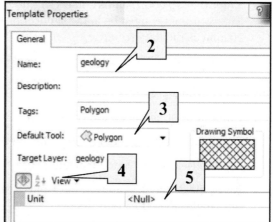

Fig. 78. Creating and modifying templates
Source: ArcMap

### *Using the sketch context menus*

Two context menus can be accessed during sketching. The Vertex menu is accessed by right-clicking on a vertex or segment of the sketch. The Sketch menu is accessed by right-clicking off the sketch.

1. To delete or edit the vertex of a sketch, right-click on a vertex to bring up the Vertex menu (Fig. 79).

2. To add a vertex to a sketch, right-click on a segment of the sketch between two vertices.

3. To bring up the Sketch menu, right-click the screen at a location off the sketch.

4. To delete the current sketch and start again, right-click to raise one of the menus and choose Delete Sketch.

5. When the sketch is complete, right-click to raise one of the menus and choose Finish Sketch, or simply double-click on the last vertex.

Fig. 79. The Vertex and Sketch menus
Source: ArcMap

### *Using Sketch menu tools*

The following functions can be accessed during a sketch by right-clicking anywhere off the sketch to open the Sketch menu and select the function.

#### Entering absolute *x-y* locations

1. Right-click off the sketch to open the Sketch menu.

2. Choose Absolute X, Y from the menu.

3. Enter the *x-y* coordinates and press Enter (Fig. 80). The coordinates must be in the same coordinate system as the data frame.

4. Finish the sketch, or switch to another sketching tool and keep sketching.

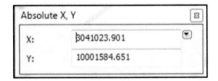

Fig. 80. Absolute X, Y
Source: ArcMap

#### Using offsets from a previous location

1. Enter at least one vertex of the sketch.

2. Right-click off the sketch to open the Sketch menu.

3. Choose Delta X, Y from the menu.

4. Enter the distance change in the *x* and *y* directions in map units (Fig. 81). Press Enter.

5. Finish the sketch, or switch to another sketching tool and keep sketching.

Fig. 81. Delta X, Y
Source: ArcMap

506

### Creating a segment in a specified direction

1. Enter at least one vertex of the sketch.

2. Open the Sketch menu and choose Direction (Fig. 82).

3. Type in the desired angle in degrees and press Enter.

4. The segment is now constrained at the desired angle. Click at the desired distance to enter the point.

5. Finish the sketch, or switch to another sketching tool and keep sketching.

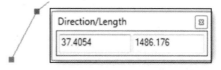

Fig. 82. Constraining the segment direction
Source: ArcMap

### Creating a segment of specific length

1. Enter at least one vertex of the sketch.

2. Open the Sketch menu and choose Length.

3. Type the length in map units in the box and press Enter (Fig. 83).

4. The segment is constrained to a set length. Click at the desired location to create the next vertex.

5. Finish the sketch, or switch to another sketching tool and keep sketching.

Fig. 83. Constraining the sketch segment to a set length
Source: ArcMap

### Creating a segment of set angle and length

1. Enter at least one vertex of the sketch.

2. Open the Sketch menu and choose Direction/Length.

3. Type in the desired angle and length and press Enter (Fig. 84).

4. Finish the sketch, or switch to another sketching tool and keep sketching.

Fig. 84. Constraining the direction and the length of a segment
Source: ArcMap

### Creating a deflected segment

Use this option to create a new segment at a specified angle from the last segment in the sketch.

1. Enter at least one segment of the sketch.

2. Open the Sketch menu and choose Deflection.

3. Enter the angle in degrees that the new segment will be deflected (Fig. 85). Press Enter.

4. The new segment is now constrained along a line. Click at the desired distance to enter the new vertex.

Fig. 85. Deflecting a vertex from the previous segment
Source: ArcMap

5. Finish the sketch, or switch to another sketching tool and keep sketching.

**Deflecting a segment from a feature**

Use this option to create a new segment at a specified angle to an existing segment.

1. Enter at least one segment (two vertices) of the sketch.

2. Right-click *on an existing feature segment* (the side of this building) to open the context menu and then choose Segment Deflection.

3. Type in the angle of deflection from the existing feature and press Enter (Fig. 86).

4. The new segment is now constrained at the specified angle (45 degrees) to the existing segment (the south side of the building). Click at the desired length to create the new vertex.

5. Finish the sketch, or switch to another sketching tool and keep sketching.

Fig. 86. Deflecting at an angle to an existing segment
Source: ArcMap

**Creating parallel or perpendicular segments**

1. Enter at least one vertex of the sketch (Fig. 87).

2. Right-click *on an existing feature segment* (the side of this building) to open the context menu and then choose Parallel or Perpendicular.

3. The new segment will be constrained in the chosen direction. Click to enter the new vertex at the desired location.

4. Finish the sketch, or switch to another sketching tool and keep sketching.

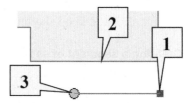

Fig. 87. Creating a segment parallel to an existing feature
Source: ArcMap

*Creating multipart features*

Multipart features include more than one contiguous feature. This technique is used to create a single feature from multiple polygon areas such as Hawaii, from multiple segments forming a single street, or from multiple points forming a single feature.

1. Sketch the first part of the feature using any of the sketching tools and context menus (Fig. 88).

2. Right-click on or off the sketch and choose Finish Part.

3. Add as many parts as desired, ending each with a Finish Part.

4. On the last part, right-click on or off the final sketch and choose Finish Sketch.

Fig. 88. Creating multipart features
Source: ArcMap

### *Creating adjacent polygons*

It is very important that adjacent polygons have identical shared boundaries. You can create adjacent polygons with coincident boundaries in two different ways by using the AutoComplete Polygons tool or by creating a large polygon boundary and cutting it up into sections.

#### Using the AutoComplete Polygons tool

1. Click the editing template for the layer in the Create Features window.

2. Choose the Polygon construction tool, or one of the other construction tools (Fig. 89).

3. Create the first polygon.

4. Choose the Auto Complete Polygon tool from the Construction tools list.

5. Sketch only the undefined boundary of the next polygon. Be sure to start and end inside the adjacent polygon or inside two adjacent polygons.

6. Finish the sketch to create the new polygon.

#### Cutting polygon features

1. Click the editing template for the layer in the Create Features window.

2. Choose one of the polygon construction tools.

3. Create the polygon around the outer boundary (Fig. 90).

4. Click the Cut Polygons tool on the Editor toolbar.

5. Sketch a line that will form the boundary between the new polygons. Important: The line must start and end **snapped to** or **outside of** the existing polygons.

6. Finish the sketch. The new polygons will be created.

**TIP:** A polygon must be selected before it can be cut with the Cut Polygons tool.

Fig. 89. Creating adjacent polygons with the Auto Complete task
Source: ArcMap

Fig. 90. Cutting a polygon
Source: ArcMap

### *Using the Attributes window*

The Attributes window is an interface for entering and editing attributes fields of features. It is convenient to dock this window during editing—under the Table of Contents is a good place.

1. Select the feature(s) to edit.

2. Click the Attributes button on the Editor toolbar. It may come up as a tab in the Create Features window.

**To edit a single feature**

3. Click the record to edit in the top panel (Fig. 91). Its attributes will be shown in the lower panel.

4. Click in the attribute field and type the new value. Press Enter or tab, or move to a different field.

5. Keep editing fields until you are satisfied.

**To edit all features**

6. Click the layer name at the top of the list. The fields on the right will go blank.

7. Click the field to edit and type the new value. The new value will replace the existing values for each record in the selected set.

**Managing records**

8. Clicking an entry in the list causes the feature to flash briefly on the map so that you can locate it.

9. Right-click one entry in the list to bring up a context menu with many useful functions (Fig. 92).

**Changing the display field**

10. Right-click one of the records in the Attributes window and choose Layer Properties.

11. Click the Display tab.

12. Choose the Display Expression Field from the drop-down box. Click OK.

13. The change will be applied to the Attributes window when you make the next selection.

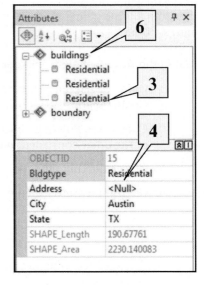

Fig. 91. The Attributes window
Source: ArcMap

Fig. 92. The context menu
Source: ArcMap

### *Using the sketching tools*

By default, a straight segment connects adjacent vertices entered by the user. The sketching tools provide additional options for entering a vertex or designating the shape of the connecting segment. Each sketching tool can be used at the beginning, middle, or end of a sketch. You can use multiple tools in succession during a single sketch. The sketching tools are selected from the Editor menu. The Straight Segment and Endpoint Arc Segment tools have their own icons on the Editor toolbar. The other tools are accessed from a drop-down on the Editor toolbar.

Source: ArcMap

510

### Straight Segment tool

1. Select an editing template.
2. Click on the Straight Segment tool if necessary (it is the default tool).
3. Enter vertices by clicking on the screen.

Source: ArcMap

### Endpoint Arc Segment tool

With this tool the user specifies the beginning, end, and radius of an arc.

1. Click on the Endpoint Arc tool. The current vertex will be the start of the arc.
2. Click on the endpoint of the arc (Fig. 93).
3. Move the mouse to visually set the desired radius of the arc. Click to enter the vertex.
4. To specify an exact radius, press the R key, type in the radius, and press Enter.

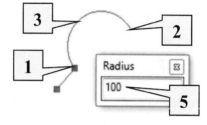

Fig. 93. The Endpoint Arc Segment tool
Source: ArcMap

### Tracing tool

1. Select the feature to be traced (Fig. 94).
2. Select the Trace tool.
3. Press the letter O key on the keyboard to specify the offset distance in map units. Type the distance into the box. Change the corner style if desired. Press Enter.
4. Click on the feature to begin tracing and move the mouse along the feature to create the trace. (If the trace appears on the wrong side, press O again and make the distance negative.)
5. Click again to finish the trace and create the sketch.
6. Finish the sketch, or switch to another sketching tool and keep sketching. The offset will remain at the same value until you change it.

Fig. 94. Tracing a feature
Source: ArcMap

**TIP:** Features have ending locations according to how they were digitized. A trace stops when it reaches the end of the feature. To continue tracing, click at the stopping point to end the current trace and then click again at the same point to begin a new trace.

### Right Angle tool

This tool is useful for creating buildings and other features with square corners.

1. Select the Right Angle tool.

2. Carefully enter the first segment along the edge of the feature (Fig. 95). All corners will be perpendicular to this segment.

3. Enter the vertices of the other corners. Each one will be perpendicular to the previous segment.

4. Instead of adding the final vertex with a click, right-click OFF the sketch and choose Square and Finish from the Sketch menu.

Fig. 95. Creating features with square corners
Source: ArcMap

### Midpoint tool

The Midpoint tool is useful for entering a vertex at the halfway point of a specified distance. Imagine that you wish to split an existing parcel exactly in half.

1. Turn on vertex snapping to ensure that the exact midpoint is found.

2. Select the parcel polygon to split and click the Cut Polygons tool on the Editor toolbar.

3. Choose the Midpoint tool.

4. Click on the location representing the start of the measuring line to be halved (Fig. 96). Be sure to snap to the parcel vertex.

5. Click on the other end of the measuring line. A vertex is entered at the midpoint of the measuring line.

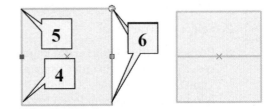

Fig. 96. Using the Midpoint tool to split a parcel in half
Source: ArcMap

6. Click to start and end another measuring line on the other side of the parcel, thereby entering a second vertex.

7. Right-click and choose Finish Sketch.

### Distance-Distance tool

1. Choose the Distance-Distance tool.

2. Click at the center of the first distance circle and move the mouse out to expand the circle (Fig. 97).

3. Press the D key to bring up the Distance window, and type in the exact distance in map units. Press Enter.

4. Click at the center of the second circle, and type D to enter a distance for it. Press Enter to close the window.

5. Move the cursor over one of the two intersection points (a blue circle will appear). Click to create the point or vertex.

6. Finish the sketch, or switch to another sketching tool and keep sketching.

Fig. 97. Finding the intersection point of two distance circles
Source: ArcMap

### Direction-Distance tool

This tool creates a new vertex using a bearing direction from a known point plus a distance from another point. Imagine digitizing a tree location that has a known bearing and distance from a building corner.

1. Select the feature template.

2. Choose the Direction-Distance tool (Fig. 98).

3. Click on the building corner to establish the rotation point of the direction segment. A temporary line appears.

4. Rotate the line to the desired direction and click, or press the D key to type a specific direction angle. Press Enter.

5. Click on the corner again to define the center of the circle.

6. Move the mouse to the desired radius and click, or press R to type a specific radius and press Enter.

7. Click on one of the two possible intersections to create the vertex.

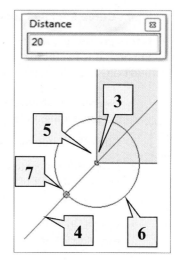

Fig. 98. Using the Direction-Distance tool
Source: ArcMap

### Finding intersections

1. Choose the Intersection tool.

2. As you move the tool over a feature edge, its extension will be indicated. Click the first line on the south side (1 in Fig. 99).

3. Click the second line on the west side. Now the point of intersection will be visible where the extensions of the two lines cross.

4. Click near the intersection to enter the point or vertex.

5. Finish the sketch, or switch to another sketching tool.

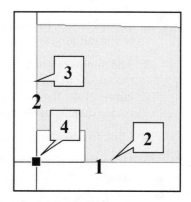

Fig. 99. Placing a vertex at the intersection of two lines
Source: ArcMap

### Arc Segment tool

1. Choose the Arc Segment tool.

2. Click the starting point of the curve (Fig. 100).

3. Click the point through which the curve will pass.

4. Visually choose the desired curve and click the endpoint to create it. Press the R key to enter a specific radius in map units and click Enter.

5. Continue adding pass-through points and endpoints to create multiple curves, if desired.

6. Finish the sketch, or switch to another sketching tool and keep sketching.

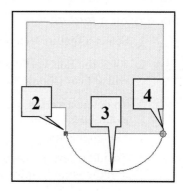

Fig. 100. Sketching a parametric curve
Source: ArcMap

### Tangent Curve Segment tool

This tool creates an arc that is tangential to the previous sketch segment and is useful for creating smooth curves for roads. At least one sketch segment must be present first.

1. Use Sketch or another sketching tool to create one or more sketch segments (Fig. 101).

2. Click the Tangent Curve Segment tool.

3. Move the mouse to the desired endpoint of the tangential curve and click to create the curve. Click R to enter a radius in the box and press Enter.

4. Finish the sketch, or switch to another sketching tool and keep sketching.

Fig. 101. Creating a tangential curve
Source: ArcMap

### Bezier Curve Segment tool

1. Choose the Bezier Curve Segment tool.

2. Add a vertex where the curve will begin (Fig. 102). The first guide will appear. Enter a second vertex to establish the length and angle of the guide.

3. Add another vertex for the second guide. Move the cursor to set the distance and angle of the curve. Make the guide longer for more curve; rotate the line to change the curve shape.

4. Click to add a vertex at the curve's end.

5. Move the cursor to modify the handle and complete the curve's shape.

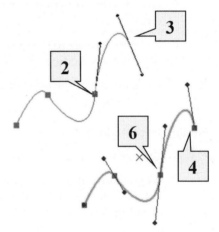

Fig. 102. Creating Bezier curves
Source: ArcMap

6. To change the shape later, select the feature with the Edit tool. Click the Edit Vertices tool on the Editor toolbar. Move the guides from the center, or rotate them from the ends, to change the curve shape.

### *Modifying a feature*

1. Select a feature to modify with the Edit tool.

2. Click the Edit Vertices tool. The sketch of the feature will appear (Fig. 103).

3. To delete a vertex, place the cursor on the vertex, right-click, and choose Delete Vertex from the Sketch menu.

4. To add a vertex, place the cursor over the spot to add the vertex, right-click, and choose Add Vertex.

5. To move a vertex, place the cursor on top of the vertex to be moved until a crosshair symbol appears. Click and drag the vertex to a new location.

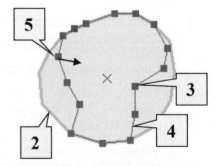

Fig. 103. Modifying a feature
Source: ArcMap

6. When done with modifying the vertices, right-click ON the sketch and choose Finish Sketch.

514

## *Reshaping a feature*

Reshaping works well for reentering part of a feature without having to redo the entire thing.

1. Turn on vertex and/or edge snapping to ensure that the start/stop points are snapped to the feature.

2. Select the feature to reshaped using the Edit tool (Fig. 104).

3. Choose the Reshape Feature tool from the Editor toolbar.

4. Click ON the feature at the start of the section to be reshaped, being sure to snap to an edge or a vertex.

5. Enter vertices defining the new shape. Use the other sketching tools and context menus as needed.

6. Double-click ON the feature at the end of the section being reshaped, or right-click and choose Finish Sketch.

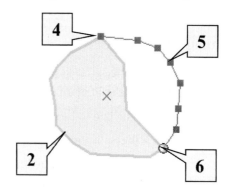

Fig. 104. Reshaping part of a polygon boundary
Source: ArcMap

**TIP:** If the start points and endpoints of the new sketch do not fall exactly on the feature, the reshaping will not be performed.

## *Flipping a line*

1. Select the line feature using the Edit tool (Fig. 105).

2. Click the Edit Vertices tool on the Editor toolbar. The sketch will appear.

3. Right-click ON the sketch and choose Flip from the Sketch menu.

4. When done, either right-click the sketch to finish it or clear the selection by clicking on the screen.

Fig. 105. Flipping a line
Source: ArcMap

### Displaying line directions

1. Click the line layer's symbol in the Table of Contents.

2. In the Symbol Selector, scroll to the bottom of the list to find the arrow symbols.

3. Choose the symbol with an arrow at the start of the line or at the end of the line (Fig. 106).

4. Change the color or width as desired. Click OK.

Arrow at End    Arrow at Start    Arrows at Start and End

Fig. 106. Arrows in the Symbol Selector for showing line directions
Source: ArcMap

## *Splitting a line*

### With the Split tool

Use the Split tool to split a single line into two pieces at the location you click.

1. Turn on vertex snapping if you wish to ensure that the line is split at an existing vertex.

2. Select the line to be split with the Edit tool.

3. Click the Split tool on the Editor toolbar.

4. Click the location on the line where the split should occur.

### Into multiple lines

This task divides a line at equal intervals defined by the user or evenly spaced along the line.

1. Select the line to divide with the Edit tool (Fig. 107).

2. Choose Editor > Split from the Editor toolbar.

3. Choose the split option based on distances, equal parts, or percentages, and enter the values.

4. The new features will replace the original feature.

## *Creating evenly spaced points*

1. Use the Edit tool to select a line along which the points will be placed (Fig. 108).

2. Choose Editor > Construct Points from the Editor toolbar. Arrows will appear on the line to show its direction.

3. Choose the template to specify the layer to which the points will be added.

4. Select the construction option as a number of evenly spaced points or by distance. Enter the values.

5. Click to place additional points at start and end, if desired.

6. Click OK.

## *Moving a feature an exact distance*

1. Select the feature to be moved (Fig. 109).

2. Choose Editor > Move from the Editor toolbar.

3. Type the change in $x$- and $y$-coordinates as an offset from the current location and press Enter.

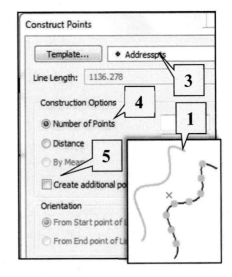

Fig. 107. Splitting a line
Source: ArcMap

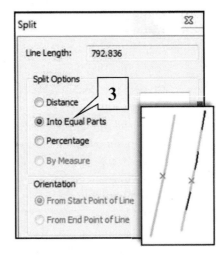

Fig. 108. Constructing evenly spaced points
Source: ArcMap

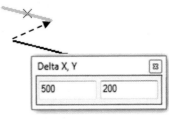

Fig. 109. Moving a line an exact offset
Source: ArcMap

### Using Copy Parallel

1. Select the feature to copy (Fig. 110).

2. Choose Editor > Copy Parallel from the Editor toolbar.

3. Choose the template specifying the layer into which the copy will be placed.

4. Type the distance offset.

5. Examine the arrows on the original line and determine if the new one should be placed to the left or right or both.

6. Change corner style if desired.

7. Usually, you will leave the boxes checked to treat the selection as a single line and to create a new feature for each selected feature. Click OK.

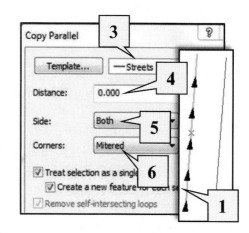

Fig. 110. Making a parallel copy
Source: ArcMap

### Merging features

1. Select at least two features to be merged (Fig. 111).

2. Choose Editor > Merge from the Editor toolbar.

3. Select the feature into which the others will be merged. The merged feature will be given the attributes of the selected feature. The value shown in the window is the primary display field for the layer. Click OK.

Fig. 111. Merging features
Source: ArcMap

### Union of features

1. Select at least two features to union (Fig. 112).

2. Choose Editor > Union from the Editor toolbar.

3. Select the template specifying the layer to which the new feature will be added.

4. The union feature remains selected and on top of the original figures, creating an overlap.

Fig. 112. Union of features
Source: ArcMap

### Intersection of features

1. Select the features to intersect (Fig. 113).

2. Choose Editor > Intersect from the Editor toolbar. (Use Customize > Toolbars > Customize to add it to the Editor menu, if needed.)

3. The new feature remains selected on top of the original polygons, creating an overlap.

Fig. 113. Intersecting two polygons
Source: ArcMap

517

## Clipping features

Overlaps of one polygon on another are considered topological errors. Clip can be used to remove the overlap.

1. Select a polygon that lies at least partially on top of another polygon (Fig. 114).

2. Choose Editor > Clip from the Editor toolbar.

3. If desired, apply a buffer to the clipping polygon by entering the value in map units.

4. Choose whether to preserve the overlapping area or to discard the overlapping area.

5. Click OK.

6. The clipping polygon will remain on top of the original feature. You may move or delete it.

Fig. 114. Clipping with a polygon
Source: ArcMap

## Buffering features

1. Select the feature(s) to be buffered. It may be a point, a line, or a polygon (Fig. 115).

2. Choose Editor > Buffer from the Editor toolbar.

3. Select the template to indicate in which layer the buffer will be placed.

4. Type the buffer distance in map units and press Enter.

5. The new buffer will remain selected on top of the original feature, creating an overlap.

Fig. 115. Buffering a polygon
Source: ArcMap

## Editing annotation

Annotation stored as features in a geodatabase must be edited in an edit session.

### Preparing to edit annotation

1. Make sure that the annotation layer is turned on in the Table of Contents.

2. Choose Editor > Start Editing from the Editor toolbar.

3. Click the annotation template and select a construction tool.

**Adding new annotation labels**

1. Click the desired construction tool in the Create Features window (Fig. 116).

2. Enter the text in the Annotation Construction window.

3. For Horizontal annotation, click on the map once to place it.

4. For Straight annotation, click once to enter the anchor point and again to establish the angle of rotation.

5. For Leader annotation, click once to place the far end of the pointer and again to define the near end and place the label.

6. For Follow Feature annotation, click on the feature to follow.

7. For Curved annotation, enter the start, mid, and endpoint of a curve.

**Make existing labels follow features**

1. Click the Edit Annotation tool and select the annotation.

2. Right-click the annotation and choose Follow > Follow Feature Options and choose Straight or Curved depending on the desired result.

3. Right-click the line to be followed and choose Follow This Feature. Place the cursor on the annotation and move it to the desired location along the line feature.

**To edit existing annotation labels**

1. Click the Edit Annotation tool and select the annotation. Colored handles appear on the annotation, as shown in Figure 117.

2. Click and drag on the red handle to enlarge or shrink the annotation.

3. Click on one of the two blue handles to rotate the annotation.

4. Click on the black handle to move the annotation to a different location.

**Modifying annotation properties**

1. Select one or more pieces of annotation using the Edit Annotation tool.

2. Right-click the feature and choose Attributes from the Context menu, or click the Attributes button on the Editor toolbar.

3. Change one or more of the attributes for the annotation and click Apply.

Fig. 116. Annotation construction tools
Source: ArcMap

Fig. 117. Action handles on selected annotation
Source: ArcMap

### *Editing with map topology*

Map topology is used to edit shared features in shapefiles or when a planar topology is not implemented. It uses the Topology toolbar.

#### Creating map topology

1. Choose Editor > More Editing Tools > Topology from the Editor toolbar. The Topology toolbar will appear.

2. Click on the Select Topology icon on the Topology toolbar.

3. Choose a Map Topology and check the boxes for the layers to participate (Fig. 118).

4. Click the Options arrow if you wish to set a different cluster tolerance from the default.

Fig. 118. Creating map topology
Source: ArcMap

#### Reshaping a common boundary

1. Turn on vertex or edge snapping, and click on the Topology Edit tool.

2. Click the shared boundary to edit. It will be selected in a purple color to show that it is shared (Fig. 119).

3. Choose the Reshape Edge tool from the Topology toolbar.

4. Click on the shared boundary at one end of the section to be reshaped.

5. Enter vertices defining the new edge.

6. End the sketch on the other end of the section and double-click to finish it.

7. The boundary will change to the shape of the sketch for both polygons.

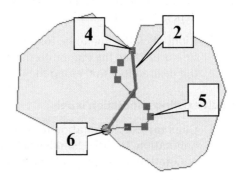

Fig. 119. Reshaping a shared boundary
Source: ArcMap

#### Moving shared nodes

1. Click the Topology Edit tool.

2. Click on a node at the intersection of the lines to select it (Fig. 120).

3. Click and drag the node to its new location.

4. As you move the node, the movement of the shared features will be shown.

5. When you release the mouse button, the node and its attached lines will go to their new places.

Fig. 120. Moving shared lines
Source: ArcMap

### Editing with planar topology

A planar topology contains rules that define how features may be spatially related to each other. You can use tools to help find and fix errors. See *Creating a planar topology*.

#### Preparing to edit

1. Click the Add Data button and navigate to the feature dataset containing the topology. Add the topology to the map. Also add the participating feature classes to be edited.

2. Choose Editor > Start Editing from the Editor toolbar.

3. Choose Editor > More Editing Tools > Topology to open the Topology toolbar.

#### Displaying topology errors

1. Double-click the topology feature class to open its properties.

2. Click the Symbology tab (Fig. 121).

3. Use the check boxes to indicate which errors will be shown.

4. Click on an error type to highlight it and see its symbols.

5. Choose to show with a single symbol or symbolized by error type.

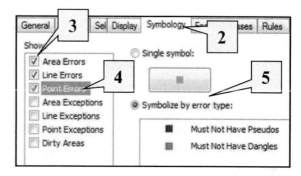

Fig. 121. Symbolizing topology errors
Source: ArcMap

6. Repeat for each error type as desired. Click OK.

#### Controlling which errors are selectable

Use this feature to be able to work on one type of error without selecting any of the others.

1. Double-click the topology feature class to open its properties.

2. Click the Selection tab (Fig. 122).

3. Check the boxes for selecting errors and exceptions as desired.

4. Check the boxes to indicate the types of errors you want to be able to select.

5. Click OK.

Fig. 122. Choosing selectable errors
Source: ArcMap

#### The Fix Topology Error tool

This tool is used to manually select errors on the screen and apply suggested fixes.

1. Pan or zoom to a set of errors on the screen, if necessary.

2. Click the Fix Topology Error tool on the Topology toolbar.

3. Click on an error to select it. If you have several errors of the same type, you can hold down the Shift key to select additional errors, and/or draw a box around the errors to select them. The selected errors will turn black.

4. Right-click a selected error to open a context menu (Fig. 123).

5. Choose an action to fix the error, Snap, Extend, or Trim. Enter the tolerance to use for the item selected. Or mark the error as an exception.

6. All of the selected errors will have the same fix applied.

**The Error Inspector**

The Error Inspector (Fig. 124) is a dockable window that allows you to search for specific errors or groups of errors and apply suggested fixes.

Fig. 123. Selecting a fix
Source: ArcMap

1. Click the Error Inspector button on the Topology toolbar. Dock the window in a convenient place, such as the bottom of the ArcMap window. Pin it if desired.

2. Set the Show drop-down to the desired error type or leave it to show all errors.

3. Check the boxes to search for errors or exceptions or both.

4. Check the box to search only in the visible extent or not.

5. Click Search Now to search for the errors.

6. Click an error to select it, or right-click an error to get a context menu.

7. Zoom to an error for a better look.

8. Select one of the fixes for the error, or mark it as an exception.

9. Continue searching for and fixing errors as needed.

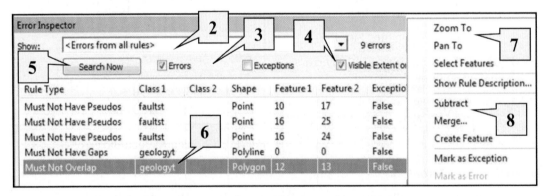

Fig. 124. The Error Inspector and the Overlap context menu
Source: ArcMap

**TIP:** You can select multiple errors in the window by clicking one, scrolling down to the last one, and holding the Shift key while you click it. Right-click one of the selected errors for the menu.

522

# MAPPING

## Maps and Symbols

### *Setting symbols for a layer*

You can set symbols in a number of ways in ArcMap. Here are a few of the most common ways—all of these assume that all features in the layer are being drawn with the same symbol.

**Changing the color of the current symbol**

1.  Right-click on the layer symbol in the Table of Contents and choose a color.

**Changing the symbol**

1.  Click on the layer symbol in the Table of Contents to open the Symbol Selector window (Fig. 125).

2.  Choose a symbol from the scroll box.

3.  Modify the symbol's color, size, thickness, outline, or other attributes by setting the options provided. Make additional changes using the Properties button.

4.  Click Edit Symbol for detailed editing of symbols, which can be saved.

5.  To load additional symbols in the scroll box, click the Style References button and choose from the list of Styles.

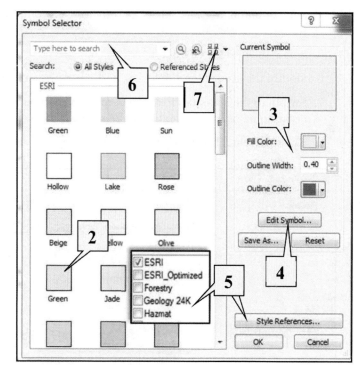

Fig. 125. The Symbol Selector
Source: ArcMap

6.  Enter a key word to search for symbols and click the magnifying glass.

7.  Use the drop-down to change the view of the symbols to a list or a compact view.

8.  Click OK when finished.

**TIP:** You can create your own symbols and groups of symbols, called styles, and work with them in the Style Manager. Read about it in the Help.

### Creating new symbols

Symbols are created from one or more layers of symbol objects, as shown in Figure 125. Users can create new layers, put predefined symbols in them, and modify their colors and other properties to create new symbols. Symbols can also be created from imported bitmap images.

1. Click the Edit Symbol button in the Symbol Selector to access the Symbol Property Editor (Fig. 126).

2. Add new layers using the + button in the Layer area of the window. Delete layers using the X. Change the order of layers with the arrow buttons.

3. Select the type of symbol character to put in the layer. Markers, arrows, and 3D symbols are some of the types available.

4. Select the font and the subset. Many different fonts are available, giving many choices for symbols and text.

5. Select the desired character.

6. Modify the size, color, thickness, and other properties of the character.

7. Switch to the Mask tab to create a mask or halo around the symbol. This tab is useful for creating halos around text to make it easier to read.

8. Click OK when finished creating the symbol.

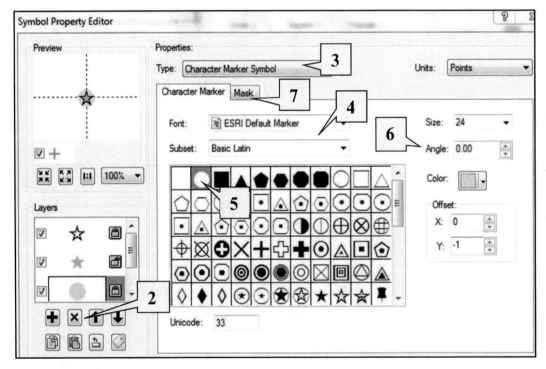

Fig. 126. Changing symbol properties
Source: ArcMap

**TIP:** You can save a symbol as part of a style, and manage them using the Style Manager. See Help for details.

## *Creating maps based on attributes*

Several different types of maps are available, but they all start the same way.

1. Right-click the layer name in the Table of Contents and choose Properties. Click the Symbology tab.

2. In the Show box, click on a heading to expand it and select the type of map desired.

3. Set the other map settings for each map type as described below.

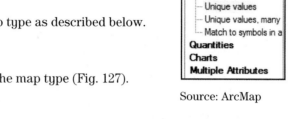

Source: ArcMap

### Unique values map

1. Choose Categories: Unique values for the map type (Fig. 127).

2. Select the Value field on which the map will be based. It should contain categorical or ordinal data.

3. Choose a Color Scheme.

4. Click Add All Values to add the categories and symbols.

5. Double-click a symbol in the list to change its properties using the Symbol Selector.

6. Highlight a row by clicking, if desired, and choose Remove to place it in the <all other values> group.

7. Use the check box to choose to display the other values or not.

8. Click OK.

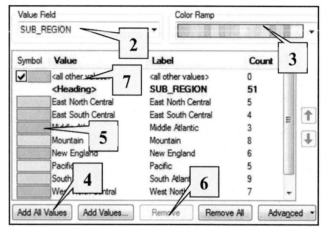

Fig. 127. Making a unique values map
Source: ArcMap

### Graduated color maps

Graduated color maps are for polygons.

1. Choose Quantities: Graduated colors for the map type.

2. Set the Value field on which the map will be based (Fig. 128). It must contain numbers.

3. Select a normalization field if one is desired.

4. Choose a Color Ramp using the drop-down box.

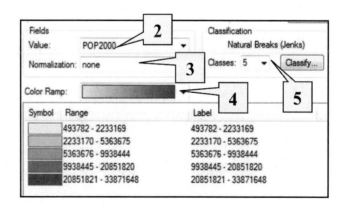

Fig. 128. Making a graduated color map
Source: ArcMap

5. Change the number of classes or classification scheme, if desired. See ***Classifying data***.

6. Click OK.

**Graduated symbols map**

Graduated symbols maps are for point or line data.

1. Choose Quantities: Graduated symbol for the map type.

2. Set the Value field on which the map will be based (Fig. 129). It must contain numeric data.

3. Select a normalization field if one is desired.

4. Set the sizes of the smallest and largest symbols.

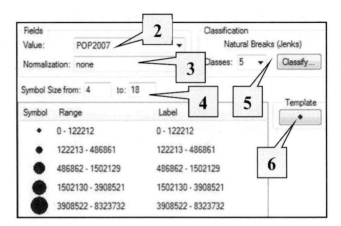

Fig. 129. Making a graduated symbols map
Source: ArcMap

5. Change the number of classes or classification scheme, if desired. See ***Classifying data***.

6. Click the Template button to open the Symbol Selector and change the shape or color of the symbol being used.

7. Click OK.

**Proportional symbols map**

A proportional symbols map is called an unclassed map.

1. Choose Quantities: Proportional symbols for the map type.

2. Set the Value field on which the map will be based (Fig. 130). It must contain numeric data.

3. Select a normalization field if one is desired.

4. Set the units of the values, if known.

Fig. 130. Making a proportional symbols map
Source: ArcMap

5. Click the Min Value button to define the shape and color of the symbol in the Symbol Selector. Set the size to be the smallest desired symbol. The larger symbols are based on a proportion of the data values to the smallest data value.

6. Click the Exclude button to define an expression excluding certain values from the map, for example, NoData values such as –99.

7. Click Rotation to rotate the symbol according to an angle specified by a field.

8. Select the number of symbols to display in the legend.

9. Click OK.

## Dot density map

A dot density map is also unclassed.

1. Choose Quantities: Dot density for the map type.

2. Select a value field and click the > button to move it to the active list (Fig. 131). More than one field may be chosen.

3. Use the < button to remove a single field from the list or << to remove them all.

4. Right-click the symbol to change its color, or select a Color Scheme to assign colors based on a ramp or symbol set.

Fig. 131. Making a dot density map
Source: ArcMap

5. If more than one field are used, select one and use the arrows to change its position.

6. Set the dot symbol size.

7. Set the equivalence value for the dot. In this case, one dot represents 1 million people.

8. Use these buttons to set the color and outline symbol for the background.

9. Click the Exclusion button to define an expression excluding certain values from the map, for example, NoData values such as –99.

10. Click OK.

## Chart map

A chart map is useful for comparing multiple fields. Pie, bar, or stacked chart maps can be made. The directions assume a pie chart, but the others are similar.

1. Choose Chart: Pie for the map type (Fig. 132).

2. Select two or more value fields and click the > button to move them to the list to be used.

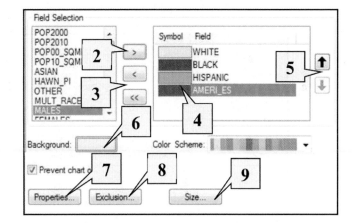

Fig. 132. Making a chart map
Source: ArcMap

3. Use the < button to remove a single field from the list or << to remove them all.

4. Right-click the symbol to change its color, or select a Color Scheme to assign colors based on a ramp or symbol set.

5. If more than one field are used, select one and use the arrows to change its position.

6. Set the background color.

7. Click Properties to set other chart properties such as 2D or 3D symbols.

8. Click the Exclusion button to define an expression excluding certain values from the map, for example, NoData values such as –99.

9. Click the Size button to set a fixed chart size, or set the chart size to vary based on an attribute field.

10. Click OK.

### Classifying data

For quantities data, both feature and raster, ArcMap applies a default classification strategy. Use the following procedure to modify the classification.

1. On a map symbology tab, a classification box will appear if the map type uses a classification.

2. Use the drop-down box to change the number of classes.

3. For other changes, click the Classify button to open the Classification dialog box (Fig. 133). Examine the classification statistics.

4. To exclude certain records, such as those containing zeros or NoData values, click the Exclusion button and enter an expression defining the records to be excluded. (Bug in 10.1–10.2.2: exclusion doesn't work the first time; reopen the window and repeat.)

5. Change the classification method and number of classes or class values.

6. The Sampling button is used to reduce the number of features used to calculate statistics, which is useful when data sets are very large. The default sample is 10,000 records.

7. To manually set class breaks, either click and drag on the blue lines in the graph or type the break values directly into the box on the right.

8. Click OK when done setting the classes.

Classification
Natural Breaks (Jenks)
Classes: 5 ▼ [Classify...]

Source: ArcMap

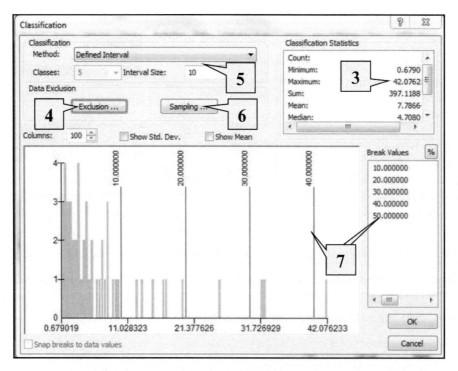

Fig. 133. Setting the classification
Source: ArcMap

528

### Modifying the appearance of the legend

Right-clicking the legend area provides a context menu with several options for modifying the legend (Fig. 134).

1. Use Flip Symbols to reverse the order of the symbols only.

2. Select a symbol by clicking on it (or multiple symbols using Ctrl-click) and choose Properties for Selected Symbols to change their appearance with the Symbol Selector. Use Properties for All Symbols to launch the Symbol Selector and modify the appearance of all of the symbols at once, such as changing them all from red to green.

**TIP:** In Versions 10.4 and 10.5, the menu for Properties for All Symbols does not always appear. If it does not, select the symbols and use the Properties for Selected Symbols instead.

3. Use Reverse Sorting to sort the classes and symbols in the reverse direction.

4. Select one or more classes and delete them with Remove Class(es).

5. Select two or more classes and merge them with Combine Classes.

6. Click Format Labels to apply numeric formatting options to the labels, such as changing the number of decimal places or including thousands separators.

7. Click Edit Description to enter or change a descriptive sentence or paragraph that can be used in the legend.

8. Click on any Range (except the lowest value of the first range) or Label in the legend to type in new values.

Fig. 134. Modifying the legend
Source: ArcMap

9. To create a new color ramp, set the first and last symbols to the desired end colors, and then choose the Ramp Colors option.

# Labels and Annotation

### Using graphic text

Graphic text is placed on a map document interactively and becomes a graphic on the map that is unique to the map document. Most graphic text uses strings typed by the user.

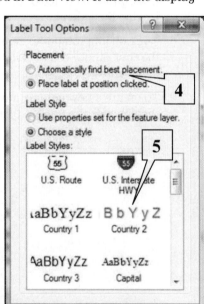

1. Make sure that you are in Layout view.

2. Look for one of the labeling tools on the Draw toolbar. Click on the black arrow for a drop-down menu to choose one of the tools.

Source: ArcMap

#### Adding text to the map

1. In Layout view, click the New Text tool.

2. Click the desired location on the map and enter the text. Press Enter when finished.

3. Newly entered text is always selected, as shown by the dashed blue box around it. At this point you can change its font, size, or color using the menus on the Draw toolbar, or you can use the cursor to click and drag it to a new location.

**TIP:** To delete text, click the black arrow on the Draw toolbar, select the text, and press the Delete key. To delete all graphic elements, choose Edit > Select All Elements from the main menu bar and press the Delete key. Be careful! This option also deletes annotation.

#### Labeling a feature with an attribute

The Label tool, unlike the other graphic text tools, must be used in Data view. It uses the display expression in the layer properties for the text string.

1. Open the layer properties and click the Display tab. Set the Display Expression Field to the desired attribute to appear in the label. Click OK.

2. Make sure that you are in Data view.

3. Choose the Label tool from the labeling drop-down button. The Label Tool Options window will appear (Fig. 135).

4. Choose whether to *Automatically find best placement* or to *Place label at position clicked*.

5. Also choose whether to use the Label properties already set on the Labels tab or to choose a symbol from the window.

6. Click on the feature to be labeled. ArcMap will label the topmost layer clicked if more than one layer is present.

Fig. 135. Label Tool Options
Source: ArcMap

**Splining text along a line**

Splining text makes it follow along a linear feature such as a road or stream.

1. Set the font size and style options desired using the Draw toolbar.

2. Choose the Splined Text tool from the labeling drop-down button.

3. Click vertices to define the line along which the text will appear. Double-click to end the line.

4. Type the text in the box and press Enter.

5. Use the Draw toolbar to modify the position or font characteristics as needed.

**Adding a callout label**

A callout places text in a box with a pointer to the feature of interest.

1. Choose the Callout tool from the labeling drop-down button.

2. Click on the feature to be labeled and drag the cursor to define the direction and length of the callout pointer.

3. Enter the text into the box and press Enter.

4. Click and drag on the text box to change its location, if desired. Click and drag on the blue dot to change the location of the pointer. Use the tools on the Draw toolbar to modify the callout's font, style, size, and so on.

**Creating wrapped text boxes**

| | |
|---|---|
| **A** | Polygon Text |
| **A** | Rectangle Text |
| **A** | Circle Text |

Source: ArcMap

1. Choose one of the wrapped text tools: the Polygon Text, the Rectangle Text, or the Circle Text.

2. For the circle or rectangle text tool, click and drag to draw the circle/box, releasing the mouse when it reaches the desired size and shape. For the polygon tool, click on each vertex to define the desired shape. Double-click when finished.

3. Double-click inside the empty shape to open the text Properties box (Fig. 136). Type the text to be displayed. Do not use the Enter key unless you wish to enforce a new line within the text.

4. Set the symbol, spacing, or other options, if necessary.

5. Use the other tabs to change the margins, columns, frame border, size, position, or area background of the text box as needed.

6. Click OK to place the text.

7. To modify the text later, double-click it to open its properties box.

Fig. 136. Wrapped text properties window
Source: ArcMap

531

### Using dynamic labels

Dynamic labels are used to quickly label all the features in a layer.

1. Open the layer Properties and click the Labels tab.

2. Check the *Label features in this layer* box (Fig. 137).

3. Make sure the method is set to *Label all the features the same way.*

4. Choose the Label Field. Click the Expression button to enter a VBA script.

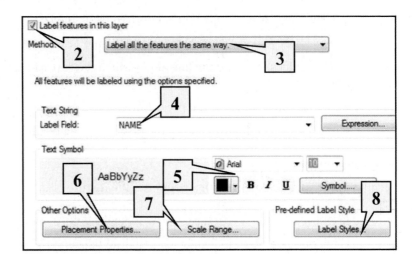

Fig. 137. The Label Properties window settings
Source: ArcMap

5. Edit the font settings, or select a predefined text symbol by clicking the Symbol button and choosing a predefined symbol style.

6. For detailed control of label placement, click the Placement Properties button.

7. Set the scale range, if desired, by using the label's scale range or by typing in new values. If the map scale is outside the specified range, the labels will not be drawn.

8. Select a label style, if desired. A label style includes BOTH a text symbol and predefined label placement options.

9. Click OK to place the labels.

**TIP:** Turn labels on and off for a layer by right-clicking the layer name in the Table of Contents and choosing the Label Features option. If the menu choice is checked, the labels are on, and choosing it will turn them off. If it is unchecked, choosing it will turn them on.

### Creating label classes

Label classes are used to assign different label styles to groups within a layer, such as giving towns small labels and large cities large labels. We will designate towns as having fewer than 100,000 people. The instructions below use this example, but they can be customized for as many classes as you wish.

1. Open the layer Properties and click the Labels tab.

2. Check the box to *Label features in this layer* (Fig. 138).

3. Make sure the method is set to *Define classes of features and label each class differently.*

4. The current class is the Default class. Uncheck the box to *Label features in this class.*

5. Click Add to add a new class. Name it Towns. Click OK.

When the class name is showing in the Class box, then all the settings in the window will be applied to that class. We continue now by defining the Towns class and setting the label properties.

6. Click the SQL Query button.

7. Enter an expression that defines the Towns class. In this case we use:
   [POP2010] <100000. See *Entering an SQL query*.

8. Make sure that the box is checked to *Label features in this class*.

9. Select the font to be used for this class.

10. Set the placement properties or scale range for this class, if desired.

Fig. 138. Creating a label class
Source: ArcMap

You are finished defining and setting properties for the Towns class. The next step is to add the Cities class.

11. Repeat Steps 5–10 for the Cities class. This time the expression will read:
    [POP2010] >= 30000.

12. Continue until all classes have been added.

13. Click OK.

**TIP:** If you see double labels on your features, you probably forgot to uncheck the box for the Default class. If you don't see any labels, or see only some classes, check your SQL queries.

## Using the Label Manager

The Label Manager makes it easier to work with labels from many layers at one time. You can choose layers to modify with a single click rather than having to open and close label properties for each one.

1. Choose Customize > Toolbars from the main menu bar and select the Labeling toolbar.

Source: ArcMap

2. Click the Label Manager button on the Labeling toolbar.

3. Use the check boxes to turn labels for layers on and off (Fig. 139). Notice that both default labels and label classes are shown in this view.

4. Click one of the label classes to highlight it. While it is highlighted, the settings in the window refer to that class.

5. Make changes to the settings. When you click Apply, the changes are applied to the highlighted class.

6. Use the Options button to modify the display of the label class tree.

7. Click OK when finished setting the properties of the labels.

Fig. 139. Using the Label Manager
Source: ArcMap

### Using the Labeling toolbar

1. Choose Customize > Toolbars > Labeling from the main menu bar to open the toolbar.

2. Choose Labeling > Options to set several options such as the color of unplaced labels (Fig. 140).

Fig. 140. Labeling toolbar buttons
Source: ArcMap

3. Choose the Label Manager button to open the Label Manager.

4. Use the Label Priority Ranking button to assign priorities for labels. Labels with a higher rank will have priority and will be drawn if they conflict with labels of a lower rank.

5. Use the Label Weights button to control which labels will be placed when there are overlaps between features and labels.

6. Use the Lock Labels toggle button to lock/unlock the labels against changes.

7. Use the Pause Labeling button to suspend drawing of labels if you accidentally set properties that produced too many labels.

8. Use the View Unplaced Labels button to show the unplaced labels in red so that you can see which ones are missing.

## *Creating annotation*

These directions show how to create annotation stored as text graphics in a map. Consult the Help files for information on creating annotation as a geodatabase feature class.

1. Use the Layer Properties to create dynamic labels for all the desired layers. Take care in setting the properties, weights, and so on because these will control the labels that appear.

2. Annotation will be created for all layers in the data frame with dynamic labels turned on. Turn off any dynamic labels for layers not to be converted.

3. Right-click the data frame name and choose Convert Labels to Annotation. The dialog box will appear (Fig. 141).

4. Examine which layers will be converted to ensure they are the desired ones.

5. Choose to create annotation in the map or in a geodatabase.

6. Choose whether to create annotation for all features, features in the current extent, or selected features.

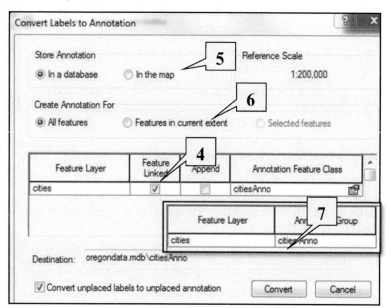

Fig. 141. Creating annotation in a geodatabase
Source: ArcMap

7. If creating geodatabase annotation, check whether to create feature-linked annotation or not. This requires an ArcGIS Standard license. Note the name of the feature class being created. If map annotation is being created, the annotation will be placed in a group.

8. Check the destination for the geodatabase annotation and change it if necessary.

9. Check the box to Convert unplaced labels if you want to place overlapping labels interactively.

10. Click Convert to create the annotation.

11. Place map overflow annotation immediately (see below). Geodatabase overflow annotation may be placed at any time.

## Placing overflow map annotation

Map annotation from overlapping dynamic labels will not be placed on the map. Instead it will be placed in an overflow window and must be placed interactively.

1. Right-click a label in the overflow list and choose a method to locate it, if necessary (Fig. 142).

2. Right-click it again and choose Add Annotation to add the label to the map. Click and drag it to adjust its location, if necessary.

3. If you decide not to place the label, delete it from the list using Delete.

4. Use Show Annotation In Extent to list only the labels in the current view. Place all of these before zooming to another location—it saves time.

## Placing overflow geodatabase annotation

Overflow annotation in a geodatabase is placed in the annotation feature class table along with the placed labels, but with an attribute field indicating that it is not placed. It can be placed at any time.

Fig. 142. Placing overflow annotation
Source: ArcMap

1. Make sure that the annotation feature class is in the Table of Contents and is turned on.

2. Start an edit session. See **Begin an editing session**.

3. Choose Editor > Editing Windows > Unplaced Annotation from the Editor toolbar.

Fig. 143. Unplaced geodatabase annotation
Source: ArcMap

4. If more than one annotation layer are present, choose the one to be searched (Fig. 143), and whether to search in the visible extent or the entire layer.

5. Click Search Now to find the unplaced labels.

6. Clicking a row will cause that annotation to flash on the screen so it can be located. Right-click an unplaced label to open the context menu.

7. Right-click the annotation and choose pan or zoom, if necessary.

8. Right-click the annotation and choose Place Annotation to put it on the map. Adjust its location or properties if needed.

9. Right-click the annotation and choose Delete if you decide not to place it.

10. Close the Unplaced Annotation window when done. You can return and do more later if any are left. Remember to save your edits, and stop editing.

### *Deleting annotation*

**Deleting map annotation**

1. Right-click the data frame and open its properties. Click the Annotation Groups tab.

2. Click on the annotation group to highlight it (Fig. 144).

3. Choose Remove Group.

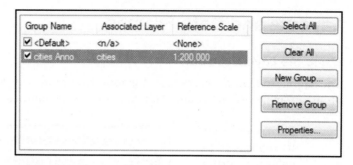

Fig. 144. Managing annotation groups
Source: ArcMap

**Deleting geodatabase annotation**

Geodatabase annotation is deleted like any other feature class.

1. In ArcCatalog or the Catalog window, navigate to the geodatabase containing the annotation.

2. Right-click the annotation feature class and choose Delete.

# Layouts and Data Frames

### *Using the Layout toolbar*

The Layout toolbar provides tools for zooming around the layout page (Fig. 145). It has no effect on the zoom in the data frame. The functions of the buttons, from left to right, are as follows:

1. The Layout Zoom In tool is used to enlarge a portion of the layout. Click the tool and then click and drag a box around the desired area.

Fig. 145. The Layout toolbar
Source: ArcMap

2. The Layout Zoom Out tool zooms out a specified distance from the layout, centered on the clicked point.

3. The Layout Pan tool moves the layout within the window. Click and drag on the layout to move it.

4. The Zoom Whole Page tool zooms so that the entire layout page can be seen.

5. The Zoom to 100% tool shows the layout at the same scale as it will be printed.

6. The Fixed Zoom In tool zooms in a specified amount, centered on the current display, when you click anywhere on the layout.

7. The Fixed Zoom Out tool zooms out a specified amount, centered on the current display, when you click anywhere on the layout.

8. The Go Back to Extent tool returns to the previous extent.

9. The Go Forward to Extent goes to the next extent. This button is only available if you have clicked the Go Back to Extent tool at least once.

10. The Zoom Control box sets a particular percent enlargement for the layout.

11. The Toggle Draft Mode allows you to display each element as a simple labeled box. This feature can make setting up the layout easier without waiting for each element or data frame to redraw each time a change is made.

12. The Focus Data Frame mode switches to Data View but lets you keep editing any text placed on the layout. It's a hybrid between Data View and Layout View.

13. The Change Layout button launches the Template window so you can add or change the layout of the map using a predefined template.

14. The Data Driven Pages button opens the Data Driven Pages toolbar, which is used for working with map tiles used to produce map books. See Help for more information.

### Setting up the map page

1. Choose File > Page and Print Setup from the main menu bar.

2. Check the printer and use Properties to change it, if necessary (Fig. 146).

3. Set the printer paper size, source, or orientation options, if necessary.

4. Set the map page width, height, and orientation.

5. To use the settings from the current printer for the map page size, check the box. Otherwise, set the page size. See the tip below.

6. If you plan to change the map size and want the map elements to be resized along with the page, check the box.

7. If desired, check the box to show the printer margins in the layout.

8. Examine the page preview to make sure that your map (the colored picture) will fit on the paper (the white area). Adjust the settings, if necessary.

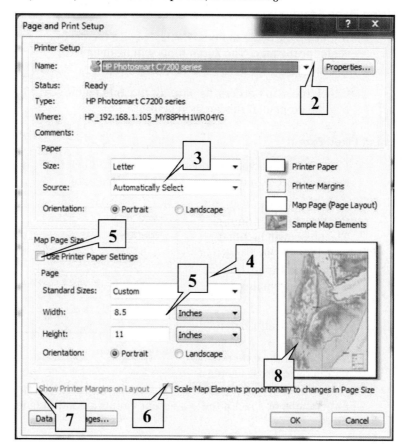

Fig. 146. The Page and Print Setup window
Source: ArcMap

**TIP:** Using the Printer Paper Settings may cause a map document to issue a warning when it is opened on a system where the original printer is not available or to make unwanted changes to the layout when the printer is changed. The author does not recommend using this check box.

### Setting the scale or extent

The Data Frame properties tab provides three options for setting the map scale of the frame: automatic, fixed scale, or fixed extent. Automatic scaling is the default. Fixed scale or fixed extent will deactivate the Zoom/Pan tools for the frame.

1. Right-click the data frame name to open its properties. Click the Data Frame tab. Choose the desired scaling method.

### Using Automatic

2. If you set the extent to Automatic, no other settings are needed. The scale will be determined using the zoom/pan tools in the data frame.

Source: ArcMap

### Using Fixed Scale

3. If you set the scale to Fixed Scale, specify the desired scale for the map. The Zoom tools will be disabled, but the Pan tool will still work.

4. Check the map borders, as parts of the data frame may be cropped if this method is used.

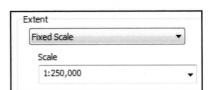

Source: ArcMap

### Using Fixed Extent

5. First, use the Zoom/Pan tools to set the Data view to the desired extent. Alternatively, determine the $x$-$y$ coordinates of the desired extent.

6. In the Data Frame tab, fill the Fixed Extent button. The boxes will be filled with the current $x$-$y$ extent of the data frame. Change the $x$-$y$ coordinates, if desired.

Source: ArcMap

7. Click the Specify Extent button for more options. You can set the extent to match the current visible extent, set it to match a particular data layer, set it to match an existing graphic such as a rectangle, or specify the $x$-$y$ extent values in longitude-latitude degrees instead of map units.

8. Click OK.

### *Setting the reference scale*

1. Use the Pan/Zoom, the Scale readout, or Bookmark tools to zoom to the desired scale.

2. Right-click the data frame name in the Table of Contents and choose Reference Scale > Set Reference Scale.

**TIP:** To set an exact scale value such as 1:24,000, right-click the data frame name, open the data frame Properties, click the General tab, and type a specific reference scale in the appropriate box.

3. To zoom to the reference scale, right-click the data frame name and choose Reference Scale > Zoom to Reference Scale.

4. To remove the reference scale, right-click the data frame name in the Table of Contents and choose Reference Scale > Clear Reference Scale.

**TIP:** Annotation, once created, retains its original reference scale set when it was created, even if the reference scale of the data frame is changed later.

### *Clipping to a layer*

1. Open the Data Frame Properties and click the Data Frame tab.

2. Under Clip Options, change the drop-down to Clip to shape (Fig. 147).

3. Click the Specify Shape button.

4. Choose Outline of Features and select the layer defining the outer boundary.

5. Choose to clip to all features or only the visible ones.

6. Alternatively, enter coordinates for a custom rectangle.

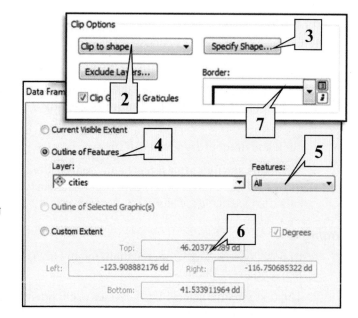

Fig. 147. Clipping layers in a data frame
Source: ArcMap

7. Select a border to outline the clip feature, if desired.

### *Composing the data frames*

The map layout has at least one data frame that can be dragged and resized on the page.

1. Make sure that the Layout view button at the bottom left of the display window is clicked.

2. Choose the Select Elements tool from the Draw toolbar or the Standard toolbar.

3. Click on a data frame to activate it. Blue handles and dashed lines will appear to indicate that it is active.

4. Click and drag on the data frame to move it to a new location (Fig. 148).

5. Click and drag a side handle to increase or decrease the size in one direction.

6. Click and drag on a corner handle to increase or decrease the size in two directions.

Layout view

Fig. 148. Click and drag the active frame to move it or change its size.
Source: ArcMap

### *Adding a north arrow*

1. Choose Insert > North Arrow from the main menu and choose the desired symbol.

2. Click the Properties button to modify the symbol. Click OK to place it.

3. If necessary, click and drag it to the desired location, or resize it.

**TIP:** In some map projections, north is not straight up. Set the north arrow marker angle manually to point true north, if necessary.

### Adding a title

1. On the main menu bar, choose File > Map Document Properties.

2. In the Title box, enter the title of the map. Click OK.

3. Choose Insert > Title from the main menu bar. The title appears at the top of the map, in a dashed blue line to indicate that it is selected.

4. The text remains selected. If necessary, change its size or font using the Draw toolbar or click and drag it to a new location.

5. To modify the text after it has been created, double-click it with the Select Elements tool, or right-click it and choose Properties. You can change the text, set the font and size, or specify a particular position for the text.

### Adding text

1. Choose Insert > Text from the main menu bar. Type the desired text in the box on the map and press Enter.

2. The text remains selected. If necessary, change its size or font using the Draw toolbar or click and drag it to a new location.

3. To modify the text after it has been created, double-click it with the Select Elements tool, or right-click it and choose Properties. You can change the text, set the font and size, or specify a particular position for the text.

### Adding graphics to layouts

The Draw toolbar (Fig. 149) provides functions for creating and modifying objects on a layout. These objects may also be created within a data frame itself, in which case they will be scaled if the map changes size. If they are in the layout, they will be unaffected by scale changes.

The Draw toolbar contains common functions found in other programs. A small black triangle on a button indicates that the button contains a menu.

Fig. 149. The Draw toolbar
Source: ArcMap

### Adding a legend

1. Use the Select Elements tool to click on the data frame containing the layers to appear in the legend. The legend is always created from, and placed in, the active frame.

2. Choose Insert > Legend from the main menu bar.

3. Choose which layers will be included in the legend. To add a layer, click it in the box on the left and click the > button. To remove a layer from the legend, click it in the box on the right and click the < button. Choose the number of columns in the legend. Click Next.

4. Modify the legend title text and formatting to desired settings. Click Next.

5. Specify a legend border, background, and drop shadow, if necessary. Click Next.

6. Click a layer to modify its symbol size and patch style. Click Next.

7. Modify the spacing, if necessary (usually not necessary). Click Finish.

8. The legend appears in the map, selected. Click and drag the legend to the desired location and resize it, if desired. Resizing will change the size of the text and boxes.

**TIP:** To edit the individual elements of a legend, right-click the legend and choose Convert to Graphics. To work with each element, choose Draw > Ungroup from the Draw toolbar.

### *Modifying a legend*

Right-click the legend and choose Properties. The window has four tabs: Legend, Items, Frame, and Size/Position.

### The General tab

1. Change the title of the legend (Fig. 150) or its symbol.

2. Choose which layers will appear in the legend by selecting one and clicking the arrows to move them back and forth.

3. Select a layer and its placement in the order of items in the legend.

4. Change the default Map Connection options if desired.

Fig. 150. The General tab
Source: ArcMap

### The Size and Position tab

The Size and Position tab specifies an exact position and size for the legend in page units (inches by default) (Fig. 151).

1. Set the position in page units, relative to the lower-right corner of the page.

2. The anchor point indicates which part of the legend sits at the XY distance. To place the lower-right corner of the legend at 3 inches from the left and 3 inches from the bottom, enter 3,3 and click the lower-left anchor point. To center the legend at 3,3, enter 3,3 and click the center anchor point.

Fig. 151. The Size and Position tab
Source: ArcMap

3. Set a specific width and height of the legend, if desired, either in page units or as a percentage of the page size.

4. If Preserve Aspect Ratio is checked, the shape of the legend will remain constant if it is resized.

**The Items tab**

This tab gives you individual control of every layer in the legend.

Fig. 152. The Items tab and the Legend Item Selector
Source: ArcMap

1. Select the layer with the properties you wish to change (Fig. 152).

2. To change the label style, choose whether it applies to the entire legend or a single layer, then set the desired font, style, size, symbol, and so on.

3. To change the headings, names, and labels included for the item, click the Style button and select one of the styles from the Legend Item Selector window.

4. Click the Properties button on the Style Selector window to edit every detail of the legend style, including the patch symbol. See ***Managing legend styles.***

5. Check the boxes for any of the desired Map Extent options.

6. To place the selected layer in a new column, check the box.

7. To set the number of columns for one or a set of items, set the column count.

## *Managing legend styles*

The Legend Item Selector allows the user to choose from a set of legend styles, which vary in what elements are shown and how. Four different elements can be included in the legend: the Layer Name, Heading, Label, and Description. Figure 153a shows how these elements might appear in a legend that contains all four elements.

The values given to the four elements are controlled in several places. Layer names and Headings are edited in the Table of Contents (Fig. 153b). Labels can be edited on the layer's Symbology tab, and Descriptions are created by right-clicking the class in the Symbology tab and entering the desired information (Fig. 153c).

The Legend Item Selector (Fig. 153d), accessed from the Items tab, can be used to select a different style with different elements present. The Properties button on it opens the Legend Item window, which allows the user to adjust every detail of the legend, if desired (Fig. 153e).

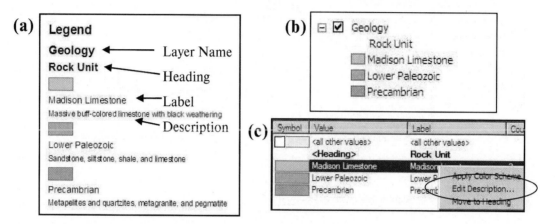

Fig. 153. Editing legends. (a) Legend items and their terms. (b) Edit layer names and headings in the Table of Contents. (c) Edit labels and descriptions on the Symbology tab. (d) The Legend Item Selector. (e) Edit every detail of the legend style.
Source: ArcMap

## The Layout tab

This tab provides general settings for the legend layout and spacing (Fig. 154).

Fig. 154. The Layout tab
Source: ArcMap

1. Change the spacing between the different parts of the legend.

2. Set the default patch style and size for the entire legend.

3. Set the desired text wrapping options for labels and descriptions.

4. Set the desired fitting strategy for the legend.

5. Choose Right to left reading for the legend, if desired.

## The Frame tab

The Frame tab can set a border around the legend (Fig. 155).

1. Choose a border style and change its color, if desired.

2. To modify the existing border, click the Border Selector button or the Edit Border button.

3. Set the gap distance between the map and the border.

4. Round the corners, if desired.

5. Set the background shade, if desired.

6. Set the Drop Shadow, if desired.

Fig. 155. The Frame tab
Source: ArcMap

### Adding neatlines, backgrounds, and shadows

Neatlines are lines that enclose one or more map elements. Many objects in layouts, including data frames, legends, and scale bars, have a tab in their properties to set up borders, backgrounds, and drop shadows. To access these tabs, open the element's properties by either double-clicking it or right-clicking it, and choose Properties. To create a neatline or shaded box as a separate object, do the following:

1. Choose Insert >Neatline from the main menu bar. The Neatline window appears (Fig. 156).

2. Choose the desired Placement option.

3. Set the border, background, and drop shadow styles.

4. To further modify the available styles, click the Border Selector or Edit Border buttons as shown.

5. Set the gap between the neatline and the elements and enter rounding, if desired. The higher the percentage, the more rounding occurs.

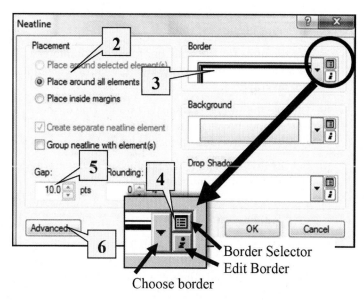

Fig. 156. Setting up neatlines and boxes
Source: ArcMap

6. The Advanced button allows customized editing of the symbols, gaps, and rounding for each border, background, and shadow.

### Adding a scale bar

1. Using the Select Elements tool, click to activate the desired frame. The scale bar will be placed in the active frame and will be sized according to the scale of the active frame.

2. Choose Insert > Scale bar from the main menu. The Scale Bar Selector window will appear (Fig. 157).

3. Choose the desired scale bar.

4. To modify the scale bar, click the Properties button to open the Scale Bar.

Fig. 157. The Scale Bar Selector
Source: ArcMap

5. The Numbers and Marks tab controls the spacing of numbers and marks on the bar (Fig. 158). These values can usually be left as defaults.

6. The Format tab controls the font of the scale text and the style of the scale bar. The defaults are usually fine.

7. The Scale and Units tab controls the length and divisions of the scale bar.

8. Set the units for the scale bar to miles, kilometers, or some other unit. The default of meters is usually not desirable.

9. Choose the *When resizing...* option. One or more of the input boxes in the window may be dimmed, depending on which option is chosen.

10. Set the division value, the number of divisions, and the number of subdivisions, as applicable.

11. Check the box to place the subdivisions before the zero point rather than in the first division.

12. The chosen units will be labeled on the scale bar (e.g., Miles). Choose the label position, change the label text and font symbol, and set the gap between the scale bar and the label, if desired.

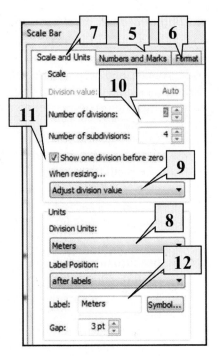

Fig. 158. Scale bar properties
Source: ArcMap

**TIP:** To change the properties of a scale bar after it has been created, double-click the scale bar using the Select Elements tool, or right-click it and choose Properties.

### Adding pictures

1. Choose Insert > Picture from the main menu bar.

2. Navigate to the folder containing the picture and click on it to select it.

3. Click Open to add the picture.

4. Resize and/or move the picture to the desired location using the Select Elements tool.

**TIP:** Information on allowed image formats can be found in the online help index by typing the entry "rasters" and choosing the subheading "formats, supported."

**TIP:** A bug in the initial release makes it difficult to resize the picture when it appears substantially larger than the layout page. A workaround is to set the layout zoom percentage to a small value so that the entire picture can be seen and resized.

### Creating a map from a template

A map template is a set of data frames, titles, styles, and other map elements that are already formatted and ready to receive the data in the data frame(s). Use a map template to quickly create a map in a standard format. You can save any map as a template to create a similar map again.

1. Click the Change Layout button in the Layout toolbar.

2. Click one of the tabs to see a choice of templates (Fig. 159). Templates you have created will be stored in the My Templates tab.

3. Use the arrow buttons to scroll the tabs to see them all.

4. Click on a template to see a preview of it.

5. Use the Browse button to navigate to another directory containing more templates saved elsewhere.

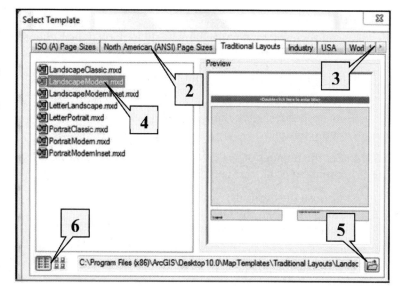

Fig. 159. Choosing a map template
Source: ArcMap

6. Click the Thumbnail button to see icons of all the templates; click the List button to go back to the original list view.

7. Click the desired template and choose Next or Finish.

**TIP:** If the template has more than one data frame, it will prompt the user to assign the data frames in the map document to the data frames in the template.

### Assigning multiple frames

8. If the data frames need assigning, click each of the frames in the list on the left (Fig. 160).

9. Use the Move Up and Move Down buttons to put them in the same order as the numbered frames in the new layout.

10. Click Finish.

11. Finally, change any titles or other map elements in the template that need to be customized.

Fig. 160. Assigning data frames to the template frames
Source: ArcMap

### *Printing a map*

1. To preview a map to see how it will look on paper, choose File > Print Preview from the main menu bar.

2. To print, Choose File > Print from the main menu. The Print window appears (Fig. 161).

3. To change the printer or its properties, click the Setup button.

4. Set the number of copies to print.

5. Choose the desired tiling options if the map is larger than the printer paper.

6. Preview the layout placement on the page to ensure that the map (color) fits the paper (white). Click OK.

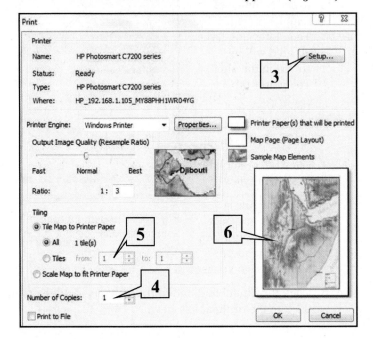

Fig. 161. The Print window
Source: ArcMap

### *Exporting a map as a picture file*

You can export a map as an image file to put it on a web page or inside another document or as a PDF to share it with others.

1. Choose File > Export Map from the main menu.

2. Navigate to a folder for the saved picture file (Fig. 162).

3. Choose the type of file and enter a name for it.

4. Click the gray arrow for more export options, such as the resolution.

5. If you plan to enlarge the map before printing, increasing the default resolution may be necessary.

6. To avoid having a white border around the picture, check the box to Clip Output to Graphics Extent.

7. Click Save.

Fig. 162. Exporting a map to a PDF
Source: ArcMap

### Creating a simple graph

This example shows the steps to create a horizontal bar graph. Other graph types will have slightly different options. Have fun experimenting.

1. Open the table from which the graph will be created. Choose Table Options > Create Graph from the main menu bar. The Graph Wizard will appear (Fig. 163).

2. Choose the graph type.

3. Set the Value field to be graphed.

4. To sort the bars, make the Y field the same as the Value field and choose Ascending or Descending.

5. Choose the X or Y label field.

6. Uncheck the Add to legend box if you don't need a legend.

7. Set the bar color. Use Match with Layer to make it the same as the map. Choose Custom to make all the bars one color of your choosing. Choose Palette to make every bar a different color.

8. Click Next.

9. Give the graph a title and footer.

10. Select a title and position for the legend if you have one.

11. Click the Left tab and provide a title for that axis.

12. Click the Bottom tab and enter a title.

13. Click Finish.

14. To place a graph on the layout, open the graph. Right-click the blue bar at the top of the graph and choose Add to Layout.

15. To edit the graph properties, right-click on the top blue bar of the graph window and choose Properties or Advanced Properties.

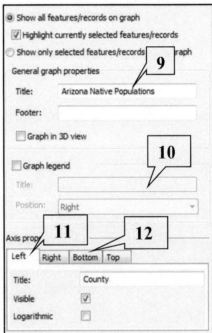

Fig. 163. The Graph Wizard
Source: ArcMap

# ANALYSIS

## Queries

### Using the Selection view

The Selection view in the Table of Contents provides quick access to many selection-related operations.

1. Click the View by Selection icon at the top of the Table of Contents (Fig. 164).

2. Right-click a layer to open a context menu with selection-related options.

3. Use the functions in the context menu for zooming, copying, creating layers, and so on. You can also open the Attribute table and the layer properties from this menu.

**TIP:** The Selection context menu can also be accessed by right-clicking a layer in Drawing Order or Source view and choosing Selection from the context menu.

4. Click the visibility icon to make the layer visible in the map or not.

5. Use the headings to determine if a layer is selectable and whether it currently has a selection.

6. Examine the values to see how many features are selected. If only a few items are selected, they will be listed in the Table of Contents.

7. Click the Clear Selection button to clear the selection for the layer.

### Setting the selectable layers

8. Click the selection icon to make the layer selectable. Click the icon again to make the layer not selectable—the icon will turn gray.

9. To make a layer the only selectable layer, right-click it and choose Make This The Only Selectable Layer.

**TIP:** The Selectable Layers setting does not apply to selections based on attributes or location.

### Clearing selections

Selections may be cleared in a number of ways. Some clear only one layer or table; others clear selections for all layers and tables, so be careful about which you use.

Fig. 164. The Selection view and context menu
Source: ArcMap

### Clearing the selection for a layer or table

1. Click the Clear selection button in the View by Selection view in the Table of Contents (Callout 7 in Fig. 164).

2. Right-click a layer in Selection view to open the context menu and choose Clear Selected Features. Features will be cleared for that layer only.

3. Right-click a layer in any other view and choose Selection > Clear Selected Features.

4. If a table is open, click the Table Options menu and choose Clear Selection.

### Clearing all selections

5. Choose Selection > Clear Selected Features from the main menu bar.

6. Click the Clear Selection button on the Tools toolbar.

### *Changing the selection method*

The selection method controls what happens to a previous selection when a new one is specified for the same layer.

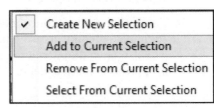

1. Choose Selection > Interactive Selection Method and choose the desired option.

Source: ArcMap

---

**TIP:** The query windows also have a drop-down bar used to change the selection method for a query made in the window.

---

### *Selecting features interactively*

Features may be selected visually by designating them on the screen using one of the interactive selection tools (Fig. 165). By default, any feature that touches the specified shape will be selected. To change the default, see *Changing the selection options*.

Fig. 165. The interactive selection tools
Source: ArcMap

1. Set the Selectable Layers, if necessary. See *Using the Selection view*.

2. If necessary, change the selection method by choosing Selection > Interactive Selection Method and picking one of the methods. See *Changing the selection method*.

3. Click the **Select by Rectangle** tool on the Tools toolbar.

4. Click on a feature in the map to select it. It will be highlighted in the selection color.

5. To add a feature to the selection, hold down the Shift key and click the feature. To remove a feature from the selection, hold down the Shift key and click the feature.

6. To add a group of features using a rectangle, click on one corner of the rectangle, drag the mouse to the desired size, and release the mouse button. All features touching the rectangle will be selected.

---

**TIP:** Hold down the Shift key while drawing the rectangle to add a group of features to the current selection.

---

554

**Using the other selection tools**

Click the desired tool and use the instructions below to specify the shape. By default, any feature touching the shape will be selected.

1. To **Select by Polygon**, click to enter each vertex and double-click to finish.

2. To **Select by Lasso**, place the cursor at the starting location, click and hold down the mouse button while drawing the shape, and release the mouse button to finish.

3. To **Select by Circle**, click in the center of the circle and drag out to the desired radius. You can also click R on the keyboard while the mouse button is down to specify a radius in map units. Release the mouse, type the radius, and click Enter.

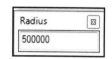

Source: ArcMap

4. To **Select by Line**, click to enter each vertex and double-click the last vertex to finish.

## *Creating a layer from selected features*

1. Perform interactive, attribute, or location queries to select the desired records.

2. Right-click the layer name and choose Selection > Create Layer from Selected Features from the context menu.

3. The new layer appears in the Table of Contents.

4. Click on the new layer name twice to give it a new name.

**TIP:** If the Table of Contents is in View by Selection mode, you can create the layer, but you must switch to another view or open the layer properties to name the new layer.

## *Creating a definition query*

A definition query is a layer property that temporarily confines the features of the layer to a subset of the stored features.

1. Open the properties for the layer, such as by right-clicking it and choosing Properties.

2. Click the Definition Query tab (Fig. 166).

3. Click the Query Builder button and enter an SQL expression to define the subset. See ***Entering an SQL query***.

4. The query will be shown on the tab. Click OK.

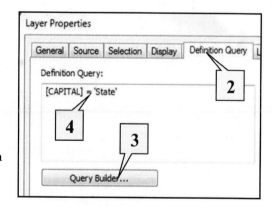

Fig. 166. Creating a definition query
Source: ArcMap

**TIP:** Always Verify the expression in the SQL Query window to ensure that it contains no errors. An expression error would cause the layer to disappear and/or generate drawing errors.

## *Entering an SQL query*

An SQL query is an expression to select features based on an attribute in the table. A similar window appears whenever an SQL expression is needed. The query is entered in the expression box in the lower part of the window.

1. Choose the selection method (Fig. 167). For a discussion of the different methods, see Chapter 8.

2. Enter a field name from the list into the expression box by double-clicking on the field name.

3. Enter an operator by clicking it once.

4. For categorical data, click Get Unique Values and double-click the desired value from the list.

5. Alternatively, enter a value by typing it into the box. If it is a text string, put single quotes around it.

6. Multiple queries use criteria from more than one field. The field name must be repeated for each query.

7. Click Verify to ensure that the query is correctly formulated, if desired.

Fig. 167. Entering a query
Source: ArcMap

8. Use the Save button to save a complex query or one that is frequently used. It will be saved with an .exp extension. Load it again later using the Load button.

9. Click Apply or OK to execute the query.

**TIP:** Many users find that typing in or editing the expression by hand tends to generate errors. Use the boxes and the buttons provided to enter expressions for quickest results.

**TIP:** Field names may be enclosed in brackets, quotes, or nothing, depending on the database and the version of ArcGIS being used.

**TIP:** When you are using multiple conditions, the field name must appear in every condition. To find rents between 500 and 1000 dollars, for example, enter [RENT] > 500 AND [RENT] < 1000. You cannot enter [RENT] > 500 AND < 1000.

## Using Select By Attributes

You can access the Select By Attributes menu in two ways.

### Selecting features from a layer

1. Choose Selection > Select By Attributes from the main menu bar.

2. Choose the layer containing the features that are to be selected (Fig. 168).

3. Check the box to show only the selectable layers in the list, if desired.

4. Choose the selection method.

5. Enter the expression. See **Entering an SQL query**.

6. Click Apply to execute the expression and create the selection.

### Selecting records from a table

1. Open the table, such as by right-clicking the table or the layer in the Table of Contents and choose Open Attribute Table.

2. Choose Table Options > Select By Attributes. The window will look similar to Figure 168, but there will be no drop-down to choose the layer.

3. Choose the selection method.

4. Enter the expression. See **Entering an SQL query**.

5. Click Apply to execute the expression and create the selection.

Fig. 168. The Select By Attributes window
Source: ArcMap

## Using Select By Location

Select By Location compares the features to be selected (from the target layer) with features from another layer (the source layer) using a spatial condition operator such as *intersects* or *contains*.

1. Choose Selection > Select By Location from the main menu bar.

2. Choose the selection method (Fig. 169).

3. Choose the target layer(s) from which to select features. Be careful—layers from previous queries remain checked until you uncheck them.

4. Choose the source layer against which the target features will be tested.

5. To use only the selected features of the source layer, check the Use selected features box.

6. Choose the spatial condition for the test.

7. If using the condition *within distance of*, or increasing the distance over which the other criteria are applied, enter a buffer amount and distance units, and check the Apply a buffer box, if necessary.

8. Click Apply to make the selection.

Fig. 169. The Select By Location window
Source: ArcMap

---

**TIP:** To use the same layer as the selection layer and the condition layer (e.g., to select all restaurants within two miles of the restaurant you own), you must first use an attribute or an interactive query to select the feature of interest—your restaurant. Then do the Select By Location and specify the same layer for Steps 3 and 4.

### Changing the selection options

The selection options control several aspects of selections and queries.

1. Choose Selection > Selection Options from the main menu bar.

2. The first box (Fig. 170) controls how shapes select features when one is using one of the interactive selection tools. By default, any feature partially or completely in the shape is selected. You may also choose to select features completely within the shape or to select features that the shape is completely within.

3. The second box sets how close you need to click to a feature before it is selected. If you often have trouble selecting points, increase this distance.

4. This box changes the selection highlight color.

5. In the fifth box, you may alter the threshold that determines how many records will be modified by a Switch Selection command. If the number exceeds the threshold, a warning prompt will be issued. This option prevents you from mistakenly doing a switch that will take a very long time to complete.

Fig. 170. Setting the selection options
Source: ArcMap

6. Finally, you can choose whether to save layers with the current selections (the default) or to clear the selections of all layers when saving the map document.

# Overlay and Spatial Joins

## *Performing a spatial join*

In a spatial join, the attributes of features in the source layer are appended to the features in the destination table. The output layer always contains the same features as the destination table. When choosing the join options in Step 4, keep in mind the cardinality of the join to pick the most appropriate one.

1. Right-click the destination layer in the Table of Contents and choose Joins and Relates > Join.

2. Choose to join to another layer based on spatial location (Fig. 171).

3. Choose the source layer from the drop-down box, or click the Browse button to choose one from the disk.

4. Choose one of the two join options and specify any summary statistics desired, if applicable.

5. Specify the shapefile or feature class file to contain the new output layer.

6. For more information on joins, click the About Joining Data button.

7. Click OK to execute the join.

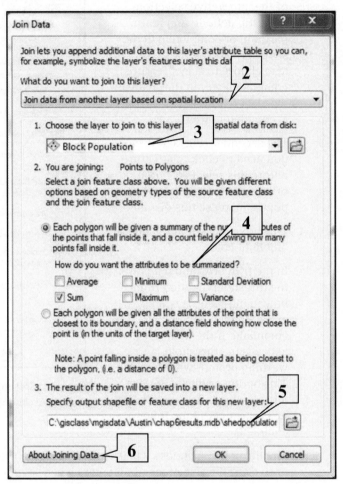

Fig. 171. Join Data window for spatial joins
Source: ArcMap

## Performing an intersection

Intersect combines the features and attributes of two input layers, preserving the areas that are common to both.

1. Choose ArcToolbox > Analysis Tools > Overlay > Intersect (Fig. 172).

2. Enter the Input Features by clicking a choice from the drop-down list or by clicking the Browse button to select a file from the disk.

3. Enter a second Input Features layer.

4. Enter the name and location of the Output Feature Class.

5. Set the JoinAttributes option to NO-FID, the best choice in most cases. ALL joins all attributes from both tables, NO-FID joins all attributes except the FIDs, and FID-ONLY joins only the FIDs.

6. Set a carefully chosen XY tolerance for help in preventing slivers.

7. Change the default output type, if desired.

8. Click OK.

Fig. 172. The Intersect tool
Source: ArcMap

**TIP:** Two input layers must be entered if the user has an ArcGIS Basic license. With an ArcGIS Advanced license, three or more layers may be entered and the intersections will be performed sequentially.

## Performing a union

A union combines the features and attributes of two polygon layers, preserving all areas from both. The Union window is nearly identical to the Intersect window shown in Figure 172.

1. Choose ArcToolbox > Analysis Tools > Overlay > Union.

2. Enter the Input Features by clicking a choice from the drop-down list or by clicking the Browse button to select a file from the disk. Both inputs must contain polygons.

3. Enter a second Input Features layer.

4. Enter the name and location of the Output Feature Class.

5. Set the JoinAttributes option to NO-FID, the best choice in most cases.

6. Set a carefully chosen XY tolerance for help in preventing slivers.

7. Check the box if gaps are to be allowed between polygons. Click OK.

Source: ArcMap

### *Creating buffers*

Buffers create polygons delineating the area within a specified distance of a set of features. Creating buffers can be a time-intensive process.

1. Choose ArcToolbox > Analysis Tools > Proximity > Buffer.

2. Select a layer for the Input Features to buffer (Fig. 173).

3. Specify the name and location for the Output Feature Class.

4. Specify the units and value for a fixed buffer distance.

5. OR choose a field from the input layer attribute table containing the buffer distances.

6. Change the Side and End type settings, if desired.

7. Set the Dissolve Type; use ALL to dissolve boundaries between overlapping buffers or NONE to leave them as is. The ALL option is usually best.

8. Choose one or more Dissolve Field(s). Buffers sharing the same value in the field will be dissolved.

9. Click OK.

Fig. 173. The Buffer tool
Source: ArcMap

### *Clipping a layer*

Clipping removes features outside a boundary specified by a second layer.

1. Choose ArcToolbox > Analysis Tools > Extract > Clip (Fig. 174).

2. Specify the Input Features layer to be clipped.

3. Specify the Clip Features layer representing the clip boundary.

4. Specify the name and location of the Output Features.

5. Set a carefully chosen XY tolerance to help avoid slivers, if desired.

6. Click OK.

Fig. 174. Clipping a layer
Source: ArcMap

## *Merging or appending layers*

Merge combines two or more feature classes to create a new feature class. Append combines two or more feature classes and places them into an existing feature class. The data type of the inputs must match (all points, for example).

1. Choose ArcToolbox > Data Management Tools > General > Merge (Fig. 175).

2. Specify the layers to be merged together.

3. Specify the output feature class to contain the merged features.

4. Modify the Field Map, if desired. The Field Map lists all unique attribute fields from the input layers. When the attributes match, the field information will be combined into the single new field. If the layers have different fields, each field will be included in the output with null values from the feature classes not containing that field. You can rename, delete, and reorder the fields in the field map.

5. Click OK.

Fig. 175. Merging layers
Source: ArcMap

**TIP:** The Append tool is nearly identical to the Merge tool window, except that there is no field map panel because the fields are defined by the existing output feature class.

## *Dissolving*

Dissolving merges features within a feature class when they share the same value of a specified attribute.

1. Choose ArcToolbox > Data Management Tools > Generalization > Dissolve.

2. Specify the Input Features layer to be dissolved (Fig. 176).

3. Specify the name and location of the Output Feature Class to be created.

4. Choose one or more attribute field(s) on which to base the dissolve. These fields will appear in the output layer; all other fields will be dropped.

5. Select a field to generate statistics from, if desired.

6. Ignore the warning, and click in the Statistic Type box to specify the type of statistic to use.

7. Select as many field-statistic combinations as desired.

8. Scroll down and uncheck the *Create multipart features* box if you wish to ensure that disjoint polygons remain separate features in the output.

Fig. 176. Dissolving a layer
Source: ArcMap

9. Check the box for *Unsplit lines* if you want lines to be dissolved only when they share an end vertex. Otherwise, disjoint lines are dissolved into a single feature.

10. Click OK to begin dissolving the layer.

# Rasters

## *Turning on Spatial Analyst*

Even if Spatial Analyst is installed and registered, it must still be turned on before use.

1. Open ArcMap, if necessary.

2. Choose Customize > Extensions from the main menu bar.

3. Check the box to turn on Spatial Analyst. Click Close.

Source: ArcMap

## *Environment settings for rasters*

Several of the Environment settings pertain specifically to rasters and should be checked before beginning analysis.

1. Choose Geoprocessing > Environments from the main menu bar (Fig. 177).

**Setting the workspace**

2. Expand the Workspace settings.

3. Set the Current Workspace to the folder or geodatabase that contains your data sets.

4. Set the Scratch Workspace to the same location, or another temporary folder.

**Setting the extent**

5. Expand the Processing Extent settings.

6. Choose one of the extent settings to control the region of the output.

7. To make the cells coincide exactly with an existing raster, select that raster in the Snap Raster box.

**Setting the cell size and mask**

8. Expand the Raster Analysis settings.

Fig. 177. Important environment settings for raster analysis
Source: ArcMap

9. Click the drop-down box to set the cell size method: the maximum or minimum of input grids, or to a specific value.

10. Use the drop-down or Browse button to select a mask raster or polygon feature class.

11. Click OK when done setting the Environment settings.

### Converting between grids and features

#### Converting features to a raster

1. Choose ArcToolbox > Conversion Tools > To Raster > Polygon to Raster (or Polyline to Raster or Point to Raster).

2. Choose the feature class to be converted (Fig. 178). If converting from a layer in ArcMap, only the selected features will be used if a selection is present.

3. Choose the field that will become the values in the output grid.

4. Specify a location and name for the output grid.

5. Select the cell assignment method to control how the attribute value is assigned to the raster.

6. Specify the cell size. Click OK.

Fig. 178. Converting polygons to a raster
Source: ArcMap

#### Converting a raster to features

1. Make sure that the raster to convert contains discrete data that will convert well to features.

2. Choose ArcToolbox > Conversion Tools > From Raster > Raster to Polygon (or Raster to Polyline or Raster to Point).

3. Select the raster to convert (Fig. 179).

4. Choose the field of the raster (usually the Value field) to become the defining attribute for the polygons.

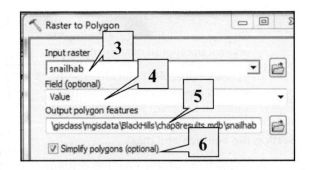

Fig. 179. Converting a raster to polygons
Source: ArcMap

5. Specify the location and name of the output feature class.

6. Check the box to *Simplify polygons*. You will usually want this option, which produces smoother boundaries that don't follow the cell edges.

### Reclassifying a raster

1. Choose ArcToolbox >
   Spatial Analyst
   > Reclass > Reclassify.

2. Choose the raster to
   reclassify (Fig. 180).

3. Choose the field in the
   raster to reclassify
   (usually Value).

**If the grid has categories**

4. Click the Unique button
   to see each value
   individually.

5. Edit the right-hand
   column to specify the
   output values.

**If the grid has many values**

6. Edit the ranges in the
   Old values column, and
   then enter the New
   values to replace them.

Fig. 180. Reclassifying a raster
Source: ArcMap

7. Use the Add Entry or the Delete Entries button to add or remove lines from the list of ranges.

8. Instead of entering ranges by hand, click the Classify button to open the Classify Values window and use it to set the ranges. See *Classifying data*.

9. Save or Load previously created classifications, if desired.

10. Specify the name and location of the output grid.

11. Check the box to convert missing values to NoData, if desired.

12. Click OK.

### Using neighborhood statistics

Neighborhood Statistics utilizes a roving window centered on a target cell. It calculates a statistic for all cells in the window and assigns the value to the target cell in the output grid. Either a block movement or a focal movement may be specified based on the tool chosen.

1. Choose ArcToolbox > Spatial Analyst > Neighborhood > Focal Statistics (or Block Statistics).

2. Choose the raster to use (Fig. 181).

3. Specify the name and the location of the output raster.

4. Choose the neighborhood shape.

5. Specify the neighborhood size. The settings will differ for different shapes.

6. Specify whether the size is in cells or map units.

7. Choose the statistic type.

8. Click OK.

Fig. 181. Calculating neighborhood statistics
Source: ArcMap

### Calculating zonal statistics

Zonal Statistics calculates statistical measures for cells within zones. Two input grids are required, one defining the zones and another containing the values to use in calculating the statistics. The output is a table containing a record for each zone and a field for each statistic.

1. Choose ArcToolbox > Spatial Analyst > Zonal > Zonal Statistics.

2. Select the raster or the feature layer that defines the zones (Fig. 182).

3. Select the field from the zone layer that defines the zones.

4. Select the raster containing the values for which to calculate statistics.

5. Specify the name and the location of the output raster.

6. Select the statistic to calculate.

7. Check the box to ignore NoData cells when calculating (usually the best option).

8. Click OK.

Fig. 182. Calculating zonal statistics
Source: ArcMap

## *Using surface functions*

The surface functions contour, slope, aspect, and hillshade all produce output grids based on an input surface and a few arguments. The general procedure is as follows:

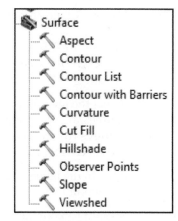

1. Choose ArcToolbox > Spatial Analyst > Surface > *function*, where *function* is the name of the desired operation.

2. Choose the surface to use as input.

3. Specify the parameters for each function as described in the following paragraphs.

4. Specify the name and the location of the output raster. (If you are contouring, the output will be a feature class.)

Source: ArcMap

To **Contour**, specify the base contour value and the contour interval. For example, a base of 10 and an interval of 20 will give the contours 10, 30, 50, and so on. You can also specify a Z factor to multiply with the surface before contouring, for example, to create foot contours on a meter surface.

For **Slope**, enter a Z factor, if desired, and specify degrees or percent for the output units.

For **Aspect**, there are no additional parameters. The resulting grid contains values between 0 and 360 degrees representing the aspect, with 0 = 360 = North. A value of –1 indicates a flat area with no aspect.

For **Hillshade**, change the default values for the azimuth and altitude of the illumination source, if desired. The azimuth is given in degrees between 0 and 360, with 0 = 360 = North. The altitude is given in degrees between 0 (horizontal) and 90 (vertical). A Z factor > 1 may be specified, which will exaggerate the topographic relief. A Z factor < 1 will subdue the topographic relief. You may also check the box to model shadows. If you choose this option, cells in the shadow of another will be set to 0 so that later you can extract these to create a map of shadows. The default option calculates the illumination of all cells, including shadowed ones. The visual difference between the two options is negligible.

For **Cut Fill**, specify the before and after surfaces. Adjust the Z factor if necessary. The output grid will have negative values for areas that have been removed and positive values where material has been added.

For **Viewshed**, specify the point or line feature class containing the observer locations. Specify a Z factor if desired.

## Calculating Euclidean distance

The Euclidean Distance tool finds the straight-line distance from each cell to a set of features.

1. Choose ArcToolbox > Spatial Analyst > Distance > Euclidean Distance.

2. Select the input raster of feature source data containing the features from which to calculate distances (Fig. 183).

3. Specify the name and location for the output raster.

4. Specify a maximum distance, if desired.

5. Specify the cell size.

6. Optionally specify a name and location for the output direction raster, indicating direction to the closest feature.

7. Click OK.

Fig. 183. Calculating Euclidean distance
Source: ArcMap

## Creating a density raster

1. Choose ArcToolbox > Spatial Analyst > Density > Point Density.

2. Specify the point layer (Fig. 184).

3. Choose an attribute of the point layer, such as population, on which the density will be based. To count the number of points, choose <None>.

4. Specify the name and the location of the output raster.

5. Enter the output cell size.

6. Enter the neighborhood shape and size settings. The settings will depend on the shape.

7. Enter the search radius in map units (maximum radius to include data points).

8. Click OK.

Fig. 184. Creating a density raster from points
Source: ArcMap

**TIP:** The Kernel Density tool spreads the values around the points using a specified function. The Line Density tool calculates line densities, such as road density in a forest.

## Interpolating between points

1. Choose ArcToolbox > Spatial Analyst > Interpolation > IDW.

2. Choose the point layer containing the data to interpolate (Fig. 185).

3. Enter the field containing the values to be interpolated.

4. Specify the name and the location of the output grid or leave it as is to create a temporary grid.

5. Enter the output cell size.

6. Enter the power exponent for the distance weighting. Entering 1 will give a linear weighting in which the influence of each point decreases linearly with distance. Entering 2 causes the influence to decrease as a square of the distance, and so on.

7. Enter the type of search radius.

A variable radius searches for a specified number of closest points, with an optional requirement to stop searching outside a specified maximum distance. A fixed search radius uses all points within a specified distance, and it has an optional requirement to limit the points used to a specified maximum number.

8. Enter the values for the search distance and the number of points.

9. Optionally specify a line feature class containing barriers across which interpolation will not occur.

10. Click OK.

Fig. 185. Interpolation using IDW method
Source: ArcMap

# Glossary

**absolute pathname** a file pathname that starts at the drive letter

**active frame** the data frame that is visible and responds to changes by the user

**adjacency** a spatial condition that quantifies whether one feature touches another

**affine transformation** a process to translate, rotate, or skew an image to fit it to a new coordinate system

**alias** an alternate name displayed for a field in a table that does not have to follow length and character restrictions

**analysis cell size** the default cell size defined for output rasters during a raster analysis session

**analysis extent** the default extent defined for output rasters during a raster analysis session

**analysis mask** a layer applied during a raster analysis session to apply NoData values outside an area of interest

**annotation** labels created from map features and stored separately for detailed editing

**append** a function that combines features from two different feature classes

**ArcToolbox** a set of functions and commands for processing spatial data

**ASCII** an international code used to store simple text letters, numbers, and symbols; synonymous with "text file"

**aspatial data** data entries that are not tied to a location on the earth's surface

**aspect** the direction of steepest slope at a location on a surface

**attribute accuracy** a measure of the type or frequency of errors occurring in the attributes of a data set

**attribute field** a column in a table containing information about spatial features

**attribute query** an operation to extract specific records of information from a file based on values in the records

**attribute table** a table containing rows of information for features

**attributes** information about map features stored in columns of a table

**automatic scaling** fitting the contents of a data frame within a specified rectangle

**azimuth** a measurement of compass direction in 360 degrees, with 0 = North

**azimuthal projection** a map projection onto a flat plane that is tangent or secant to the earth's surface

**background** where objects that are low in the visual hierarchy fade into

**band** a single array of values stored in a raster, which may store one array or many

**bilinear** a resampling method that interpolates a new cell value based on the four closest values

**binary** numeric data stored in base 2 as a series of ones and zeros; a computer's native data storage format

**block function** an analysis function that moves over a raster in adjacent, non-overlapping neighborhoods

**bookmarks** links to a particular map area and scale for quick access

**Boolean** an adjective describing data, operators, values, or files that contain or manipulate only true or false values

**Boolean operators** operators (AND, OR, NOT, XOR) that evaluate a pair of true or false propositions and return a true or false result

**Boolean overlay** evaluating rasters with Boolean operators to evaluate areas where combinations of specific conditions exist

**Boolean raster** a raster with cells that contain a 1 to indicate true or 0 to indicate false regarding a specified condition

**branding** repeating colors, symbols, fonts, images, or other design elements across multiple publications to foster recognition

**buffer** the delineated area within a specified distance of a feature

**byte** a unit of data storage in base 2 containing eight zeros or ones and holding a number from 0 to 255

**capture** using the Alt-PrntScrn keys to place a copy of the active window on the clipboard for pasting into another program

**cardinality** the relationship between records in two tables: one-to-one, one-to-many, many-to-one, or many-to-many

**cartographic generalization** simplifying map features in order to construct clearer and less cluttered maps

**categorical data** data that place objects into unranked groups; examples are land use and geology data

**cell** a square data element in a raster corresponding to one value representing conditions on the ground

**cell size** the dimensions of one square of map information in a raster data set

**central meridian** the central longitude of a map projection for which the $x$ coordinate equals zero

**chart map** a map showing several different attributes in chart form, with one chart for each feature

**choropleth map** a map in which each feature, such as a state, is colored according to the values in a data field

**class breaks** break points used to classify values in data field groups

**classification** assigning features to two or more groups based on numeric values in an attribute field

**classified** a raster display method that divides values into two or more groups based on their numeric values

**cloud** a computing system composed of warehouses of computers and hard drives managed by a company that rents processing power and disk space to clients

**clip** to remove features and portions of features that lie outside of the features of another layer

**cluster tolerance** a defined distance used in topological editing, causing vertices to be made coincident if they are closer together than the tolerance distance

**CMYK** a method of specifying color using mixtures of cyan, magenta, yellow, and black; primarily used in printing and publishing

**coded domain** a rule that permits only certain specific values to be assigned to an attribute, such as land use codes

**coincident boundary** an identical boundary shared by two adjacent polygons

**colormap** a set of defined colors matched to specific image pixel values, which determines how the image will appear

**completeness** a measure of how well a data set has captured the entire area or all representatives of a feature class

**complex edge** a network entity composed of multiple linear features that behaves as a single linear feature in the network

**conflict detection** determining which labels will overlap each other

**conic projection** a map projection derived by projecting latitude-longitude values on a paper cone covering a sphere

**connectivity** a property of linear features when they are connected to each other via junctions

**connotation** a typical emotional reaction elicited by certain colors or symbols

**containment** the property of one feature including another in whole or in part

**context menus** computer menus that appear when certain objects on the screen are clicked with the right mouse button

**continuous** data that take on a variety of values and that change rapidly across a data set, such as elevation

**contour** a line indicating a constant value of a quantity on a surface, such as an elevation contour at 2000 ft

**contrast** the difference in hue, saturation, or value between adjacent colors

**convention** a symbol for mapping that is commonly chosen and understood to have particular meaning

**coordinate pair** a single pair of $x$ and $y$ values indicating position in a planar coordinate system

**coordinate space** the range of $x$ and $y$ values onto which maps are plotted

**coordinate system** (1) a specified range of $x$-$y$ values onto which a map is plotted; (2) the definition of a coordinate space used by a map layer, including the ellipsoid, datum, and projection

**coverage** the spatial data format created for, and used by, ArcInfo

**CSDGM** (Content Standard for Digital Geospatial Metadata) a metadata standard developed by the FGDC and commonly used in the United States

**cubic convolution** a resampling method that interpolates a new cell value from the 16 closest input cells

**cut/fill** a function that determines the difference between a before and after topographic surface

**cylindrical projection** a map projection derived by projecting spherical latitude-longitude values onto a cylinder wrapped about a sphere

**dangle** a line feature that fails to connect to another line feature, leaving a gap

**data frame** a container holding layers that are viewed and analyzed together; a map view

**Data view** the data frame mode optimized for the display and analysis of map data

**datum** a combination of an earth ellipsoid and a reference point to reduce mapping discrepancies

**dBase** a database program whose file format has been adopted for the shapefile data model and tables in ArcGIS

**default** an automatic input value assumed by a program when no value has been entered by the user

**Define Projection tool** a tool to guide the user through the task of assigning a coordinate system to a spatial data layer

**defined interval** a classification method in which the user specifies a size range for all the classes

**definition query** an operation to set a map layer to display only the features whose attributes meet specific criteria

**degrees** the measurement units used in the spherical coordinate system; a circle has 360 degrees

**DEM** see *digital elevation model*

**destination table** the table that receives data from another table during a join operation

**digital elevation model (DEM)** a raster array of values representing elevations at the earth's surface

**digital raster graphic (DRG)** a scanned image of a USGS topographic map

**digitize** to convert shapes on a paper map to a digital map layer by entering vertices

**discrete** data that take on a relatively small number of distinct values

**discrete color** a display option for rasters in which every different value is assigned a random color

**Display tab** the tab in the Table of Contents that shows layers in the order in which they are drawn from bottom to top

**display units** the units in which ArcMap reports the current $x$-$y$ location of the cursor on the map

**dissolve** to combine features together when they share the same value for an attribute

**distance join** a join that combines the information from two feature tables based on the features that lie closest to each other

**divergent color set** a set of colors with a neutral middle value that grades to increasingly saturated colors of different hues above and below

**division units** the units in which a map scale bar is measured and drawn, such as miles or kilometers

**division value** the length of one section of a scale bar as given in division units, such as 100 km

**divisions** the number of sections given to a scale bar

**domain** a rule that determines the values that may be entered into an attribute

**dot density map** a map representing attribute values by a proportional number of randomly placed dots

**double-precision** a numeric value stored using 16 bytes of information

**DRG** see *digital raster graphic*

**dynamic labels** labels determined from an attribute and placed on a map automatically each time features are drawn and redrawn

**edge** a shared boundary between two features being edited using topology

**edge snapping** ensuring that new features are automatically connected to the edges of existing lines or polygons

**ellipsoid** a spheroidal volume with unequal axes, used to approximate the shape of the earth in map projections

**end snapping** ensuring that new features are automatically connected to the ends of existing line features

**enterprise GIS** a long-term GIS project developed by a large organization and involving many people over a long period of time

**Environment settings** program-level or tool-level settings that impact how tools run or set characteristics of the output, such as cell size or coordinate system

**equal interval** a classification method in which the user specifies a number of classes that have equal size ranges

**erase** an overlay function that removes features lying inside the external boundary of another polygon feature class

**Euclidean distance** the straight-line distance between two points

**event layer** a map layer of points created from a series of coordinate pairs in a table

**export** to create a new data file from all or a subset of features in an existing one, often while converting it to a different format

**expression** a statement containing field names, values, and/or functions used to extract records in a query or calculate values in a table

**extent** the range of $x$-$y$ coordinates displayed in a map or stored in a data layer

**extent rectangle** the range of $x$-$y$ coordinates occupied by the features in a data layer

**extraction** functions or commands that pull out certain records or pixels based on field values or a designated spatial area

**false easting** an arbitrary $x$-coordinate translation applied to a map projection, usually to ensure that all values are positive

**false northing** an arbitrary $y$-coordinate translation applied to a map projection, usually to ensure that all values are positive

**feature** a spatial object composed of one or more $x$-$y$ coordinate pairs and having one or more attributes in a single record of an associated table

**feature class** a set of similar objects with the same attributes stored together in a spatial data file

**feature dataset** a set of feature classes in a geodatabase that share a common coordinate system and can participate in networks and topology

**feature service** an Internet map layer in which the actual features are available for display, query, and sometimes editing

**feature template** a set of properties and attributes that stores all the information needed to edit a layer

**feature weight** the priority assigned to a layer when determining which features may be drawn on top of others

**FGDC** (Federal Geographic Data Committee) an organization that promotes and establishes standards for exchange of data

**FID** (feature ID) a unique number assigned to every feature in a spatial data file and used for identification and tracking

**field** a single column of information in a data table

**field definition** parameters specified when creating an attribute field, including type of data, field length, precision, and scale

**field length** the maximum number of characters that can be stored in a text attribute field

**filter** a moving window applied to a raster data layer that calculates new values for the center pixel based on some function of the values in the window

**fixed extent** a constraint applied to a data frame to prevent changes in scale or extent in layout mode

**fixed scale** a constraint applied to a data frame to prevent the scale of the map from being changed in layout mode

**flat file database** a database that stores data in simple files

**flipping lines** swapping the start and end nodes of a line feature so that it goes in the opposite direction

**focal function** raster analysis functions that determine new values for a target cell based on values in a moving neighborhood around the target

**folder connection** (1) a link in ArcCatalog and ArcMap that points to a folder with GIS data and serves as a shortcut to frequently used folders; (2) a link to a DBMS allowing data to be transferred

**foreground** where objects higher in the visual hierarchy appear

**GCS** see *geographic coordinate system*

**generalization** simplifying a feature in order to store it or display it at smaller scales than its source scale

**geocoding** the matching of a location stored in a table to a spatial point feature based on a reference spatial data layer, most often applied to converting addresses to locations

**geodatabases** a data model developed for ArcGIS 8 that employs recent database technology for storage and implements rules and topology

**geographic coordinate system (GCS)** a spherical coordinate system of degrees of latitude and longitude that is used to locate features on the earth's surface

**geoid** the shape of the earth as defined by mean sea level and affected by topographic and gravitational factors

**geometric accuracy** the accuracy with which the shape and position of features are represented

**geometric interval** a classification method that bases the class intervals on a geometric series in which each class is multiplied by a constant coefficient to produce the next higher class

**geometric network** a system of linear edges and point junctions used to model the flow of commodities, such as traffic or utilities

**geoprocessing** analysis of spatial data layers, such as dissolving, intersecting, and merging

**geoprocessing service** an analysis tool provided by a GIS server, in which the processing is done on the server rather than on the client's device

**georeferenced** a spatial data layer that is tied to a specific location on the earth's surface for display with other data

**georelational** a data model that links features to attributes in a separate table using a unique feature ID code

**graduated color map** a map that divides numeric data from a polygon feature class into classes based on value and displays the classes with different colors

**graduated symbol map** a map that divides numeric data from a line or point feature class into classes based on value and displays the classes with different size or thickness of symbols

**graphic text** text placed on a map layout that is not associated with a feature attribute; can be manipulated by the Drawing tools

**graticule grid** marks of latitude and longitude placed on a map boundary

**grid** (1) specific raster format native to GRID and Spatial Analyst; (2) generic term for a raster data set

**ground control points** a set of points that match easily identifiable locations on two different data layers to enable georeferencing of one layer to another

**hierarchical database** a database that stores information in tables with permanent links between them

**hillshade** a raster that displays the brightness variation of a surface as if it were illuminated by a light source at a specified azimuth and zenith angle

**histogram** a graph showing the number of pixels contained for each data value in a raster

**HSV** a method of specifying color based on hue, saturation, or value

**hue** (in HSV color model) the shade of a color, such as red, green, or blue, measured on a 0–360 scale

**image** a raster data layer, usually referring to a raster that displays brightness values, as in a photograph

**image service** an Internet GIS service that provides imagery for display or analysis

**import** to bring a copy of a data set into a geodatabase, sometimes including a conversion from a different data format

**INFO** an early database system upon which the Arc/Info software data model was based

**inside join** combining the information from two feature tables based on one feature that lies inside another

**interactive labels** simple graphic labels on a map that are placed by the user one at a time

**interactive selection** extracting one or more features by manually picking them from the screen with selection tools

**Interchange file** a file used to store a coverage for file transfer; has an extension of .e00

**interpolation** to calculate values at locations between known measurements; to populate a raster with values extrapolated from a known set of point values

**intersect** an operation to overlay two spatial data layers and find the areas common to both while discarding areas unique to either

**intersection** the property of two features touching each other in whole or in part

**interval data** values that follow a regular scale but have no natural zero point, such as temperature or pH

**ISO** (International Organization of Standards) an international body that approves standard industry practices, including metadata standards

**Item Description** brief style of metadata offering basic information about a data set

**Jenks method** a way to classify numeric data into ranges defined by naturally occurring gaps in the data histogram

**join** the temporary combination of data from two tables based on a common attribute field or location

**kernel** a moving window of values applied to a raster to calculate new values at a target in the middle of the window; see also *filter*

**key** an attribute field that is used to extract or match records in a table

**label weights** a priority rating assigned to values to determine which ones will be placed in case of an overlap

**latitude** a spherical unit measuring angular distance north or south from the equator

**latitude of origin** the reference latitude of a map projection where the $y$ value is zero

**layer** a reference to a feature class and its associated properties

**layer file** a file that stores a pointer to spatial data along with information on how to display it

**layer package** a file sent from a GIS server with the symbol properties and necessary data included for use

**layer properties** settings that control how a data set is to be displayed or viewed in ArcMap

**layout** the specification for a map page, including the map frames, legend title, scale bar, and so on; stored in a map document

**Layout view** a mode of ArcMap that is used to design and create a printed map and that allows manipulation of map layers, titles, scale bars, north arrows, and more

**legend** a map element that displays the names and symbols used to portray layers on a map

**line** a spatial feature composed of a string of $x$-$y$ coordinate vertices and used to represent linear features such as streets

**lineage** the original source and processing steps behind the production of a digital geospatial data set

**lock file** a small file created in a data set's folder to indicate that the data set is in use

**logical consistency** a measure of how well data features represent real-world features, in particular with respect to topology

**logical expression** a statement composed of field names, operators, and values that specifies criteria used to select records or values from a layer or table

**logical operators** functions, such as >, <, or =, that compare values and generate a true or false result

**longitude** a spherical unit measuring angular distance east or west from the Prime Meridian

**loops** connected circular paths in a network in which flow is indeterminate

**mantissa** the part of an exponential number that stores the significant digits

**manual class breaks** a classification scheme in which the user sets each class break to the desired value

**map algebra** a system that permits calculations and operations on entire raster arrays, such as adding two rasters together

**map document** file used by ArcMap to contain a set of data for mapping and analysis

**map elements** objects placed on a map layout, such as titles, legends, scale bars, north arrows, images, and charts

**map extent** the range of $x$-$y$ values of the area being displayed in a map

**map overlay** to combine two spatial data layers, either for display or to evaluate the relationships between them

**map package** a file sent from a GIS server with the map layout and necessary data included for use

**map scale** the ratio of feature size on a map to its size on the ground

**map service** an Internet map that is available for viewing but that cannot be manipulated, symbolized, or saved by the user

**map topology** temporary spatial relationships developed between features during editing in ArcMap to facilitate editing of features with common nodes or boundaries

**map units** the units of the coordinate system in which a map is stored or displayed

**mask** a raster layer applied during analysis to nullify unwanted cells, such as those outside a study area boundary

**measurement grid** a data frame grid that provides measured $x$-$y$ coordinate system values around the perimeter of the frame

**merge** (1) to combine two or more map features into one feature; (2) to combine two or more data layers into a single layer

**merge policy** a rule that states how the attributes of a feature should be handled if it is merged with another feature

**metadata** information stored about data to document its source, history, management, uses, and more

**metadata standard** a set of requirements laying out the types of information and organization to be used for creating metadata

**metadata template** a partially completed metadata document containing information repeated for many data sets and used to aid in metadata creation

**model** a sequence of steps or calculations used to convert raw data into useful information; a scheme used to understand and predict processes in the real world based on the manipulation of data

**Model Builder** a graphical interface used to combine existing tools to create new tools and scripts

**Modifiable Areal Unit Problem (MAUP)** statistical and visual issues caused by aggregating measured data using arbitrary areal units such as political boundaries

**modify** an editing technique used to edit individual vertices to change the shape of a feature

**multipart feature** a single feature composed of nontouching units, such as a single state feature composed of the seven Hawaiian islands

**NAD** see *North American Datum*

**natural breaks** a data classification scheme that divides values based on natural groupings or gaps between values

**nearest neighbor** a resampling method for discrete data that grabs the closest value to a cell center

**neatline** a line used to enclose one or more map elements in a rectangle

**negative space** blank areas in a design composition

**neighborhood functions** calculation of a value for a target cell or feature based on surrounding values from a defined region

**network or network topology** an association of linear edges and connecting points, used to model the flow of a commodity, such as traffic or utilities

**NoData** a special value used to designate that a data value is absent or unknown

**node** the beginning and endpoint of a line feature

**nominal data** values that name or identify an object, such as a street name

**normalized data** to divide the values of an attribute field by the total of the field or by the values in another field

**North American Datum (NAD)** a combination of a spheroid and reference point that is used to minimize map distortion in North America

**object-oriented** a programming and database approach that treats software and model elements as objects with defined properties and relationships

**oblique projection** a map projection in which locations on a sphere are projected to a cylinder or cone of paper at an arbitrary angle

**OID** (ObjectID) unique number identifying a row in a table or a feature in a geodatabase feature class

**ordinal data** data values that indicate a rank or ordering system

**origin** the (0,0) point of a coordinate system

**orthographic projection** a map projection in which locations on a sphere are projected onto a planar surface

**orthophoto** an aerial photograph that has been geometrically corrected to match a map base

**overlap** a spatial condition that quantifies whether one feature covers all or part of another feature

**overshoot** a type of dangle in which a line crosses too far over a line that it is supposed to meet exactly

**pan** to move the display window to another part of the map without changing the map scale

**parameter** (1) specific value associated with map projections that define how it appears; (2) a variable that serves as an input to a model

**parametric arc** a line feature composed of a smooth curve derived from a given radius for each segment

**pathname** a list of the folders that must be traversed to locate a particular file, such as c:\mgisdata\usa\states.shp

**pixel** a square data element in a raster corresponding to one value representing conditions on the ground

**planar topology** an association of feature classes in a feature dataset, established by rules regarding the spatial relationships between features, such as not overlapping each other

**point** a one-dimensional feature defined by a single $x$-$y$ coordinate pair

**point snapping** ensuring that new vertices are automatically connected to existing point features

**polygon** a closed, two-dimensional area feature defined by three or more $x$-$y$ coordinate pairs

**positional accuracy** a measure of the likelihood that features on a map are actually in the locations specified

**precision** the number of digits allotted to store a numeric value

**Prime Meridian** the line of zero longitude on the earth, passing through Greenwich, England

**process step** data assembly and geoprocessing function performed in generating a GIS data set

**profile** a formal customization of an international standard developed and maintained by a community of users

**Project tool** a tool that converts a feature class from one coordinate system to another

**projection** a mathematical transformation that converts spherical units of latitude and longitude to a planar $x$-$y$ coordinate system

**project-oriented GIS** a GIS project with fairly limited objectives and a finite life span

**proportional symbol map** a map that displays attribute values with marker or line symbols that are proportional in size relative to the value of the feature

**proximity** the property of one feature being close to another feature

**pseudonode** a type of topology error in which only two lines meet at a coincident node instead of three or more

**pyramids** a set of rasters with different resolutions that is calculated from a raster and used to speed displays at smaller scales

**quantile** a classification method that divides the data into the specified number of quantiles, so that each class has the same number of features

**quantities data** numeric attribute data

**query** an operation to extract records from a database according to a specified set of criteria

**range domain** a rule that stipulates the largest and smallest values that can be stored in a particular attribute

**raster** a data set composed of an array of numeric values, each of which represents a condition in a square element of ground

**raster model** a data model that uses rasters or numeric arrays to represent real-world features

**ratio data** data having a regular scale of measurement and a natural zero point, such as precipitation or population

**read access** granting permission to a user to view a file but not to change or save it

**reclassify** to replace sets or ranges of values in a raster with different sets or ranges of values

**record** a row in a table containing one object

**rectify** to rotate, resize, or warp an image to match a map base using a selected set of ground control points

**reference grid** a data frame grid that provides lettered and numbered squares on a map for use with an index of the features

**reference latitude** the latitude of origin; see also *latitude of origin*

**reference scale** the scale at which text or symbols appear at their assigned size

**relate** a temporary association between two tables based on a common field whereby fields may be selected based on whether they match selected records in the other table

**relational database** a database that stores information in tables and constructs temporary relationships between them

**relative pathname** path to a file that starts in the current folder

**resampling** to change the resolution of a raster using a defined strategy to convert values from one cell grid to another

**reshape** an editing technique used to re-enter part of an existing line or polygon

**resolution** (1) the ground area represented by one cell value in a raster; (2) the default storage precision of a vector data set

**RGB** a method of specifying color using mixtures of red, green, and blue on scales of 0–255 each

**RGB composite** an image displayed by assigning one band of brightness information to each red, green, and blue color gun in a display monitor

**RMS error (RMSE)** the root mean squared differences between a set of original and translated points

**rubbersheeting** a process to deform a raster to fit it to a new coordinate system

**Rule of Joining** each record in the destination table must have one and only one matching record in the source table

**rule of thirds** guideline to place important parts of a graphic composition at the intersections of lines dividing the page in thirds

**saturation** (in HSV color model) the intensity of a color measured on a scale of 0–100

**scale** (1) the ratio of the size of features in a map to their size on the ground; (2) the number of decimal places allotted to an attribute field for storing numbers

**scale range** the range of scales for which a data layer will be displayed, set by the user to avoid clutter, or the display of layers at inappropriate scales

**schema** the layout plan of a geodatabase, including the tables, the fields, and the relationships

**script** a program including ArcGIS commands and functions

**secant projection** a map projection in which spherical coordinates are projected onto a surface that intersects the sphere along two great circles

**select** to extract one or more features or records from a layer or table in preparation for another operation; to perform a query

**Select By Attributes** to choose a subset of features or table objects based on the values in one or more attribute fields

**Select By Location** to choose a subset of features based on their spatial relationship to other features

**selected set** the set of features that has been extracted prior to another operation; the result of a query

**selection method** a setting that controls what happens to previously and newly selected objects when a new selection is made

**Selection tab** the tab in the Table of Contents that shows the selection state and number of selected features for each layer

**shapefile** the spatial data model developed for, and used by, ArcView 3 and later versions

**shared features** features that are linked to or that share a boundary with other features

**short integer** an attribute field definition that uses up to five bytes to store a binary integer and that can store values up to about 62,000

**simple edges** linear elements in a network that always end at junctions and behave as separate entities

**simple join** a join that combines two layers according to common locations when a one-to-one spatial relationship exists between the layers

**single symbol** a map type in which every feature in a layer is displayed with the same symbol

**single-precision** a numeric value stored using eight bytes of information

**sketch** a provisional figure created during editing; when finished, it becomes a feature in a data layer

**Sketch menu** a context menu accessed by right-clicking *off* a sketch during editing

**slice** to divide the values in a raster into a specified number of even classes

**sliver** small polygon created during map overlay due to slight boundary differences in the inputs; usually considered an error

**slope** the drop in elevation for a specified horizontal distance, expressed as an angle or a percentage

**snap tolerance** a distance set during editing such that new vertices close to an existing feature are snapped to it

**snapping** ensuring that features within a specified distance are automatically adjusted to meet at exactly the same location to avoid gaps between features

**source** (1) a spatial data file that provides the features for a map layer; (2) the original information used to develop a spatial data set

**source scale** the original analog scale or resolution at which a digital data set is converted to digital form

**source table** the table that provides the information to be appended to another table in a join

**spaghetti model** a model that stores spatial features as a series of *x-y* coordinates and does not store topological relationships between features

**Spatial Analyst** a program extension to ArcMap that is used to analyze raster data

**spatial data** information that is tied to a specific location on the earth's surface

**spatial join** a function that combines the attributes of features in two layers based on containment or distance

**spatial operators** a set of functions that evaluate spatial relationships, including intersection, containment, and proximity, between sets of features

**spatial query** an operation to extract records from a data layer based on location relative to another data layer

**spatial reference** the complete description of how the spatial data are stored for a feature class that includes the coordinate system, the X/Y Domain, and the precision

**SPCS** see *State Plane Coordinate System*

**spheroid** ellipsoid; see also *ellipsoid*

**split policy** a rule that states how attributes of a feature should be handled if it is split

**SQL** see *Structured Query Language*

**standalone table** a table of information not linked to spatial data features

**standard deviation** a classification scheme in which the class breaks are based on the standard deviation values of the data being mapped

**standard parallel** parameter of a map projection indicating the latitude(s) at which the projection surface lies tangent or secant to the sphere

**State Plane Coordinate System (SPCS)** a group of projections defined for different regions of the United States and designed to minimize map distortions

**stereographic projection** a map projection in which locations on a sphere are projected onto a planar surface

**stream digitizing** method of entering vertices automatically during editing, instead of clicking to create each one

**street type** the part of an address that indicates the type of street, such as St, Rd, Ave, and so on

**stretch** to spread the values of an image to cover the entire range of symbols available, often supplemented by ignoring the tails of the distribution

**stretched** a display method that spreads the data values over the entire range of symbols available

**Structured Query Language (SQL)** an established syntax for creating logical expressions to extract records from a database according to specified criteria

**style** a collection of map symbols and colors that are stored together and used together

**Style Manager** a window used to create symbols and manage them within collections of symbols called styles

**subdivisions** the number of units into which a single division of a scale bar is split, usually appearing on the left end

**subtypes** different categories created for a feature class, with each category having its own symbol and default values

**suffix direction** a direction appended to the end of a street name to indicate a part of the city, such as Main St *North*

**suitability analysis** evaluating a landscape to determine which areas are best fit for a given purpose, based on a set of factors

**Summarize** a function that groups records in a table according to one categorical field and calculates statistics for each group

**summarized join** a join that combines the records of two attribute tables with a one-to-many cardinality by assigning each output record a single value derived statistically from the many input values

**surface analysis** functions designed for application to rasters representing a three-dimensional surface, such as elevation

**symbol** a shade, line, or marker with specified shape and color used to display map features

**table** data stored as an array of rows and columns, with each row representing an object or a feature and each column representing an attribute or a property of that object

**Table of Contents** the ArcMap window that lists the data frames and layers in the map document

**tangent projection** a map projection in which spherical coordinates are projected upon a surface that lies tangent to the sphere along one line of latitude

**templates** a map design that can be saved and applied to many different map documents

**temporal accuracy** a determination of the time period for which a data set is considered valid

**thematic accuracy** the degree to which attribute values represent the true properties in the real world

**thematic mapping** displaying the features of a spatial data layer based on values in its attribute table

**thematic raster** a raster that contains categorical or nominal data values, such as land use codes or soil types

**themes** terminology in ArcView 3 used to indicate a data layer containing similar features, such as roads or states

**thumbnail** a small snapshot showing the appearance of a data layer and displayed in ArcCatalog to aid the user in finding data

**TIN** (Triangulated Irregular Network) a data model for storing surfaces as triangular facets with varying orientations

**Tobler's Law** a statement of a common property of spatial data in which values measured at distances close to each other are more likely to be similar than values at greater distances

**topological model** a data model that stores spatial relationships between features in addition to their $x$-$y$ coordinates

**topology** the spatial relationships between features, such as which are connected or adjacent to each other

**trace** to follow an existing linear feature when creating a new feature that follows an existing one

**trace solvers** programs that analyze network flow by tracing paths in the network

**transformation** conversion of one geographic coordinate system (GCS) and datum to another

**transverse projection** a map projection in which spherical coordinates are converted to locations on a cylinder or cone tangent to the sphere along a line of longitude

**undershoot** a type of dangle in which a line falls short of meeting another line

**union** an operation to create a new feature or set of features by combining all the areas from two input layers

**unique values map** a map in which each attribute value is assigned its own symbol

**Universal Transverse Mercator (UTM)** a family of map projections defined for 60 zones around the world and based on a transverse cylindrical projection

**unprojected data** a spatial data layer stored in a geographic coordinate system with units of degrees of latitude and longitude

**UTM** see *Universal Transverse Mercator*

**validation** a step performed on a planar topology to test the features against the defined topology rules and identify errors

**value** (in HSV color model) the darkness or lightness of a color on a scale of 0–100

**vector** a spatial data storage method in which features are represented by one or more pairs of *x-y* coordinate values forming points, lines, or polygons

**Venn diagram** a diagram used in set theory to portray sets and subsets

**vertex** (pl. **vertices**) a point at which the segments of a line or polygon feature change direction

**Vertex menu** a short context menu accessed by clicking *on* the sketch during editing

**vertex snapping** ensuring that new features are automatically connected to the vertices of existing line features

**viewshed** the area on a three-dimensional surface that is visible from a specified point or set of points

**visual center** where the center of a graphic composition appears to be, about 5% above the geometric center

**visual hierarchy** the order in which objects in a design are perceived

**web map** an interactive map based solely on Internet data services and accessed from many platforms, including desktop software, web pages, and mobile devices

**weighted overlay** an analysis that uses a set of reclassified ranked condition rasters and weights to calculate a suitability index

**world file** a text file containing the georeferencing information for a raster

**write access** granting permission to a user to view and edit and save a file

**XML** (eXtensible Markup Language) a tagged text presentation format used to store metadata

**X/Y Domain** the maximum allowable range of values stored by a feature class

**XY tolerance** a setting used for geoprocessing that defines the minimum allowable distance between vertices

**zenith angle** an angular measure of the distance of an object above the horizon

**zipped** refers to a set of files that have been compiled into a single file and compressed, used to make file transfers and downloads easier

**zonal statistics** raster analysis functions that calculate statistics from one raster for zones defined by a different raster

**zone** the combined areas of a raster that share the same attribute value, such as all areas with the commercial land use code

# Selected Answers

*Some problems can be done several ways. Your answers may sometimes be slightly different.*

*Because of the small size, legends and other items are omitted from most map answers. Your submissions should be better; check with your instructor to determine what you need to submit.*

## Ch. 1 Tutorial

1. Wallowa County
2. 36 rows
3. Wheeler County
4. NAME is the field.
5. USDA Natural Resources Conservation Service.

## Ch. 1 Exercises (Hints or Maps)

1. Use the Add Data button to view the geodatabase contents.
2. Open the layer properties and examine the Source tab.
3. Open the table and query for STATE_NAME = 'California'.
4. Open the lakes table and sort on the Area field.
5. Use Find to search the table.
6. Open the table and use Statistics.
7. Open Help and search for *feature class*; choose feature class basics.
8. See map.

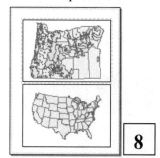

Source: ArcMap

9. Answers vary; see example map.

Source: ArcMap

10. See map.

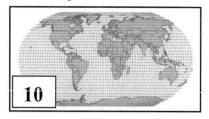

Source: ArcMap

## Ch 1. Challenge

Answers will vary.

## Ch. 2 Tutorial

1. Cell size 1000 meters; 696 columns and 495 rows
2. 2 coverages, 1 table, 1 raster, 1 layer file, and 7 shapefiles
3. ST = 'NJ'
4. Congressional districts
5. STATE_ABBR = 'NJ'
6. Feature classes are stored on disk and can be managed by ArcCatalog. Layers exist in ArcMap and are managed by setting properties in the Table of Contents.
7. From viewing the metadata on the web site
8. Yes, it is in NJ State Plane, but in units of feet, not meters.
9. Geographic Coordinate System WGS 1984

## Ch. 2 Exercises (Hints)

1. Use the Catalog tab to examine the geodatabase.
2. Examine the raster properties in the Catalog tab.
3. Examine the Item Description in the Catalog tab.
4. Where is the data it uses?

## Ch. 2 Exercises (Hints), con't.

7. Some outputs may be empty because no features exist in the home state.

## Ch. 2 Challenge

Answers will vary.

## Ch. 3 Tutorial

1. GCS_WGS_1984, degrees

2. About –80° and 25°

3. Both units are decimal degrees.

4. About –5550 and 1740 miles

5. Mercator units are meters.

6. South America mostly negative; Asia mostly positive

7. Because the states datum is NAD 1983 and the data frame datum is WGS 1984.

8. No, it will not choose an automatic transformation, because there are multiple choices.

9. Central meridian is –96° and the latitude of origin is 40°. Standard parallels are at 20° and 60°.

10. Yes, because the latitude of origin lies midway between the northern and southern borders of the continental United States.

11. Secant conic, because it has two standard parallels.

12. Geographic coordinate system (GCS)

## Ch. 3 Exercises (Hints)

1. Examine the layer properties Source tab.

2. Examine the layer properties Source tab.

3. View the data frame properties General tab and Coordinate System tab.

4. (a–d) Add the utmzones and spcszones to the map and see where the areas fall. (e–f) Search the data frame coordinate system tab for these locations.

5. Modify the central meridian of Illinois State Plane E or W.

6. Modify the parallels of Colorado State Plane Central.

7. Use Define projection to set the appropriate coordinate system.

8. Answers will vary.

## Ch. 3 Challenge

Answers will vary.

## Ch. 4 Tutorial

1. ELEVATION is interval; TYPE is categorical.

2. I for Interstates, U for US highways, S for state highways

3. Ratio data; a graduated color map

4. Interval data; stretched or classified

5. Ordinal data, unique values with monochromatic ramp

## Ch. 4 Exercises

1. Volcanic hazards frame

Source: ArcMap

2. Farms frame

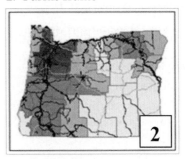

Source: ArcMap

3. Housing frame

Source: ArcMap

## Ch. 4 Exercises, con't.

4. Physiography frame

Source: ArcMap

## Ch. 4 Challenge

Answers will vary.

## Ch. 5 Tutorial

1. Mercator and Web Mercator

2. Annotation has a reference scale.

3. Left to right: subdivision, division, division units

## Ch. 5 Exercises

1. Answers should look like the map shown underneath the question.

2. Answers will vary.

## Ch. 5 Challenge

Answers will vary.

## Ch. 6 Tutorial

1. Population does not require decimal values; population density does.

2. Not normal; about 38 million people

3. 190 Democratic districts

4. DISTRICTID

5. Republicans

6. Standalone table

7. Nine states have one district.

8. Republicans control nearly six times the area.

9. One-to-one relationship

10. One-to-many relationship

11. Districts is source, states is destination.

12. 21 New England reps

13. STATION_NAME and STATION

## Ch. 6 Exercises (Hints and Maps)

1. (a) Select zero magnitude quakes and examine in table; view statistics. (b) Sort each set of quakes by year and compare. (c) Select smaller quakes and view on a map.

2. Use Select By Attributes and Statistics.

3. Summarize by Sub_Region and sort results.

4. Add a field and calculate percentage of Blacks; sort table on number and percent.

5. Look for a code.

6. Query for states that lost population and create a layer from the selected features. See map.

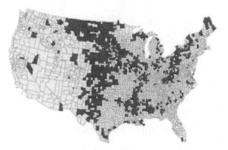

Source: ArcMap

7. Summarize on STATE_NAME field.

8. Relate US States to 114th Congress.

9. Select on CAPITAL field and get statistics.

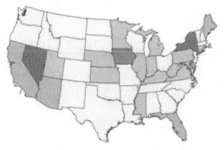

Source: ArcMap

10. Add new field and calculate percentage. See map.

## Ch. 6 Challenge

Answers will vary. See map for salaries from 2012–2013 National Center for Education Statistics (NCES) data (top) and normalized by COLI (bottom).

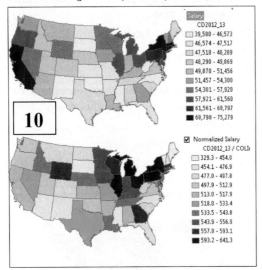

Source: ArcMap

## Ch. 7 Tutorial

1. NAD 1983 State Plane Texas Central
2. ~1200 feet; 1:200,000
3. Angle –115 degrees or 245 degrees.
4. Map units, in this case, feet

## Ch. 7 Exercises

Answers will vary but should show the digitized features for the requested areas.

## Ch. 7 Challenge

Answers will vary.

## Ch. 8 Tutorial

1. 15 counties, about 314,000 people
2. 9 cities
3. 303 counties
4. 133 counties
5. 34 cities
6. 889 counties
7. Hughes County contains Pierre, SD.
8. 3641 cities
9. 93% of cities
10. Brazos, Canadian, Pecos, Red, and Rio Grande

11. The Flathead and the Salt
12. 19 target counties, 902,000 people

## Ch. 8 Exercises

1. Select By Attributes and sort.
2. Select By Attributes and view in table, calculate % with a calculator.
3. Use Select By Attributes and do the 3 queries in steps.
4. Select By Location
5. Use Select By Location first, then Select By Attributes.
6. Use Select By Location and Statistics.
7. Select Crater Lake and then use Select By Location twice, once for volcanoes and once for interstates.
8. Several queries required, both attribute and spatial. See map.

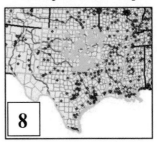

Source: ArcMap

9. Query for capitals within 50 miles and then switch the selection.
10. Select cities using attribute and spatial queries and export to new table. Add field and calculate funding amount.

## Ch. 8 Challenge

Answers will vary.

## Ch. 9 Tutorial

1. 591 out of 1360 wells
2. About 880 feet
3. FACILITY_N
4. NAD 1983 Texas State Plane Central, in feet
5. 16,477 streets
6. Min. 836, max. 167,563, avg. 32,719
7. Walnut, Little Walnut, Williamson
8. South Brushy, Walter E. Long Lake, Barton Creek

## Ch. 9 Exercises

1. Summarized join of wells to destination watersheds

2. Simple join of zoningoverlays to destination wells, and sort on depth

3. Summarized join of wells to destination zoneoverlays

4. Simple join of restrooms to destination bbq_pits

5. Export elementary schools; summarized join of schools to destination pools.

6. What kinds of pools are they?

7. Export post offices; summarized distance join of blockpop to destination offices.

8. Simple join of airports to destination cities.

9. Summarized join of cities to destination airports.

10. Summarized join of lakes to destination parks.

## Ch. 9 Challenge

Source: ArcMap

## Ch. 10 Tutorial

1. ALL, NO-FID, ONLY-FID. ALL or NO-FID is best.

2. 4.9 km$^2$

3. 32.2 km$^2$

4. About 15%

5. Opening new map document sets them back to defaults.

## Ch. 10 Exercises

1. Method: Intersect roads and geology; use statistics.

2. See map.

Source: ArcMap

3. See map.

Source: ArcMap

4. See map.

Source: ArcMap

5. The answer is the table.

| sumshedrange | |
| --- | --- |
| SHEDNAME | Area |
|  | 29,813,100 |
| Whitewood | 9,055,570 |
| False Bottom | 8,385,018 |
| Bear Butte | 1,695,465 |

Source: ArcMap

## Ch. 10 Exercises, con't.

6. See map.

7. See map.

Source: ArcMap

8. See map.

Source: ArcMap

9. See map.

Source: ArcMap

10. See map. Off-limits area is 34.3 km$^2$.

Source: ArcMap

## Ch. 10 Challenge

Source: ArcMap

## Ch. 11 Tutorial

1. About 13%

2. 354,549 cells × 30 m × 30 m/1 million m$^2$/km$^2$, or about 319 km$^2$

3. About 51°

4. North is red; south is light blue.

5. Stores 0–3, indicating the number of points from which the cell can be seen

6. Floating-point raster, because the precipitation values have decimals

7. Pleasant Valley, about 165 million m$^3$

## Ch. 11 Exercises

1. See map.

2. See map.

Source: ArcMap

## Ch. 11 Exercises, con't.

3. See map.

4. Area is 40.4 km$^2$. See map.

Source: ArcMap

5. 12 sites. See map.

Source: ArcMap

6. Highest is the Madison Limestone; lowest is the Upper Mesozoic.

| NAME | MEAN |
|---|---|
| Cenozoic | 5.2 |
| Upper Mesozoic | 3 |
| Lower Mesozoic | 7.6 |
| Upper Paleozoic | 12.5 |
| Madison Limestone | 15.4 |
| Precambrian | 11.7 |
| Lower Paleozoic | 11.9 |

Source: ArcMap

7. See map.

8. See map.

Source: ArcMap

9. Highest is Anchor Hill at 1723 meters; lowest is Sly Hill at 1183 meters.

10. See map.

Source: ArcMap

## Ch. 11 Challenge

Answers will vary, but might look something like this one. The gray "off limits" areas should be the same in all answers.

Source: ArcMap

## Ch. 12 Tutorial

1. The cluster tolerance must be smaller than the stream tolerance.

2. If G were done next, it would leave a hole for polygon F.

## Ch. 12 Exercises

Answers will vary.

## Ch. 12 Challenge

Answers will vary.

## Ch. 13 Tutorial

1. NAD 1927 UTM Zone 13N

2. To save space or to be consistent with previous data or policies

3. Create a Float Double domain also.

## Ch. 13 Exercises

Answers will vary.

## Ch. 13 Challenge

Answers will vary.

## Ch. 14 Tutorial

1. Publication date is 6/30/2010.

2. Originated by National United States Atlas and USGS; published by ESRI

3. Two steps in 2004 and 2006

4. The National Atlas

5. Source scale 1:2 million

6. Contact ESRI to purchase the software.

7. Produced by ArcGIS Content Team; follows NAP of ISO 19115 2003

8. Depth unit is kilometers.

## Ch. 14 Exercises

Answers will vary.

## Ch. 14 Challenge

Answers will vary.

# Index

## A

Absolute pathname, 50
Access constraints and use, 415
Accuracy
   attribute, 411, 413, 415
   geometric, 16, 25
   in maps, 15
   positional, 411, 412–413
   temporal, 411, 414–415
   thematic, 16, 25
   vs. precision, 17
Active frame, 20
Adding territory, digitizing by, 366–368
Adjacency, 11, 351
Adjacent polygons, 207, 211–212, 223
Adjust division value, 154
Adjust number of divisions, 154
Adjust width, 154
Aerial photographs, 12, 16–17, 23, 115
Affine (first-order) transformation, 79
Aggregation, 142
Alias, 183
Alignment of map elements, 145
Analysis
   data analysis functions, 2
   raster, 319–347
   spatial analysis functions, 294–296
   suitability, 291, 320
   surface, 325
   viewshed, 325
Annotation, 154–155
   editing, 402–407
   in geodatabases, 388–389
   working with, 162–165
Annotation Groups tab, 164
Append function, 295–296
ArcCatalog, 21, 52, 59–61, 84–85
   using ArcMap to export data, 66–68
   using ArcToolbox, 65–66
   Catalog Tree, 59, 64
   Catalog window in ArcMap, 64–65
   creating and organizing data, 61–64
   using data from ArcGIS Online and
      Internet, 68–69
   using ESRI Data and Maps, 71
   getting data from Internet, 69–71
   to manage files, 52–53
   program, 52

ArcEditor, 19
ArcGIS, 2, 18–19, 81, 119–120, 152, 176
   annotation, 154–155
   using ArcCatalog to manage files, 52–53
   ArcMap documents, 19–20
   using ArcToolbox, 53–55
   attribute data for, 183–187
   choosing selection method, 240–242
   CS, 83
   data files in, 20–23
   documenting data, 55–56
   editing, 207–212
   editing in, 357–360
   geodatabase model, 352
   interactive selection, 238
   Internet data services, 23–24
   labeling, 154–155
   labeling coordinate systems, 83–85
   map documents and pathnames, 49–51
   map documents and shared data, 48–49
   map scales, 152–153
   maps in, 152
   and metadata, 418–422
   on-the-fly projection, 85
   processing layers with selections, 237–238
   projecting data, 85
   scale bars, setting up, 153–154
   selecting by attributes, 239
   selecting by location, 240
   Selection window, 242
   sources for GIS data, 56
   text files, 154–155
   troubleshooting coordinate system
      problems, 85–87
   understanding file locks, 51–52
ArcGIS 9 editors, 420
ArcGIS Advanced, 19
ArcGIS Basic, 19
ArcGIS Desktop, 19
ArcGIS/NAP metadata maps, 422
ArcGIS Online, 5, 8, 24, 46, 68–69
ArcGIS Standard, 19
Arc/Info, 18–19
ArcMap, 14, 82, 83–85, 151, 208
   adding data to, 27–28
   data frame coordinate systems, 36–38
   documents, 19–20
   exercises, 42
   to export data, 66–68
   using internet map services, 38–41
   layer files, 33–35
   map symbols, 31–33

# Length Conversions

Multiply by the factor indicated to convert the column units to the units in the row.

| | Inches | Centimeters | Feet | Yards | Meters | Miles | Kilometers | Nautical miles |
|---|---|---|---|---|---|---|---|---|
| Inches | 1 | 0.3937 | 12 | 36 | 39.37 | 69,840 | 39,370 | 60,964 |
| Centimeters | 2.540 | 1 | 30.48 | 91.44 | 100 | 27,496 | $1.00 \times 10^5$ | $1.548 \times 10^5$ |
| Feet | 0.0833 | 0.0328 | 1 | 3 | 3.281 | 5,280 | 3,281 | 5,080 |
| Yards | 0.0278 | 0.0109 | 0.333 | 1 | 1.094 | 1,940 | 1,094 | 1,693 |
| Meters | 0.0254 | 0.0100 | 0.3048 | 0.914 | 1 | 1,610 | 1,000 | 1,852 |
| Miles | $1.432 \times 10^{-5}$ | $3.637 \times 10^{-5}$ | $1.718 \times 10^{-4}$ | $5.155 \times 10^{-4}$ | $6.21 \times 10^{-4}$ | 1 | 0.6215 | 1.151 |
| Kilometers | $2.539 \times 10^{-5}$ | $1.00 \times 10^{-5}$ | $3.048 \times 10^{-4}$ | $9.144 \times 10^{-4}$ | 0.001 | 1.609 | 1 | 1.852 |
| Nautical miles | $1.640 \times 10^{-5}$ | $6.458 \times 10^{-6}$ | $1.968 \times 10^{-4}$ | $5.905 \times 10^{-4}$ | $5.40 \times 10^{-4}$ | 0.869 | 0.5400 | 1 |

# Area Conversions

Multiply by the factor indicated to convert the column units to the units in the row.

| | Inches | Centimeters | Feet | Yards | Meters | Miles | Kilometers | Hectares | Acres |
|---|---|---|---|---|---|---|---|---|---|
| Inches | 1 | 0.1550 | 144 | 1,296 | 1,550 | $4.878 \times 10^9$ | $1.550 \times 10^9$ | $1.550 \times 10^7$ | $6.272 \times 10^6$ |
| Centimeters | 6.452 | 1 | 929 | 8,361 | 10,000 | $7.560 \times 10^8$ | $1.00 \times 10^{10}$ | $1.000 \times 10^8$ | $4.046 \times 10^7$ |
| Feet | 0.0069 | $1.076 \times 10^{-3}$ | 1 | 9 | 10.76 | $3.387 \times 10^7$ | $1.076 \times 10^7$ | $1.076 \times 10^5$ | 43,560 |
| Yards | $7.716 \times 10^{-4}$ | $1.196 \times 10^{-4}$ | 0.1111 | 1 | 1.196 | $3.764 \times 10^6$ | $1.196 \times 10^6$ | 11,960 | 4,840 |
| Meters | $6.442 \times 10^{-4}$ | $1.00 \times 10^{-4}$ | 0.0929 | 0.8361 | 1 | $2.591 \times 10^6$ | 1,000,000 | 10,000 | 4,047 |
| Miles | $2.050 \times 10^{-10}$ | $1.323 \times 10^{-9}$ | $2.952 \times 10^{-8}$ | $2.657 \times 10^{-7}$ | $3.859 \times 10^{-7}$ | 1 | 0.3863 | $3.859 \times 10^{-3}$ | $1.563 \times 10^{-3}$ |
| Kilometers | $6.449 \times 10^{-10}$ | $1.00 \times 10^{-10}$ | $9.290 \times 10^{-8}$ | $8.361 \times 10^{-7}$ | $1.00 \times 10^{-6}$ | 2.589 | 1 | 0.0100 | $4.047 \times 10^{-3}$ |
| Hectares | $6.452 \times 10^{-8}$ | $1.00 \times 10^{-8}$ | $9.290 \times 10^{-6}$ | $8.361 \times 10^{-5}$ | 0.0001 | 259 | 100 | 1 | 0.4047 |
| Acres | $1.594 \times 10^{-7}$ | $2.471 \times 10^{-8}$ | $2.296 \times 10^{-5}$ | $2.066 \times 10^{-4}$ | $2.471 \times 10^{-4}$ | 640 | 247 | 2.471 | 1 |